The Arboviruses: Epidemiology and Ecology

Volume I

Editor

Thomas P. Monath, M.D.
Director
Division of Vector-Borne Viral Diseases
Centers for Disease Control
Public Health Service
U.S. Department of Health and Human Services
Fort Collins, Colorado

CRC Press is an imprint of the
Taylor & Francis Group, an **informa** business

CRC Press
Taylor & Francis Group
6000 Broken Sound Parkway NW, Suite 300
Boca Raton, FL 33487-2742

Reissued 2019 by CRC Press

A Library of Congress record exists under LC control number:

Publisher's Note
The publisher has gone to great lengths to ensure the quality of this reprint but points out that some imperfections in the original copies may be apparent.

Disclaimer
The publisher has made every effort to trace copyright holders and welcomes correspondence from those they have been unable to contact.

ISBN 13: 978-0-367-23529-1 (hbk)
ISBN 13: 978-0-367-23532-1 (pbk)
ISBN 13: 978-0-429-28022-1 (ebk)

Visit the Taylor & Francis Web site at http://www.taylorandfrancis.com and the CRC Press Web site at http://www.crcpress.com

FOREWORD

The term "arbovirus" is used to describe a diverse array of viruses which share a common feature, namely transmission by arthropod vectors. This ecological grouping now includes over 500 viruses, most belonging to five families — the Togaviridae, Flaviviridae, Bunyaviridae, Reoviridae, and Rhabdoviridae. Over 100 of these agents have been associated with naturally acquired disease in humans and/or domestic animals, and among these approximately 50 of the most important pathogenic viruses have been selected for detailed review under this title.

The complexity of arbovirus ecology requires a fundamental understanding of the influence of each of the multiple components (virus, vector, viremic host, clinical host, and environment) on infection and transmission cycles. The first volume, devoted to the variables which affect arbovirus transmission, provides this background and also contains guidelines for the design of future epidemiological investigations. Armed with these general principles, the student, teacher, or research worker will be able to structure specific knowledge about individual arbovirus infections found in Volumes II through V.

Recent textbooks are available which provide comprehensive coverage of the clinical aspects, pathogenesis, virological characteristics, and molecular biology of arboviruses. The intent of this book is different, for it focuses on the epidemiology and ecology of the arboviruses, the risk factors underlying the appearance of disease in the community, and the roles of arthropod vector and vertebrate hosts in virus transmission. Emphasis is placed on the field and laboratory evidence for involvement of vector and host species and on the ecological dynamics which determine their ability to spread infection. Elements of transmission cycles which are susceptible to surveillance, field investigation, prevention, and control are elucidated.

A number of arboviruses which have caused human disease only on rare occasions are not included in the book or are mentioned in passing within chapters on related diseases. Although these viruses (for example, Spondweni, Ilheus, Rio Bravo, Usutu, Orungo, Wanowrie) are inherently interesting and may, with changing ecologic conditions, turn out to be medically important, little or no information about their epidemiology (insofar as it relates to clinical hosts) is available. Finally, the scope of this book has been limited strictly to arthropod-borne infections and other viruses sometimes considered under the aegis of arbovirology (e.g., rodent-borne viral hemorrhagic fevers) are not included.

The compilation of a book of this scope required sacrifices in time and energy by a large number of contributors, all of whom faced multiple other commitments. This sacrifice will, I expect, be partially compensated by the availability of a useful compendium of collective knowledge.

Thomas P. Monath
October 1986

THE EDITOR

Thomas P. Monath is Director of the Division of Vector-Borne Viral Diseases, Centers for Disease Control, and is an affiliate faculty member of the Department of Microbiology, College of Veterinary Medicine, Colorado State University.

He received his undergraduate and M.D. degrees from Harvard University and his clinical training in Internal Medicine at the Peter Bent Brigham Hospital, Boston. In 1968 he joined the U.S. Public Health Service, serving as Medical Officer in the Arbovirology Unit, Centers for Disease Control, Atlanta, and later as Chief of the Arbovirus Section. Between 1970 and 1972, he was assigned to the Virus Research Laboratory of the Rockefeller Foundation, University of Ibadan, Nigeria, where he conducted field research on the epidemiology of yellow fever and Lassa fever. Since 1974, Dr. Monath has been Director of the Division of Vector-Borne Viral Diseases, Fort Collins, Colorado. In 1984 — 1985 he spent a sabbatical year in the Gastroenterology Unit of the Massachusetts General Hospital.

Dr. Monath is a Fellow of the American College of Physicians, the Infectious Disease Society, and the Royal Society of Tropical Medicine and Hygiene. He is a member of the American Society of Virologists, the American Society of Tropical Medicine and Hygiene, and the Association of Military Surgeons. He serves on the Editorial Boards of the *American Journal of Tropical Medicine and Hygiene, Acta Tropica,* and the *Journal of Virological Methods.* Dr. Monath is a member of the Committee on Research Grants, Board of Science and Technology for International Development, National Research Council, and is currently Chairman of the AIBS Infectious Diseases and Immunology Peer Review Panel to the U. S. Army Medical Research and Development Command. He is a member of the World Health Organization Expert Committee on Virus Diseases and the Pan American Health Organization Scientific Advisory Committee on Dengue, Yellow Fever, and *Aedes aegypti.* He has served as Chairman of the Executive Council of the American Committee on Arthropod-Borne Viruses, as a Councilor of the American Society of Tropical Medicine and Hygiene, and as a member of the Directory Board, International Comparative Virology Organization and the U. S.-Japan Cooperative Medical Research Program Panel on Virus Diseases.

Dr. Monath has authored or coauthored over 140 scientific publications in the field of virology and is editor of the book, *St. Louis Encephalitis,* published by the American Public Health Association. His main research interests are the ecology, epidemiology, and pathogenesis of arbovirus infections.

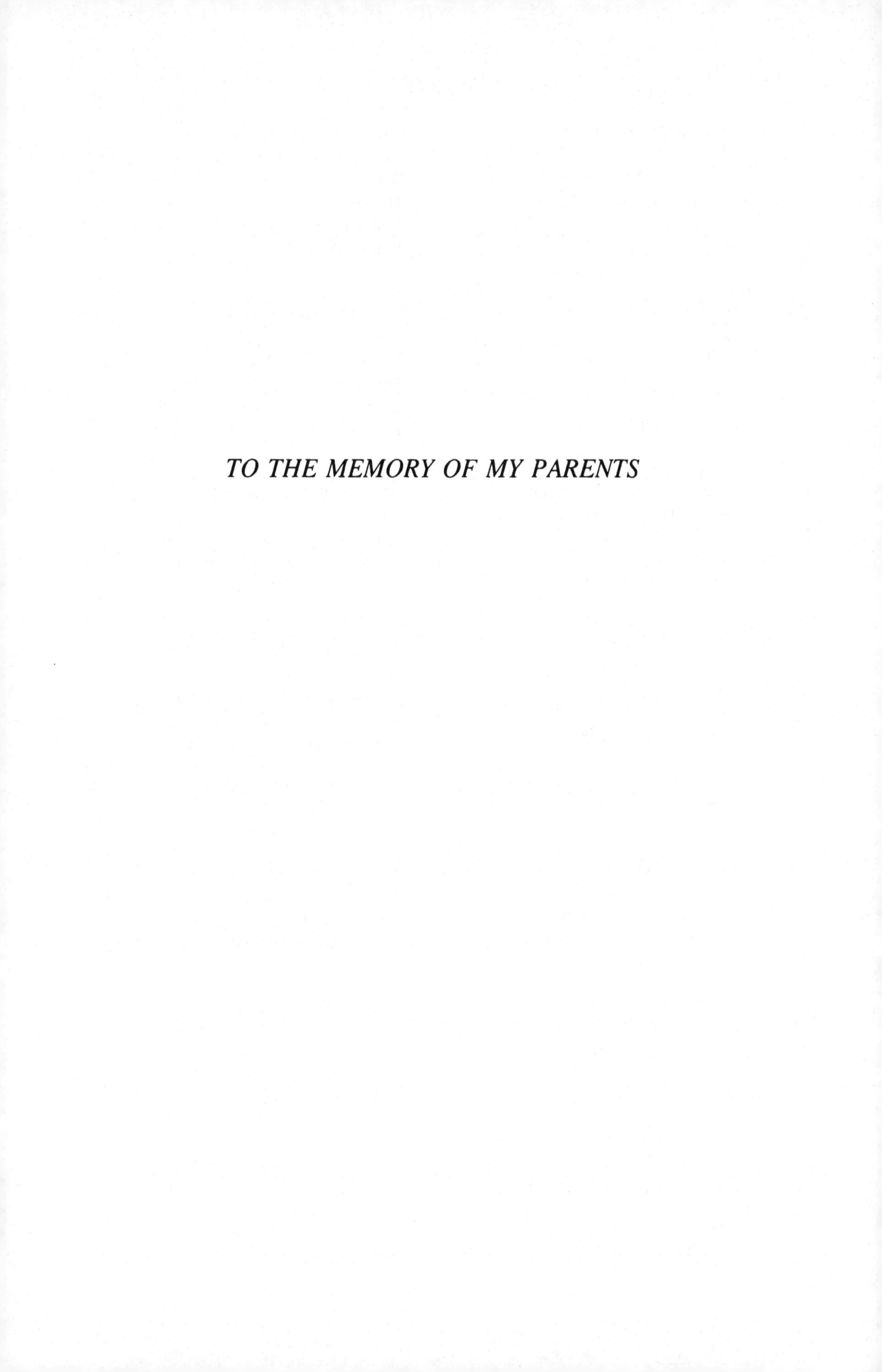

TO THE MEMORY OF MY PARENTS

CONTRIBUTORS

Barry J. Beaty, Ph.D.
Professor
Department of Microbiology and
Environmental Health
Colorado State University
Fort Collins, Colorado

Paul Brès, M.D.
Former Chief
Virus Diseases Unit
World Health Organization
Geneva, Switzerland

Charles H. Calisher, Ph.D.
Chief
Arbovirus Reference Branch
Division of Vector-Borne Viral Diseases
Centers for Disease Control
Fort Collins, Colorado

Frederick L. Dunn, M.D., Ph.D.
Professor
Department of Epidemiology and
International Health
University of California
San Francisco, California

John D. Edman, Ph.D.
Professor
Department of Entomology
University of Massachusetts
Amherst, Massachusetts

Dana Alan Focks, Ph.D.
Research Entomologist
Agricultural Research Service
U. S. Department of Agriculture
Gainesville, Florida

Michael B. Gregg, M.D.
Deputy Director
Epidemiology Program Office
Centers for Disease Control
Atlanta, Georgia

James L. Hardy, Ph.D.
Professor of Medical Virology
Department of Biomedical and
Environmental Health Sciences
University of California
Berkeley, California

Nick Karabatsos, Ph.D.
Research Microbiologist
Arbovirus Reference Branch
Division of Vector-Borne Viral Diseases
Centers for Disease Control
Fort Collins, Colorado

Carl J. Mitchell, Sc.D.
Chief
Vector Virology Laboratory
Division of Vector-Borne Viral Diseases
Centers for Disease Control
Fort Collins, Colorado

Thomas P. Monath, M.D.
Director
Division of Vector-Borne Viral Diseases
Centers for Disease Control
Fort Collins, Colorado

Paul Reiter, M.Phil., D.Phil.
Research Entomologist
Dengue Branch, San Juan Laboratories
Division of Vector-Borne Viral Diseases
Centers for Disease Control
San Juan, Puerto Rico

John T. Roehrig, Ph. D.
Chief
Immunochemistry Branch
Division of Vector-Borne Viral Diseases
Centers for Disease Control
Fort Collins, Colorado

Thomas W. Scott, Ph.D.
Assistant Professor
Department of Entomology
University of Maryland
College Park, Maryland

Daniel E. Sonenshine, Ph.D.
Professor
Department of Biological Sciences
Old Dominion University
Norfolk, Virginia

Andrew Spielman, Sc.D.
Professor
Department of Tropical Public Health
Harvard School of Public Health
Boston, Massachusetts

Dennis W. Trent, Ph.D.
Chief
Molecular Virology Branch
Division of Vector-Borne Viral Diseases
Centers for Disease Control
Fort Collins, Colorado

Michael J. Turell, Ph.D., M.P.M.
Entomologist
Department of Arboviral Entomology
Disease Assessment Division
USAMRIID, Fort Detrick
Frederick, Maryland

TABLE OF CONTENTS

Volume I

Chapter 1

IMPACT OF ARBOVIRUSES ON HUMAN AND ANIMAL HEALTH

Paul Brès

TABLE OF CONTENTS

I. INTRODUCTION

Arboviruses (arthropod-borne viruses) belong to different taxonomic families and are transmitted from infected to susceptible vertebrates by certain species of mosquitoes, ticks, sand flies *(Phlebotomidae),* or biting midges *(Culicoides, Ceratopogonidae)* in which they multiply. Initially, arboviruses were given the name of the illness which they caused, e.g., bluetongue (a disease of sheep), yellow fever, dengue and chikungunya (which breaks the bones); later, it became necessary to add a geographic connotation to differentiate the causative virus, e.g., eastern or western equine encephalitis virus. Ultimately, arboviruses were only given the name of the place where they were isolated for the first time even if they caused a disease, e.g., Mayaro, Rocio.

The vector/vertebrate interaction gives unique epidemiologic characteristics to the impact of arboviruses on human and domestic animal communities. This impact may be revealed by a review of the clinical importance of the diseases, the risks of infection, the impact on the community, the intervention of health services, and the future trends of these diseases.

II. CLINICAL IMPORTANCE OF ARBOVIRAL INFECTIONS

Of the 504 arboviruses registered in 1985 in the *Catalogue of Arthropod-Borne and Selected Vertebrate Viruses of the World,* approximately 100 are known to infect humans and 40 to infect livestock. Some of these viruses cause only subclinical infections detected by the presence of antibodies. Such infections do not provoke an evident effect on the community health, but they indicate that a virus is circulating and that viremic subjects may be virus reservoirs for insect or tick vectors. These viruses will not be dealt with here. Arboviruses causing overt infections important for humans are indicated in Table 1, and those of veterinary importance are listed in Table 2. Some viruses causing mild symptoms and whose occurrence is not frequent have not been included in the tables.

A. Disease in Humans

Human arbovirus diseases can be classified according to the predominant syndrome caused: systemic febrile illness, encephalitis, or hemorrhagic fever. However, the clinical spectrum associated with arboviruses causing encephalitides and hemorrhagic fevers is quite wide, and these agents can also cause systemic febrile illnesses and even subclinical infections.

1. Systemic Febrile Illnesses

These infections cover a wide clinical pattern ranging from simple fever to a characteristic dengue-like syndrome.

Table 1
MOST IMPORTANT ARBOVIRUSES CAUSING DISEASES AND EPIDEMICS IN MAN

Family/virus	Abbreviation	Epidemics	CFR[a](%)
Systemic Febrile Illness			
Togaviridae			
Chikungunya	CHIK	*	
Mayaro	MAY	*	
O'nyong-nyong	ONN	*	
Ross River	RR	*	
Sindbis	SIN	*	
Flaviviridae			
Dengue-1—4	DEN-1—4	*	
Wesselsbron	WSL		
West Nile	WN	*	
Bunyaviridae			
Bunyamwera	BUN		
Bwamba	BWA		
Germiston	GER		
Group C (10 viruses)		*	
Ilesha	ILE		
Sandfly fever (Naples, Sicilian)	SFN-SFS	*	
Rift Valley fever	RVF	*	
Tataguine	TAT		
Reoviridae			
Colorado tick fever	CTF		
Encephalitis			
Togaviridae			
Eastern equine encephalitis	EEE	*	50—75
Venezuelan equine encephalitis	VEE	*	0.1—20[b]
Western equine encephalitis	WEE	*	5—10
Flaviviridae			
Japanese encephalitis	JBE	*	30—40[c]
Kyasanur Forest disease	KFD		5
Louping ill	LI		Rare
Murray Valley encephalitis	MVE	*	20—70
Rocio	ROC	*	13
St. Louis encephalitis	SLE	*	4—20
Tick-borne encephalitis	TBE		
Far eastern			30
Central European			1—10
West Nile	WN		Rare
Bunyaviridae			
California encephalitis (La Crosse)	LAC		1
Hemorrhagic Fever			
Togaviridae			
Chikungunya	CHIK	*	Rare
Flaviviridae			
Dengue-1—4	DEN 1—4	*	3—12
Kyasanur Forest disease	KFD		5
Omsk hemorrhagic fever	OMSK		1—2
Yellow fever	YF	*	5—20

Table 1 (continued)
MOST IMPORTANT ARBOVIRUSES CAUSING DISEASES AND EPIDEMICS IN MAN

Family/virus	Abbreviation	Epidemics	CFR[a](%)
Bunyaviridae			
Crimean-Congo hemorrhagic fever	CCHF		15—20
Rift Valley fever	RVF		1—5

[a] CFR = case-fatality rate.
[b] High in children.
[c] Less in children.

Table 2
MOST IMPORTANT ARBOVIRUSES CAUSING DISEASES IN DOMESTIC ANIMALS

Family/virus	Animal[a]	Disease
Togaviridae		
Eastern equine encephalitis (EEE)	E	Encephalitis
Getah (GEV)	E	Febrile systemic, diphasic; rash
Semliki Forest (SF)	E	Encephalitis (in Senegal)
Venezuelan equine encephalitis (VEE)	E	Febrile systemic, encephalitis
Western equine encephalitis (WEE)	E	Encephalitis
Flaviviridae		
Japanese encephalitis (JE)	P	Abortions
	H	Subclinical, encephalitis
Louping ill (LI)	S	Febrile systemic, encephalitis
Wesselsbron (WSL)	S	Hemorrhagic fever
Bunyaviridae		
Akabane (AKA)	C,S,G	Congenital abnormalities
Crimean-Congo hemorrhagic fever (CCHF)	C,S,G	Mild illness
Nairobi sheep disease (NSD)	S,G	Hemorrhagic gastroenteritis
Rift Valley fever (RVF)	S,G,C	Bloody diarrhea, abortions
Reoviridae		
African horse sickness (AHS)	E	Febrile, pulmonary, cardiac, illness
Bluetongue (BT)	S	Hemorrhages, edemas, lameness
Rhabdoviridae		
Bovine ephemeral fever (BEF)	C	Febrile systemic, respiratory
Vesicular stomatitis (VS)	C,E,S,P	Vesicles in the mouth and on feet
Iridoviridae		
African swine fever (ASF)	P	Diarrhea, pneumonia, hemorrhages

[a] C = cattle; E = equine; G = goat; P = pig; S = sheep.

The pattern of mild systemic febrile illness generally consists of fever of 1 to 3 days' duration, headache, and some degree of arthralgia and myalgia, occasionally with rash, and, is caused by most viruses listed in Table 1. The differential diagnosis is extremely difficult and, unless special attention is given to such cases, they are readily confused with malaria, influenza, or fever of unknown origin.

Typical dengue, caused by four serotypes of the virus, is a characteristic syndrome with an abrupt onset of fever, headache, photophobia, retroorbital and low back pains, rapidly followed by general malaise, crippling myalgias and arthralgias, and prostration; there is often a transient improvement on day 3 to 4 and a relapse (saddleback fever), during which lymphadenopathy and rash may appear. There are no fatalities. The convalescence phase is marked by prolonged asthenia.

A few viruses listed in Table 1, while generally causing dengue like syndromes, exhibit an additional clinical note. West Nile virus is responsible for a high frequency of inapparent infection in children, a typical dengue syndrome in adults, and meningeal symptoms and sometimes encephalitis in the elderly. Ross River and chikungunya virus diseases are recognized by recurrent and persistent arthralgias of the extremities and even arthritis. Dengue fever may be complicated by myocarditis or encephalopathy and Oropouche virus disease by meningitis. Colorado tick fever may be accompanied in children by an hemorrhagic or encephalitic syndrome. During the epidemic of Rift Valley fever in Egypt in 1977—1978 and previous epidemics in South Africa, complications such as retinitis, encephalitis, and hemorrhages were not uncommon. These features indicate an underlying neurotropic or viscerotropic potential of these viruses.

2. *Encephalitis*

Certain arboviruses possess a marked neurotropism. A prodromal phase, marked by some of the systemic symptoms described above, may or may not precede the onset of neurologic signs. The full-blown encephalitic syndrome comprises a sudden rise of fever, vomiting, stiffness of the neck, dizziness, drowsiness, disorientation, confusion, and progression to stupor and coma. The arboviruses most frequently causing encephalitis are listed in Table 1. The case-fatality rate varies considerably according to the etiologic agent, the age of the affected individual, and nonbiological factors (surveillance, hospitalization, and treatment). Recovery may be uneventful or leave sequelae such as psychiatric symptoms in adults or mental retardation in children. A myelitic component may be evident in some infections. For example, paralysis and paralytic sequelae of the shoulder girdle or upper limbs are typical of the Far Eastern type of tick-borne encephalitis. Meningoencephalitis may sometimes be caused by several arboviruses, listed in Table 1, as an exceptional complication of systemic illness; Rift Valley and Colorado tick fevers have been mentioned in this regard. The arboviral encephalitides have to be differentiated by laboratory examinations from other viral, bacterial, rickettsial, and parasitic infections affecting the central nervous system.

3. *Hemorrhagic Fever*

After a systemic prodromal phase of about 3 days duration and a short period of remission of a few hours, the hemorrhagic syndrome may suddenly appear. Responsible viruses are listed in Table 1. Two general patterns are found. One form is typical of dengue hemorrhagic fever with a sudden cardiovascular collapse caused by plasma leakage and hemoconcentration with cold and clammy extremities, petechiae, hemorrhages, and shock, which rapidly becomes irreversible. The second form is characterized by sudden onset of high fever, flushing of the face and chest, and conjunctival infection. After a short remission period, characteristic symptoms appear on the 3rd to 4th day with petechiae, echymoses, bleeding from the nose and the gums, hematemesis, melena, metrorrhagia, abortion, proteinuria, azotemia, sudden cardiovascular collapse, and terminal shock. This clinical picture is characteristic of yellow fever, Crimean-Congo hemorrhagic fever and Omsk hemorrhagic fever and may be confused with similar hemorrhagic fevers transmitted by rodents (Argentine and Bolivian hemorrhagic and Lassa fever) as well as those caused by Marburg and Ebola viruses.

B. Disease in Domestic Animals

As in man, subclinical infections are frequent in domestic animals and occur either as isolated cases which are rarely diagnosed or during epizootics, when they are detected if systematic serological surveys are carried out. As in humans, overt infections also consist of systemic, encephalitic, or hemorrhagic syndromes caused by viruses listed in Table 2.

The encephalitic syndrome is seen mainly in equines. The responsible viruses also cause human disease; however, the severity of the disease may not be identical in both hosts. Eastern equine encephalitis virus has a high case-fatality rate in horses (90%) and man (50 to 75%), but western equine encephalitis virus is less often fatal in humans (5 to 10%) than in horses (20 to 30%). Japanese encephalitis occurs less frequently in equines than in man, and West Nile virus is a rare cause of encephalitis in both hosts.

The features of the human hemorrhagic fevers are found in African swine fever, epizootic hemorrhagic disease of deer, bluetongue of ovines, Wesselsbron virus disease in ovines, Rift Valley fever in sheep, and Nairobi sheep disease, with hemorrhages, abortion, and death of newborns.

Other viruses cause systemic syndromes with specific clinical and pathological features. African horse sickness is characterized by edema and cardiac failure. Rift Valley fever, Wesselsbron, and Akabane viruses are associated with fetal malformations such as arthrogryposis and hydrancephaly. Vesicular stomatitis virus causes lesions in the mouths, teats, and hooves of affected cattle and horses, and Getah virus produces a characteristic skin eruption in horses. Japanese encephalitis is an important cause of stillbirth and reduced fecundity in swine.

III. THE RISKS OF INFECTION

By definition, arboviruses are transmitted to man and domestic animals by the bite of infected arthropods. The risk of infection in natural conditions varies according to the geographic distribution of the viruses.

A. Transmission by Arthropods

1. Two Cycles of Transmission

Arboviruses can be transmitted to man and domestic animals by infected arthropods (mosquitoes, ticks, sand flies, biting midges) in two different contexts: the sylvatic or jungle cycle and the urban type of transmission cycle. Arboviruses are maintained by vector arthropods in a determined area (ecological niche) by a transmission chain from infected to receptive vertebrates. In the first context, humans and domestic animals may be bitten by an infected arthropod by intrusion in an ecologic niche at some distance from their dwellings. This tangential mode of infection results in sporadic cases or a small cluster of cases contaminated at the same site, usually in forests. It is also possible that infected arthropods escape from their natural niche and come in contact with humans and domestic animals in nearby villages. Similarly, an infected wild animal may come close to human dwellings and be bitten by arthropods which also bite humans or domestic animals and transfer the virus to them. This mode of infection generally occurs in villages close to the site of arbovirus maintenance and results in sporadic, endemic/enzootic, or even epidemic/epizootic cases.

In the second context, a person or domestic animal infected as described above becomes an amplifying host in the transmission chain and initiates an "urban" cycle (in a village as well as in a town) involving a domestic arthropod vector capable of transferring the virus to other persons or domestic animals. Such cases occur as epidemics or epizootics.

Yellow fever is a good illustration of these two modes of transmission, but many other arboviruses exhibit one or both patterns.

2. Factors Influencing the Frequency of Transmission

The frequency of transmission of an arbovirus to humans and domestic animals is related to the vector population density, which is influenced by temperature and humidity and may be constantly, seasonally, or periodically favorable. It is also related to the feeding preferences of the vector(s), the efficiency of transmission by the vector(s) (vector competence), the flight range of the vector(s), age structure of the vector population, and other factors. Similar elements of the risk of transmission apply to vertebrate hosts (abundance, susceptibility and level of viremia, attractiveness to vectors, distribution, and mobility). The dynamics of these factors in transmission cycles are considered in other chapters on individual diseases.

B. Unusual Modes of Transmission

Apart from biological transmission by the usual arthropod vectors (multiplication of the virus in the arthropod with transmission after an extrinsic incubation period), other blood-sucking species, such as tabanids, can sometimes spread arboviruses immediately by "mechanical" feeding. This requires interruption of the blood meal on one infected vertebrate and completion of it on another vertebrate host. This transmission mechanism may probably play some role when arthropods are particularly numerous in urban epidemics, and it has recently been demonstrated experimentally with Rift Valley fever virus.[1]

Direct transmission from vertebrate to vertebrate may occur in exceptional circumstances. Laboratory infections, documented for at least 58 arboviruses, have sometimes been the result of transmission by aerosol or bite from an infected vertebrate. Nosocomial fatal cases of Crimean-Congo hemorrhagic fever have been reported in surgical and intensive care units,[2]

Other unusual modes of transmission may also be found with animal arboviruses. African swine fever has been described as a "facultative" arbovirus because, in addition to ticks, it can be transmitted by direct contact and ingestion of contaminated meat and swill. Venezuelan equine encephalitis may occasionally be transmitted by the aerosol route.

C. Geographic Risks

The risk of infection by arboviruses also depends on their geographic distribution. This distribution may be limited to a small ecological niche or extended to a region or even to several zoogeographic regions. In general, the distribution of an arbovirus is determined by the presence of the vector(s) and host(s) involved in transmission. Viruses transmitted by ticks or those having terrestrial small vertebrate hosts tend to have more restricted distributions than viruses transmitted by flying insects or having migratory avian hosts. However, "exotic" arboviruses may be imported into areas where they do not occur naturally.

1. Local Risks

The ecologic niche of certain arboviruses may be restricted to a small area within a country or, at most, extended to a few neighboring countries. In India, for example, Kyasanur Forest disease, a tick-borne flaviviral infection, is restricted to Karnataka State and parts of adjacent states into which it is extending at present. Omsk hemorrhagic fever, also caused by a tick-borne flavivirus is limited to the Omsk and Novosibirsk regions of the U.S.S.R.

The area of arbovirus activity is frequently smaller than the area infested by its vector. The reasons may be complex. The virus may never have been introduced into suitable areas past the periphery of its range. Ecological factors, such as climate or vertebrate host density, may restrict viral transmission. The virus may be localized to areas inhabited by particular subspecies or even biotype of the vector, which is not different morphologically, but which fulfills the required physiologic and behavioral conditions for transmission of that virus.

2. Regional Risks

The area of disease risk may involve one or several large geographic regions (Africa and the Middle East, North America, Central and South America, Europe and continental Asia,

Table 3
REGIONAL PREVALENCE OF THE MOST IMPORTANT ARBOVIRUSES OF MAN AND DOMESTIC ANIMALS[a]

Region	Arboviruses of man	Arboviruses of domestic animals
Africa and the Middle East	CCHF, CHIK, DEN, ONN, RVF, SIN, WN, YF	AHS, AKA, BEF, BT, CHF, NSD, RVF, SF, WSL
North America	CE, CTF, DEN, EEE, SLE, VEE, WEE	BT, EEE, VEE, VS, WEE
Central and South America	DEN-DHF,[b] EEE, MAY, ORO, ROC, SLE, VEE, YF	ASF, BT, EEE, VEE, VS, WEE
Europe and continental Asia	CHF, LI, OMSK, SFN, SFS, SIN, TBE,[c] WN	ASF, BT, LI
India and Southeast Asia, including China and Japan	CHIK, CHF, DEN-DHF, JE, KFD, SFN, SFS	AKA, BT, BEF, CHF, JE
Oceania	DEN-DHF, MVE, RR	AKA, BT, BEF, GET, MVE

[a] The entire region may not be affected.
[b] DEN-DHF = dengue hemorrhagic fever.
[a] TBE = tick-borne encephalitis, including strains of central and far-eastern Europe.

India, and Southeast Asia, Oceania) if the principal vector is present or if alternate vectors can transmit the virus in different areas. Dengue virus is transmitted by *Aedes aegypti* in may areas and, in addition, by *Ae. albopictus* in Asia. The regional risk is higher in terms of frequency and diversity of arboviral infections in tropical and subtropical zones than in temperate regions because of the greater variety of arboviruses, the rich vector and host faunas, and the favorable temperature conditions for insect, tick, and virus development.

The regional distribution of the most important arbovirus disease of humans and domestic animals is indicated in Table 3. A few of these viruses are responsible for epidemics of regional and even interregional importance. The majority of epidemics and epizootics are caused by alpha- and flaviviruses.[3] In Africa and the Middle East, arboviruses are responsible for recurring epidemics (yellow fever, Rift Valley fever, chikungunya), recent outbreaks in previously unaffected areas (dengue), past outbreaks with future epidemic potential (O'nyong-nyong, West Nile), and nosocomial epidemics (Crimean-Congo hemorrhagic fever).[4] According to official notifications, yellow fever causes about 50 to 100 cases a year, but these figures are underestimated and larger figures are obtained when epidemiological investigations are carried out. From 1960 to 1962, the yellow fever epidemic in Ethiopia caused 100,000 cases and 30,000 deaths in a population of 1 million; however, only 2000 deaths were officially notified. Outbreaks of dengue-2 or -3 have occurred recently in East and Central Africa, and sylvatic circulation of dengue-2 virus has been recognized in West Africa; the outbreaks have been characterized by mild disease in the local population and typical forms in expatriates. In the past, dengue caused a severe epidemic in Durban, South Africa in 1927, with hemorrhagic illness reported.

In North America, arthropod-borne encephalitides caused 7,928 cases in the U.S. between 1955 and 1985. In tropical America, Venezuelan, eastern, and western equine encephalitis and Rocio virus have caused important outbreaks.[5] The recent recurrence of pandemic dengue has raised a growing concern in the region.[6] This concern was justified by the 1981 epidemic of dengue hemorrhagic fever in Cuba with 350,000 cases, of which 116,143 hemorrhagic cases were hospitalized (about 30%) and 158 were fatal. The morbidity rates per 10,000 were 28 in persons over 15 years, 28 between ages 5 and 14, 24 between ages 1 and 4, and 24 for infants under 1 year of age. Jungle yellow fever occurs annually in the tropical zone; the recent reinvasion of urban centers by *Ae. aegypti* raises a considerable risk of reurbanization of the disease.

In Europe and continental Asia tick-borne encephalitis clinically more severe in the east than in the west, causes more public health concern in endemic countries than Sindbis-related viruses, West Nile, and phlebotomus fever viruses. Crimean-Congo hemorrhagic fever virus occurs in the U.S.S.R., but has also caused nosocomial outbreaks in Pakistan. The dengue 1 epidemic in Athens and Piraeus of 1927—1928 attacked 90% of the population of all age groups, and and large number of cases exhibited features of hemorrhagic fever. In India and Southeast Asia, Japanese encephalitis is endemic and epidemic. More than 10,000 cases occur annually in China alone.[8] The epidemiology of the disease has changed since the late 1970s, with the appearance of recurrent severe epidemics in northern India, Nepal, and Thailand involving thousands of cases. The frequency of dengue is increasing in the region, and dengue hemorrhagic fever is becoming endemic — not only in large cities but also in rural areas including tropical China, where dengue had been absent for 40 years.[9] During the interval 1980 to 1984, the average reported number of cases of dengue hemorrhagic fever was 39,218 in Thailand, 8413 in Indonesia, and 2087 in Burma, with case-fatality rates between 0.7 and 3.7%.[10] In Oceania (Australia, New Zealand, Papua-New Guinea, and the Pacific Islands), dengue and sometimes dengue hemorrhagic fever is also present. Ross River fever (epidemic polyarthritis) spread from Australia to Papua-New Guinea, Samoa, Fiji, and other islands in 1979 with more than 30,000 cases.

The distribution pattern and risks for the most important domestic animal pathogens are as follows. The tropical climate in the major part of Africa favors many arthropod-borne diseases. African swine fever is widespread in wild warthogs, and the disease is mild in domestic pigs. So are African horse sickness and bluetongue which are epizootic only in imported livestock. In contrast, Rift Valley fever has caused considerable economic losses from South Africa to Egypt, where the 1977—1978 epizootic was unexpected.[11] Other viruses affecting livestock are Wesselsbron and Nairobi sheep disease. In Europe, louping ill, a disease of sheep, is limited to the U.K. Epizootics of African swine fever have followed importation of the disease into Europe and the Americas. In Asia, Japanese encephalitis causes fetal wastage in swine. Akabane virus, which causes abortion and fetal abnormalities in cattle, sheep, and goats, exists in Japan and Australia. Ibaraki virus, antigenically related to bluetongue, caused severe epizootics in Japan. Bluetongue is increasingly recognized as a cause of clinical or subclinical infections in temperate, tropical, and subtropical regions. Bovine ephemeral fever occurs in Japan and Australia in addition to Africa.

3. Exotic Arbovirus Diseases

An exotic disease is one which has been imported into a place where it does not naturally occur, not necessarily from a tropical country, but often so. The present rapidity of air travel permits persons (or animals) to reach an unaffected area while still circulating virus in their blood. This has little public health importance unless there is a risk of initiation of transmission cycle by a local vector. However, in the case of dengue fever, the principal vector, *Ae. aegypti,* has a worldwide distribution, and air travel has resulted in dissemination of dengue serotypes and initiation of epidemics. Dengue outbreaks have also crossed international boundaries in the form of an epidemic wave; during the summer of 1980, the virus spread from Mexico to Texas, resulting in the first indigenous transmission in the U.S. since 1942.[12] *Ae. aegypti*-infested countries of Africa, the Americas, the Middle East, and the Far East are also potentially at risk of introduction of yellow fever, with millions of possible victims. Other disabling diseases such as chikungunya, West Nile, and Rift Valley fever also present a potential risk of spread outside of present endemic zones because of the wide distribution of their vectors.

Invasion of new territories by trading of animals may occur when protective rules are infringed. African swine fever was exported during the past decade from its natural focus in central Africa into Spain and Portugal, where it became enzootic, and in the south of

France, Holland, Cuba, Hispañola, and Brazil, resulting in quarantine and the slaughter of millions of pigs. Rift Valley fever was probably introduced into Egypt in 1977 by smuggling infected camels over the Sudanese border.

Movement of infected vectors is also a possible cause of dissemination. Midges can be carried over long distances on the wind (40 to 700 km at 4000 ft altitude), and this may explain the spread of African horse sickness in 1959 from Africa to Iran, Afghanistan, and Pakistan. Birds have also been incriminated as means of transport of infected ticks along their migration routes, possibly explaining the spread of Crimean-Congo hemorrhagic fever virus over central Asia and Africa. Migratory birds have been found to be viremic with eastern equine encephalitis and Mayaro viruses far from enzootic areas.

IV. IMPACT ON THE COMMUNITY

The effects of arbovirus infections on the human community and domestic animals can be estimated by the attack rate, the severity of cases, and the amount of economic losses.

A. Effects of Febrile Systemic Infections

Arbovirus infections characterized by nonspecific systemic syndromes or by a low incidence frequently remain undiagnosed even in countries where good laboratory facilities exist for their diagnosis. Their impact on the community is considered practically nil. For this reason, these viruses have not been included in Table 1. Although the medical importance of these viruses may be ignored for some time, it may increase due to improved surveillance of changing ecologic conditions. An example is Tataguine virus, which was isolated from mosquitoes in Africa in 1966 and was found only 7 years ago to cause a febrile systemic infection.[13] In a study of the etiology of cases of febrile illness in the Central African Republic from 1966 to 1979, arboviruses were implicated in 7.3% of the cases, the most frequent infecting agents being chikungunya, Ilesha, Tataguine, and Bwamba viruses.[14]

Outbreaks with a high incidence of febrile systemic infections are characteristic of certain arboviruses such as chikungunya, O'nyong nyong, dengue, Oropouche, Mayaro, Rift Valley fever, sandfly fever, Ross River, Venezuelan equine encephalitis, Sindbis, and West Nile fever. Attack rates may reach 25% or more. Although lethal infections are absent or rare, outbreaks result in widespread social disruption and loss of productive work. Certain arboviruses also have had a significant military impact; dengue and phlebotomus fever caused severe outbreaks in troops during World War II.

B. Effects of Encephalitides and Hemorrhagic Fevers

Even if they are not numerous, encephalitic and hemorrhagic fever cases strike the public attention. Approximate case-fatality rates are indicated in Table 1. The impact of these severe infections is directly related to their lethality. Yellow fever, which is responsible for outbreaks affecting thousands of persons, kills approximately 20% of victims who develop jaundice. Japanese encephalitis results in a similar case fatality rate. The encephalitides place an additional burden on society and the individual, namely neuropsychiatric sequelae which occur in a high proportion of persons surviving severe infections with Japanese, Rocio, and the equine encephalitides.

Epidemics of encephalitis or hemorrhagic fever tend to frighten the population and may even cause a general panic. This happened in Bangkok in 1958 when more than 8000 children under 8 years of age were affected for the first time by dengue hemorrhagic fever over a period of 3 months. Exaggerated description of events by the media may aggravate the panic in such episodes. In contrast, education of the population in Cuba alleviated public concern over the large outbreak of dengue hemorrhagic fever in 1981.

Paradoxically, some severe epidemics may be ignored by health authorities in developing countries either by lack of surveillance facilities in remote areas or for political reasons. In some instances, epidemics of yellow fever in Africa have escaped detection, and most of the time official notifications have been grossly underestimated by 10 or 100 times.[4]

C. Economic Losses

In addition to human suffering, epidemics of arboviruses cause economic damage due to loss of manpower, premature death, disruption of public services, and overload on health facilities. Unfortunately, specific studies are not undertaken on such losses, but they are certainly very high in any epidemic. Estimates of the costs resulting from the 1977 dengue epidemic in Puerto Rico[15] ($6.4 to 16.7 million) and the 1980 dengue hemorrhagic fever epidemic in Thailand ($6.97 million) have been published. The economic cost of epidemic St. Louis encephalitis (Dallas, 1966) was estimated at $796,500 ($4631 per patient).[16] These estimates have not included indirect losses in tourism, school absenteeism, and other factors.

Concerning domestic livestock, economic losses would seem easier to estimate, but here also the data are sparse. However, a few example can be given. During an epizootic in the U. S. in 1938, 184,000 horses were infected with eastern or western equine encephalitis virus with a respective case-fatality rate of 90 and 20 to 30%. Venezuelan equine encephalitis caused an epizootic in Colombia from 1967 to 1968 which killed 27,000 donkeys and 40,000 horses and mules (250,000 to 500,000 persons were infected). By 1969, it had spread to Ecuador and 1000 to 1500 burros died (31,000 human cases), and it killed 3000 equines in Peru. During 1970, it caused the death of approximately 10,000 equine animals in Mexico. In 1971, it spread to Texas where thousands of horses died.[17] The direct cost of the epizootic in south Texas in 1971 (for spraying, vaccines, surveillance, quarantine) was estimated at $20 to 30 million; indirect costs in lost revenues were even higher. Getah virus hit 39% of race horses in a training center in Japan in 1978. Transmitted by *Culicoides,* African horse sickness spread from Africa to the Middle East in the 1960s, where it caused a loss of 300,000 horses, mules, and donkeys and had severe effects upon agriculture and transportation. A Rift Valley fever epizootic caused about 20,000 deaths and 60,000 abortions in cattle in Zimbabwe in 1978; over 100,000 sheep and cattle had already died in the years 1950 to 1951. From 1977 to 1978, an unprecedented epizootic in Egypt caused severe losses in domestic animals and about 1 million human cases. Akabane virus caused widespread losses in cattle in Japan from 1972 to 1974.[18] African swine fever eradication programs in Europe and the Americas have been especially costly. Between 1957 and 1985, such programs have resulted in the slaugther of over 2 million swine at a cost of over $50 million.

V. INTERVENTION OF HEALTH SERVICES

Health services may be faced with two different situations: sporadic or endemic occurrence of arbovirus diseases and sudden outbreaks. The former calls for preventive measures and the latter for emergency measures followed by prevention of recurrence. The possible spread of arbovirus diseases across national borders may require international measures. Although of great interest for health planners, very little information is available on the cost/benefit aspects of such measures.

A. Emergency Measures

The strategy of interventions to protect human populations is different than that for domestic animals, but they may have to be coordinated when the same virus affects both humans and animals simultaneously.

1. Protection of the Human Population

Emergency measures begin with field investigations which must be well organized and coordinated between epidemiologists, entomologists, and virologists to define the nature and the extent of the outbreak.[19] Recent progress has improved the rapidity of laboratory techniques and made it possible to use them in the field. Although the lack of diagnostic laboratory services provides an obstacle to surveillance and epidemiological investigations in developing countries, much technical assistance can be obtained from the reference laboratory network organized under the aegis of the American Committee for Arthropod-borne Viruses and the World Health Organization (WHO).[20]

Medical care of patients with arbovirus diseases is solely supportive because of the lack of specific antiviral drugs, at least for the present time. Intensive supportive care is necessary to decrease the case-fatality rate of the encephalitides and hemorrhagic fevers. Where well-equipped hospitals do not exist in developing countries, some improvisation following the general principles of intensive care may significantly improve patient survival. This was illustrated during the Rocio encephalitis epidemic in Brazil in 1973, when the case-fatality rate dropped dramatically after the institution of better supportive care.

Immunization in emergency situations is only feasible with yellow fever vaccine, a live attenuated virus which requires a single inoculation. However, protective antibodies do not fully develop before 5 to 7 days postinjection, and this constitutes a period during which cases can still appear and the virus can be transmitted to other persons. Killed vaccines which require two to three injections cannot be expected to have a rapid effect on an outbreak.

Quarantine measures to prevent movement of the population out of an epidemic focus are not indicated unless there are arthropod vectors in other places that could be infected by biting a viremic patient. This measure may be applied to nonvaccinated persons during an outbreak of yellow fever, but it has limited indications in other diseases.

Vector control is the ultimate measure against diseases transmitted by mosquitoes, sand flies, and midges. Entomologic investigations are necessary to determine the responsible vector species, its behavior, the best choice of insecticides, and the strategy for their application. In an established outbreak, it is necessary to interrupt transmission by killing the infected adult vectors by use of insecticide space sprays. Spraying is generally more effective against domestic vectors than against forest vectors. The control of tick-borne diseases requires knowledge of the complicated transmission cycles which, furthermore, may be different from one country to another.[21]

2. Protection of Domestic Animals

The important losses in domestic animals during epizootics point out many problems, particularly in developing countries[22] where the necessary resources in manpower, equipment, and finance are often inadequate and require international aid.[23]

Rapid identification of the causative agent is essential, as for outbreaks in humans. Veterinary laboratories are often insufficient in developing countries, but the particular clinical and pathological features of the disease and background knowledge of the distribution and activity of the causative virus in the region usually permit a presumptive diagnosis and institution of control measures.

Once the etiology has been established, vaccines may be utilized. Vaccines exist for the majority of domestic animal arbovirus diseases either as live attenuated viruses or inactivated viruses. Generally, the use of live vaccines is possible during epizootics, taking advantage of their rapidity of action and knowing that a possible reversion of the strain to virulence in a few animals would not be as catastrophic as during a nonepizootic period. Epizootic spread of Venezuelan equine encephalitis has been terminated by use of the attenuated live vaccine TC-83. The existence of multiple serotypes having incomplete cross-protection complicates the use of vaccines (e.g., for bluetongue and African horse sickness).

Vector control raises the same logistic problems as in epidemics affecting humans. Changes in farming methods can reduce morbidity when the ecology of the virus is well known. For example, the breeding time of piglets, which are amplifying hosts of Japanese encephalitis, has been changed not to coincide with the period when the vector, *Culex tritaeniorhynchus,* is abundant.

Slaughtering animals is the ultimate method required to stop the spread of viruses which are transmitted directly from animal to animal and for which no vaccines are available, e.g., African swine fever.

B. Preventive Measures

Preventive measures consist of surveillance, immunization, and vector control. The principles are almost identical for human and animal diseases.

1. Surveillance

The objective of surveillance is to provide an early warning of the threat of epizootics/epidemics. An essential component is a system for prompt reporting of clinical cases. However, for those arboviruses which have a transmission cycle involving wild vertebrate hosts, surveillance aimed at detecting virus transmission prior to spill-over to clinical hosts provides the most sensitive predictive capability. Effective systems have been based on the use of sentinel animals, such as chickens, for the North American arboviral encephalitides. Surveillance of vectors and/or wild reservoir animals is generally difficult, but has been successfully applied in some cases, e.g., St. Louis encephalitis. Surveillance of wild nonhuman primates for detection of antibody or (in tropical America) illness has provided useful information about yellow fever activity.

Maintenance of surveillance systems is generally costly, labor-intensive, and requires a long-range commitment on the part of the agencies involved and economic planners. Political support and funding for such programs may be difficult to maintain during interepidemic periods.

2. Immunization

In humans, preventive immunization against yellow fever with the attenuated 17D vaccine strain has obvious advantages, and immunity is probably lifelong. There is no risk of vaccine virus reversion and transmission. Live, attenuated dengue vaccines are presently in the developmental stage. Inactivated vaccines against tick-borne encephalitis have been successfully employed in Europe and the U.S.S.R. Other inactivated vaccines which have been developed, but less widely used (or not used), include those to Kyasanur Forest disease, Omsk hemorrhagic fever, and chikungunya. Preventive immunization against Japanese encephalitis is discussed in Section D, "Costs/Benefit of Interventions". Inactivated viruses are used for immunization against some other viruses such as the equine encephalitides and Rift Valley fever, but their use is generally restricted to persons with high-risk occupational exposures such as laboratory workers or veterinarians. In the case of Venezuelan equine encephalitis and Rift Valley fever, vaccination of animals is beneficial to humans since it decreases the virus reservoir.

In domestic animals, killed virus vaccines are preferable for prevention in nonepizootic periods if the live vaccine is pathogenic or may revert to virulence. Inadequate inactivation of vaccines may, however, be a problem. Several outbreaks of Venezuelan equine encephalitis probably occured following use of insufficiently inactivated vaccines. The duration of immunity is shorter with killed vaccines, and boosters are necessary which increase the difficulty and cost of use in domestic livestock.

3. Vector Control

In certain situations, it is appropriate to undertake adulticide spraying to reduce vector populations as a preventive measure. For example, heavy precipitation or natural disasters, such as floods, can dramatically increase the density of vector mosquitoes (e.g., *Cx. tarsalis*, which transmits western equine and St. Louis encephalitis viruses); if the virus is found to be present, preventive vector control may be warranted.

In the case of *Ae. aegypti*-borne viruses (dengue, yellow fever, chikungunya), long-term preventive measures are aimed at reducing vector densities below a level at which epidemic transmission may occur (possibly below a Breteau index of 5). Attempts to accomplish this by use of insecticides alone have not been successful, and an integrated approach which includes emphasis on breeding source reduction is necessary. Preventive control of forest vectors is not practical.

C. International Measures

The need for international emergency aid to countries affected by epidemics or epizootics has been mentioned previously. In addition, international agreements have been reached on measures to be enforced in order to limit the spread of arboviruses (among other communicable diseases of humans and animals) across national borders.

1. Human Diseases

Yellow fever is the only arbovirus subject to the International Health Regulations adopted by the WHO Member States.[24] The Regulations provide for the implementation of four obligations. Outbreaks, whether their etiology is presumptive or confirmed, must be notified without any delay to WHO which, in turn, informs Member States by telex and via the *Weekly Epidemiological Record.* In the endemic zone defined by WHO, airports must be maintained as mosquito-free areas, and facilities must exist for diagnosis, isolation, and treatment of patients. Before they enter countries where the vector *Ae. aegypti* exists, persons who have traveled within 6 days in an infected area must present a valid vaccination certificate, otherwise, they could be isolated during the 6-day incubation period. Aircraft, ships, and vehicles arriving from infected areas must be disinsected. Difficulties in the application of these regulations arise through disagreements on the definition of an area infected by urban or forest transmission of the virus. In addition, the stringent anti-*Ae. aegypti* regulations are, in practice, inefficient, in part due to the multiplicity of uncontrolled private means of transport. There are no similar regulations for other arboviruses, and WHO simply recommends that rapid notification be given of the occurrence of outbreaks of any disease which could spread to other countries.

2. Animal Diseases

Information on the occurrence of epizootics is centralized worldwide by the Food and Agricultural Organization (FAO) of the United Nations (UN) and the Office International des Epizooties (OIE). Member Countries are kept informed through the *Monthly Epizootic Circular* of the OIE or by direct contact in case of emergency and are expected to provide monthly reports. Several other international organizations are also involved in these activities.

The OIE has promoted an interdependent arrangement among its Member Countries for international protection against animal diseases.[25] Difficulties met in the application of these measure include the weakness of surveillance, the illegal movements of animals across the borders of certain countries, and reluctance to declare diseases calling for quarantine or slaughtering measures which represent severe economic losses for the farmers.

D. Costs/Benefits of Interventions

Human and animal suffering deserve a priority consideration over any economic analysis. Very few data on the economic cost of arbovirus diseases are available to decision makers who have to weigh the respective advantages and disadvantages of emergency measures vs.

preventive measures. The conditions are so different for each arbovirus and in each country that no precise method of evaluation can be proposed. The theoretical considerations given below may provide a basis for future attempts to evaluate the cost/benefit ratios of emergency and preventive measures in practical cases of human or animal arboviral diseases.

1.Human Diseases

a. Emergency Measures

Expenses incurred during an outbreak of arboviral disease include:

1. Investigations — costs of investigative field teams and laboratory support which are proportional to the extent of the investigated area, the density of the population at risk, and the number of patients
2. Patient care — out- and in-patient diagnosis and treatment costs which are proportional to the number of patients and severity of the disease
3. Immunization — costs of an emergency mass campaign (using a single vaccine) which are proportional to the populaton at risk inside and outside the focus
4. Vector control — cost of investigations, insecticides, equipment, spraying operations, transport which are proportional to the size and accessibility of the epidemic area and surrounding areas at risk
5. Indirect costs/socioeconomic losses — absenteeism, premature death, diversion of resources of cooperating official and private services (public works, police, etc.), disruption of country economy (trade, tourism, agriculture, etc.)

b. Preventive Measures

Expenses include the costs of surveillance, immunization, and vector control programs.

Preventive immunization is feasible and cost-effective for yellow fever since the vaccine is relatively inexpensive (less than $0.20 U.S. per dose), and immunity is probably lifelong. In contrast, use of the inactivated Japanese encephalitis vaccine is more problematic. The attack rate of Japanese encephalitis is about 1:1000, and cases are scattered over rural areas or suburbs.[20] The Japanese encephalitis vaccine is expensive (two shots required at $2 U.S. per dose); immunization of 1000 persons to prevent one case is hardly feasible in certain countries. Nevertheless, vaccination of school children in Japan, the Republic of Korea, and China has been widely practiced, and some locally produced vaccines are significantly less expensive.

In spite of appearances, preventive measures are not necessarily less expensive than emergency measures, and their use is always complicated by the general unpredictability of epidemics. The effectiveness of mass vaccination against Japanese encephalitis in Korea in the 1960s could not be assessed because the level of endemic virus activity regressed spontaneously in the control population.

A cost/benefit evaluation of the *Ae. aegypti* eradication program carried out in the western hemisphere since 1947 to prevent dengue and yellow fever was undertaken in 1970.[26] The conclusions underscored the difficulty in arriving at exact figures and estimated that there was a slight advantage for eradication over casual vector control and yellow fever vaccination and this might increase if dengue was to become preeminent. Since then, reinvasion by *Ae. aegypti* has further progressed, and dengue epidemics and the threat of urbanization of yellow fever have increased.

2. Domestic Animal Diseases

The theoretical considerations expressed above are also applicable here. Losses of animals can be estimated by their market value, but during the epizootic in Colombia in 1967, a horse was worth six to eight months' earnings of a peasant, a figure which should be entered into the statistics. The invasion of Venezuelan equine encephalitis in Texas in 1971 killed thousands of horses and caused 88 human cases. Expenses for controlling the outbreak by animal vaccination and insecticide spraying were in excess of $20 million, and animal losses could have been avoided if preventive measures had been taken in advance, as recommended by arbovirologists.[3]

VI. FUTURE TRENDS

Certain natural and socioeconomic processes may increase the risk of arbovirus diseases, whereas others may decrease it. It is difficult to predict what will be the final issue which, furthermore, may vary between geographic regions.

A. Factors Favoring Increased Incidence of Arboviral Diseases

Growing urbanization is one of the consequences of the population explosion in the developing countries. Uncontrolled development of towns in warm climates has resulted in the absence of piped water, the storage of water at home, the absence of drainage of sewage, and the accumulation of garbage which multiply the breeding sites of *Ae. aegypti* and several *Culex* vectors.[27] A number of cities in developing tropical countries now have several million inhabitants who are at risk of *Ae. aegypti*-transmitted dengue. The invasion of yellow fever into such areas, especially in the densely populated Far East, would be a catastrophe, resulting in thousands of deaths. Urbanization in the Amazon Basin has led to the occurrence of recent outbreaks of Oropouche virus disease transmitted by a peridomestic culicoid midge.

Increasing air travel enhances the risk of spread of certain arboviruses, enabling viremic persons to cross international borders. Among many examples, dengue and Ross River fever have been disseminated in this way during the past decade. The trade of draught and food animals presents a similar danger which is enhanced by the increasing use of air transport. Aircraft, ships, and other vehicles can also transport vectors which may be able to breed in new ecologic niches and transmit exotic diseases or introduce infected vectors able to start an epidemic/epizootic. The recent introduction and establishment of *Ae. albopictus* into North America serves as an example; should this effective rural vector of dengue viruses subsequently be disseminated to tropical America, it would greatly increase the danger of epidemics and complicate control strategies. The number of viremic subjects or infected vectors which are necessary to start an epidemic or an epizootic is still unknown and should be studied.

Man-made environmental changes may interfere with arbovirus ecology. Deforestation undertaken in certain zones exposes the workers to close contact with the forest cycles of enzootic viruses, such as yellow fever. However, clearing the forest may also reduce the impact of some diseases since amplifying vertebrate hosts and sylvan mosquito vectors may disappear. The development of irrigation has increased the impact of many arboviruses. The occurrence of western equine and St. Louis encephalitis in the western U.S. is closely linked to irrigated agriculture and, in the Far East, the population density of *Cx. tritaeniorhynchus*, the vector of Japanese encephalitis, is associated with the extension of rice fields. Irrigation may also change the species composition of mosquito populations in an area; in Nigeria, for example, *Mansonia uniformis*, a vector of chikungunya, Ndumu, Wesselsbron, Spondweni, and Pongola viruses was replaced by *Anopheles gambiae*, a vector of O'nyong-nyong, Bwamba, Nyando, and malaria.[28]

Increasing insecticide resistance diminishes the possibilities of chemical vector control. The high cost of developing new insecticides has discouraged manufacturers. The withdrawal of DDT, the occurrence of resistance to other insecticides, the increasing costs of manpower, and the senescence of vector control programs have contributed to the reinvasion of the western hemisphere by *Ae. aegypti*. The extensive trade of used automobile tires has also been incriminated. Groot described the progression of *Ae. aegypti* reinvasion in Colombia after 1966 (the country had been previously freed from the mosquito). In 1971—1972, 450,000 cases of dengue-2 occurred, with *Ae. aegypti* indexes of more than 40% in certain towns; dengue-2 virus caused not less than 200,000 cases in 1974; and dengue-1 was disseminated throughout all the country in 1978 and caused more than 770,000 cases.[29] Ten years had sufficed for the mosquito to reinvade all places occupied before 1952 and to reach new places. Emergency control measures taken in 1978 against the vector and use of the 17D vaccine avoided the urbanization of three serious outbreaks of jungle yellow fever in 1978—1979 which occurred in the immediate neighborhood of towns infested with *Ae. aegypti*. The reinvasion by *Ae. aegypti* has further progressed in the western hemisphere, and the threat of dengue hemorrhagic fever and urban yellow fever is continuously increasing.[30]

B. Factors Which May Improve the Situation

Several factors should decrease the risks of arboviral outbreaks, at least in the long term. Important progress has been made during the last decade in the understanding of the ecology of many arboviruses. Molecular techniques are available to characterize virus variants and to uncover the origin of epidemics/epizootics. Rapid laboratory diagnostic techniques, based on the detection of IgM antibodies or antigen by monoclonal antibodies, and nucleic acid probes have been developed which improve surveillance and predictability of outbreaks. Improvements in vaccine technology open perspectives for the deployment of live attenuated vaccines against dengue and Japanese encephalitis. Genetic engineering provides avenues for the eventual development of safe and inexpensive recombinant or synthetic vaccines or live attenuated strains with appropriate deletions. It may eventually be possible to insert dengue immunoprotective epitopes in the yellow fever 17D virus genome.

In more immediate perspectives, a concerted effort is still necessary to improve the situation in developing countries. This is the task of international organizations through multilateral and bilateral aid. In the case of human diseases, aid to developing countries is mainly directed toward the control of epidemics. In animal diseases, aid to developing countries is directed toward improving husbandry, training of personnel, surveillance, and prevention of outbreaks. Both developed and developing countries have common interests in the field of arbovirus diseases.

REFERENCES

1. **Hoch, A. L., Gargan, T. P., II, and Bailey, C. L.,** Mechanical transmission of Rift Valley fever virus by hematophagous diptera, *Am. J. Trop. Med. Hyg.*, 34, 188, 1985.
2. **Baskerville, A., Satti, A., Murphy, F. A., and Simpson, D. I. H.,** Congo-Crimean haemorrhagic fever in Dubai: histopathological studies, *J. Clin. Pathol.*, 34, 871, 1981.
3. **Shope, R. E.,** Medical significance of togaviruses: an overview of diseases caused by togaviruses in man and in domestic and wild vertebrate animals, in *The Togaviruses*, Schlesinger, R. W., Ed., Academic Press, New York, 1980, chap. 3.
4. **Monath, T. P.,** Impact of arthropod-borne virus diseases on Africa and the Middle-East, in *Applied Virology*, Kurstack, E., Al-Nakib, E., and Kurstak, C., Eds., Academic Press, New York, 1984, chap. 24.
5. **Monath, T. P.** Arthropod-borne encephalitides in the Americas, *Bull. WHO*, 57, 513, 1979.
6. **Ehrenkranz, N. J., Ventura, A. K., Cuadrado, R. R., Pond, P. D., and Porter, P. D.,** Pandemic dengue in Caribbean countries and the southern United States — past, present and potential problems, *N. Engl. J. Med.*, 285, 1460, 1971.
7. **Anon.,** Program for dengue elimination and *Aedes aegypti* eradication in Cuba, *Epid. Bull. PAHO*, 3, 7, 1982.

8. **Umenai, T., Krzysko, R., Bektimirow, T. A., and Assaad, F. A.,** Japanese encephalitis: current worldwide status, *Bull. WHO*, 63, 625, 1985.
9. **Halstead, S. B.,** Dengue haemorrhagic fever — a public health problem and a field for research, *Bull. WHO*, 58, 1, 1980.
10. *WHO Wkly. Epidemiol. Rec.*, 61, 20, 1986.
11. **Brès, P.,** Prevention of the spread of Rift Valley fever from the African continent, in *Rift Valley Fever*, Vol. 3, Swartz, T. A., Klingberg, M. A., and Golblum, N., Eds., S. Karger, Basel, 1981, 178.
12. Centers for Disease Control, Imported dengue fever — United States, 1984, *Morbid. Mortal. Wkly. Rep.*, 34, 488, 1985.
13. **Fagbami, A. H. and Tomori, O.,** Tataguine virus isolations from humans in Nigeria, 1971-1975, *Trans. R. Soc. Trop. Med. Hyg.*, 75, 788, 1981.
14. **Georges, A. J., Saluzzo, J. F., Gonzalez, J. P., and Dussarat, G. V.,** Arboviruses en Centafrique: incidence et aspects diagnostiques chez l'homme, *Med. Trop. (Marseilles)*, 40, 561, 1980.
15. **Van Allmen, S. D., Lopez-Correa, R. H., Woodall, J. P., Morens, D. M., Chiriboga, J., and Casta-Velez, A.,** Epidemic dengue fever in Puerto Rico, 1977. A cost analysis, *Am. J. Trop. Med. Hyg.*, 28, 1040, 1979.
16. **Schwab, P. M.,** Economic cost of St. Louis encephalitis epidemic in Dallas, Texas, 1966, *Publ. Health Rep.*, 83, 860, 1968.
17. **Lord, R. D.,** History and geographic distribution of Venezuelan equine encephalitis, *Bull. PAHO*, 8, 100, 1974.
18. **Parsonson, I. M., Della-Porta, A. J., and Snowdon, W. A.,** Devolopmental disorders of the fetus in some arthropod-borne virus infections, *Am. J. Trop. Med. Hyg.*, 30, 660, 1981.
19. **Murphy, F. A.,** Control and eradication of exotic virus affecting man, *Prog. Med. Virol.*, 25, 69, 1979.
20. WHO Scientific Group, *Arthropod-Borne and Rodent-Borne Viral Diseases*, Tech. Rep. Ser. No. 719, World Health Organization, Geneva, 1985.
21. **Hoogstraal, H.,** Established and emerging concepts regarding tick-associated viruses, and unanswered questions, in *Arboviruses in the Mediterranean Countries*, Vesenjak-Hirjan, J., Ed., Gustav Fischer Verlag, Stuttgart, 1980, 49.
22. **Henderson, W. M.,** Identification of existing and prospective problems of disease control, in *Virus Diseases of Food Animals*, Vol. 1, Gibbs, E. P. J., Eds., Academic Press, New York, 1981, chap. 1.
23. **Gibbs, E. P. J.,** Organizations, development aid and animal disease, in *Virus Diseases of Food Animals*, Vol. 1, Gibbs, E. P. J., Ed., Academic Press, New York, 1981, chap. 3.
24. *International Health Regulations (1969)*, 3rd annotated ed., World Health Organization, Geneva, 1983.
25. **Watson, W. A. and Brown, A. C. L.,** Legislation and control of virus diseases, in *Virus Diseases of Food Animals*, Vol. 1, Gibbs, E. P. J., Ed., Academic Press, New York, 1981, chap. 14.
26. **Anon.,** *The Prevention of Diseases Transmitted by Ae. aegypti in the Americas — A Cost Benefit Study*, Arthur D. Little, Inc., Cambridge, Mass., 1972,
27. **Surtees, G.,** Urbanization and the epidemiology of mosquito-borne disease, *Abstr. Hyg.*, 46, 121, 1971.
28. **Surtees, G.,** Effects of irrigation on mosquito populations and mosquito-borne diseases in man, with particular reference to ricefield extension, *Int. J. Environ. Stud.*, 1, 35, 1970.
29. **Groot, H.,** The reinvasion of Colombia by *Aedes aegypti*: aspects to remember, *Am. J. Trop. Med. Hyg.*, 29, 330, 1980.
30. **Groot, H.,** *Aedes aegypti*: a sword of Damocles over tropical America, *Bull. PAHO*, 15, 267, 1981.

Chapter 2

ARBOVIRUS SEROGROUPS: DEFINITION AND GEOGRAPHIC DISTRIBUTION

Charles H. Calisher and Nick Karabatsos

TABLE OF CONTENTS

I. INTRODUCTION AND DEFINITIONS

The viruses causing bluetongue, Nairobi sheep disease, yellow fever, and African horse sickness were isolated in 1901, 1910, 1927, and 1932, respectively.[1] These are all arthropod-borne viruses, but no relationships between them and other arthropod-borne viruses existed, other than obvious ecological and epidemiological ones, until Smithburn[2] demonstrated in 1942 an antigenic relationship between West Nile, St. Louis, and Japanese encephalitis viruses. Later, Sabin and co-workers[3-7] demonstrated hemagglutination (HA) by these viruses, as well as by Russian spring-summer encephalitis, dengue, and western equine encephalitis viruses. HA by these viruses was shown to be pH- and species-specific in regard to the erythrocytes used. At first, these findings were taken to indicate that all arthropod-borne viruses possessed HA activity. Casals and Brown[8] greatly extended the list of arthropod-borne viruses with known HA activity and, using hemagglutination-inhibition (HI) tests, were able to demonstrate antigenic relationships and differences between many of these viruses. Although investigators, such as Havens et al.[9] and Clarke and Theiler[10] demonstrated cross-reactivities between certain arthropod-borne viruses, it was Casals and co-workers[11,12] who provided the most comprehensive and unifying data regarding these relationships. First, Casals and Brown[8] designated serogroups A and B, then Casals demonstrated cross-reactivity with certain other arthropod-borne viruses and established serogroups Bunyamwera,[13] C,[14] California,[15] Guama,[16] Capim,[17] and many others. Placement of viruses into serogroups was difficult, since not all the viruses were found to have HA activity. Viruses were considered "arthropod-borne viruses" based on circumstances of isolation, inactivation by sodium deoxycholate or ether, and other minimally descriptive methods.

The thoroughness of Casals and co-workers set an example that led to a second phase of serogroup definitions and descriptions. Using the HI test, Casals[11] observed that, within serogroups, certain viruses are more closely related to each other than they are to the remaining members of the group. Casals termed this collection of closely related viruses an "antigenic complex". Whereas HI was used to place hemagglutinating viruses in a serogroup, complement-fixation (CF) and neutralization (N) tests were used as confirmatory procedures; these tests were also used to study arthropod-borne viruses that did not possess HA activity. Intragroup relationships heretofore not detected by HI were found; CF tests revealed finer distinctions within serogroups, and N tests were shown to be even more discriminating. We now know that HA and N are properties of external viral glycoproteins and that CF antibody may be directed against a variety of antigenic entities, including the nucleocapsid and nonstructural proteins.

Thus, a serogroup can be defined as two or more viruses, distinct from each other by quantitative serologic criteria (fourfold or greater differences between homologous and heterologous titers of both sera) in one or more test but related to each other or to other viruses by some serologic method. Arboviruses that are very closely related but distinct from each other may constitute an antigenic complex. Individual agents, antigenically related but easily separable (fourfold or greater differences between homologous and heterologous titers of both sera) by one or more serologic test are considered viruses or types. Subtypes are virus isolates separable from each other by at least a fourfold difference between the homologous and heterologous titers of one but not both of the two sera tested. Varieties are those isolates differentiable only by the application of special tests or reagents (kinetic HI, monoclonal antibody assays, etc.).

In present practice, the first discovered virus of a newly recognized serogroup lends it name to the antigenic cluster. However, before a virus can be assigned to an antigenic group, it must be shown to be serologically related to, but clearly distinguishable from, a previously identified virus.

When one or, at most, a very few viruses assigned to a serogroup cross-react to a limited extent with one or more members of another serogroup, a classification problem exists. This was found to be the case when Guaroa virus, assigned to the California serogroup, was shown to be a serorelative of certain members of the Bunyamwera serogroup.[15] It soon became apparent that low-level but repeatable cross-reactivity occurred between viruses belonging to not only the California and Bunyamwera serogroups but serogroups C, Guama, Patois, Anopheles A, Bwamba, Capim, Koongol, Olifantsvlei, Simbu, Tete, and Gamboa.[17-20] Casals[18] solved this dilemma by establishing the Bunyamwera supergroup.

Subsequent to the classic serologic studies of Casals, electron microscopy was introduced to investigate viral morphology, site of replication and formation, and site of liberation of these viruses from infected cells.[21] Assignment of viruses to taxons based upon antigenic relationships has been supported and extented by morphologic criteria. Most recently, molecular analyses have substantiated the previous classification schemes, and a clearer view of the taxonomy of the arboviruses has emerged.

It is apparent that the term arbovirus has considerable merit, insofar as virus ecology is concerned. However, the *International Catalogue of Arboviruses*[1] lists more than 500 viruses, most with antigenic relationships to others, some without such relationships, that are antigenically, taxonomically, and ecologically distinct. Many of the viruses registered in the *International Catalogue of Arboviruses* do not meet the definition of arboviruses, i.e., viruses "maintained in nature by a biological transmission cycle between susceptible vertebrate hosts and hematophagous arthropods";[22] in part, this is due to lack of information regarding transmission, but other viruses are certainly not arthropod-borne. A number of viruses registered in the *International Catalogue of Arboviruses* are there merely by virtue of their having been isolated by arbovirologists. The term "arbovirus" is used here to denote an ecologic description, but in the interest of accuracy, mention of these viruses employs taxonomic descriptions. Differences in antigenic, biochemical, and morphological characteristics are used to separate the arboviruses into families, genera, serogroups, complexes, viruses, subtypes, and varieties in an increasing order of relatedness. In this chapter, we present information regarding more than 655 registered and unregistered viruses.

Until recently, the family Togaviridae was composed of four genera: *Alphavirus, Flavivirus, Rubivirus,* and *Pestivirus*. The International Committee on Taxonomy of Viruses has now reclassified this family, based on additional molecular data regarding strategy of replication, dividing the Togaviridae into two families: Togaviridae (genera *Alphavirus, Rubivirus, Pestivirus,* and *Arterivirus* and Flaviviridae (genus *Flavivirus)*[23,24] Because invertebrate hosts are not known for viruses of the *Rubivirus* (rubella virus), *Pestivirus* (mucosal disease-bovine virus diarrhea virus, hog cholera virus, and border disease virus), or *Arterivirus* (equine arteritis virus) genera, they are not covered in this chapter.

Many arboviruses are closely related to each other morphologically, biophysically, biochemically, genetically, and antigenically. Classification schemes are not only useful in bringing together these viruses, they are also useful in separating them, which is epidemiologically functional. It is important to be able to distinguish virus serotypes, subtypes, and varities from each other because several may coexist within the same horizontal geographic boundaries or even vertically in rain forests, and some of them cause serious human or livestock diseases, while other, closely related but distinct viruses do not. For example, only one virus has been assigned to the Venezuelan equine encephalitis complex (family Togaviridae, genus *Alphavirus)* and six subtypes of this virus have been identified. These subtypes are rather easily distinguished by the use of standard CF tests, not a difficult chore for almost any reasonably prepared virus laboratory. However, six varieties of subtype I are known, three of which have been responsible for epizootics in equines and epidemics in humans. The other three have not been implicated in such outbreaks. These variants can be distinguished from one another by the use of antisera raised specifically to individual E2 glycoproteins, infection-immune rodent sera, monoclonal antibodies selected for epitope specificity, and other reagents used in simple or timed (kinetic) HI or N tests or enzyme-linked immunosorbent assays (ELISA). Therefore, it is a relatively straightforward task to assess the disease potential of a virus isolate belonging to this antigenic complex, to determine the efficacy of vector control operations in areas where both epizootic and enzootic viruses have been occurring, and to otherwise perform epidemiologic investigations and other important disease control strategies.

Close antigenic relationships between and among other viruses led to difficulties in distinguishing the etiology of infection in both individuals and populations at risk to these viruses. Chikungunya and o'nyong nyong viruses cause similar clinical manifestations in humans, and both viruses occur in Africa; Bwamba virus, which can cause a febrile illness in humans, occurs in areas of Africa where Pongola virus, not known to cause human illness, has been isolated; and La Crosse and snowshoe hare viruses occur sympatrically in the northeastern U.S., the former being the etiologic agent of a disease with significant clinical and epidemiologic impact, whereas the latter has only rarely been shown to cause human disease. Similar supportive examples of the need for laboratory confirmation and applications to epidemiologic studies are the closely related western equine encephalitis and Highlands J viruses and Hantaan and Seoul viruses.

The obverse side of the same epidemiologic coin is the cross-reactivity of antibodies elicited in response to infections with viruses of the family Flaviviridae, genus *Flavivirus*. Because of the complex and extensive antigen sharing among them, it is often exceedingly difficult to determine the etiologic agent in infection caused by these viruses. A prerequisite for any laboratory intending to undertake such determinations is an extensive battery of these viruses and antigens. Where the question narrows to that of determining whether St. Louis encephalitis or West Nile virus is the culprit, such a determination can be made on epidemiologic grounds, since these two viruses are found in distinct geographic areas. However, when two or more flaviviruses coexist in the same geographic areas and when the patient has a history of yellow fever vaccination or extensive travel, then the epidemiologist has a need to subtype a virus isolate or perform multiple and complex serodiagnostic procedures to distinguish the etiology of the infection. Probably the most clear-cut case in point is the need to determine the etiologic agent of infections causing dengue fever. In areas hyperendemic for dengue, it is imperative to follow the sequence of infections with the four dengue viruses in a community, since current evidence suggests that dengue hemorrhagic fever with shock syndrome, the malignant form of this usually benign disease, is due to a particular sequence of infections with different dengue types. Application of monoclonal antibody preparations have been and will be even more extensively useful in answering such epidemiologic questions.

In addition to classification schemes presented in this chapter, we have also attempted to summarize information regarding the geographic distributions of these viruses. The known distributions correspond to virus isolations and have not taken into account evidence obtained from serosurveys.

II. FAMILY TOGAVIRIDAE, GENUS *ALPHAVIRUS* (FORMER GROUP A ARBOVIRUSES)[23,25]

Alphaviruses are spherical particles, 60 to 70 nm in diameter. They contain one segment of single-stranded, positive-sense RNA, molecular weight 4×10^6, and three structural proteins. Two of these are envelope glycoproteins (E1 and E2) with molecular weights of 50 to 59×10^3, and one is a nonglycosylated capsid protein with a molecular weight of 30 to 34×10^3; Semliki Forest virus possesses three envelope glycoproteins. At present, more than 37 viruses, subtypes, and varieties have been assigned to the genus *Alphavirus*. All but two have been isolated from mosquitoes. Fort Morgan virus, which has been isolated only from passerine birds and the nest bug *(Oeciacus vicarius)*, is an example of a virus whose distribution is restricted by its vector. Alphaviruses of the eastern and western equine encephalitis complexes appear to utilize avians as principal vertebrate hosts in their maintenance cycles in nature; this probably accounts for their widespread distribution. Subtypes of Venezuelan equine encephalitis virus, on the other hand, appear to be restricted principally to small mammals in discrete enzootic foci in the Americas. Certain Semliki Forest virus complex members, Semliki Forest, chikungunya, Ross River, and Mayaro viruses, also are widely distributed. At least one of these agents (Mayaro virus) has been associated with avian hosts. Two of the four antigenic subtypes of Getah virus (Sagiyama and Bebaru) appear to be geographically isolated, but little is known of their natural histories. Individual members have been isolated on six continents (Table 1). A provisional classification scheme for these viruses has been published.[26] Triniti virus may be a member of the family Togaviridae, but there is insufficient information regarding it to warrant further (genus) placement.

III. FAMILY FLAVIVIRIDAE, GENUS *FLAVIVIRUS* (FORMER GROUP B ARBOVIRUSES)[24]

Flaviviruses are spherical particles 40 to 50 nm in diameter. They contain one segment of single-stranded positive-sense RNA, molecular weight about 4×10^6, and three structural proteins. One is an envelope glycoprotein (E) with a molecular weight of 51 to 59×10^3, another is the core protein (C) with a molecular weight of 13 to 16×10^3, and the third is a membrane-like protein (M), molecular weight 7 to 9×10^3. More than 70 viruses have been assigned to the flavivirus genus. Some are transmitted principally by mosquitoes, some by ticks, and some have not been associated with an arthropod vector. When divided by vector or nonvector associations, flaviviruses generally show serologically distinct differences, possibly as a consequence of adaptation to vector type.[27] The mosquito-borne flaviviruses include some of great epidemiologic significance, including yellow fever, West Nile, St. Louis encephalitis, Murray Valley encephalitis, Japanese encephalitis, Rocio, and the dengue viruses. It is interesting to note that the viruses causing encephalitides are more closely related to each other antigenically than they are to yellow fever or dengue viruses, which cause diseases distinct from each other and which are antigenically distinct from each other.

The tick-borne flaviviruses, including Russian spring-summer encephalitis, Central European encephalitis, Omsk hemorrhagic fever, and Kyasanur Forest disease viruses, are of considerable importance and concern from central Europe to eastern Siberia. They are only distantly related to the mosquito-borne flaviviruses, but that relationship is noteworthy on phylogenetic grounds. It is intriguing to speculate that evolutionary pressures have created a divergence of the virus-vector relationships, perhaps from a common original relationship. In fact, West Nile virus replicates nearly as well in ticks as in its usual arthropod vector (*Culex* mosquitoes). Powassan, a tick-borne flavivirus from Canada, the northern U.S., and the U.S.S.R., is widely divergent antigenically from the mosquito-borne flaviviruses but has been isolated from *Anopheles hyrcanus* mosquitoes and from larvae of *Aedes togoi* mosquitoes in the U.S.S.R.

The vector-unassociated flaviviruses (Table 4), not closely related to any other flavivirus(es), are connected to each other by virtue of host and geography (Rodentia, New World; Chiroptera, Old World). This set of viruses represents divergent evolution, indicating relatively restricted epidemiologic spread (vertebrate to vertebrate), whereas the mosquito- and tick-borne viruses represent more moderate divergent evolution and relatively unlimited geographic spread. Given the fairly discrete geographic foci and econiches in which the vector-unassociated viruses are found, such antigenic dissimilarities are not surprising.

Table 1
GEOGRAPHIC DISTRIBUTION OF VIRUSES BELONGING TO THE FAMILY TOGAVIRIDAE, GENUS *ALPHAVIRUS* (GROUP A ARBOVIRUSES)[1,23,25,26]

Complex	Virus	Subtype	Variety	Distribution
Eastern equine encephalitis	Eastern equine encephalitis		2	North America,
				South America
Middelburg	Middelburg			Africa
Ndumu	Ndumu			Africa
Semliki Forest	Semliki Forest			Africa, U.S.S.R.
	Chikungunya	1 — chikungunya	Several	Africa, Asia
		2 — o'nyong-nyong		Africa
	Getah	1 — Getah		Australia, Asia
		2 — Sagiyama		Japan, Okinawa
		3 — Bebaru		Malaysia
		4 — Ross River		Australia, Oceania
	Mayaro	1 — Mayaro		South America
		2 — Una		South America
Venezuelan equine encephalitis	Venezuelan equine encephalitis	1 — Venezuelan equine encephalitis	A-B	South America
			C	Colombia, Venezuela
			D	Colombia, Panama
			E	Panama, Mexico
			F[a]	Brazil
		2 — Everglades		Florida (U.S.)
		3 — Mucambo	Mucambo	Brazil
			Tonate	French Guiana
			(71D-1252)[a]	Peru
		4 — Pixuna		Brazil
		5 — Cabassou		French Guiana
		6 — (AG80-663)[a]		Argentina
Western equine encephalitis	Western equine encephalitis	Several		North America;
				South America
	Y 62-33[a]			U.S.S.R.
	Highlands J			U.S.(eastern)

	Fort Morgan		U.S.(western)
	Sindbis	1 — Sindbis	Africa, Asia, Europe, Australia
		2 — Babanki	Cameroon, Senegal, Central African Republic, Ivory Coast
		3 — Ockelbo[a]	Europe
		4 — Whataroa	New Zealand
		5 — Kyzylagach	Azerbaijan Rep. (U.S.S.R.)
	Aura		Brazil, Argentina
Barmah Forest	Barmah Forest		Australia

[a] These viruses are not in the published or working *International Catalogue of Arboviruses*.[1]

Antigenic relationships betwen the flaviviruses may be summarized as:

1. Those that are mosquito-borne are far more similar to each other than to the tick-borne or vector-unassociated flaviviruses, and those that are tick-borne are far more similar to each other than to the mosquito-borne or vector-unassociated flaviviruses
2. Those that are vector-unassociated are nearly as dissimilar from one another as they are from the mosquito- and tick-borne members of the group

Koutango (Table 2), a vector-unassociated virus, is closely related antigenically to one or more mosquito-borne flaviviruses, suggesting common ancestry.

Classification schemes based on studies by de Madrid and Porterfield,[28] Varelas-Wesley and Calisher,[29] and Calisher, Karabatsos, Dalrymple, and Shope (unpublished data) provide information regarding geographic distribution of these viruses (Tables 2 to 5). Simian hemorrhagic fever virus has been assigned to this family, but information regarding it is insufficient, and it has not been placed in a genus.

IV. FAMILY BUNYAVIRIDAE

When Casals[18] demonstrated antigenic relationships between viruses belonging to different serogroups, he united them in the so-called Bunyamwera supergroup solely on the basis of intergroup antigenic relationships. Additional morphologic and molecular evidence supported this rationale and extended it to include in the family (Bunyaviridae) five genera having certain characteristics in common (single-stranded RNA, three circular RNA segments; spherical and enveloped 90 to 100 nm diameter virions). Members of these genera do not share antigenic relationships with members of the other four genera. These five genera are *Bunyavirus, Phlebovirus, Nairovirus, Uukuvirus,*[30] and the recently proposed *Hantavirus.*[31] At present, more than 265 viruses are included in these five genera. Viruses that share antigenic, morphologic, replicative, or molecular characteristics with members of a recognized genus are placed in that genus on a provisional basis. In addition, 38 viruses have been provisionally placed in the family Bunyaviridae but have not been assigned to a genus. Thus, more than 300 viruses are members of the family Bunyaviridae.

A. Genus *Bunyavirus*[30]

Members of the genus *Bunyavirus* possess negative-sense RNA replication strategy. Total molecular weight of the RNA segments is 4.78 to 5.9×10^6. The 3′ terminal sequence is UCAUCACAUG. Two of the four structural proteins are glycosylated (G1 and G2) with molecular weights of 108 to 120 and 29 to 41×10^3, respectively. Another is a nucleocapsid protein (N), molecular weight 19 to 25×10^3. A minor large protein (L) with a molecular weight of 145 to 200×10^3 has also been recognized. Within this genus (the former Bunyamwera supergroup), 16 serogroups, containing more than 153 viruses, are recognized. The bunyaviruses are found worldwide; are transmitted by mosquitoes and culicoids; have as their principal vertebrate hosts rodents and other small mammals, birds, or ungulates; and usually exist in silent sylvatic transmission cycles. At least 30 of these viruses or their subtypes and varieties have been reported to cause disease in humans or animals of veterinary importance, and three (Oropouche, La Crosse, and Bwamba viruses) have caused epidemics in humans. The geographic distributions of these viruses and their classification scheme have been established (Calisher et al., unpublished data) (Table 6). We have schematically demonstrated the antigenic interrelationships within the genus (Figure 1). No Bunyamwera serogroup viruses have been isolated in Australia, and only Batai virus has been isolated in Asia and Europe, probably because birds and migrating mammals are not involved in the mosquito-rodent cycles of these viruses in nature. Bunyamwera serogroup viruses are com-

Table 2
GEOGRAPHIC DISTRIBUTION OF VIRUSES BELONGING TO THE FAMILY FLAVIVIRIDAE, GENUS *FLAVIVIRUS* (GROUP B ARBOVIRUSES)[1]

Complex	Virus	Subtype	Variety	Distribution
St. Louis encephalitis	St. Louis encephalitis		3	North and South America
	Alfuy			Australia
	Japanese encephalitis			Asia
	Kokobera			Australia, New Guinea
	Koutango[a]			Senegal, Central African Republic
	Kunjin			Australia, Indonesia, Malaysia
	Murray Valley encephalitis			Australia, New Guinea
	Stratford			Australia
	Usutu	Usutu		Africa
		Yaounde		Cameroon, Central African Republic
	West Nile			Africa, Asia, Europe
Uganda S	Uganda S			Uganda, Nigeria
	Banzi			Africa
	Bouboui			Central African Republic, Senegal, Cameroon
	Edge Hill			Australia
Dengue	Dengue-1			North and South America, Asia, Africa, Australia, Oceania
	Dengue-2			North and South America, Asia, Africa, Australia, Oceania
	Dengue-3			North and South America, Asia, Australia, Tahiti
	Dengue-4			North and South America, Africa, Asia, Oceania
Ntaya	Ntaya			Uganda, Cameroon, Central African Republic
	Tembusu	Tembusu		Malaysia, Thailand
	Israel turkey meningoencephalitis	Israel turkey meningoencephalitis		Israel, South Africa
		Bagaza		Central African Republic, Cameroon, Senegal

[a] Koutango virus has not been isolated from naturally infected mosquitoes.
[b] This virus is not in the published or working *International Catalogue of Arboviruses.*[1]

Table 3
GEOGRAPHIC DISTRIBUTION OF TICK-BORNE FLAVIVIRUSES[1]

Complex	Virus	Subtype	Variety	Distribution
Russian spring-summer encephalitis	Russian spring-summer encephalitis	Russian spring-summer encephalitis		U. S. S. R., Europe
		Omsk hemorrhagic fever		Western Siberia
		Central European encephalitis[a]		Europe
		Kyasanur Forest disease		India
		Langat	Langat	Asia
			Negishi	Japan
			Carey Island	Malaysia
		Louping Ill		Scotland, England, Ireland
	Powassan			Canada; northern U. S., U. S. S. R.
	Karshi			Southern U. S. S. R.
	Royal Farm			Afghanistan
Tyuleniy	Tyuleniy			Eastern U. S. S. R., Oregon (U. S.)
	Saumarez Reef			Australia
	Meaban			France

[a] This virus is not in the published or working *International Catalogue of Arboviruses*.[1]

Table 4
GEOGRAPHIC DISTRIBUTION OF VECTOR-UNASSOCIATED FLAVIVIRUSES[1]

Complex	Virus	Subtype	Variety	Distribution
Modoc	Modoc			Western U.S.
	Cowbone Ridge			Florida (U.S.)
	Jutiapa			Guatemala
	Sal Vieja			Texas (U.S.)
	San Perlita	San Perlita	San Perlita	Texas (U.S.)
			(MA 387-72)[a]	Texas (U.S.)
Rio Bravo	Rio Bravo			Western U.S., Sonora (Mexico)
	Apoi			Japan
	Bukalasa bat[a]			Uganda
	Dakar bat			Africa
	Entebbe bat			Uganda
	Saboya			Senegal

[a] These viruses are not in the published or working *International Catalogue of Arboviruses.*[1]

Table 5
GEOGRAPHIC DISTRIBUTION OF FLAVIVIRUSES NOT ASSIGNED TO AN ANTIGENIC COMPLEX[1]

Virus	Distribution
Aroa	Venezuela
Bussuquara	Brazil, Colombia, Panama
Cacipacore	Brazil
Gadgets Gully	Macquarie Island (Australia)
Ilheus	North and South America
Jugra	Malaysia
Kadam	Uganda, Saudi Arabia
Kedougou	Senegal, Central African Republic
Montana Myotis Leukoencephalitis	Montana (U.S.)
Naranjal	Ecuador
Phnom-Penh bat	Cambodia, Malaysia
Rocio	Brazil
Sepik	New Guinea
Sokuluk	U.S.S.R.
Spondweni	Africa
Tamana bat[a]	Trinidad
Wesselsbron	Africa, Asia
Yellow fever	Africa, South America
Zika	Africa, Malaysia

[a] This virus is not in the published or working *International Catalogue of Arboviruses.*[1]

monly isolated in the Americas and in Africa.[1] The relative insularity of the North American Bunyamwera serogroup viruses may, in some way, be related to the distributions of their principal vertebrate hosts and the competence of their vectors. It is interesting to speculate that Batai and Northway viruses may represent phylogenetic links between Bunyamwera serogroup viruses in Africa and in North America. Rabbits, for example, are viremic or produce antibody after inoculation with Batai, Northway, Tensaw, or Cache Valley viruses,[1] whereas horses and perhaps hares are not susceptible to these viruses.[1] Two other Bunyamwera serogroup viruses from North America, Lokern and Main Drain viruses, replicate well in hares and have been isolated most frequently from *Culicoides* sp., but not from mosquitoes; Main Drain virus has been isolated from the brain of an encephalitic horse.[33]

Table 6
GEOGRAPHIC DISTRIBUTION OF VIRUSES OF THE FAMILY BUNYAVIRIDAE, GENUS *BUNYAVIRUS* (BUNYAMWERA SUPERGROUP)[1]

Complex	Virus	Subtype	Variety	Distribution
		Bunyamwera Serogroup		
Bunyamwera	Bunyamwera			Africa
	Batai (Calovo)			Europe, Asia
	Birao			Central African Republic
	Bozo[a]			Central African Republic
	Cache Valley	Cache Valley	Cache Valley	Canada,U. S.
			Tlacotalpan	Mexico
		Maguari	Maguari	South America
			CbaAr 426[a]	Argentina
			AG83-1746[a]	Argentina
		Playas		Ecuador
		Xingu[a]		Brazil
	Germiston			Africa
	Ilesha			Africa
	Lokern			Western U. S.
	Mboke			Cameroon
	Ngari			Senegal, Burkina Faso, Central African Republic, Madagascar
	Northway			Alaska (U. S.), Canada
	Santa Rosa			Mexico
	Shokwe			Africa
	Tensaw			Southern U. S.
Kairi	Kairi			South America
Main Drain	Main Drain			Western U. S.
Wyeomyia	Wyeomyia			South America, Panama
	Anhembi	Anhembi		Brazil
		Iaco		Brazil
		BeAr 328208[a]		Brazil
		Macaua		Brazil

		Sororoca		Brazil
		Taiassui[a]		Brazil
		Tucunduba[a]		Brazil
Anopheles A Serogroup				
Anopheles A	Anopheles A			Colombia
	CoAr 3624[a]			Colombia
	ColAn 57389[a]			Colombia
	Las Maloyas			Argentina
	Lukuni			Trinidad, Brazil
	Trombetas[a]			Brazil
Tacaiuma	Tacaiuma	Tacaiuma		Brazil
		SPAr 2317[a]	SPAr 2317	Brazil
			Virgin River	Arizona (U. S.)
		H-32580[a]		Brazil
	CoAr 1071[a]	CoAr 1071	CoAr 1071	Colombia
			CoAr 3627	Colombia
Anopheles B Serogroup				
Anopheles B	Anopheles B			Colombia
	Boraceia			Brazil
Bwamba Serogroup				
Bwamba	Bwamba	Bwamba		Africa
		Pongola		Africa
Group C Serogroup				
Caraparu	Caraparu	Caraparu	2	South America, Panama
		Ossa		Panama
	Bruconha[a]			Brazil
	Vinces			Ecuador
	Apeu			Brazil
Madrid	Madrid			Panama

Table 6 (continued)
GEOGRAPHIC DISTRIBUTION OF VIRUSES OF THE FAMILY BUNYAVIRIDAE, GENUS *BUNYAVIRUS* (BUNYAMWERA SUPERGROUP)[1]

Complex	Virus	Subtype	Variety	Distribution
Marituba	Marituba	Marituba		Brazil
		Murutucu		South America
		Restan		Trinidad, Surinam
	Nepuyo	Nepuyo	Nepuyo	North and South America
			63U11[a]	Mexico
		Gumbo Limbo		Florida (U. S.)
Oriboca	Oriboca	Oriboca		South America
		Itaqui		Brazil
		California Serogroup		
California encephalitis	California encephalitis	California encephalitis		Western North America
		Inkoo		Finland
		La Crosse	La Crosse	U. S.,
			snowshoe hare	Canada
		San Angelo		Western U. S.
		Tahyna	Tahyna	Europe, U. S. S. R.
			Lumbo[a]	Africa
Melao	Melao	Melao	Melao	Brazil, Trinidad, Panama
			AG83-497	Argentina
		Jamestown Canyon	Jamestown Canyon	U. S.
			Jerry Slough	California (U. S.)
			South River[a]	N. J., N. Y., Pa. (U. S.)
		Keystone		Southern U. S.
		Serra do Navio		Brazil
Trivittatus	Trivittatus			U. S.
Guaroa	Guaroa			Colombia, Brazil, Panama

		Capim Serogroup		
Capim	Capim			Brazil
Guajara	Guajara	Guajara	Guajara	Brazil, Panama
			GU71U350[a]	Guatemala
Bush Bush	Bush Bush	Bush Bush		Trinidad, Brazil
		Benfica		Brazil
		GU71U344[a]		Guatemala
	Juan Diaz			Panama
Acara	Acara			Brazil, Panama
	Moriche			Trinidad
Benevides	Benevides			Brazil
		Gamboa Serogroup		
Gamboa	Gamboa			Panama, Surinam
	Pueblo Viejo	Pueblo Viejo		Ecuador
		75V-2621[a]		Ecuador
Alajuela	Alajuela[a]			Panama
	San Juan	San Juan		Ecuador
		78V-2441[a]		Argentina
		75V-2374[a]		Ecuador
		Guama Serogroup		
Guama	Guama			South America
	Ananindeua			Brazil
	Mahogany Hammock			Florida (U.S.)
	Moju			Brazil
Bertioga	Bertioga			Brazil
	Cananeia			Brazil
	Guaratuba			Brazil
	Itimirim			Brazil
	Mirim			Brazil
Bimiti	Bimiti			Northern South America
Catu	Catu			Northern South America
Timboteua	Timboteua			Brazil

Table 6 (continued)
GEOGRAPHIC DISTRIBUTION OF VIRUSES OF THE FAMILY BUNYAVIRIDAE, GENUS *BUNYAVIRUS* (BUNYAMWERA SUPERGROUP)[1]

Complex	Virus	Subtype	Variety	Distribution
		Koongol Serogroup		
Koongol	Koongol			Australia, New Guinea
	Wongal			Australia
		Minatitlan Serogroup		
Minatitlan	Minatitlan			Mexico, Guatemala
	Palestina			Ecuador
		Olifantsvlei Serogroup		
Olifantsvlei	Olifantsvlei	Olifantsvlei	Olifantsvlei	Africa
			Bobia	Central African Republic
Botambi	Botambi			Central African Republic
	Dabakala			Ivory Coast
	Oubi			Ivory Coast
		Patois Serogroup		
Patois	Patois			Mexico, Central America
	Babahoyo			Ecuador
	Shark River			Florida (U.S.)
	Abras			Ecuador
Zegla	Zegla			Mexico, Central America
	Pahayokee			Florida (U.S.)

Simbu Serogroup				
Simbu	Simbu			Africa
Akabane	Akabane			Japan, Australia, Kenya
	Yaba-7[a]			Nigeria
Manzanilla	Manzanilla	Manzanilla		Trinidad
		Ingwavuma		Africa, India, Thailand, Cyprus
		Inini		French Guiana
		Mermet		Central U.S.
	Buttonwillow			Western U.S.
	Nola			Central African Republic
	Oropouche	Oropouche		Trinidad, Brazil
		Facey's Paddock[a]		Australia
		Utinga		Brazil
		Utive[a]		Panama
	Sabo	Sabo		Nigeria
		Tinaroo		Australia
Sathuperi	Sathuperi	Sathuperi		India, Nigeria
		Douglas		Australia
Shamonda	Shamonda			Nigeria
	Sango	Sango		Nigeria, Kenya
		Peaton		Australia
Shuni	Shuni			Nigeria, South Africa
	Aino	Aino	Aino	Japan, Australia
			Kaikalur	India
Thimiri	Thimiri			India, Egypt
Tete Serogroup				
Tete	Tete	Tete		South Africa, Nigeria
		Bahig		Egypt, Cyprus, Italy
		Matruh		Egypt, Italy
		Tsuruse		Japan
	Batama			Central African Republic

Table 6 (continued)
GEOGRAPHIC DISTRIBUTION OF VIRUSES OF THE FAMILY BUNYAVIRIDAE, GENUS *BUNYAVIRUS* (BUNYAMWERA SUPERGROUP)[1]

Complex	Virus	Subtype	Variety	Distribution
		Turlock Serogroup		
Turlock	Turlock			North and South America
	Umbre			India, Malaysia, Australia
	Lednice			Czechoslovakia, Romania
M'Poko	M'Poko	M'Poko		Central African Republic
		Yaba-1[a]		Nigeria
		No Serogroup Assigned		
	Kaeng Khoi			Thailand

[a] These viruses are not in the published or working *International Catalogue of Arboviruses*.[1]

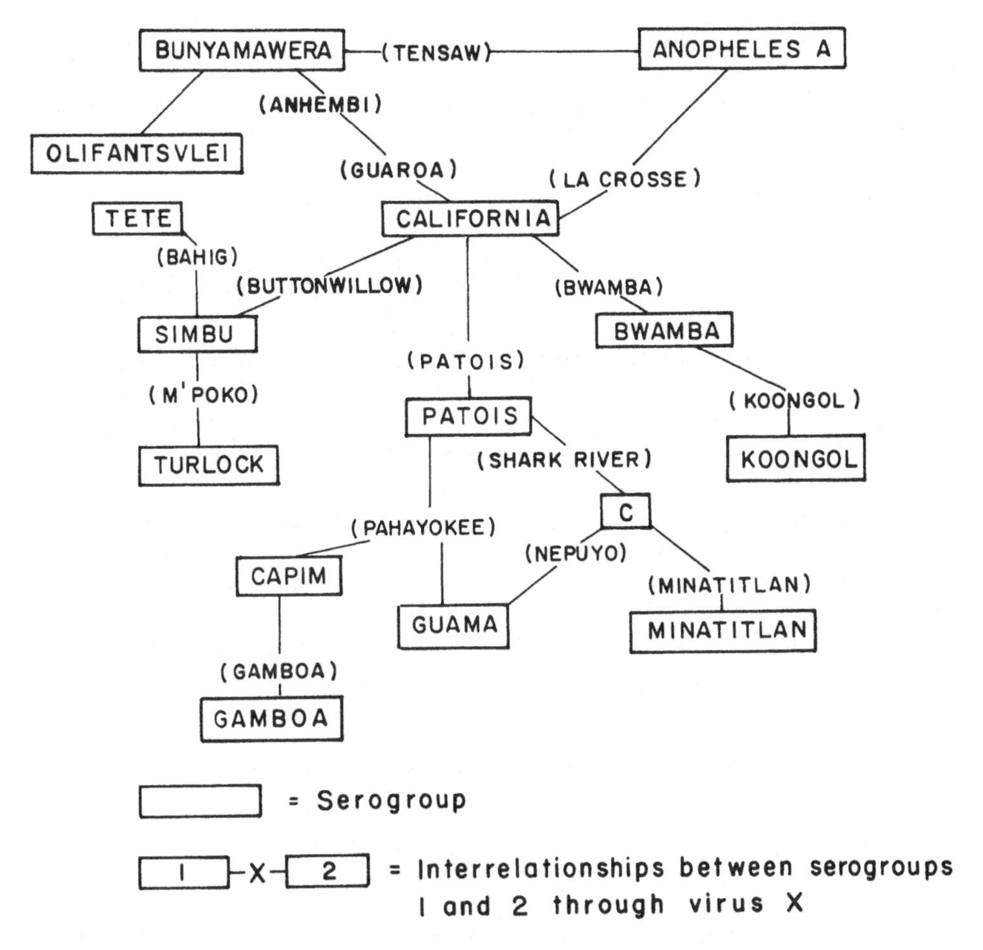

FIGURE 1. Schematic representation of antigenic interrelationships within the genus *Bunyavirus* (Bunyamwera supergroup).[17,32] (Anopheles B virus, not shown to be antigenically related to any of the viruses of the Bunyamwera supergroup, has been placed in the genus *Bunyavirus* on molecular grounds.[49])

Restriction of a virus to a vertebrate-vector pairing with defined geographic distribution may lead to natural isolation and, therefore, genetic stability in divergent evolution. For example, African Bunyamwera serogroup viruses have been found in distinct or overlapping geographic areas and ecosystems, but the South American members appear to coexist in horizontally or vertically contiguous, but not identical, ecosystems. One mechanism for separate maintenance of sympatric, closely related serotypes may be differences in vector susceptibility to them. Woodall[34] has suggested that the group C bunyavirus Itaqui is transmitted mainly by *Culex vomerifer,* a species apparently resistant to infection by Oriboca virus. Apeu and Marituba, also group C viruses, have been isolated from marsupials but not from rodents, whereas Caraparu and Murutucu viruses, present in the same area of Brazil, have been isolated frequently from rodents. Ultimately, cross-protection tests in vertebrates and the application of more sophisticated analyses, such as the use of western blotting techniques, will be necessary if we are to ascertain possible reasons for the coexistence of such closely related viruses in certain areas.

Like the Bunyamwera serogroup members, bunyaviruses of the California serogroup are transmitted between small mammals by mosquitoes, principally of the genus *Aedes*. The vector and host relationships of each virus appear to be quite restricted, possibly as a consequence of transovarial transmission in the arthropod vector; thus, the California serogroup viruses are geographically distributed in relation to the range of their vectors and hosts.[35] For example, Keystone virus has been isolated from rabbits and from *Sigmodon hispidus* (cotton rats) from Georgia and Florida.[35] Because *S. hispidus* rats only infrequently occur outside the southeastern U.S. and because of their association with the implicated principal mosquito vectors, *Ae. atlanticus/tormenter* and *Ae. infirmatus*, the distribution of Keystone virus is limited to this area. With other mosquitoes and other mammalian hosts, similar associations could be shown for others of the California serogroup. In fact, such postulations have been made for all of the North American members of this serogroup.[35]

The California and group C viruses are, in many respects, quite similar to those of the Bunyamwera serogroup; mammal-feeding mosquitoes transmit virus to small mammal hosts within geographic foci determined largely by the distribution and limited movements of the vertebrate hosts. Potentially competitive serotypes are excluded by natural, selective disadvantages imposed by vector-host restriction and ability to induce cross-protection. Recombination, when it occurs, takes place only between closely related serotypes. Shope and Causey[36] have shown that six of the group C bunyaviruses form three indistinguishable antigen pairs in complement-fixation tests. Karabatsos and Shope[37] extended these studies and suggested that because the complement-fixing antigen common to the members of the pairs is not an antigen shared by all members of the serogroup, ''pairing'' might have resulted from natural genetic recombination. The apparent plethora of such bunyavirus variants may indicate the natural order of things, in which continuous attempts to gain competitive advantages are made and then discarded or preserved. Nevertheless, one is continually impressed with the stability of viral antigenicity rather than its lability.

A clear-cut case in point may be made for bunyaviruses of the Australian region. About 100 million years ago during the Cretaceous period, the early mammals of Africa diverged into at least two groups, the marsupials and the placentals. Because Australia and Africa were connected by land at that time, movement of marsupials to Australia occurred. Later, Australia drifted to its present position surrounded by water and, therefore, became relatively isolated from the other continents. The evolution of marsupials in Australia has led to the appearance of these mammals in econiches filled by placental mammals in other regions (i.e., there are, in Australia, squirrel-, rabbit-, and wolf-like marsupials, as well as kangaroos, which correspond to the grazing mammals of other areas).

Fenner and Myers have suggested that the rabbit-myxoma virus host-parasite relationship may have originated in the Pleistocene period when the modern leporids were evolving and spreading throughout the world. These authors also inferred that leporiviruses might have evolved in conjunction with the evolution of these mammals.[38]

The principal vertebrate hosts of the Bunyamwera and California serogroup viruses are placental mammals. The absence of these viruses in Australia may indicate that they arose somewhat later than the occurrence of the drift of the Australian continent, perhaps fewer than 10 million years ago. The current situation is probably not one of phylogenetic stagnation but merely a point in time of a continuing occurrence.

Simbu-, Tete-, and Turlock-serogroup viruses have been isolated from resident and migrating birds, which may explain their relatively widespread distributions. Other bunyaviruses are confined to single continents because they replicate in mosquitoes and rodents, marsupials, bats, and other mammals but not in birds. Gamboa-serogroup viruses represent a third type. They replicate in and are transovarially transmitted by *Aedeomyia squamipennis* mosquitoes, which feed principally on birds. However, the geographic distribution of these viruses appears to be limited to the distribution of the arthropod vector, not the vertebrate host.

B. Genus *Phlebovirus*[30]

Members of the genus *Phlebovirus* possess an ambisense RNA replication strategy. Total molecular weight of the RNA segments is 5.1 to 5.8 $\times$ 10^6. The 3′ terminal sequence is

Table 7
GEOGRAPHIC DISTRIBUTION OF VIRUSES OF THE FAMILY BUNYAVIRIDAE, GENUS *PHLEBOVIRUS* (PHLEBOTOMUS FEVER SEROGROUP)[1]

Complex	Virus	Subtype	Distribution
Sandfly fever Naples	Sandfly fever Naples		Italy, Africa, Asia
	Karimabad		Iran
	Tehran		Iran
	Toscana		Italy
Bujaru	Bujaru		Brazil
	Munguba		Brazil
Rift Valley fever	Rift Valley fever (Zinga)	Rift Valley fever	Africa
		Belterra[a]	Brazil
	Icoaraci		Brazil
Candiru	Candiru		Brazil
	Alenquer		Brazil
	Itaituba	Itaituba	Brazil
		Oriximina	Brazil
	Nique		Panama
	Turuna		Brazil
Punta Toro	Punta Toro	Punta Toro	Panama
		Buenaventura	Colombia, Panama
Frijoles	Frijoles		Panama
	Joa[a]		Brazil
Chilibre	Chilibre		Panama
	Cacao		Panama
Salehabad	Salehabad		Iran
	Arbia		Italy
(No complex assigned)	Sandfly fever Sicilian		Italy, Africa, Asia
	Aguacate		Panama
	Anhanga		Brazil
	Arumowot		Africa
	Caimito		Panama
	Chagres		Panama
	Corfou		Greece
	Gabek Forest		Africa
	Gordil		Central African Republic
	Itaporanga		Brazil, Trinidad, French Guiana
	Pacui		Brazil, Trinidad
	Rio Grande		Texas (U.S)
	Saint-Floris		Central African Republic
	Urucuri		Brazil

[a] These viruses are not in the published or working *International Catalogue of Arboviruses.*[1]

UGUGUUUCG. Two of the four structural proteins are glycosylated (G1 and G2) with molecular weights of 55 to 70 and 50 to 60 $\times$ 10^3, respectively. Another is a nucleocapsid protein (N), molecular weight 20 to 30 $\times$ 10^3. A minor large protein (L) with a molecular weight of 145 to 200 $\times$ 10^3 has also been recognized. A single serogroup, the Phlebotomus fever serogroup, constitutes this genus. Of the 37 members, none occur in Australia, one has been found in the U.S., three occur only in Europe, three only in Asia, five only in Africa, and two in Africa, Asia, and Europe (Table 7). The last two, sandfly Sicilian and sandfly Naples viruses, have caused epidemics, and Rift Valley fever virus,[39] although limited to the African continent, is widespread there and has caused extensive and serious epizootics from South Africa to Egypt.[40] The other phleboviruses occur in South or Central America, and their ecologic characteristics suggest a relationship between arboreal or ground-dwelling mammalian hosts and virus distribution; alternatively, the geographic distributions of these viruses may be simply restricted by the distribution of their arthropod vectors.

C. Genus *Nairovirus*[30]

RNA replication strategy of viruses belonging to the genus *Nairovirus* is unknown. However, total molecular weight of the three RNA segments is 6.2 to 7.5 $\times$ 10^6. The 3′ terminal sequence is AGAGUUUCU. Two of the four structural proteins are glycosylated (G1 and G2) with molecular weights of 72 to 84 and 30 to 40 $\times$ 10^3, respectively. Another is a nucleocapsid protein (N), molecular weight 48 to 54 $\times$ 10^3. A minor large protein (L) with a molecular weight of 145 to 200 $\times$ 10^3 has also been recognized. Six serogroups, containing 28 viruses and subtypes, comprise this genus (Table 8). With the exception of Belem virus, all the nairoviruses have been isolated from ticks; Dugbe and Ganjam viruses of the Nairobi sheep disease serogroup have also been isolated from culicine mosquitoes, and strains of Dugbe, Nairobi sheep disease, and Crimean-Congo hemorrhagic fever viruses have been obtained from *Culicoides* spp. as well. With certain notable exceptions (Nairobi sheep disease and Crimean-Congo hemorrhagic fever), little is known of the vertebrate hosts of the nairoviruses; however, at least one member each of the Sakhalin and Hughes serogroups have been isolated from seabirds, probably accounting for the relatively widespread distributions of these viruses.

D. Genus *Uukuvirus*[30]

RNA replication strategy of viruses belonging to the genus *Uukuvirus* is unknown. However, total molecular weight of the three RNA segments is 3.4 to 4.4 $\times$ 10^6. The 3′ terminal sequence is UGUGUUUCUGGAG. Two of the four structural proteins are glycosylated (G1 and G2) with molecular weights of 70 to 75 and 65 to 70 $\times$ 10^3, respectively. Another is a nucleocapsid protein (N), molecular weight 20 to 25 $\times$ 10^3. A minor large protein (L) with a molecular weight of 180 to 200 $\times$ 10^3 has also been recognized. The uukuviruses have been recovered primarily from ticks, and birds are their principal vertebrate hosts. None of these viruses is known to cause disease in humans or livestock. The genus contains one serogroup with 11 members (Table 9). With the exception of Uukuniemi virus, which is distributed throughout Europe, viruses belonging to this genus (serogroup) appear to occur focally in areas of Europe, Asia, Africa, North America, and Australia (Macquarie Island).

E. Genus *Hantavirus*[31]

Members of the proposed genus *Hantavirus* possess negative-sense RNA replication strategy. Total molecular weight of the RNA segments is about 4.5 $\times$ 10^6. The 3′ terminal sequence is AUCAUCAUCUG. Two of the four structural proteins are glycosylated (G1 and G2) with molecular weights of 68 to 72 and 54 to 60 $\times$ 10^3, respectively. Another is a nucleocapsid protein (N), molecular weight 50 to 53 $\times$ 10^3. A minor large protein with a molecular weight of about 200 $\times$ 10^3 has also been recognized. At present, only one serogroup comprises this proposed genus. Seven Hantaan serogroup viruses,[31,41] have been found in Asia, North and South America, and Europe (Table 10). There is a strong likelihood that one or more members of this serogroup will be found in Australia and that many more members of the serogroup will be isolated in the near future. Existing information suggests that the host associations of the hantaviruses are species-specific. Hantaan virus has been recovered from *Apodemus* spp. field mice, Seoul, Tchoupitoulas, Girard Point, and Sapporo rat viruses from rats, Prospect Hill virus from *Microtus* sp. meadow voles, and Puumala virus from *Clethrionomys* spp. bank voles. Several isolates of hantaviruses have not been definitively typed, so these apparent host and geographic associations may not be borne out, or they may be extended.

Table 8
GEOGRAPHIC DISTRIBUTION OF VIRUSES OF THE FAMILY BUNYAVIRIDAE, GENUS *NAIROVIRUS*[1]

Serogroup	Complex	Virus	Subtype	Distribution
Crimean-Congo hemorrhagic fever	Crimean-Congo hemorrhagic fever	Crimean-Congo hemorrhagic fever		Africa, U.S.S.R., Europe, Asia
		Hazara		Pakistan
		Khasan		U.S.S.R.
Dera Ghazi Khan	Dera Ghazi Khan	Dera Ghazi Khan		Pakistan
		Abu Hammad		Egypt, Iran
		Abu Mina[a]		Egypt
		Kao Shuan		Taiwan, Australia
		Pathum Thani		Thailand
		Pretoria		South Africa
Hughes	Hughes	Hughes		Florida (U.S.), Trinidad, California (U.S.), Cuba
		Farallon[a]		California (U.S.)
		Fraser Point[a]		Australia
		Punta Salinas		Peru
		Raza[a]		Mexico
		Sapphire II[a]		Southwestern U.S.
		Soldado		Trinidad, Europe, Africa, Seychelles
		Zirqa		Persian Gulf
Nairobi sheep disease	Nairobi sheep disease	Nairobi sheep disease (Ganjam)		Africa, India
		Dugbe		Africa
Qalyub	Qalyub	Qalyub		Egypt
		Bandia		Senegal
		Omo		Ethiopia
Sakhalin	Sakhalin	Sakhalin	Sakhalin	Eastern U.S.S.R.
			Tillamook[a]	Oregon (U.S.)
		Kachemak Bay[a]		Alaska (U.S.)
		Clo Mor		Scotland
		Avalon (Paramushir)		U.S.S.R., Newfoundland (Canada)
		Taggert		Macquarie Island (Australia)

[a] These viruses are not in the published or working *International Catalogue of Arboviruses*.[1]

Table 9
GEOGRAPHIC DISTRIBUTION OF VIRUSES OF THE FAMILY BUNYAVIRIDAE, GENUS *UUKUVIRUS* (UUKUNIEMI SEROGROUP)[1]

Complex	Virus	Subtype	Distribution
Uukuniemi	Uukuniemi	Uukuniemi	Europe, U.S.S.R.
		Oceanside[a]	Oregon (U.S.)
	Grand Arbaud		France
	Manawa		Pakistan
	Murre[a]		Alaska (U.S.)
	Ponteves		France
	Precarious Point		Macquarie Island (Australia)
	Zaliv Terpeniya		U.S.S.R.
	EgAn-1825-61[a]		Egypt
	Fin V-707[a]		Norway
	UK FT/254[a]		Scotland

[a] These viruses are not in the published or working *International Catalogue of Arboviruses.*[1]

Table 10
GEOGRAPHIC DISTRIBUTION OF VIRUSES OF THE FAMILY BUNYAVIRIDAE, GENUS *HANTAVIRUS* (HANTAAN SEROGROUP)[1,30,31]

Complex	Virus	Subtype	Distribution
Hantaan	Hantaan	Hantaan	Korea, China, U.S.S.R.
		Seoul	Asia
		Tchoupitoulas[a]	Louisiana (U.S.)
		Girard Point[a]	Pennsylvania (U.S.)
		Sapporo rat[a]	Japan
	Prospect Hill		Maryland (U.S.)
Puumala	Puumala		Finland, Sweden, U.S.S.R.

[a] These viruses are not in the published or working *International Catalogue of Arboviruses.*[1]

F. Bunyavirus-Like Viruses[1,30]

Thirty-eight viruses have morphologic and molecular characteristics in common with members of the family Bunyaviridae; usually, the common denominator is morphology. In the absence of adequate molecular studies or antigenic relatedness with a recognized member of one or another of the serogroups within the family, these bunyavirus-like viruses have been placed provisionally within the family but are denoted as "possible members of the family".[30] Included among these are 22 viruses belonging to 8 serogroups and 16 ungrouped viruses (Table 11). Viruses of the Bhanja, Kaisodi, and Upolu serogroups are principally tick-borne, while those of the Bakau, Mapputta, Nyando, and Resistencia serogroups appear to be principally mosquito-borne. Matariya serogroup viruses have been isolated only from birds. Knowledge of the geographic distributions of all these viruses is quite limited, as is that of the ungrouped mosquito-borne and ungrouped tick-borne bunyavirus-like viruses.

Table 11
GEOGRAPHIC DISTRIBUTION OF POSSIBLE MEMBERS OF THE FAMILY BUNYAVIRIDAE, GENUS UNASSIGNED[1]

Serogroup	Complex	Virus	Subtype	Distribution
Bakau	Bakau	Bakau		Malaysia, Pakistan
		Ketapang		Malaysia
Bhanja	Bhanja	Bhanja		Asia, Africa, Europe
		Forecariah		New Guinea
		Kismayo[a]		U. S. S. R.
Kaisodi	Kaisodi	Kaisodi		India
		Silverwater		Canada, Wisconsin and Alaska (U. S.)
		Lanjan		Malaysia
Mapputta	Mapputta	Mapputta		Australia
		Maprik	Maprik	New Guinea
			GanGan	Australia
		Trubanaman		Australia
Matariya	Matariya	Matariya		Egypt
		Burg el Arab		Egypt
		Garba		Central African Republic
Nyando	Nyando	Nyando		Kenya, Central African Republic, Senegal
		Eret-147[a]		Ethiopia
Resistencia	Resistencia	Resistencia		Argentina
		Barranqueras		Argentina
		Antequera		Argentina
Upolu	Upolu	Upolu		Australia
		Aransas Bay		Texas (U. S.)
(Serogroup undetermined)		Bangui		Central African Republic
		Belmont		Australia
		Bobaya		Central African Republic
		Caddo Canyons[a]		Oklahoma (U. S.)
		Enseada		Brazil
		Issyk-Kul (Keterah)		U. S. S. R., Malaysia
		Kowanyama		Australia
		Leanyer		Australia
		Lone Star		Kentucky (U. S.)
		Pacora		Panama
		Razdan		U. S. S. R.
		Sunday Canyon		Texas (U. S.)
		Tamdy		U. S. S. R.
		Tataguine		Africa
		Wanowrie		India, Egypt, Sri Lanka, Iran
		Witwatersrand		Africa

[a] These viruses are not in the published or working *International Catalogue of Arboviruses*.[1]

V. FAMILY REOVIRIDAE GENUS *ORBIVIRUS*[1,25]

Orbiviruses are icosahedral particles, 60 to 80 nm in diameter. They contain 10 or 12 segments of double-stranded RNA. Molecular weight of the 10-segment RNA of the prototype virus of the genus, bluetongue virus, is about 12×10^6. Seven structural proteins have been found, with molecular weights ranging between 32 and 155×10^3. Polypeptide 2 contains the main antigenic determinant for neutralization. Thirteen serogroups comprise the orbiviruses. Certain members of the bluetongue, epizootic hemorrhagic disease, and

Eubenangee serogroups have been shown to share intergroup antigenic relationships.[42] This, perhaps, indicates the existence of a "bluetongue supergroup". The Palyam serogroup, which may share antigens with these viruses,[42] could be included in such a supergroup. Kemerovo serogroup viruses have been studied thoroughly, and assignment of individual viruses to antigenic complexes has been suggested.[43] Members of most serogroups in this genus are almost exclusively transmitted by either culicoids (African horse sickness, bluetongue, epizootic hemorrhagic disease, Wallal), phlebotomine flies (Changuinola), or ticks (Colorado tick fever,* Kemerovo). Corriparta serogroup viruses are transmitted by mosquitoes and Eubenangee, Palyam, and Warrego serogroup viruses have been isolated from mosquitoes and culicoids. The equine encephalosis serogroup is comprised of four recognized serotypes. Vector associations have not been defined, but the geographic distribution of these four viruses appears to be limited to South Africa. Seven ungrouped mosquito-borne, one ungrouped tick-borne, and one ungrouped virus with unknown vector has been placed in the genus on morphologic grounds. The orbiviruses have a wide geographic distribution (Table 12).

VI. FAMILY RHABDOVIRIDAE

Two genera have been established within this family. All rhabdoviruses from vertebrates and invertebrates are bullet-shaped and contain one segment of single-stranded, negative-sense RNA and four structural proteins. Vesiculoviruses are 178 to 188 × 60 to 75 nm in size, and their four structural proteins (L, G, N, and M) weigh 150, 70 to 80, 50 to 62, and 20 to 30 × 10^3, respectively. Lyssaviruses are 180 × 60 to 75 nm in size, and their four structural proteins (L, G, N, and M2) weigh 190, 65 to 80, 58 to 62, and 22 to 25 × 10^3, respectively. The genus *Vesiculovirus* includes the vesicular stomatitis serogroup (Table 13), while the genus *Lyssavirus* includes rabies and the rabies-related viruses (Table 14).[25] The rhabdoviruses mentioned here either are known to be transmitted by arthropods or have not been associated with an arthropod vector.

Vesicular stomatitis serogroup viruses are transmitted by culicine mosquitoes or phlebotomine flies. It has been speculated that transmission of vesicular stomatitis viruses by plant-feeding insects may be important in establishing epizootics.[44,45] This hypothesis suggests that vesicular stomatitis virus (New Jersey) is a plant virus that also infects mammals. Vesicular stomatitis (Indiana) and (New Jersey) serotypes are distributed throughout the western hemisphere, while other members of the serogroup appear to be more localized in areas of the Americas and elsewhere.

Whereas rabies and rabies-related viruses are transmitted vertebrate-to-vertebrate, at least two members of this serogroup, kotonkan and obodhiang viruses, have been isolated from *Culicoides* spp. midges and unengorged *Mansonia uniformis* mosquitoes, respectively. With the exception of rabies virus, distributed essentially worldwide, other lyssaviruses are found only in Africa; most appear to be restricted to relatively discrete geographic foci, but the isolation of Kolongo virus from birds suggests a more complex natural cycle than has been recognized.

Viruses belonging to nine other serogroups have been placed in the family Rhabdoviridae, based on virus morphology. Geographic distribution of the rhabdoviruses not assigned to a genus are summarized in Table 15.

Hart Park-serogroup rhabdoviruses appear to be maintained in nature in a cycle involving culicine mosquitoes and birds. Hart Park virus is found throughout essentially all subarctic western North America, while Flanders virus is found throughout essentially all subarctic eastern North America. Little is yet known about Mosqueiro virus.

* Taxonomic status being reevaluated. May be placed in a separate genus at a later date.

Table 12
GEOGRAPHIC DISTRIBUTION OF ORBIVIRUSES (FAMILY REOVIRIDAE, GENUS *ORBIVIRUS*)[1]

Serogroup	Complex	Virus	Subtype	Distribution
African horse sickness		African horse sickness (9 or 10 serotypes)		Africa, Asia
Bluetongue		Bluetongue (23 serotypes)		Africa, Asia, Europe, North America, Australia
Epizootic hemorrhagic disease		Epizootic hemorrhagic disease	Epizootic hemorrhagic disease (N.J.)[a]	North America, Nigeria, South Africa
			Epizootic hemorrhagic disease (Alberta)	Canada
		Ibaraki		Japan, Australia, Indonesia, Taiwan
Eubenangee		Eubenangee		Australia
		Ngoupe		Central African Republic
		Pata		Central African Republic.
		Tilligerry		Australia
Palyam		Palyam		India
		Abadina[b]		Nigeria
		Bunyip Creek		Australia
		CSIRO village		Australia
		D'Aguilar		Australia
		Kasba		India
		Kindia		New Guinea
		Marrakai		Northern Australia
		Nyabira[b]		Zimbabwe
		Petevo		Central African Republic
		Vellore		India
Umatilla		Umatilla		Western U.S.
		Llano Seco		California (U.S.)
		Netivot[b]		Israel
Kemerovo	Kemerovo	Kemerovo		Africa, Asia
		Lipovnik		Czechoslovakia
		Tribec	Tribec	Czechoslovakia, Italy, Romania, U.S.S.R.
			Brezova[b]	Czechoslovakia

Table 12 (continued)
GEOGRAPHIC DISTRIBUTION OF ORBIVIRUSES (FAMILY REOVIRIDAE, GENUS *ORBIVIRUS*)[1]

Serogroup	Complex	Virus	Subtype	Distribution
	Great Island	Great Island		Newfoundland (Canada)
		Bauline		Newfoundland (Canada)
		Cape Wrath		Scotland
		Fin V-808[b]		Finland
		Kenai[b]		Alaska (U. S.)
		Mill Door/79[b]		Scotland
		Mykines		Denmark
		Nugget		Macquarie Island (Australia)
		Okhotskiy		Eastern U.S.S.R.
		Poovoot[b]		Alaska (U. S.)
		Tindholmur		Denmark
		Yaquina Head		Oregon and Alaska (U. S.)
	Chenuda	Chenuda		Egypt, South Africa, U.S.S.R.
		Baku		Azerbaijan (U.S.S.R.)
		Essaouira[b]		Morocco
		Huacho		Peru
		Mono Lake		California (U.S.)
		Sixgun City		Texas and Colorado (U.S.)
	Wad Medani	Wad Medani		Africa, Asia, Jamaica
		Seletar		Singapore, Malaysia
Changuinola		Changuinola		Panama
		Almeirim		Brazil
		Altamira		Brazil
		Caninde		Brazil
		Gurupi		Brazil
		Irituia		Brazil
		Jamanxi		Brazil
		Jari		Brazil
		Monte Dourado		Brazil
		Ourem		Brazil
		Purus		Brazil
		Saraca		Brazil

Corriparta	Corriparta	Corriparta	Australia
		Bambari[b]	Brazil
	Acado		Ethiopia
	Jacareacanga	Jacareacanga	Brazil
		BeAr 263191[b]	Brazil
Wallal	Wallal		Australia
	Mudginbarry[b]		Australia
Warrego	Warrego		Queensland (Australia
	Mitchell River		Queensland (Australia)
Equine encephalosis	Bryanston[b]		South Africa
	Cascara[b]		South Africa
	Gamil[b]		South Africa
	Kaalplaas[b]		South Africa
Colorado tick fever	Colorado tick fever		North America
	Eyach		West Germany, France
	S6-14-03[b]		California (U.S.)
(Serogroup undetermined)	Chobar Gorge		Nepal
	Ife		Nigeria, Cameroon
	Ieri		Trinidad, Brazil
	Japanaut		New Guinea
	Lebombo		South Africa, Nigeria
	Orungo		Uganda, Nigeria, Central African Republic, Senegal, Ivory Coast
	Ndelle		Cameroon
	Paroo River		New South Wales (Australia)

[a] Two varieties.
[b] These viruses are not in the published or working *International Catalogue of Arboviruses*.[1]

Table 13
GEOGRAPHIC DISTRIBUTION OF VIRUSES OF THE FAMILY RHABDOVIRIDAE, GENUS *VESICULOVIRUS* (VESICULAR STOMATITIS SEROGROUP)[1]

Complex	Virus	Subtype	Variety	Distribution
Vesicular stomatitis (Indiana)	Vesicular stomatitis (Indiana)			North, South, and Central America
	Carajas[a]			Brazil
	Cocal			Trinidad, Brazil, Argentina
	Maraba[a]			Brazil
	Perinet			Madagascar
	Vesicular stomatitis (Alagoas)			Brazil
Vesicular stomatitis New Jersey	Vesicular stomatitis (New Jersey)	Vesicular stomatitis (New Jersey)	2	U.S., Central and South America
Piry	Piry			Brazil
Chandipura	Chandipura	Chandipura	2	India, Nigeria
Isfahan	Isfahan			Iran, U.S.S.R.
	Jurona			Brazil
	La Joya			Panama
	Yug Bogdanovac			Yugoslavia
	Calchaqui[a]			Argentina
Bahia Grande	Bahia Grande[a]			Texas and New Mexico (U. S.)
	Reed Ranch[a]			Texas (U. S.)
	Muir Springs[a]			Colorado and North Dakota (U. S.)
(No complex assigned)	Malpais Spring[a]			New Mexico (U. S.)

[a] These viruses are not in the published or working *International Catalogue of Arboviruses*.[1]

Table 14
GEOGRAPHIC DISTRIBUTION OF VIRUSES OF THE FAMILY RHABDOVIRIDAE, GENUS *LYSSAVIRUS* (RABIES SEROGROUP)[1]

Complex	Virus	Subtype	Variety	Distribution
Rabies	Rabies[a]	Rabies	Several	Worldwide
	Lagos bat	Lagos bat	2	Nigeria, South Africa, Central African Republic
	Mokola[a]			Nigeria, Central African Repulic, Zimbabwe, Cameroon
	Kolongo[a]			Central African Republic
	Duvenhage[a]			South Africa
Obodhiang	Obodhiang[a]			Sudan
Kotonkan	Kotonkan			Nigeria

[a] These viruses are not in the published or working *International Catalogue of Arboviruses*[1]

Table 15
GEOGRAPHIC DISTRIBUTION OF VIRUSES ASSIGNED TO THE FAMILY RHABDOVIRIDAE (NOT ASSIGNED TO A GENUS)[1]

Serogroup	Virus	Subtype	Variety	Distribution
Hart Park	Hart Park			Western North America
	Flanders			Eastern North America
	Mosqueiro			Brazil
Sawgrass	Sawgrass			Florida (U. S.)
	New Minto			Alaska (U. S.)
	Connecticut			Connecticut (U. S.)
Kwatta	Kwatta	Kwatta	2	Surinam, Brazil
Timbo	Timbo			Brazil
	Chaco			
	Sena Madureira			Brazil
Mossuril	Mossuril			Mozambique, Central African Republic, South Africa
	Bangoran			Central African Republic
	Barur			India, Kenya, Somalia
	Charleville			Australia
	Cuiaba			Brazil
	Kamese			Uganda, Central African Republic
	Kern Canyon			California and Texas (U. S.)
	Marco			Brazil
Malakal	Malakal			Sudan
	Puchong			Malaysia
Bovine ephemeral	Bovine ephemeral fever			Australia, Africa, Asia
	Kimberley			Australia
	Adelaide River			Australia
	Berrimah			Australia
	Oak-Vale[a]			Australia
Tibrogargan	Tibrogargan			Australia
	Bivens Arm[a]			Florida (U. S.)
	Coastal Plains[a]			Australia
Le Dantec	Le Dantec			Senegal
	Keuraliba			Senegal

Table 15 (continued)
GEOGRAPHIC DISTRIBUTION OF VIRUSES ASSIGNED TO THE FAMILY RHABDOVIRIDAE (NOT ASSIGNED TO A GENUS)[1]

Serogroup	Virus	Subtype	Variety	Distribution
(Serogroup undetermined)	Almpiwar			Australia
	Aruac			Trinidad
	Boteke			Central African Republic
	Gossas			Senegal
	Gray Lodge			California (U. S.)
	Inhangapi			Brazil
	Joinjakaka			New Guinea
	Klamath			Oregon and Alaska (U. S.)
	Kununurra			Australia
	Mount Elgon bat			Kenya
	Navarro			Colombia
	Ngaingan			Australia
	Nkolbisson			Cameroon
	Oita 296[a]			Japan
	Ouango			Central African Republic
	Parry Creek[a]			Australia
	Porton[a]			Sarawak, Indonesia
	Rochambeau			French Guiana
	Sandjimba			Central African Republic
	Sripur			India
	Xiburema			Brazil
	Yata			Central African Republic

[a] These viruses are not in the published or working *International Catalogue of Arboviruses.*[1]

Members of the Sawgrass serogroup are transmitted by ticks to rabbits in such disparate areas as Florida (Sawgrass virus), Alaska (New Minto virus), and Connecticut (Connecticut virus).

Kwatta virus has been isolated from culicine mosquitoes and has been isolated only in Surinam; a closely related virus has been isolated from birds in Belem, Brazil.

Timbo serogroup viruses have not been associated with a vector and have been isolated from lizards and in Brazil only.

Rhabdoviruses in the Mossuril serogroup represent an interesting mix of antigenically related viruses with apparently different ecologies. Bangoran and Mossuril viruses are transmitted in culicine mosquito-to-bird cycles, while Kamese virus has been isolated only from culicine mosquitoes; Barur virus has been isolated from culicine and anopheline mosquitoes, ticks, and rodents; Charleville virus has been isolated from phlebotomine flies and a lizard; Cuiaba virus has been isolated from a toad; Marco virus has been isolated only from a lizard, and Kern Canyon virus has been isolated only from bats.

Although placed in what is known as "minor antigenic groups", the viruses are not "minor" in terms of importance or interest, only in the size of the serogroups. For example, the bovine ephemeral fever serogroup contains at least one virus of economic importance, bovine ephemeral fever virus. It is most likely transmitted to cattle by *Culicoides* sp. midges, although it has also been recovered from mosquitoes. The four other members of this serogroup and those of the Tibrogargan serogroup probably are maintained in nature in similar associations, although only Kimberly, Tibrogargan, and Bivens Arm viruses have been isolated from midges; all but Tibrogargan and Bivens Arm viruses have been isolated from cattle. With the exception of bovine ephemeral fever virus, found in Australia, Africa, and Asia, viruses of both the bovine ephemeral fever and Tibrogargan serogroups have been found only in Australia: Bivens Arm virus has been isolated from *Culicoides insignis* in Florida.

Vector associations have not been determined for Le Dantec virus, isolated from the serum of a human, and for Keuraliba virus, on several occasions isolated from rodents. These two viruses have been found only in Senegal. Malakal serogroup viruses have been isolated only from *Mansonia uniformis* mosquitoes and, thus far, only from Sudan and Malaysia.

As with the rhabdoviruses assigned to serogroups, the antigenically ungrouped rhabdoviruses exhibit a diversity of geographic distributions, vectors, and vertebrate host associations. Most have been isolated from culicine mosquitoes and two, Inhangapi and Sripur viruses, have been recovered from phlebotomine files; a few rhabdoviruses have not been associated with an arthropod vector.

Rabies, Mokola, and Duvenhage viruses have caused disease in humans. In fact, Mokola and Duvenhage viruses have been shown to infect animals and humans and to have produced disease in spite of a history of rabies vaccination. Other rhabdoviruses implicated in human disease include the vesiculoviruses vesicular stomatitis (Indiana), vesicular stomatitis (New Jersey), vesicular stomatitis (Alagoas), Piry, and Le Dantec viruses. Neutralizing antibodies to Isfahan virus, a member of the *Vesiculovirus* genus, have been detected in sera of humans residing in the U.S.S.R. and Iran.

While this information may simply reflect the intensity of the efforts to isolate viruses belonging to one serogroup or another, the presence of rhabdoviruses on five continents and the fact that they have been isolated from lizards, a toad, rodents, and bats and from culicine and anopheline mosquitoes, phlebotomine flies and ticks, indicates the adaptability of these viruses and suggests that additional studies will provide fascinating information.

VII. FAMILY ARENAVIRIDAE, GENUS *ARENAVIRUS*[25]

Arenaviruses are spherical to pleomorphic particles, 50 to 300 nm in diameter. They contain five RNA segments. Two of these are virus-specified (segments L and S) with molecular weights of 1.1 to 1.6 and 2.1 to 3.2 $\times$ 10^6, respectively. Three other RNAs are of host cell origin. The RNAs are single-stranded and have an ambisense replication strategy. Three stuctural proteins are known: a major nonglycosylated protein with a molecular weight of 63 to 72×10^3 has been associated with the nucleocapsid, and one or two glycosylated proteins have also been recognized. One, with molecular weight about 34 to 44×10^3, is consistently detected in arenaviruses; the other, with molecular weight about 54 to 72×10^3, has been detected in some, but not all, arenaviruses. A single genus constitutes the family Arenaviridae, and all are members of the Tacaribe serogroup (Table 16). Although isolated from bats and arthropods on rare occasions, Tacaribe serogroup viruses are principally rodent-borne, a probable explanation for their restricted distributions. Junin, Machupo, and Lassa viruses cause severe disease in humans, and stringent containment facilities are required to work safely with them. Consequently, they have been studied by arbovirologists and are registered in the *International Catalogue of Arboviruses.*[1] Most arenaviruses have been isolated only in South America, Lassa, Toure, Ippy (Africa), Tamiami (Florida) and lymphocytic choriomeningitis viruses being exceptions.

Knowledge of zoogeographical distributions of animal species may provide some insight into the relationship of geologic change to viral phylogeny. Perhaps the clearest example of this is the relationship between the South American cricetid rodents and arenaviruses of the Tacaribe complex. Ancestral Cricetidae from North America invaded South America during the Miocene-Pliocene epoch[46] and evolved into a wide variety of forms. In concert with the divergent evolution of hosts, one may speculate that an evolution of new arenaviruses

Table 16
GEOGRAPHIC DISTRIBUTION OF VIRUSES OF THE FAMILY ARENAVIRIDAE, GENUS *ARENAVIRUS* (TACARIBE SEROGROUP)[1]

Complex	Virus	Subtype	Variety	Distribution
Tacaribe	Tacaribe			Trinidad
	Amapari			Brazil
	Flexal			Brazil
	Junin			Argentina
	Latino			Bolivia, Brazil
	Machupo			Bolivia
	Parana			Paraguay
	Pichinde			Colombia
	Tamiami			Florida (U.S.)
Lymphocytic choriomeningitis	Lymphocytic choriomeningitis[a]	Lymphocytic choriomeningitis	2	Worldwide
	Lassa	Lassa	2	Africa
	Toure			Senegal
	Ippy			Central African Republic

[a] This virus is not in the published or working *International Catalogue of Arboviruses*[1]

occurred. Of interest is the apparently unique occurrence of a member of the Tacaribe complex (Tamiami virus) in the cricetid rodent, *Sigmodon hispidus*, in Florida. As Arata and Gratz[47] have pointed out, the evolution of *S. hispidus* as a South American species is certain from the Pleistocene fossil record, and this rodent is, therefore, a recent invader back into North America. Like the arenaviruses, many bunyavirus groups are associated with neotropical cricetid and hystricomorph rodents and marsupials. These groups contain members found both in Florida and South and Central America. Thus, Gumbo Limbo (group C), Mahogany Hammock (Guama group), and Shark River and Pahayokee (Patois group) bunyaviruses have been isolated in Florida; the South American corollaries of these viruses are Nepuyo, CoAn 49888, Patois, and Zegla viruses, respectively. It may be that as with the arenavirus Tamiami virus, North American bunyaviruses have been brought from South America or perhaps the viruses had an earlier distribution in Florida separated from the neotropical habitat by geologic changes.

Another member of the Tacaribe serogroup, lymphocytic choriomeningitis virus, is not registered in the *International Catalogue of Arboviruses;* however, it has been found essentially worldwide.

VIII. OTHER VIRUSES

Twelve of the 13 viruses belonging to the families Coronaviridae, Nodaviridae, Poxviridae, Orthomyxoviridae, Iridoviridae, Herpesviridae, and Paramyxoviridae have been registered in the *International Catalogue of Arboviruses.*[1,25] These viruses are from diverse vertebrate and arthropod species. With the exceptions of Nodamura and African swine fever viruses, the isolation of these viruses from arthropods probably represents an incidental presence in them. Again, registration of viruses, such as the coronaviruses Bocas and Tettnang, poxviruses Cotia, Oubangui, Salanga, and Yoka the orthomyxoviruses Dhori and Thogoto, herpesvirus Agua Preta, and the paramyxovirus Nariva, may merely indicate the wide-ranging interests of arbovirologists. In fact, both Bocas and Tettnang have been isolated in mice inoculated with tissues of bats, mosquitoes, and ticks; serologic relationships between these

Table 17
GEOGRAPHIC DISTRIBUTION OF OTHER VIRUSES REGISTERED IN THE *CATALOGUE OF ARBOVIRUSES*[1]

Family	Virus	Subtype	Distribution
Coronaviridae	Bocas		Wisconsin (U. S.), Panama, Colombia
	Tettnang		West Germany, Czechoslovakia, Egypt
Nodaviridae	Nodamura		Japan
Poxviridae	Cotia		Brazil, French Guiana
	Oubangui		Central African Republic
	Salanga		Central African Republic
	Yoka[a]		Central African Republic
Orthomyxoviridae	Dhori		India, Egypt, Portugal, U. S. S. R.
	Thogoto	Thogoto	Africa, Europe, Iran
		SiAr 126[a]	Italy
Iridoviridae	African swine fever		Africa, Portugal, Spain
Herpesviridae	Agua Preta		Brazil
Paramyxoviridae	Nariva		Trinidad, Colombia

[a] This virus is not in the published or working *International Catalogue of Arboviruses.*[1]

Table 18
UNCLASSIFIED AND ANTIGENICALLY UNGROUPED, MOSQUITO-ASSOCIATED VIRUSES[1]

Virus	Distribution
Andasibe	Madagascar
Arkonam	India
Gomoka	Central African Republic
Itupiranga	Brazil
Minnal	India, Australia
Okola	Cameroon
Para	Brazil, Argentina
Picola	Australia
Tanga	Tanzania
Tembe	Brazil
Termeil	Australia
Venkatapuram	India
Wongorr	Australia
Yacaaba	Australia

Table 19
UNCLASSIFIED AND ANTIGENICALLY UNGROUPED, TICK-, *CULICOIDES* SP.-, and *PHLEBOTOMUS* SP.-ASSOCIATED VIRUSES[1]

Virus	Distribution
Aride	Seychelles
Batken	U. S. S. R.
Chim	U. S. S. R.
Estero Real	Cuba
Lake Clarendon	Australia
Matucare	Bolivia
Slovakia	Czechoslovakia

viruses and mouse hepatitis virus have been demonstrated.[48] Alternatively, two or more of these 13 viruses may be true arboviruses (Dhori and Thogoto) or may someday be shown to be antigenically related to an arbovirus. Further studies of these viruses are needed. Their geographic distributions are summarized in Table 17.

IX. UNCLASSIFIED VIRUSES[1]

Forty-three viruses registered in the *International Catalogue of Arboviruses* and two unregistered viruses have not been characterized sufficiently by physicochemical, electron microscopic, biological, or molecular methods to warrant placement in a taxon. These viruses are listed in Tables 18 to 21. For most, serologic relationships with other unclassified viruses have not been shown; however, there are five serogroups of unclassified viruses, viruses

Table 20
UNCLASSIFIED AND ANTIGENICALLY UNGROUPED VIRUSES WITHOUT RECOGNIZED ARTHROPOD VECTORS[1]

Virus	Distribution
Araguari	Brazil
Fomede	Guinea
Kammavanpettai	India
Kannamangalam	India
Landjia	Central African Republic
Mapuera	Brazil
Mojui dos Campos	Brazil
Sakpa	Central African Republic
Santarem	Brazil
Sebokele	Central African Republic
Sembalam	India
Thottapalayam	India

Table 21
UNCLASSIFIED VIRUSES BELONGING TO ANTIGENIC SEROGROUPS[1]

Serogroup	Virus	Distribution
Marburg	Marburg	South Africa
	Ebola	Zaire, Sudan
Nyamanini	Nyamanini	South Africa, Egypt, Nigeria
	Midway[a]	Central Pacific islands, Japan
Quaranfil	Quaranfil	Africa, Afghanistan, Iran
	Johnston Atoll	Central Pacific islands. Australia, New Zealand
Tanjong Rabok	Tanjong Rabok	Malaysia
	Telok Forest	Malaysia
Yogue	Yogue	Senegal, Central African Republic
	Kasokero[a]	Uganda

[a] These viruses are not in the published or working *International Catalogue of Arboviruses.*[1]

that are related antigenically to each other, but which have not been placed in a virus family (Table 21). A family name has been proposed for members of the Marburg serogroup (Table 21), Marburg and Ebola viruses, two important human pathogens.[50] However, pending official recognition of this name, they have been placed among the unclassified viruses for the purposes of this chapter. It is the ongoing task of reference laboratories to attempt classification of such viruses, but new ones are found essentially as fast as previously recognized unclassified viruses are characterized. Within this category may very well be viruses belonging to recognized families, viruses representing heretofore unrecognized families, duplicates of viruses already registered in the *International Catalogue of Arboviruses,* and nonviral agents.

ADDENDUM

Continuing biochemical and antigenic studies of viruses place viral taxonomy in a constant state of revision. This chapter is no exception. As an example, in determinations of the antigenic relationships among rhabdoviruses, we have found cross-reactions between rabies-related and bovine ephemeral fever-related viruses, which suggest placement of the latter in the genus *Lyssavirus*. Also, Klamath, Ngaingan, Nkolbisson, Sandjimba, Sripur, and others indicated in Table 15 as "(serogroup undetermined)" have been shown to be related to other ungrouped rhabdoviruses or to viruses placed in one or the other genus of rhabdoviruses. Certain unclassified and antigenically ungrouped viruses (Tables 18 to 20) and unclassified viruses belonging to serogroups (Table 21) have now been shown to have antigenic relatives. It was not possible to include all our findings on a continuing basis and still meet publication deadlines.

Therefore, certain tables in this chapter should not be considered as any more than provisional. For additional information regarding current classification status of any of the viruses listed, please contact one of the authors.

ACKNOWLEDGMENTS

This chapter is a summary of the combined and often concerted efforts of numerous investigators. For example, the Subcommittee on Interrelationships Among Catalogued Arboviruses (SIRACA) of the American Committee on Arthropod-borne Viruses, has, under the successive leadership of Jordi Casals and Robert E. Shope, been responsible for concentrating studies of antigenic relationships performed in many laboratories. Our tables, while not an official report of the deliberations of SIRACA, are reflections of the general sense of these deliberations, with our interpretations superimposed upon them and with our opinions having to suffice where classification schemes are not available. The mention of viruses and virus names that are neither registered in the *International Catalogue of Arboviruses*[1] nor published is not intended to constitute priority of publication. Less generally, we thank Robert B. Tesh, Robert E. Shope, and Andrew J. Main, Yale Arbovirus Research Unit, Yale University, New Haven, Conn.; Barry M. Gorman, Queensland Institute of Medical Research, Brisbane, Australia; and all members of SIRACA, past and present, who contributed so much in fact and in conversations. Without the abilities of Carol J. Frank, who typed the manuscript, this chapter would be incomplete still.

REFERENCES

1. **Karabatsos, N., Ed.,** *International Catalogue of Arboviruses Including Certain Other Viruses of Vertebrates,* 4th ed., American Society of Tropical Medicine and Hygiene, San Antonio, Tex. 1985.
2. **Smithburn, K. C.,** Differentiation of the West Nile virus from the viruses of St. Louis and Japanese B encephalitis, *J. Immunol.*, 44, 25, 1942.
3. **Sabin, A. B. and Buescher, E. L.,** Unique physico-chemical properties of Japanese B encephalitis virus hemagglutinin, *Proc. Soc. Exp. Biol. Med.*, 74, 222, 1950.
4. **Sabin, A. B.,** Hemagglutination by viruses affecting the human nervous system, *Fed. Proc.*, 10, 573, 1951.
5. **Chanock, R. M. and Sabin, A. B.,** The hemagglutinin of St. Louis encephalitis virus, *J. Immunol.*, 70, 271, 1953.
6. **Chanock, R. M. and Sabin, A. B.,** The hemagglutinin of western equine encephalitis virus: recovery, properties and use for diagnosis, *J. Immunol.*, 73, 337, 1954a.

7. **Chanock, R. M. and Sabin, A. B.,** The hemagglutinin of West Nile virus: recovery, properties and antigenic relationships, *J. Immunol.*, 73, 352, 1954b.
8. **Casals, J. and Brown, L. V.,** Hemagglutination with arthropod-borne viruses, *J. Exp. Med.*, 99, 429, 1954.
9. **Havens, W. P., Jr., Watson, D. W., Green, R. H., Lavin, G. I., and Smadel, J. E.,** Complement-fixation with the neurotropic viruses, *J. Exp. Med.*, 77, 139, 1943.
10. **Clarke, D. H. and Theiler, M.,** The hemagglutinins of Semliki Forest and Bunyamwera viruses, their demonstration and use, *J. Immunol.*, 75, 470, 1955.
11. **Casals, J.,** The arthropod-borne group of animal viruses, *Trans. N. Y. Acad. Sci.*, 2, 19 and 219, 1957.
12. **Clarke, D. H. and Casals, J.,** Techniques for hemagglutination and hemagglutination-inhibition with arthropod-borne viruses, *Am. J. Trop. Med. Hyg.*, 7, 561, 1958.
13. **Casals, J. and Whitman, L.,** A new antigenic group of arthropod-borne viruses, the Bunyamwera group, *Am. J. Trop. Med. Hyg.*, 9, 73, 1960.
14. **Casals, J. and Whitman, L.,** Group C, a new serological group of hitherto undescribed arthropod-borne viruses. Immunological studies, *Am. J. Trop. Med. Hyg.*, 10, 250, 1961.
15. **Whitman, L. and Shope, R. E.,** The California complex of arthropod-borne viruses and its relationship to the Bunyamwera group through Guaroa virus, *Am. J. Trop. Med. Hyg.*, 1, 691, 1962.
16. **Whitman, L. and Casals, J.,** The Guama group: a new serological group of hitherto undescribed viruses. Immunological studies, *Am. J. Trop. Med. Hyg.*, 10, 259, 1961.
17. **Bishop, D. H. L. and Shope, R. E.,** Bunyaviridae, in *Comparative Virology*, Fraenkel-Conrat, H. and Wagner, R. R., Eds., Plenum Press, New York, 1979, 1.
18. **Casals, J.,** New developments in the classification of arthropod-borne animal viruses, *Ann. Microbiol.*, 11, 13, 1963.
19. **Calisher, C. H., Sasso, D. R., Maness, K. S. C., Gheorghiu, V. N., and Shope, R. E.,** Relationships of Anopheles A group arboviruses, *Proc. Soc. Exp. Biol. Med.*, 143, 465, 1973.
20. **Casals, J.,** Arboviruses: incorporation in a general system of virus classification, in *Comparative Virology*, Maramorosch, K. and Kurstak, E., Eds., Academic Press, New York, 1971, 307.
21. **Murphy, F. A., Harrison, A. K., and Whitfield, S. G.,** Bunyaviridae: morphologic and morphogenetic similarities of Bunyamwera serologic supergroup viruses and several other arthropod-borne viruses, *Intervirology*, 1, 297, 1973.
22. WHO Scientific Group, Arboviruses and human disease, *WHO Tech. Rep. Ser.*, 369, 1, 1967.
23. **Westaway, E. G., Brinton, M. A., Gaidamovich, S. Ya., Horzinek, M. C., Igarashi, A., Kaariainen, L., Lvov, D. K., Porterfield, J. S., Russell, P.K., and Trent, D. W.,** Togaviridae, *Intervirology*, 24, 125, 1985.
24. **Westaway, E. G., Brinton, M.A., Gaidamovich, S. Ya., Horzinek, M. C., Igarashi, A., Kaariainen, L., Lvov, D. K., Porterfield, J. S., Russell, P. K., and Trent, D. W.,** Flaviviridae, *Intervirology*, 24, 183, 1985.
25. **Matthews, R. E. F.,** Classification and nomenclature of viruses. Fourth report of the International Committee on Taxonomy of Viruses, *Intervirology*, 17, 1, 1982.
26. **Calisher, C. H., Brandt, W., Casals, J., Shope, R. E., Tesh, R. B., and Wiebe, M. E.,** Recommended antigenic classification of registered arboviruses. I. Togaviridae, Alphaviruses, *Intervirology*, 14, 229, 1980.
27. **Chamberlain, R. W.,** Epidemiology of arthropod-borne togaviruses: the role of arthropods as hosts and vectors and of vertebrate hosts in natural transmission cycles, in *The Togaviruses*, Schlesinger, R. W., Ed., Academic Press, New York, 1980.
28. **de Madrid, A. T. and Porterfield, J. S.,** The Flaviviruses (Group B arboviruses): a cross-neutralization study, *J. Gen. Virol.*, 23, 91, 1974.
29. **Varelas-Wesley, I. and Calisher, C. H.,** Antigenic relationships of flaviviruses with undetermined arthropod-borne status, *Am. J. Trop. Med. Hyg.*, 31, 1273, 1982.
30. **Bishop, D. H. L., Calisher, C. H., Casals, J., Chumakov, M. P., Gaidamovich, S. Y., Hannoun, C., Lvov, D. K., Marshall, I. D., Oker-Blom, N., Pettersson, R. F., Porterfield, J. S., Russell, P. K., Shope, R. E., and Westaway, E. G.,** Bunyaviridae, *Intervirology*, 14, 125, 1980.
31. **Schmaljohn, C. S., Hasty, S. E., Dalrymple, J. M., LeDuc, J. W., Lee, H. W., von Bonsdorff, C.-H., Brummer-Korvenkontio, M., Vaheri, A., Tsai, T. F., Regnery, H. L., Goldgaber, D., and Lee, P. W.,** Antigenic and genetic properties of viruses linked to hemorrhagic fever with renal syndrome, *Science*, 227, 1041, 1985.
32. **Calisher, C. H.,** Antigenic relationships of the arboviruses: an ecological and evolutionary approach, in *Proc. Int. Symp. New Aspects in the Ecology of Arboviruses*, Labuda, M. and Calisher, C. H., Eds., Slovak Academy of Sciences, Bratislava, Czechoslovakia, 1980.
33. **Emmons, R. W., Woodie, J. D., Laub, R. L., and Oshiro, L. S.,** Main Drain virus as a cause of equine encephalomyelitis, *J. Am. Vet. Med. Assoc.*, 183, 555, 1983.

34. **Woodall, J. P.**, personal communication, 1978.
35. **Sudia, W. D., Newhouse, V. F., Calisher, C. H., and Chamberlain, R. W.**, California group arboviruses: isolation from mosquitoes in North America, *Mosq. News*, 31, 576, 1971.
36. **Shope, R. E. and Causey, O. R.**, Further studies on the serological relationships of group C arthropod-borne viruses and the application of these relationships to rapid identification of types, *Am. J. Trop. Med. Hyg.*, 11, 293, 1962.
37. **Karabatsos, N. and Shope, R. E.**, Cross-reactive and type-specific complement-fixing structures of Oriboca virions, *J. Med. Virol.*, 3, 167, 1979.
38. **Fenner, F. and Myers, K.**, Myxoma virus and myxomatosis in retrospect: the first quarter century of a new disease, in *Viruses and Environment*, Kurstak, E. and Maramorosch, K., Eds., Academic Press, New York, 1978.
39. **Shope, R. E., Peters, C. J., and Walker, J. S.**, Serological relation between Rift Valley fever virus and viruses of Phlebotomus fever serogroup, *Lancet*, 1, 886, 1980.
40. **Meegan, J. M., Hoogstraal, H., Khalil, G. M., and Adham, F. K.**, Symposium on Rift Valley Fever. The Rift Valley Fever epizootic in Egypt 1977-78. Description of the epizootic and virological studies, *Trans. R. Soc. Trop. Med. Hyg.*, 73. 618, 1979.
41. **Lee, H. W., Lee, P. W., and Johnson, K. M.**, Isolation of the etiologic agent of Korean hemorrhagic fever, *J. Infect. Dis.*, 137, 298, 1978.
42. **Gorman, B. M., Taylor, J., and Walker, P. J.**, Orbiviruses, in *The Reoviridae*, Joklik, W. K., Ed., Plenum Press, New York, 1983.
43. **Main, A. J., Shope, R. E., and Wallace, R. C.**, Cape Wrath: a new Kemerovo group orbivirus from *Ixodes uriae* in Scotland, *J. Med. Entomol.*, 13, 204, 1976.
44. **Jonkers, A. H.**, The epizootiology of the vesicular stomatitis viruses: a reappraisal, *Am. J. Epidemiol.*, 86, 286, 1967.
45. **Johnson, K. M., Tesh, R. B., and Peralta, P. H.**, Epidemiology of vesicular stomatitis virus: some new data and a hypothesis for transmission of the Indiana serotype, *J. Am. Vet. Med. Assoc.*, 155, 2133, 1969.
46. **Hershkovitz, P.**, Mice, land bridges and Latin American faunal interchange, in *Ectoparasites of Panama*, Wenzel, R. L. and Tipton, V. J., Eds., Field Museum of National History, Chicago, 1966.
47. **Arata, A. A. and Gratz, N. G.**, The structure of rodent faunas associated with arenaviral infections, *Bull. WHO*, 52, 621, 1975.
48. **Smith, A. L., Casals, J., and Main, A. J.**, Antigenic characterization of Tettnang virus: complications caused by a passage of the virus in mice from a colony enzootically infected with mouse hepatitis virus, *Am. J. Trop. Med. Hyg.*, 32, 1172, 1983.
49. **Klimas, R. A., Ushijima, H., Clerx-Van Haster, C. M., and Bishop, D. H. L.**, Radioimmune assays and molecular studies that place Anopheles B and Turlock serogroup viruses in the *Bunyavirus* genus (Bunyaviridae), *Am. J. Trop. Med. Hyg.*, 30, 876, 1981.
50. **Kiley, M. P., Bowen, E. T. W., Eddy, G. A., Isaacson, M., Johnson, K. M., McCormick, J. B., Murphy, F. A., Pattyn, S. R., Peters, D., Prozesky, O. W., Regnery, R. L., Simpson, D. I. H., Slenczka, W., Sureau, P., van der Groen, G., Webb, P. A., and Wulff, H.**, Filoviridae: a taxonomic home for Marburg and Ebola viruses?, *Intervirology*, 18, 24, 1982.

Chapter 3

VIRUS VARIATION AND EVOLUTION: MECHANISMS AND EPIDEMIOLOGICAL SIGNIFICANCE

Barry J. Beaty, Dennis W. Trent, and John T. Roehrig

TABLE OF CONTENTS

I. INTRODUCTION

The unifying feature of arbovirus biology is the development of complex arthropod-vertebrate virus transmission/maintenance cycles. In such cycles, virus genetic information must be successfully expressed in two phylogenetically unrelated biological systems. It is remarkable that viruses as diverse in physicochemical and morphogenetic properties as the togaviruses, flaviviruses, bunyaviruses, orbiviruses, and rhabdoviruses have evolved transmission cycles which involve continuous bidirectional bridging of the phylogenetic gap separating vertebrate from invertebrate.[1] Biological and ecological impediments to the success of arbovirus cycles must be compensated for by alternate factors associated with vector transmission. Vectors undoubtedly contribute to arbovirus cycles in a variety of ways; dissemination, maintenance, overwintering and amplification of viruses are apparent vector contributions. In addition, the possibilities for multiple virus infections of arthropods and the persistent nature of vector infections may provide unique opportunities for the generation of genetic diversity in arbovirus populations. Thus, virus interactions with both vector and vertebrate host may be determinants of the genotypic and phenotypic composition of arbovirus populations.

There are now 504 viruses registered in the *International Catalogue of Arboviruses;*[2] this number is testimony to their evolutionary success. The majority of the viruses have been serologically classified in the families Togaviridae, Flaviviridae, Reoviridae (orbiviruses), Bunyaviridae, and Rhabdoviridae. It is not the intent of this chapter to review the serologic relationships, geographic distribution, and vertebrate/invertebrate interactions which have been used to classify and to derive putative evolutionary relationships between the taxa. The reader is referred to Chapter 2 and to other articles for a thorough discussion of these matters.[3-12] Rather, mechanisms which could account for arbovirus genetic diversity are described, selected studies which describe the molecular bases of antigenic and biologic diversity are analyzed, and possible implications of such genetic diversity are discussed.

The vast majority of the arboviruses possess RNA genomes. Indeed, African swine fever is the only DNA virus that is unequivocally considered to be an arbovirus.[2] Similarly, the majority of plant viruses that are transmitted by vectors also contain RNA genomes.[13,14] It is curious that RNA viruses have been so successful in exploiting arthropod-associated transmission cycles. Perhaps the genomic plasticity, biochemical compatibility of replication, and basic biology of RNA viruses are necessary to permit replication in both vertebrate and vector hosts. Transmission from invertebrate to vertebrate may upset RNA virus population equilibrium and result in selection of genotypes more capable of replication in the vector or vertebrate.[13] In the biological setting, the interplay of phenotypic expression of evolving RNA with the host immune system, with virus receptors in vertebrate and invertebrate systems and with vector specificities for vertebrate host, add multiple dimensions to the evolution and selection of these viruses as they cycle in nature.

Elucidation of the evolutionary patterns of arboviruses is important. Genotypic changes, which result in altered phenotypes, could exhibit enhanced virulence or altered tropisms in vertebrates. Resultant antigenic changes could also provide a mechanism for viruses to elude existing immunologic barriers in the vertebrate populations, resulting in serious epidemiologic consequences. Possibilities for altered virus expression in nature are futher enhanced for arboviruses by vector transmission. Should a new virus efficiently infect, replicate in, and be transmitted by alternate vector species, the virus could then be expressed in a new arbovirus cycle involving new vertebrate host species. Vectors that feed upon other than the preferred vertebrate host conduct ongoing arthropod experimentation to test viruses in new systems.[3,4] Adaptation of the virus to the new vertebrate hosts and generation of sufficient viremia titers to infect attendant vector species could lead to establishment of a new arbovirus cycle.

II. MECHANISMS FOR GENERATION OF GENETIC DIVERSITY

The RNA genome of arboviruses can be either segmented or nonsegmented. Therefore, at least two distinct processes may function, either independently or synergistically, in the generation of genetic diversity. Genomic change could result from either intramolecular events or segment reassortment. These possibilities represent the principal mechanisms by which variation occurs in influenza viruses: genetic drift and shift, respectively.[15-19]

A. Evolutionary Mechanisms for Arboviruses with Nonsegmented Genomes

1. General Considerations

Arboviruses with nonsegmented genomes include the flaviviruses, alphaviruses, and rhabdoviruses. With these viruses, genomic changes most probably arise intramolecularly as a consequence of nucleotide sequence deletions, inversions, and substitutions. Compared to DNA viruses, genomes of RNA viruses mutate at a very high frequency, presumably because of the fidelity of viral replication enzymes and the lack of proofreading enzymes.[13,20] Nucleotide substitution error rates have been demonstrated to be as frequent as 10^{-3} to 10^{-4} in an RNA bacteriophage.[21] In contrast, the spontaneous mutation frequency reported for a variety of eukaryotic genomes ranges from approximately 10^{-6} to 10^{-9}.[20,22-24] The relative evolutionary potential of RNA and DNA genomes is illustrated by a comparison of spontaneous mutation rates in the temperate bacteriophage lambda. When replicating as a prophage (integrated into the bacterial genome), the nucleotide substitution error rate is 20 to 100 times lower than when the lambda phage replicates lytically.[22] As many as 50% of these nucleotide substitutions may be silent; base substitutions in the third position of codons do not usually result in corresponding amino acid changes.[24] Such changes are phenotypically and presumably biologically neutral.

High frequency mutation rates have been reported for influenza and vesicular stomatitis (VS) viruses. The mutation rate for nucleotide substitutions in seven genes of influenza virus averaged 1.1×10^{-2}.[23] In persistent VS infections of cell cultures and nude mice, base substitution rates per average site per year can equal or exceed 10^{-3}.[13,20,24,25] These high frequency mutation rates for VS are associated with the generation of defective interfering (DI) particles which compete with wild type (*wt*) virus. Sequencing studies have revealed a stepwise accumulation of DI particle-resistant mutants of VS during persistent infections.[13] Concomitantly, new DI particles with altered interference capabilities appear and disappear. Most DI particle-resistant mutants have altered base substitutions in genes coding for proteins involved in replication and encapsidation.[13] Thus, the *wt* virus apparently must evolve in order to compete with newly generated DI particles for replicative enzymes and substrates. These circumstances are conducive to rapid accumulation of mutations in the virus genome. In contrast to the high frequency mutation rates observed during persistent infections, VS maintained by lytic passage remains genetically stable as evidenced by the similar oligonucleotide fingerprints of VS maintained for many years in different laboratories.[26]

Assuming that the rate of base substitution for arboviruses in general may be as frequent as 10^{-3}, then a single progeny RNA of 12 kb would likely differ from the parent genome by at least 1 base. Thus, an arbovirus isolate represents a population of RNA molecules that differ slightly from one another. Although many of the nucleotide changes from the average sequence are silent, these mutants may be at a selective disadvantage because of the need to preserve the secondary structure of the parental RNA, the replication efficiency, and genome function. For arbovirus evolution, this means that selective pressures of different stringencies function at the genome level to preserve genome integrity: relatively moderate selective pressure against silent mutations, stronger pressure against mutations in codons which affect protein function, and absolute selection against lethal mutations.[11]

The role of intramolecular recombination in the evolution of arboviruses remains to be determined. Intramolecular recombination has been demonstrated for certain RNA viruses;[27-30] however, little effort has been devoted to determine if the phenomenon occurs with arboviruses. Opportunities for dual arbovirus infection of both vertebrate and invertebrate do exist, suggesting that intramolecular recombination could occur.

2. Variation in Alphaviruses and Flaviviruses

The genomic organization and antigenic structures of viruses in the familes Togaviridae (alphaviruses) and Flaviviruses have been extensively studied. Representatives of these two groups will be used to illustrate current knowledge and hypotheses concerning putative evolutionary processes of viruses with nonsegmented genomes.

The alphavirus genus of the family Togaviridae contains 25 members, many of which have geographic variants or subtypes.[7,31] Inclusion in this genus is based primarily on serological relationships defined by virus neutralization (N), hemagglutination inhibition (HI), and complement fixation (CF). The alphaviruses have been separated into six complexes with members of each complex being serologically related.[32] These antigenic relationships imply phenotypic similarities derived from amino acid sequences which define both common and unique antigenic epitopes and protein structure.[33] Morphologically the alphaviruses are similar in appearance, containing an outer envelope with two glycoproteins, El and E2, surrounding a nucleocapsid which encloses the viral genomic RNA. [33,34] Polyacrylamide gel electrophoresis (PAGE) and isoelectric focusing (IEF) of purified structural proteins reveal that the corresponding proteins of each alphavirus have similar molecular weights and IEF points.[33] In the alphavirus genome, the structural protein genes are located at the 3′ one third of the genome immediately following a nontranslated section of 150 to 450 bases in the order 5′C,E2,E3,E2,6K,E1-3′.[11] Analysis of alphavirus genomic RNA species by oligonucleotide fingerprinting has demonstrated that alphaviruses within a serotype maintain relative genetic stability over time and throughout their respective geographic distributions.[11,35]

Isolates of Ross River virus from various localities in eastern Australia, Samoa, Fiji, and the Wallis and Cook Islands have been examined at the genome level. Restriction endonuclease digests of cDNA of 14 isolates of Ross River identify three genetic types.[36] There is a 1 to 5% nucleotide sequence diversity between the types and small, but significant, differences among isolates within the type. All isolates obtained during the Ross River epidemic in the Pacific islands are virtually identical, suggesting that extensive evolution did not accompany virus spread.

Analysis by oligonucleotide fingerprinting of genomic RNA from isolates of western equine encephalitis (WEE) virus reveals a surprising conservation in genetic structure.[37] WEE virus isolates from North and South America obtained over a 34-year-period share 75 to 100% of their long oligonucleotides. Highlands J virus isolates obtained over a 21-year-period share from 75 to 100% of their long oligonucleotides. Similarly, fingerprint analysis of various Venezuelan equine encephalitis (VEE) virus strains (subtypes IA and IB), including isolates made during the 1971 epidemic in Central America, Mexico, and Texas, reveal amazing genetic homology. The viruses share 95 to 100% of their oligonucleotides.[38,39] Eastern equine encephalitis (EEE) virus isolates obtained from Florida, New Jersey, and Michigan over a 25-year-period, including the prototype EEE virus, share 78 to 95% genomic homology.[129] Sindbis virus isolates from a broad geographic area are classified into five distinct genetic varieties based on oligonucleotide fingerprints as well as nucleic acid hybridization and tryptic peptide analyses.[40,41]

Reasons for the relative genetic stability of the alphaviruses must be determined by biological constraints involving virus replication and assembly. These constraints exert sufficient pressures to insure survival of the virus. Nonetheless, adequate flexibility is retained to permit evolution in genomic regions not critical to biological function. As with polio and measles viruses, the neutralization antigenic epitopes of alphaviruses appear to be maintained.

This would imply that although mutations in the genome coding for the outer viral proteins do occur, these mutations, which result in changes in critical virion structures involved in virus attachment and neutralization, are lethal.[11] Genetic events leading to antigenic variants that result in changes in serotypes could arise by point mutations, short deletions, or insertions in genes coding for the surface proteins. Intramolecular recombination between alphaviruses has not been demonstrated. If such events do occur, selection of such recombinants has been constrained within closely related viruses. Such a mechanism may explain the evolution of the VEE virus complex. The neutralizing epitope of VEE subtype IAB is conserved and shared by subtypes IC and II.[32]

In the alphavirus genome, strict nucleotide conservation has been maintained in critical regions.[11] For example, in proteins El and E2, the positions of all cysteines are conserved (Figure 1). Amino acids important in the secondary structure of the alphavirus glycoproteins are also conserved at many other positions. This suggests that the three-dimensional configuration of the glycoprotein is similar for all alphaviruses. Hydrophobicity plots for VEE and other alphaviruses are also similar, suggesting that although specific sequence diversity occurs, characteristic hydrophilic domains are conserved (Figure 2).

However, considerable variation in the primary amino acid sequence of each of the structural proteins has evolved. This variability is especially evident in certain positions of the genome.[11] The greatest divergence is found in regions near the N-terminus of the capsid protein and among the transmembranal domains (Figure 1). The most conserved domains lie in the amino terminal third of the El protein and in the carboxyl half of the capsid protein. Examination of amino acid sequence data for Sindbis,[35] Semliki Forest,[42] Ross River,[43] VEE,[44] and EEE[129] viruses reveals that E2, which contains the epitopes reactive with neutralizing antibody, is the least conserved glycoprotein among the viruses. Since E2 evolves more rapidly than El, it is probably involved more directly in strain diversity. More will be said about the role of antibody selection on the antigenic composition of arbovirus populations later.

The amino acid sequence of the nonstructural alphavirus proteins is more homologous, suggesting that these proteins evolve more slowly than the structural proteins.[11] The putative molecular basis for this phenomenon has been discussed for other virus families.[45] Differential rates of evolution may occur in the nonstructural genes of a virus. In Sindbis and Middelburg viruses, which share a genomic homology of 73%, the nonstructural protein nsP3 is more divergent in amino acid sequence than is nsP4.[11]

Codon usage for VEE virus is not random (Table 1).[46] The frequency of dinucleotide CG usage in the alphavirus genome is more similar to that of the mosquito host than to the vertebrate.[11] This codon usage preference is a general feature of the alphaviruses and may reflect an adaptation of the viruses to replicate efficiently in their mosquito hosts.

Although the alphaviruses have probably evolved from a single parent, extensive sequence divergence has occurred during evolution. Many protein sequence domains have been preserved, but amino acid codon usage has been virtually randomized. Outside of the regions of sequence conservation, there are no common nucleotide sequences that suggest any evolutionary relationships between viruses. The extensive divergence is presumably due in part to the rapid evolutionary potential of RNA genomes discussed previously.

The *Flavivirus* genus of the family Flaviviridae contains 66 members.[2] These viruses have a single stranded plus-sense RNA genome which encodes for three structural proteins and at least five nonstructural proteins.[47] The viral envelope contains a single protein (E) which is usually glycosylated and a small membrane protein (M) which is the cleavage product of the larger protein designated pM.[10] The genome of approximately 11 kdaltons is enclosed within a capsid structure composed of a single polypeptide designated C. The flavivirus genome is organized similarly to that of picornaviruses. The structural proteins are located at the 5′ one third of the genome, there is no subgenomic message, and there is one single

```
      Capsid
VEE  MFPFQPMYPMQPMPYRNPFAAPRRPWFPRTDPFLAMQVQELTRSMANLTFKQRRDAPPEGPSAKKPKKEASQKQKGGGQGKKKKNQGKKKAKTGPPNPKA  C 100
SIN  .NRGFFNML-GRR.FPA.T.MW.PRRRRQAA.MP.------RNGL.SQIQQLTTAVSALVIGQATRPQPPRPRPPPR-.K.QAPK.PP.PK.PKTQEK.K  C  92
SF   .NYIPTQTFYGRRWRPR.A.R.WPL---QAT.VAPVV----PDFQ.QQMQQLISAVNALTMRQNAIAPARPP.P.KKKTT.P.PKTQP..INGKTQQQ.K  C  93
RR   .NYIPTQTFYGRRWRPR.AFR.WQVSM-QPT.TMVTPMLQAPDLQ.QQMQQLISAVSALTTKQNVKAPKGQRQK.QQKPKE..E..K..PTQKKKQQQ.P  C  99

VEE  QNGNKKKTNKKPGKRQRMVMKLESDKTFPIMLE-GKINGYACVVGGKLFRPMHVEGKIDNDVLAALKTKKASKYDLEYADVPQNMRADTFKYTHEKPQGY  C 199
SIN  KQPA----KP........AL...A.RL.DVKN.D.DVI.H.LAME..VMK.L..K.T..HP..SK..FT.S.A..M.F.QL.V...SEA.T..S.H.E.F  C 188
SF   KDKQAD.KK......E..C..I.N.CI.EVKH.-..VT....L..D.VMK.A..K.V...AD..K.AF..S......C.QI.VH..S.AS.......E.H  C 192
RR   KPQA---KK....R.E..C..I.N.CI.EVK.D-..VT....L..D.VMK.A..K.T...PD..K.TY..S......C.QI.VH.KS.AS.......E.H  C 195

                                                                                 E3
VEE  YSWHHGAVQYENGRFTVPKGVGAKGDSGRPILDNQGRVVAIVLGGVNEGSRTALSVVMWNEKGVTVKYTPENCEQWS---LVTTMCLLANVTFPCAQPP  E3 20
SIN  .N........SG....I.R...GR......M..S..........AD..T.......T..S..K.I.T...GT.E./.AAP...A....GNVS...DR..  E3 23
SF   .N........SG....I.T.A.KP.......F..K..........A..........T..-.DMVTRV...GS.E./.A-P.I.A..V...A...F...  E3 22
RR   .N........SG....I.T.A.KP.......F..K..........A...A.......T.T-.DMVTRV...GT.E./.A-A.M--..I...TS...SS..  E3 20

                                            E2
VEE  ---ICYDRKPAETLAMLSVNVDNPGYDELLEAAVKCP--GRKRRSTEELFKEYKLTRPYMARCIRCAVG-SCHSPIAIEAVKSDGHDGYVRLQTSSQYG  E2 54
SIN  ---T..T.E.SRA.DI.EE..NHEA..T..N.ILR.GSS..SK./.VIDG.T---..S..LGT.SY.HHTEP.F..VK..Q.WDEAD.NTI.I....A.F.  E2 52
SF   CVPC..ENNAEA..R..ED...R...YD..Q..LT.RNQT.H../.VSQH.NV..A.....I.Y.AD.GA.H.....V.....R.EAT..MLKI.F.A.I.  E2 55
RR   CYPC..EKQ.EQ..R..ED..NR...Y.....SMT.RNNS.H../.VT.H.NV..A.....L.Y.AD.GD.YF.Y..V...KIRDEAP..MLKI.V.A.I.  E2 55

VEE  LDSSGNLKGRTMRYD---MHGTIKEIPLHQVSLHTSRPCHIVDGHGYFLLARCPAGDSITMEFKKD-SVTHSCSVPYEVKFNPVGRELYTHPPEHGVEQA  E2 150
SIN  Y.Q..AASANKY..MSLKQDH.V..GTMDDIKIS..G..RRLSYK......K..P...V.VSIVSS-NSAT..TLARKI.PKF....K.DL..V..KKIP  E2 151
SF   I.K.D.HDYTKI..A---DGHA.ENAVRSSLKVA..GD.FVHGTM.H.I..K..P.EFLQVSIQDTRNAVRA.RIQ.HHDPQ.....KF.IR.HY.K.IP  E2 152
RR   ..KA.THAHTKI..M---AGHDVQ.SKRDSLRVY..AA.S.HGTM.H.IV.H..P..YLKVS.EDAD.HVKA.K.Q.KHDPL.....KFVVR.HF...LP  E2 152

VEE  CQVYAHDAQNRGAYVEMHLPGSEVDSSLVSLSGSSVTVTPPVGTSALVECECGGTKISKTINKTKQFSQCTKKEQCRAYRLQNDKWVYNSDKLPKAAGAT  E2 250
SIN  .T..DRLKETTAG.IT..R.RPHAYT.YLEE.SGK.YAK..S.KNITY..K..DY.TGTVSTR.EITG-..AIK..V..KSDQT...F..PD.IRHDDH.  E2 250
SF   .TT.QQITALIVEEID..M.PDTP.RT.L.QQSGN.-KITVG.KKVKYNKT..TGNVGT.NSDMTINT-.-LI...HVSVTDHK..QF..PFV.R.DEPA  E2 249
RR   .TS.QLTTAPTDEEID..T.PDIP.RT.L.QTAGN.-KITAG.RTIRYNKT..RDNVGT.STDKTINT-.-.ID..H.AVTSH...QFT.PFV.R.DQTA  E2 249

VEE  LKGKLHVPFLLADGKCTVPLAPEPMITFGFRSVSLKLHPKNPTYLTTRQLADEPHYTHELISEPAVRNFTVTEKGWEFVWGNHPPKRFWAQETAPGNPHG  E2 350
SIN  AQ....L..K.IPST.M..V.HA.NVIH..KHI..Q.DTDHL.L....R.GAN.EP.T.W.VGKT......DRD.L.YI....E.V.VY...S...D...  E2 350
SF   R...V.I..P.DNIT.R..M.R..IVIH.K.E.T.H...DH..LFSY.T.GED.Q.HE.WVTAAVE.TIP.PVD.M.YH...ND.V.L.S.L.TE.K...  E2 349
RR   RR..V....P.TNVT.R....RA.DV.Y.KKE.T.R...DH..LFSY.S.GA...PYE.WVDKFSE.IIP...E.I.YQ...N..V.L...L.TE.K...  E2 349

                                                                                        6K
VEE  LPHEVIIHYYHRYPMSTILGLSICAAIATVSVAASTWLFCRSRVACLTPYRLTPNARIPFCLAVLCCARTARAETTWESLDHLWNNNQQMFWIQL----  6K 22
SIN  W...IVQ.....H.VY...AVASATVAMMIG.TVAVLCA.KA.RE.....A.A...V..TS..L...V.S.N./..FT.TMSY..S.S.PF..V..----  6K 22
SF   W..QIVQY..GL..AA.VSAVVGMSLL.LI.IF..CYMLVAA.SK.....A...G.AV.WT.GI....PR.H./ASVA.TMAY..DQ..AL..LEFAAPV  6K 26
RR   W...I.QY..GL..AA..AAV.GASLM.LLTL..TCCMLATA.RK.....A...G.VV.LT.GL....PR.N./ASFA.TMAY..DE.KTL..MEFAAPA  6K 26
```

```
                                           ┌→E1
VEE  -LIPLAALIVVTRLLRCVCCVVPFLVMAGA--A-APA|YEHATTMPSQAGISYNTIVNRAGYAPLPISITPTKIKLIPTVNLEYVTCHYKTGMDSPAIKC  E1  61
SIN  -C.....F..LM.--C.S.-CL....V...YL.KVD./......V.NVPQ.P.KAL.E.......NLE..VMSSEVL.ST.Q..I..KFT.VVP..K...  E1  61
SF   AC.LIITYCLRNV--L.C.K-SLSFLVLLSLG.T.R./...S.V..NVV.FP.KAHIE.P..S..TLQMQVVETS.E..L....I..E...VVP..YV..  E1  61
RR   AALA.L.CCIKSL--I.C.K-PFSFLVLLSLG.S.K./...TA.I.NVV.FP.KAHIE.N.FS.MTLQLEVVETS.E..L....I..E...VVP..F...  E1  61

VEE  CGSQFCIPTYRPDEQCKVFTGVYPFMWGGAYCFCDTENIQVSKAYVMKSDDCLADHAEAYKAHTASVQAFLNITVGEHSIVTTVYVNGETPVNFNGVKLT  E1 161
SIN  ...L..Q.AAHA.YT....G..........Q....S..S.M.E...EL.AV.AS...Q.I.V...AMKVG.R.VY.NTTSFLD.....V..GTSKDL.VI  E1 161
SF   ..AS..STKEK..Y....Y...............S....L.E...DR..V.RH...S........LK.KVRVMY.NVNQTVD.....DHA.TIG.TQFI  E1 161
RR   ..TS..SSKEQ..Y....Y...............S....L.E...DR..V.KH...S........LK.TIR.SY.TINQT.EAF....HA..VG.S.FI  E1 161

VEE  AGPLSTAWTPFDRKIVQYAGEIYNYDFPEYGAGQPGAFGDIQSRTVSSSDLYANTNLVLQRPKAGAIHVPYTQAPSGFEQWKKDKAPSLKFTAPFGCEIY  E1 261
SIN  ...I.ASF....H.V.IHR.LV..........MK........ATSLT.K..I.S.DIR.IK.S.KNV.......S....M..NNSGRP.QE......K.A  E1 261
SF   F....S......N...V.KD.VF.Q...P..S....R..........E.N......A.K.A..SP.MV......T....KY.L.E.GTA.NTK.....Q.K  E1 261
RR   F..I....S...N...V.KDDV..Q...P..S....R..........E.K......A.K.S..SP.VV.....PT....KY.L.E.GS..NTK.....K.K  E1 261

VEE  INPIRAFNCAVGSIPLAFDIPDALFTRVSETPTLSAAECTLNECVYSSDFGGIATVKYSASKSGKCAVHVPSGTATLKEAAVELTEQGSATIHFSTANIH  E1 361
SIN  V..L..VD.SY.N..ISI...N.A.I.T.DA.LV.TVK.EVS..T..A....M..IQ.VSDRE.Q.P..SH.S....Q.ST.HVL.K.AV.V.....SPQ  E1 361
SF   ...V..M.....N..VSMNL..SA...IV.A..IIDIT..VAT.TH......VI.LT.KTN.N.D.S..SH.NV...Q..TAKVKTA.KV.L.....SAS  E1 361
RR   ...V..MD.......VSM....SA....VDA.AVTDLS.QVVV.TH......V..LS.KTD.P......SH.NV...Q..T.DVK.D.KV.V.....SAS  E1 361
                                         ___________________________________________
VEE  PEFRLQICTSYVTCKGDCHPPKDHIVTHPQYHAQTFTAAVSKTAWTWLTSLLGGSAVIIIIGLVLATIVAMYVLTNQKHN                       E1 442
SIN  AN.IVSL.GKKT..NAE.K..A....ST.HKND.E.Q..I...S.S..FA.F..ASSLL....MIFACSM.LTS.RR                          E1 439
SF   .S.VVSL.SARA..SAS.E.......PYAAS.SNVVFPDM.G..LS.VQKIS..LGAFA.GAILVLVV.TC-IGLRR                          E1 438
RR   .A.KVSV.DAKT..TAA.E.......PYGAS.NNQVFPDM.G..M..VQR.AS.LGGLAL.AV.VLVL.TC-ITMRR                          E1 438
```

FIGURE 1. The deduced amino acid sequences of translated regions of VEE virus (Trinidad strain),[44] Sindbis virus,[35] Semiliki Forest virus,[42] and Ross River virus[43] 26S mRNAs. Deletions (dashes) have been introduced to maximize homologies. Solid dots indicate amino acid identity with the amino acid sequence of VEE. Asn-linked potential glycosylation sites are stippled. Solid bars indicate probable transmembrane domains of the VEE E1 and E2 envelope glycoproteins.

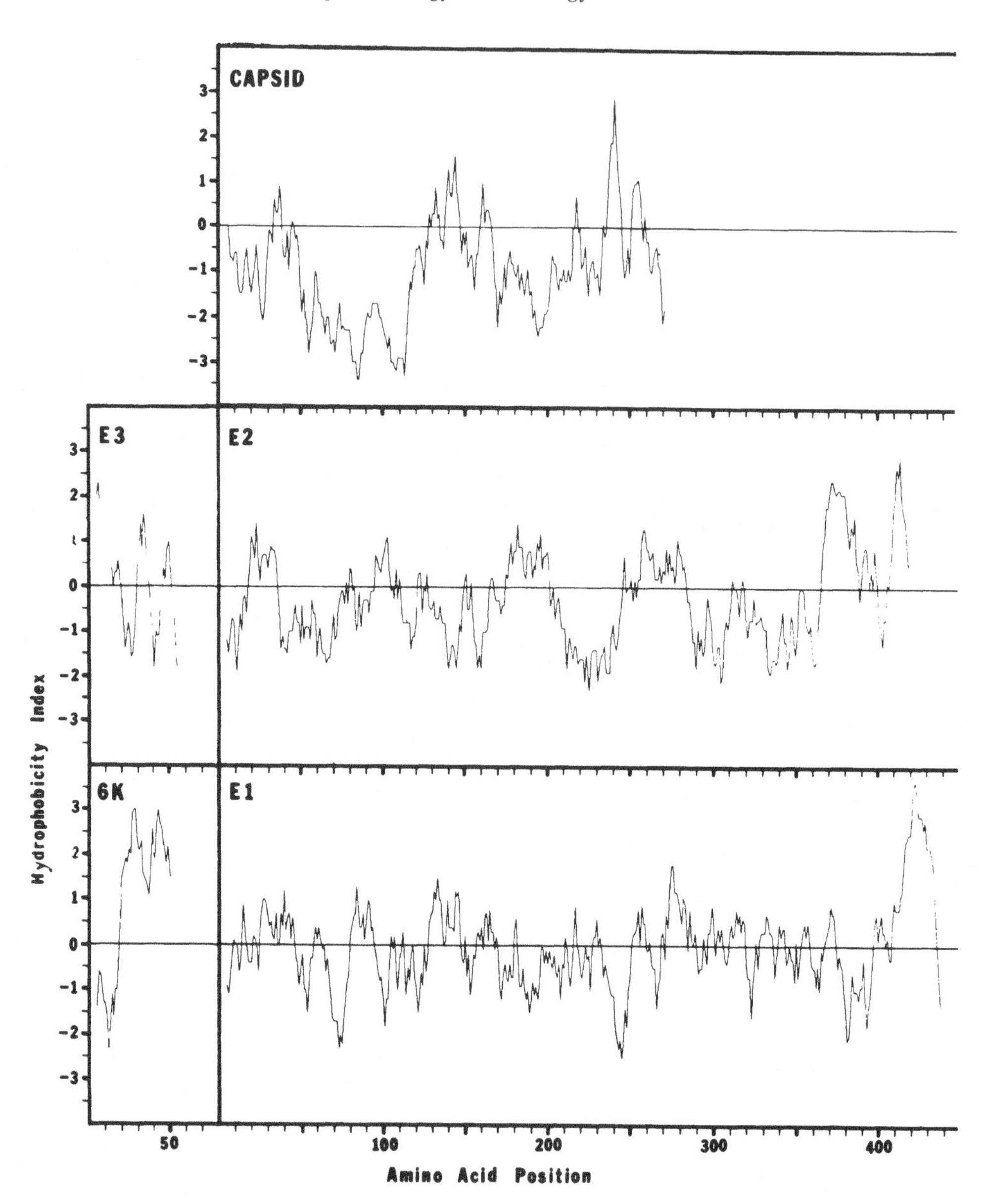

FIGURE 2. Moving average hydrophobicity plots (for peptide segments nine amino acids in length) of the proteins encoded by VEE virus (Trinidad strain) 26S RNA. Positive values indicate hydrophobicity; negative values indicate hydrophilic amino acid regions. Potential Asn-linked glycosylation sites are indicated by solid circles.

long reading frame.[10] The structural genes of yellow fever,[47] Murray Valley encephalitis,[48,49] West Nile,[50,51] and St. Louis encephalitis[52] are arranged in the sequence 5′ noncoding region, C, pM/M,E-3′. Nucleotide and amino acid sequence data for all of the nonstructural genes of yellow fever virus, Ns1, Ns2a, Ns2b, and Ns3 of Murray Valley encephalitis, and Ns1 of St. Louis encephalitis virus agree, such that following the structural genes, the nonstructural genes are organized along the genome in the order 5′-Ns1,Ns2a,Ns2b,Ns3,Ns4a, Ns4b,Ns5, noncoding region-3′.

Biochemical analyses of gene organization and virus replication demonstrate that the flaviviruses are similar in their morphogenesis, morphology, and replication strategy. The flaviviruses are antigenically more diverse than the alphaviruses, and the family has been

Table 1
CODON USAGE IN THE TRANSLATED REGION OF VEE, STRAIN TRD, 26S SUBGENOMIC RNA

Phe	UUU	20[a]	Ser	UCU	11	Tyr	UAU	31	Cys	UGU	11
	UUC	23		UUC	18		UAC	22		UGC	32
Leu	UUA	8		UCA	21	Ocher	UAA	0	Opal	UGA	1[b]
	UUG	14		UCG	9	Amber	UAG	0	Trp	UGG	17
Leu	CUU	11	Pro	CCU	24	His	CAU	17	Arg	CGU	1
	CUC	10		CCC	15		CAC	23		CGC	12
	CUA	12		CCA	28	Gln	CAA	22		CGA	1
	CUG	31		CCG	25		CAG	26		CGG	7
Ile	AUU	17	Thr	ACU	21	Asn	AAU	18	Ser	AGU	3
	AUC	21		ACC	37		AAC	31		AGC	14
	AUA	14		ACA	30	Lys	AAA	40	Arg	AGA	17
Met	AUG	28		ACG	11		AAG	46		AGG	14
Val	GUU	17	Ala	GCU	17	Asp	GAU	21	Gly	GGU	11
	GUC	23		GCC	38		GAC	22		GGC	19
	GUA	8		GCA	34	Glu	GAA	27		GGA	30
	GUG	44		GCG	19		GAG	35		GGG	25

[a] Number of times codon is used.
[b] The translated polyprotein precursor is terminated by an opal codon.

divided into seven serologic complexes based on cross reactivities in neutralization tests. Monoclonal antibody analysis of the E glycoprotein have revealed group reactive, complex reactive, and type specific epitopes.[53,54]

The sequences of the N-terminal 40 amino acids of yellow fever, West Nile, Murray Valley encephalitis, and St. Louis encephalitis viruses are 50 to 60% conserved between viruses belonging to different serogroups, and 75 to 80% conserved in viruses belonging to the same serogroup.[47,48,52] The entire amino acid sequence of the E protein of yellow fever and Murray Valley encephalitis is 45% conserved. The amino acid sequences of the E glycoprotein of St. Louis encephalitis, Murray Valley encephalitis, and West Nile viruses are 60% conserved; specific hydrophilic regions within these proteins exhibit homologies of 80 to 90%. The amino acid sequence conservation seen in the flavivirus E glycoprotein is not unlike that observed with the alphaviruses.[10,11]

Given the high mutability of RNA, the comparative stability of the flavivirus genome in nature is surprising. Oligonucleotide mapping of viral RNAs representing isolates of a specific virus over time and from different geographic regions has revealed the presence of genotypic groups or topotypes.[55,56] The evolution of St. Louis encephalitis (SLE) virus has been examined by comparing oligonucleotide fingerprint maps of viruses isolated over the geographic range of the virus.[55,56] SLE virus isolates from the Mississippi River basin are genotypically distinct from strains isolated from the western U.S. Likewise, strains isolated in South America differ from the topotypes prevalent in North America. SLE virus strains isolated in 1933 in St. Louis have 75% fingerprint homologies with strains isolated in 1975 from that same region, indicating that although genetic changes had occurred over a 40 year period, much of the genome sequence has been conserved. For North American SLE virus strains, it is clear that genotype is a determinant of biological characteristics of the virus such as viremia in sparrows, virulence for adult mice, and mosquito vector associations involved in the transmission cycle.[55,56,57]

Analysis of yellow fever virus isolates from Africa and the Americas reveals that geographically isolated and epidemiologically unrelated viruses have unique oligonucleotide fingerprints.[58,59] These observations suggest that the yellow fever genome evolves slowly and that individual virus topotypes are stable antigenically, biologically, and epidemiologically. Passage of yellow fever virus more than 200 times, which resulted in attenuation of the Asibi virus strain, alters the genome by less than 1% based on fingerprint similarities.[60] This slow rate of evolution may explain why yellow fever is the only flavivirus for which a useful attenuated vaccine is available.

Each of the four serotypes of dengue (DEN) has a distinct oligonucleotide fingerprint.[61] Geographic varieties of DEN-1 and -2 viruses can be distinguished on the basis of oligonucleotide fingerprints.[62,63] Isolates of DEN-1 virus from the same geographic are more similar to each other than to other isolates. On this basis of percent of shared oligonuclotides, three geographical topotypes of DEN-1 have been proposed: Caribbean, African, and Pacific-Southeast Asian.[63]

Analysis of DEN-2 virus isolates from endemic regions throughout the world reveals that isolates from a geographic region associated with recognized epidemics have similar oligonucleotide fingerprint maps. These studies have now established the existence of at least 12 genetic varieties of DEN-2 virus.[130] Viruses isolated in the same DEN outbreak often have fingerprint similarities of 90% or more and, during extended periods of time, continue to exhibit genetic homogeneity. Examination of DEN-2 virus isolates made in Thailand over a 20 year period confirms the genetic stability of the four virus topotypes and further supports the concept that evolution of the DEN virus genome proceeds gradually within a defined geographic locale.[130] Conservation of the genome is generally reflected in antigenic epitopes present on the DEN-2 virion E glycoprotein. Antigenic signature analysis of the different genetic topotypes revealed striking antigenic differences in nine different epitopes; in contrast, viruses with similar fingerprints were shown to have similar antigenic signatures.[64]

A unique evaluation of genetic relatedness among DEN viruses has been obtained by cDNA-RNA hybridization.[65] Using Sl nuclease resistance of the hybrids as a measure of genetic relatedness, a close relationship is observed between DEN-1 and DEN-4 viruses and between DEN-2 and Edge Hill viruses. These relationships have been confirmed by the use of monoclonal antibodies. This suggests that these viruses share complex-specific sequences coding for determinants in the E glycoprotein. Whether the viruses have evolved in parallel or separately, with one being the progenitor of the other, remains to be determined.

Genetic variation of Japanese encephalitis (JE) virus isolates from various geographic areas has also been examined by oligonucleotide fingerprint analyses.[66] Isolates from the same geographic area in the same year are similar but differ from those from other areas. Nakayama (Japan, 1935) and Peking (China, 1949) strains have similar fingerprints, suggesting that early Japanese strains of JE may be epidemiologically related to the China strains. Recent isolates from Thailand and Vietnam are similar but differ from current isolates from Japan. Genotypes of JE virus isolates from human brain, mosquitoes, and sentinel pigs in Thailand have been compared.[67] Fingerprints of JE virus isolates from southern Thailand, where encephalitis does not occur, are very different from those isolated from northern Thailand, where the disease does occur. JE virus strains that produce fatal encephalitis in humans tend to have a highly conserved genotype. In contrast, strains isolated from mosquitoes and swine are genotypically divergent.

Doublet frequencies in flavivirus RNA differ from other eukaryotic RNA viruses; there is a pronounced difference in the frequency of UA and CG doublets.[68] Based upon the relative frequencies of the CG doublets, the origins of alpha- and flaviviruses have been hypothesized.[10,69] Alphaviruses, all of which replicate well in both vertebrate and invertebrate hosts, have a CG content more like the invertebrate. The flavivirus RNA CG content is more like that of the mammalian host, and perhaps significantly, certain flaviviruses are not transmitted by vectors. Although the functional significances of the CG complex usage is not clear, these observations suggest that alphaviruses are truly invertebrate viruses that have evolved to replicate in mammals and that the flaviviruses are mammalian viruses that have evolved to replicate in an invertebrate vector. The unique structure of the flavivirus genome (lack of 3′-poly A sequences, type 1,5′-cap structure, no subgenomic RNA for translation of structural proteins, and unique morphogenesis mechanisms) certainly indicate the alpha- and flaviviruses represent different branches of the RNA virus evolutionary tree. The presence of conserved sequences of the 3′-ends of both plus and minus strands of flavivirus RNA is similar to that observed for the negative stranded viruses. Thus, the flaviviruses may be on a branch of the evolutionary tree which contains the negative stranded bunyaviruses.[10]

In summary, alpha- and flaviviruses apparently differ in their evolutionary potential. In each geographic region, strains of flaviviruses evolve independently. In contrast, alphavirus evolution appears to be more constrained, and the alphavirus genotypes are more preserved over the entire geographic range. Apparently, as with the alphaviruses, flaviviruses have evolved from a common ancestor. Evolutionary divergence of the flaviviruses, reflected in sequence conservation, antigenic epitope patterns, and biological transmission cycles, has been limited by the basic constraints on virus replication and transmission, although apparently not to the same extent as for the alphaviruses. Factors which permit evolution of the flavivirus genome, as reflected by the changes in the envelope protein, undoubtedly permit mutations in other genes. However, methods for rapid analysis of other structural and nonstructural flavivirus genes are not yet available. Nonetheless, one would predict that biological constraints placed on each gene differ and that, as with the alphaviruses, sequence changes in certain genes are less likely to produce lethal mutants.

The comparative genetic stability of the alpha- and flavivirus genome is surprising and must in part relate to the natural cycle of transmission between invertebrate and vertebrate hosts. It is noteworthy in this regard that each of the nonarthropod-borne flaviviruses is antigenically distinct. In contrast, the arthropod-borne flaviviruses exhibit a plethora of serologic cross reactions.[6]

B. Evolutionary Mechanisms for Arboviruses with Segmented Genomes

1. Intramolecular Changes

Arboviruses that have segmented genomes apparently also evolve by genetic drift. For example, no two isolates of La Crosse (LAC) virus obtained from nature have identical genomes as evidenced by oligonucleotide fingerprinting.[70-72] This is true for viruses isolated from different geographic locales at the same or different times, for viruses isolated from the same locale at different times, and for viruses isolated from the same locale at the same time. This genotypic diversity supports the hypothesis that bunyavirus evolution occurs principally through genetic drift via the accumulation of point mutations, sequence deletions, and inversions.

2. Segment Reassortment

Arboviruses with segmented genomes (bunyaviruses, orbiviruses) have even greater evolutionary potential due to the capability for RNA segment reassortment. Two evolutionary mechanisms, intramolecular changes and reassortment, function to promote genetic diversity for segmented viruses. These mechanisms may well complement one another.[73] The greater evolutionary potential of the segmented genome is evidenced by the fact that a majority of registered arboviruses are either bunya- or orbiviruses.[2] For example, 222 of the 427 viruses registered in the major arbovirus families (Alphaviridae, Flaviviridae, Bunyaviridae, Reoviridae, and Rhabdoviridae) are classified as bunyaviruses or bunyavirus-like.[2] Reassortment of virus genomes provides a unique capability for viruses to assume new antigenic and biologic phenotypes, which could have serious epidemiologic consequences. For example, exchange of genetic material between strains of influenza viruses can yield new epidemiologically successful variants, capable of causing worldwide pandemics of disease.[19]

***a. Reassortment of Arboviruses* In Vitro**

High frequency reassortment of RNA segments of bunyaviruses has been demonstrated in vitro.[9,74,75] Eight (2^3) possible genotypes can result from dual infection of a single cell. Dual infections with certain California and Bunyamwera group viruses, respectively, result in the recovery of the six reassortant genotypes as well as parental genotypes.[9,75] However, reassortment within the groups is not completely random. Among the Bunyamwera group viruses, reassortment is restricted at the gene product level rather than due to incompatibility of genome subunits.[75] Such genetic incompatibilities suggest that reassortment in nature is not random and that the genotypes resulting from these events could be predicted. Understanding the molecular basis of this phenomenon may permit predictive capability concerning the epidemic potential of these viruses. When bunyaviruses containing nonhomologous genome subunit combinations are used as parent viruses, reassortment is random.[75] Nonetheless, reassortment is apparently gene pool specific; serologically related viruses reassort, whereas unrelated viruses do not. California group viruses reassort with other members of the serogroup, but do not reassort with Bunyamwera group viruses.[9,12]

High frequency reassortment of orbiviruses, which have ten RNA segments, has been demonstrated in vitro.[76] The genomic possibilities resulting from dual infection and RNA segment reassortment of orbiviruses are great; 1024 (2^{10}) possible genotypes could result.

b. Reassortment of Arboviruses in Nature

Bunyavirus RNA segment reassortment does occur in nature. Initial evidence was based on the observations that two isolates of a bunyavirus serotype (defined by HI or N tests) may express CF antigens of a totally different virus.[9] Since bunyavirus glycoproteins are known to function in HI and N tests, and the nucleocapsid protein is a major antigen involved in the CF reaction, RNA segment reassortment is a logical explanation for viruses in which HI/N and CF antigens do not co-segregate. The group C (Bunyavirus genus) viruses illustrate the phenomenon. From the small Utinga Forest near Belem, Brazil, six group C viruses have been isolated which segregate into three groups by HI/NT: Caraparu-Apeu, Murutucu-Marituba, and Oriboca-Itaqui.[77,78] However, by CF the viruses segregate into different pairs: Caraparu-Itaqui, Apeu-Marituba, and Oriboca-Murutucu. These differences are most easily explained by ancestral RNA segment reassortment.

Genotypic analyses also provide evidence for the reassortment of alternate arboviruses in nature. Shark River and Pahayokee viruses (Patois serogroup) have been shown by viral L-, M-, and SRNA oligonucleotide fingerprint analyses to represent naturally occurring reassortant viruses.[79] Similar analyses have demonstrated that intertypic reassortment has occurred between strains of the LAC virus in the midwestern U.S.[70,71] RNA segment reassortment apparently also occurs between orbiviruses in nature. Reassortant bluetongue viruses have been isolated from sheep.[80,81] Oligonucleotide fingerprint analyses of a variety of BTV isolates suggest that evolution of orbiviruses occurs both by reassortment and by intramolecular changes.[80-82]

In summary, precedents exist for arbovirus RNA segment reassortment in nature. However, the importance of reassortment and the extent to which genomic reassortment can lead to changes in the epidemiology or epizootiology of a virus remain a matter of speculation.

III. SITE OF VIRUS EVOLUTION

In theory, evolution of arboviruses could occur in either the vertebrate or the vector.

A. Virus Evolution in the Vertebrate

Arbovirus infections of vertebrates typically are acute in nature and infection results in long-term immunity. These characteristics would seem to limit opportunities for arbovirus evolution. However, there is evidence that evolution of viruses does occur in vertebrates.

1. Nonsegmented Genome Viruses

Nonsegmented genome viruses such as VS, which have been maintained for many years by lytic passage, demonstrate little genetic diversity.[26] In contrast, persistent VS infections of cell culture and nude mice result in rapid, significant oligonucleotide changes.[25] If acute VS infection in vertebrates is more analogous to lytic passage than to persistent infections in vitro, little genetic change would be predicted from a normal, acute infection of immunologically competent vertebrates. However picornavirus infection of man and animals, which are acute in nature, can result in the generation of multiple virus genotypes.[83-87] A mutation rate of 1 to 2% of the total genome nucleotide bases has been observed for poliovirus during human passage over a 13 month period.[83,84] Likewise, the genome of foot and mouth disease virus is highly mutable during epizootic transmission.[85-87] In theory, intramolecular recombination of arboviruses could also occur in the vertebrate host. In this regard, two serotypes of DEN (1 and 4) have been isolated from one patient.[88] Thus, the vertebrate host, even during acute infection situations, may serve as a site for generation of genetic diversity in nonsegmented genome arbovirus populations.

During persistent infections in vertebrates, alternate mechanisms may promote antigenic variation in nonsegmented genome viruses. The observation that lactic dehydrogenase elevating virus (LDV) can establish a persistent, relatively benign infection in mice is of interest. LDV is structurally similar to the arthropod-borne flaviviruses.[89,90] This persistent infection appears to be mediated by the formation of circulating infectious virus-antibody immune complexes that fail to be cleared from the host.[91,92] The serum from these animals is infectious, and transfer of infected blood results in a rapid establishment of a persistent infection. Similar persistent infections with other arboviruses have yet to be identified. Persistent virus infection could help to explain the phenomena of arbovirus overwintering and persistently high and long-lasting IgM titers in flavivirus vaccines.[93] These types of infected animals could serve as reservoirs for variant virus production. Persistent infections with prolonged viremia would be especially important for viruses transmitted by blood-feeding insects. Depending upon the subsequent selection conditions, significant perturbation of the composition of the virus pool could occur leading to the production of new antigenic virus entities.

2. Segmented Genome Viruses

Opportunities for evolution of viruses with segmented genomes would seem to be limited in the vertebrate. The acute nature of virus infections in vertebrates and the production of antibody, which could limit subsequent infections with serologically related viruses, would limit the possibility of dual infection and subsequent segment reassortment in vertebrate hosts. The orbiviruses, which cause lengthy viremias in vertebrates, may well be an exception. Two serotypes of BTV have been isolated from one bovine,[94] and new genotypes of BTV, presumably resulting from reassortment, have been isolated from sheep.[80,81]

The potential for segment reassortment of bunyaviruses in vertebrates has been investigated; however, little positive evidence for reassortment has been obtained.[131] Simultaneous peripheral, intraperitoneal (i.p.) or intracranial (i.c.) inoculation of suckling mice with two *ts* mutants of LAC virus from different complementation/recombination groups does not result in the recovery of *wt* virus. More biologically relevant studies have also been conducted. When mosquitoes infected with the respective *ts* mutant viruses are permitted to engorge on suckling mice, *wt* virus cannot be detected in any of the mice. Thus, the two *ts* mutant viruses do not infect the same cell or if reassortment does occur, it is at low frequency and is undetectable in the assay systems.

In contrast, reassortment of Thogoto viruses (an arbovirus containing seven RNA segments) has been demonstrated in dually infected hamsters.[132] When *ts* mutants of Thogoto virus from distinct complementation groups are inoculated into hamsters, approximately 10%

of the viruses recovered are *wt* phenotype, indicating high frequency reassortment. Why reassortment can be demonstrated in the Thogoto infected hamster but not in the LAC virus infected mice is not known. However, viremia titers in LAC infected mice seldom exceed 4.0 $\log_{10}$ $TCID_{50}/m\ell$, whereas in the hamster Thogoto viremias were greater than 7.0 $\log_{10}$ $TCID_{50}$. This suggests that greater viremia might have resulted in more frequent dual infection of cells.

B. Virus Evolution in the Vector

Conceptually, the arthropod would seem to provide greater opportunity for viral evolution than the vertebrate. In contrast to the generally acute nature of arbovirus infections in vertebrates, arboviruses characteristically establish persistent infections in arthropods. Certain arboviruses are transovarially transmitted,[95] resulting in virus persistence in arthropod tissues for long periods of time. Thus, the vector would seem to be an excellent site for arbovirus intramolecular genomic changes to accrue; there would be ample opportunity for the generation and accumulation of spontaneous virus mutations. The unique life histories of arthropods are also conducive to evolution of segmented genome viruses via reassortment. Adult mosquitoes may feed several times during their relatively brief life span,[96] providing multiple opportunities for dual infections. In the case of LAC and certain other arboviruses, mosquitoes can also be dually infected by transovarial or venereal routes.[97] Such multiple infection strategies and persistent infections provide ample opportunity for evolution of viruses in nature. Chances for dual virus infection may be further enhanced in more long-lived vector species. Ixodid ticks, which can live for many years, may exhibit different host feeding preferences at different life stages, thereby enhancing opportunities for dual infection of vectors. Argasid ticks can be extremely long lived and will ingest many blood meals during their life span. In both these groups, the ticks may live longer than their vertebrate hosts, providing an extraordinary persistent infection. Thus, because of their respective biologies, arthropods provide unique opportunities to serve as a site for virus evolution, both by generation of intramolecular changes in virus genomes and for dual infection and subsequent virus reassortment.

1. Virus Reassortment in Vectors

Simultaneous parenteral infection of *Aedes triseriatus* mosquitoes with homologous or heterologous LAC-snowshoe hare (SSH) *ts* mutants results in high frequency reassortment (Table 2).[98] Restriction of reassortment between SSH-II and LAC-I genotypic mutants occurs in cell culture, and similar restrictions have been noted for other bunyaviruses.[9,75] Since these restrictions also occur in the vector, they may be epidemiologically important. When *Aedes triseriatus* mosquitoes are permitted to ingest two bunyaviruses simultaneously or sequentially by interrupted feeding, vectors become dually infected and subsequently serve as a site for RNA segment reassortment (Table 3).[99] Of mosquitoes simultaneously infected with two *ts* mutant viruses, 25% yield *wt* virus. When mosquitoes sequentially ingest two *ts* mutant viruses, 20% of the mosquitoes subsequently yield *wt* virus (Table 3). In both of these cases, reassortant viruses can be isolated from mice fed upon by mosquitoes. Thus, new virus genotypes are generated in the dually infected mosquitoes, and the new virus can be transmitted. Such a sequence of events would be necessary for expression of new virus genotypes in nature. It is important to note that all of the mosquitoes in these studies that did not contain *wt* virus (Table 3) were probably dually infected with the two *ts* mutant viruses; all control mosquitoes that ingested one or the other *ts* mutant virus did become infected. However, in only 25% of the mosquitoes did both viruses infect the same cell, resulting in reassortant viruses. Perhaps with different virus combinations, different vector mosquitoes, or different maintenance conditions, more of the dually infected mosquitoes would yield reassortant viruses.

Table 2
VIRUSES RECOVERED FROM *AEDES TRISERIATUS* MOSQUITOES DUALLY INFECTED PARENTERALLY AND FROM MICE ON WHICH THE INFECTED MOSQUITOES WERE ALLOWED TO FEED[a]

	Mosquitoes		Mice	
Virus cross	*%ts*	*%wt*	*%ts*	*%wt*
SSH I-1 × SSH II-22	95	5	90	10
LAC I-20 × LAC II-4	45	55	20	80
SSH I-1 × LAC II-5	35	65	55	45
SSH II-21 × LAC I-20	100	0	100	0

[a] Viruses recovered from dually infected mosquitoes that had been inoculated with *ts* viruses of LAC or SSH virus (mutants LAC I-20, SSH I-1, SSH II-22, etc.) or viruses obtained from moribund or dead mice. Samples were plated on BHK-21 cells at 33°C; virus plaques were picked and reassayed at both 33 and 40°C to score for *ts* and *wt* viruses. The results for each cross represent the averages of analyses of several mosquitoes recovered after 7-, 14-, and 21-day mosquito feedings.[98]

Table 3
DUAL ORAL INFECTION OF *AEDES TRISERIATUS* MOSQUITOES AND GENERATION OF REASSORTANT VIRUSES[a]

	Infection rates	
Infection protocol	**33°C assay**	**40°C assay**
Simultaneous		
LAC*ts*I-16 + LAC*wt*	15/15 (100%)	15/15 (100%)
LAC*ts*I-16 + LAC*ts*II-5	8/8 (100%)	2/8 (25%)
Interrupted feeding		
LAC*ts*I-16 then LAC*wt*	19/19 (100%)	18/19 (95%)
LAC*ts*I-16 then LAC*ts*II-5	4/20 (20%)	
	20/20 (100%)	
Total feeding		
LAC*ts*I-16 + LAC*wt*	34/34 (100%)	33/34 (97%)
LAC*ts*I-16 + LAC*ts*II-5	28/28 (100%)	6/28 (21%)

[a] Infection rates are expressed as the number of mosquitoes that were found to contain virus (>10 pfu), as detected by plaque assay at 33 or 40°C, divided by the number tested. Virus-blood meals on which the mosquitoes were initially allowed to feed (partially for the intrupted feeding protocol) contained 6.5 to 7.3 log of each the indicated viruses per milliliter. At 2 hr postingestion, the mosquitoes in the interrupted feeding experiment were allowed to engorge to completion on blood-virus mixtures containing 6.5 to 7.3 log/mℓ of the second virus (LAC*ts*II-5 or LAC*wt*, virus). All mosquitoes vere held for 14 days, titurated, then assayed for virus.[99]

Table 4
SUPERINFECTION OF MOSQUITOES PREVIOUSLY INFECTED PARENTERALLY WITH *ts* MUTANT VIRUSES[a]

Temp sensitive virus	Wild type virus	Growth at 33°C	Growth at 40°C	Approx % inhibition
L-I-16	LAC	Yes	No	100
	SSH	Yes	No	100
	TAH	Yes	No	100
	TVT	Yes	No	100
	WN	Yes	Yes	0
	VS	Yes	Yes	0
LAC-II-5	GRO	Yes	No	100
	TAH	Yes	No	100
	TVT	Yes	Yes	0
	WN	Yes	Yes	0
	Vs	Yes	Yes	0

[a] Groups of 4 mosquitoes (minimally) were parentally infected with 1.4 $\log_{10}$ pfu of a *ts* mutant of LAC virus (LAC-I-16 contains a *ts* lesion in the mRNA segment; LAC-II-5 contains the lesion in the lRNA segment). Mosquitoes were subsequently injected 7 days later with 2 — 4 $\log_{10}$ pfu of *wt* virus (SSH = snowshoe hare; TAH = tahyna; TVT = trivittatus; GRO = Guaroa; WN = West Nile; VS = vesicular stomatitis virus). After a further 7 days, mosquitoes were assayed at permissive and nonpermissive temperatures.[100]

Colorado tick fever viruses can also reassort in arthropods.[133] When ticks are inoculated with two strains of Colorado tick fever virus, which exhibit distinct electrophoretic RNA profiles, high frequency reassortment occurs. Interestingly, certain reassortant genotypes predominate. Whether these preferred genotypes reflect selective advantages or random reassortment events remains to be determined. This is a significant issue. If genes which code for virulence factors in vertebrates are preferentially expressed in vector species, reassortment in vectors could be epidemiologically important.

2. Interference to Superinfection in Vectors

Opportunities for simultaneous infection of vector species via ingestion of two or more viruses in one blood meal is probably restricted by the acute nature of virus infections in the vertebrate host. Prospects for dual infection would be greater if the female mosquito became infected in succeeding blood meals or if transovarially infected females ingested an alternate virus and became superinfected. A possible impediment to these sequential modes of infection would be virus interference to superinfection. Interference to bunyavirus superinfection of *Ae. triseriatus* mosquitoes has been demonstrated.[100] Mosquitoes parenterally infected with an LAC *ts* mutant virus and subsequently challenged with a California group virus, Guaroa (a bunyavirus), West Nile (a flavivirus), or VS (a rhabdovirus) are refractory to superinfection with California serogroup viruses but not to superinfection with unrelated viruses (Table 4).

However, in nature many species of mosquitoes exhibit a behavior, designated interrupted feeding, that could preclude interference. If defensive behavior of the vertebrate host inter-

Table 5
INTERFERENCE TO LAC VIRUS ORAL SUPERINFECTION OF *AEDES TRISERIATUS* MOSQUITOES

Time until ingestion of challenge virus	Infection rates[a]	
	33°C assay	40°C assay
Simultaneous	15/15 (100%)	15/15 (100%)
2 — 8 hr	26/26 (100%)	25/26 (96%)
1 — 2 days	29/29 (100%)	14/29 (48%)
7 — 28 days	18/18 (100%)	0/18 (0%)

[a] Infection rates are expressed as the number of mosquitoes containing >10 pfu of virus detected by plaque assay at 33 or 40°C divided by the number tested. Mosquitoes initially engorged partially or fully on a meal containing 6.5 to 7.8 log of LAC *ts* mutant I-16 per milliliter. Either simultaneously or at the indicated times postingestion, mosquitoes engorged to repletion on a blood meal containing 7 to 7.8 log/mℓ of *wt* LAC virus. After 14 days extrinsic incubation, mosquitoes were triturated, then assayed for virus.[99]

rupts feeding, the vector will likely finish engorgement promptly on an alternate host.[96] Under such circumstances, it would be possible for a mosquito to ingest a blood meal from two different vertebrate hosts viremic with two different viruses in a period of time short enough that the first virus would not interfere with infection by the second virus.

Mosquitoes infected orally become resistant to superinfection; however, there is a period of time in which dual infection can occur. When *Ae. triseriatus* mosquitoes are permitted to engorge a blood meal containing an LAC *ts* mutant virus and an LAC *wt* virus, 100% become infected with *wt* virus. Presumably most of these mosquitoes also contain *ts* virus, because approximately 100% of the control mosquitoes that only ingest *ts* virus become infected. When mosquitoes ingest challenge virus from 30 min to 8 hr after the *ts* mutant, 96% become superinfected (Table 5). When mosquitoes ingest *wt* virus from 1 to 2 days postinfection, 48% become superinfected. Those mosquitoes that ingest challenge virus 7 or more days after the initial virus are refractory to superinfection (Table 5).

3. Superinfection of Transovarially or Transstadially Infected Vectors

Alternate scenarios exist for natural dual infection of vectors in nature. Female mosquitoes infected either transovarially[95] or venereally[97] could possibly become superinfected with an alternate virus. Since California group viruses are efficiently transovairally transmitted and may overwinter in eggs of the vector, superinfection of transovarially infected females may be a likely mechanism for dual infection with the bunyaviruses. Similarly, acarines infected during larval or nymphal stages could become superinfected during ingestion of virus during the nymphal or adult stages, respectively. Preliminary evidence suggest that *Ae. triseriatus* females that are transovarially infected with a *ts* mutant virus can be superinfected orally with *wt* LAC virus.[134] Most females contain *wt* virus; however, the titer of *wt* virus in the superinfected females is quite low compared to the titer of the *ts* virus. This would suggest that a minimal number of cells become infected with the challenge virus. Whether such low level infections are biologically important remains to be determined.

4. Summary of Mechanisms for Generation of Genetic Diversity in Arthropods

These studies demonstrate that the vector can be a potential site for generation of genetic diversity for the segmented arboviruses. Dual infection and subsequent segment reassortment of the LAC virus in its vector species clearly demonstrates the potential for antigenic shift to occur in the bunyaviruses. There would seem to be great potential for other segmented genome arboviruses to generate genetic diversity via this mechanism in their respective vector species. However, virtually nothing is known concerning the biological, temporal, and phylogenetic constraints on reassortment for other arbovirus-vector systems. In addition, the evolutionary importance of these events remains to be determined.

The role of the vector in the generation of genetic diversity via intramolecular events also remains to be defined. Persistent bunyavirus infections of mosquito cell cultures results in the production of multiple size classes of the respective RNA species, suggesting DI (defective interfering) particle accumulation.[101] Thus, it is likely that the molecular phenomena that promote rapid evolution in vitro (i.e., accumulations of *ts* mutants, DI particles, RNA species, etc.) occur in vivo in the arthropod. Further, since dual infection of arthropod vectors can occur, intramolecular recombination may be likely.

If one assumes that evolutionary processes do occur in the arthropod, it then becomes important to determine if preferred genotypes result from vector passage and to determine how vector passage exerts selective pressure on the virus population. For example, studies with LAC and SSH viruses and their reassortant progeny viruses have demonstrated that the presence of the M RNA segment of LAC virus was essential for efficient oral infection of and transmission by *Ae. triseriatus* mosquitoes.[102,103] Viruses with SSH or SSH-like M RNA genomes would be unlikely to be perpetuated in an arbovirus cycle in which *Ae. triseriatus* is the vector. Determination of the molecular basis of such vector-virus restrictions may permit a predictive capability concerning the epidemic potential of evolutionary events in nature.

It is curious that if the vector is such a permissive site for generation of genetic diversity, the arboviruses have not taxonomically degenerated into a confusing array of antigenically distinct viruses. Apparently, constraints exist on the genetic variation tolerated. Such constraints could operate at the virus, vertebrate, or vector level in the arbovirus cycle. Likely determinants for conservation of structural components of the virion would be receptor and maturational events in both vertebrate and invertebrate cells. Some constraint on protein structure may exist at the level of the arthropod-virus interaction; certain protein structures may be required for vector infection, replication, and transmission.

It is unlikely that vector-virus interactions are the sole constraints on the evolutionary potential of arboviruses. The vertebrate and the vertebrate immune response are undoubtedly also major protagonists in the determination of the genetic and antigenic composition of arbovirus populations.

IV. THE IMMUNE RESPONSE AND DETERMINATION OF ARBOVIRUS PHENOTYPES

Thus far, a number of mechanisms have been delineated by which stable genetic variation can be introduced into a virus population. Without appropriate selective pressures, which enrich for these variant populations, many of these genotypic changes would fail to be phenotypically expressed. One of the most potent mechanisms for producing phenotypic diversity is variant selection by the immune system of the vertebrate host. Immune selection can occur in two ways. One mechanism is the selective depletion of virus subpopulations expressing targeted antigenic determinants. This depletion can occur by antibody-mediated neutralization of virus infectivity, antibody-complement-mediated lysis of both the virus and virus-infected cells, or specific cytotoxic T-cell lysis of virus-infected cells. Changes in these targeted antigenic determinants enable the variant to escape the host's preexisting antiviral immunity. The second mechanism is selective enrichment of virus subpopulations expressing "protective" antigenic determinants. This enrichment can occur by antibody-mediated enhancement of virus replication or by formation of infectious virus-antibody complexes which subsequently escape immune clearance. While it should be remembered that immune selection in the host occurs within the framework of a diverse polyclonal antibody response, the advent of monoclonal antibody (MAb) technology has allowed the production of detailed antigenic maps of viral proteins. These maps are being used to delineate epitopes that serve as targets for immune interaction that could result in the production of stable virus populations altered in phenotypic and genotypic characteristics.

A. Selection of Antigenic Variants by Neutralizing Antibody

Selection of variants possessing an altered N-site is conceptually the easiest to understand. If a virus displaying an altered N-site infects an immune vertebrate host, the preexisting antibody repertoire will be unreactive at this important determinant, and therefore the immune response will be less efficient at limiting virus dissemination. While N-variants are not difficult to isolate in the laboratory, results derived from experiments using these variants have been difficult to interpret. This difficulty is based in the lack of understanding of the structure and function of virus N-sites, as well as the interaction of these sites with neutralizing antibody. MAb analysis of the structural proteins of alphaviruses have identified at least eight epitopes on the E2-glycoprotein and at least eight epitopes on the El-glycoprotein.[104,107] Not all of these epitopes elicit neutralizing antibody. The highest titer neutralizing antibody defines E2-epitopes. Recent observations with the members of the VEE virus complex indicate that those MAbs which react with the epitope at the hub of the N-site (E2c) appear to affect N by blocking virus binding to cell receptors (Table 6). MAbs defining epitopes more distal from the central N-site are less efficient at neutralizing virus and blocking virus binding to cells. These results indicate that the N-site of the VEE virus is intimately associated with virus adsorption to cells. It appears that variation at the N-site has twofold importance. Initially, N-site variation would render the virus resistant to N. Second, this variation may reduce the ability of the virus to adsorb to susceptible cells. While variation at the N-site would appear to be the easiest mechanism of variant selection, N-site epitopes within the VEE virus complex have undergone only limited variation.[108] Other epitopes not involved in virus N demonstrate more variation. This may indicate that other mechanisms of variant selection are more biologically significant for alphaviruses. This is not surprising if the alphavirus N-site is the part of the glycoprotein that adsorbs to cells. A mutation in this site may so significantly alter the glycoprotein structure that it would most likely be lethal. In contrast, high titer flavivirus neutralizing antibodies are usually type specific, which indicates great antigenic variation in these epitopes.[54,109-112] These results are consistent with preliminary observations that flavivirus N probably does not involve blocking of virus binding to cells.[135] If flavivirus N and binding to cells is not intimately associated, then the flavivirus glycoprotein receptor domains may be more pliable with regard to structure-function relationships. This hypothesis may account for the wide variety of closely related, but antigenically distinct flaviviruses.

B. Selection of Variants by Antibody-Complement-Mediated Mechanisms

The role that complement plays in protecting immune animals from virus infection has been extensively studied with the alphaviruses.[113,114] Passive protection studies using mice depleted of complement could detect no difference in the ability of neutralizing antiviral poly- or monoclonal antibodies to protect from a lethal virus challenge. Observations with Sindbis virus and WEE virus indicated that non-neutralizing anti-El MAbs were protective even though these epitopes were expressed on the surface of the infected cell; they were

Table 6
EFFECTS OF ANTIBODY BINDING ON ATTACHMENT OF VEE VIRUS (TC-83) TO VERO CELLS

		Assay	
MAb	Epitope	% Inhibition[a]	N titer[b]
5B4D-6	E2a	0	0
3B4C-4	E2c	89	5.0
1A4A-1	E2c	82	5.0
1A5C-3	E2d	0	0
1A3A-5	E2e	0	0
1A4D-1	E2f	—[c]	2.4
1A3A-9	E2g	30	3.3
1A3B-7	E2h	76	3.3
3B2D-5	E1a	0	0
3B2A-9	E1b	33	3.0
5B6A-6	E1c	0	0
3A5B-1	E1d	0	0
6B6C-1[d]	—	0	0
Anti-E1[e]	—	3	0
Anti-E2	—	61	4.3

[a] Results are average of multiple samples (n = 5). Antibody was adjusted to 1 mg/mℓ and was included in the test at 10 μg/test. 5 μg of virus was used.
[b] Results reported as $\log_{10}$ N titer.
[c] Binding of 1A4D-1 enhances virus binding to cell by 1.5- to 2-fold.
[d] Mab, 6B6C-1, is an anti-SLE virus antibody. It is used here as a negative control.
[e] Rabbit anti-E1 and anti-E2 monospecific antisera.

apparently not accessible on the mature virion.[106-115] Similar results have been observed with MAbs specific for the NS-1 (NV-3) nonstructural protein of yellow fever virus.[116] This protein is not present in the mature virion. In each case the proposed mechanism of protection is complement-mediated lysis of virus-infected cells. More work needs to be done to characterize further the epitopes recognized in this phenomenon. However, significant variation in these epitopes would render this potentially important immune mechanism useless. Such a variant could readily escape the preexisting immune response, which would lead to amplification of this phenotype in the virus population.

C. Selection of Antigenic Variants by Virus-Specific Cytotoxic T-Cells

Very little is know about the interaction of cytotoxic T-lymphocyte subsets and arbovirus-infected cells. Alphavirus immunity can be transferred by transplantation of immune lymphocytes.[117,118] Populations of cytotoxic T-cells can apparently recognize alphavirus group-reactive determinants.[119-121] These studies implicate the El as the target for cross-reactive cytotoxic T-cells. The actual role that virus-specific cytotoxic T-lymphocytes play in protection from arbovirus disease is unclear. It is probably reasonable to assume that these immune mechanisms will be important for arbovirus pathogens. Antigenic variation of virus encoded T-cell receptors expressed on the surface of the infected cell would allow these viruses to escape from this important arm of the host immune response, resulting in a modification of the antigenic properties of the virus pool.

D. Selection of Variants by Antibody-Mediated Enhancement of Virus Growth

It has been known for a number of years that some DEN infections are more severe in patients with preexisting, heterologous anti-DEN antibody.[122] The observation that DEN virus replicated well in cells that possessed antibody Fc surface receptors (e.g., blood monocytes) led to the formulation of the hypothesis that DEN virus infection of these cells was being mediated by initial complexing to antibody, followed by adsorption via the Fc receptor. This intriguing hypothesis led many investigators to attempt identification of the antigenic determinant(s) involved in binding enhancing antibody. As with other flaviviruses, DEN virus N antibody is primarily type specific, therefore antibody-mediated immune enhancement was thought to be a function of cross-reactive DEN epitopes. Recent studies mapping DEN surface E-glycoprotein epitopes have been unable to completely explain this phenomenon. Studies with West Nile virus, JE virus, and yellow fever virus indicate that all antivirus antibodies can enhance virus infection on macrophage or macrophage-like continuous cell lines if used in the appropriate concentrations.[109,123] With DEN virus, however, only more cross-reactive antibodies could mediate immune enhancement.[124] It is apparent that the phenomenon of immune enhancement is complex. Current observations could depend on both the virus or cell types used in the immune enhancement assays. Recent studies with VEE viruses identified an anti-E2 antibody with low level N activity that appeared to both enhance and stabilize the binding of virus to susceptible VERO cells.[135] Because VERO cells have no identifiable Fc receptors, binding of this antibody probably modifies the overall glycoprotein structure, rendering it more efficient in binding to cells. Fc receptors do not appear to be necessary for this type of immune enhancement. Whether non-Fc-mediated immune enhancement is occurring during flavivirus infection is unknown. Enhancement mediated by reorganization of proteins following antibody binding might explain some of the conflicting observations on immune enhancement of flaviviruses. Regardless of mechanism, it appears that immune enhancement is epitope dependent. Virus strains possessing these epitopes would have a distinct selective advantage over similar strains mutant in the "enhancement epitopes". Under such circumstances, enrichment of variant virus subpopulations possessing enhancement epitopes could readily occur, and a new antigenically distinct virus population would result.

E. Summary of Immune Mechanisms in Antigenic Variability of Arboviruses

A number of immune mechanisms by which subpopulations of mutant viruses could be amplified have been discussed. The importance of these mechanisms in nature is currently unknown. The effects that antigenic diversity has on the phenotypic markers of adsorption, replication, pathogenesis, and virulence are also unknown. It will be important to determine whether those antigenic variants produced in the laboratory resemble naturally occurring viruses. It is obvious that rampant antigenic variation does not occur. In fact, many of the arbovirus populations are extremely stable (e.g., EEE virus). This antigenic stability probably results from the exquisite checks and balances between immune response and the structural constraints on the association of virus with its cellular receptors in vertebrate and invertebrate hosts. With such potent control mechanisms in place, it is probably not unusual that significant virus antigenic variation occurs only infrequently.

V. EPIDEMIOLOGIC SIGNIFICANCE

One can only speculate concerning the epidemiologic implications of arbovirus evolution. Dramatic changes in the epidemiology of certain arboviruses may or may not be attributable to evolutionary events. There are examples of disease outbreaks or altered arbovirus epidemiology in which virus variation could have played an important role. For example, o'nyong-nyong (ONN), an alphavirus closely related to chikungunya (CHIK), appeared in an explosive epidemic involving several million human cases in East Africa.[126] The virus was repeatedly isolated from man and *Anopheles* species mosquitoes until it apparently vanished; only one isolation of ONN has been made since. One can speculate that ONN somehow evolved from CHIK, a virus associated with *Aedes* and *Culex* species mosquitoes. Molecular changes in the CHIK genome may have resulted in a virus, ONN, capable of eluding existing herd immunity barriers to CHIK and capable of being transmitted by alternate vector species. Thus, the putatively newly evolved virus could exploit a newly susceptible vertebrate population using newly available vector species. Genomic hybridization analysis of CHIK and ONN, using stringent hybridization conditions, revealed that the two are significantly more similar than the other alphaviruses.[127] Recent dramatic changes in the epidemiology of other arboviruses, such as the appearance of Ross River virus outside of Australia, the expansion of Japanese encephalitis throughout India, and the emergence of Rift Valley fever virus from sub-Saharan Africa, remain unexplained. While such changes in the epidemiology of arboviruses may be due to improved surveillance capability or introduction of the virus into a new region, the possibility also exists that evolutionary events have permitted these viruses to exploit new ecological situations.

There is little doubt that specific virus genotypes are associated with specific biologic and epidemiologic situations. As was described previously, oligonucleotide fingerprints of isolates of SLE virus from western and midwestern regions of the U.S., respectively, are similar to each other but differ from fingerprints of isolates from other geographic areas. Thus, distinct genotypes are correlated with the unique epidemiology of the virus in the respective geographic areas, involving different preferred invertebrate and presumably vertebrate hosts. Further, the specific genotypes of SLE differ dramatically in certain biologic characteristics, such as duration of viremia in avians and virulence in mice.[55] It is noteworthy that in light of the genetic heterogeneity of isolates of LAC virus described previously, the two isolates of LAC virus obtained from brain specimens at autopsy have virtually identical oligonucleotide fingerprints.[9] These two viruses were isolated from brain tissues of children whose infections were separated by 50 mi and 20 years. Isolates of JE virus obtained from brains of patients in Thailand also have similar oligonucleotide fingerprints.[67] Further, virus isolates from serious cases of DEN in Thailand exhibit similar fingerprints.[130] Thus, one wonders if certain virus genotypes are indeed more virulent and if infection with these specific genotypes results in serious disease. Alternatively, the specific genotypes could be selected for during the pathogenesis of the virus in the host. Considering the plasticity of the RNA virus genome and the realization that only a single nucleotide change, even in a noncoding region of the virus genome, can lead to neurovirulence,[128] such a hypothesis does not seem unreasonable.

At this time, it would seem that only one conclusion can be drawn about the epidemiologic significance of arbovirus evolution. The advent and application of modern molecular approaches, such as recombinant DNA technology, genomic analyses, and MAbs, may finally provide definitive answer to these interesting and important questions concerning virus epidemiology, biology, and pathology.

REFERENCES

1. **Schlesinger, R. W.,** in *The Togaviruses: Biology, Structure and Replication,* Schlesinger, R. W., Ed., Academic Press, New York, 1980, 1.
2. **Karabatsos, N.,** *International Catalogue of Arboviruses,* 3rd ed., American Society of Tropical Medicine and Hygiene, San Antonio, Tex., 1986.
3. **Darlington, C. D.,** Origin and evolution of viruses, *Trans. R. Soc. Trop. Med.,* 54, 90, 1960.
4. **Mattingly, P. F.,** Ecological aspects of the evolution of mosquito-borne virus diseases, *Trans. R. Soc. Trop. Med.,* 54, 97, 1960.
5. **Smith, C. E. G.,** Factors in the past and future evolution of the arboviruses, *Trans. R. Soc. Trop. Med.,* 54, 113, 1960.
6. **Calisher, C. H.,** Antigenic relationships of the arboviruses: an ecological and evolutionary approach, in *Proc. Int. Symp. New Aspects in Ecology of Arboviruses,* Institute of Virology, Slovak Academy of Sciences, Prague, 1979, 117.
7. **Porterfield, J. S.,** Antigenic characteristics and classification of togaviridae, in *The Togaviruses: Biology, Structure and Replication,* Schlesinger, R. W., Ed., Academic Press, New York, 1980, 13.
8. **Mitchell, C. J.,** Mosquito vector competence and arboviruses, in *Current Topics in Vector Research,* Vol. 1, Harris, K. F., Ed., Praeger, New York, 1983, 63.
9. **Bishop, D. H. L. and Shope, R. E.,** Bunyaviridae, in *Comprehensive Virology,* Vol. 14, Fraenkel-Conrat, H. and Wagner, R. R., Eds., Plenum Press, New York, 1979, 1.
10. **Rice, C. M., Strauss, E. G., and Strauss, J. H.,** Structure of the flavivirus genome, in *The Togaviruses and Flaviviruses,* Schlesinger, S. and Schlesinger, M., Eds., Plenum Press, New York, 1986, 1.
11. **Strauss, E. G. and Strauss, J. H.,** Structure and replication of the alphavirus genome, in *The Togaviruses and Flaviviruses,* Schlesinger, S. and Schlesinger, M., Eds., Plenum Press, New York, 1986, 68.
12. **Bishop, D. H. L.,** The genetic basis for describing viruses as species, *Intervirology,* 24, 79, 1985.
13. **Holland, J. J.,** Continuum of change in RNA virus genomes, in *Concepts in Viral Pathogenesis,* Notkins, A. L. and Oldstone, M. B. A., Eds., Springer-Verlag, New York, 1984, 137.
14. **Nahmias, A. J. and Reanney, D. C.,** The evolution of viruses, *Ann. Rev. Ecol. Syst.,* 8, 29, 1977.
15. **Young, J. F., Desselberger, U., and Palese, P.,** Evolution of human influenza viruses in nature: sequential mutations and the genomes of new H1N1 isolates, *Cell,* 18, 73, 1979.
16. **Ortin, J., Najera, R., Lopez, C., Davila, M., and Domingo, K.,** Genetic variability of Hong Kong (H3N2) influenza viruses: spontaneous mutations and their location in the viral genome, *Gene,* 11, 319, 1980.
17. **Palese, P. and Young, J. F.,** Variation of influenza A, B, and C viruses, *Science,* 215, 1468, 1982.
18. **Webster, R. G., Laver, W. G., Air, G. M., and Schild, G. C.** Molecular mechanisms of variation in influenza virus, *Nature (London),* 296, 115, 1982.
19. **Laver, W. G., Webster, R. G., and Chu, C. M.,** Summary of a meeting on the origin of pandemic influenza viruses, *J. Infect. Dis.,* 149, 108, 1984.
20. **Holland, M., Spindler, K., Morodyski, F., Grabau, E., Nichol, S., and VandePol, S.,** Rapid evolution of RNA genomes, *Science,* 215, 1577, 1982.
21. **Domingo, E., Sabo, D., Taniguchi, T., and Weissman, C.,** Nucleotide sequence heterogeneity of an RNA phage population, *Cell,* 13, 735, 1978.
22. **Drake, J.,** Comparative rates of spontaneous mutation, *Nature (London),* 221, 1132, 1979.
23. **Hayashida, H., Toh, H., Kikuno, R., and Miyata, T.,** Evolution of influenza virus genes, *Mol. Biol. Evol.,* 2, 289, 1985.
24. **Jukes, T. H.,** Silent nucleotide substitutions and the molecular evolutionary clock, *Science,* 210, 973, 1980.
25. **Holland, J. J., Grabaw, E. A., Jones, C. L., and Semler, B. L.,** Evolution of multiple genome mutation during long term persistent infection by vesicular stomatitis virus, *Cell,* 16, 495, 1979.
26. **Clewley, J. P., Bishop, D. H. L., Kang, C.-Y., Coffin, F., Schnitzlein, M. W., Reichmann, M. E., and Shope, R. E.,** Oligonucleotide fingerprints of RNA species obtained from rhabdoviruses belonging to the vesicular stomatitis virus subgroup, *J. Virol.,* 23, 152, 1977.
27. **King, A., MacCahon, D., Slade, W., and Newman, J.,** Recombination in RNA, *Cell,* 29, 921, 1982.
28. **Tolskaya, E., Ramonova, L., Koliniskova, M., and Agol, V. I.,** Intertypic recombinants in poliovirus: genetic and biochemical studies, *Virology* 124, 121, 1983.
29. **Fields, S. and Winter, G.,** Nucloetide sequences of influenza virus segments 1 and 3 reveal mosaic structure of a small RNA virus RNA segment, *Cell,* 28, 303, 1982.
30. **McCahon, D., King, A., Roc, D., Slade, W., Newman, J., and Cleary, A.,** Isolation and biochemical chracterization of intertypic recombinants of foot-and-mouth disease virus, *Virus Res.,* 3, 87, 1985.
31. **Calisher, C. H., Shope, R. E. Brandt, W., Casals, J., Karabatsos, N., Murphy, F. A., Tesh, R. B., and Wiebe, M. E.,** Proposed antigenic classification of registerd arboviruses. I. Togaviridae, alphavirus, *Intervirology,* 14, 229, 1980.

32. **Karabatsos, N.,** Antigenic relationships of group A arboviruses by plaque reduction neutralization testing, *Am. J. Trop. Med. Hyg.*, 24, 527, 1975.
33. **Simons, K., Garoff, H., and Helenius, A.,** Alphavirus proteins, in *The Togaviruses: Biology, Structure, and Replication*, Schlesinger, W., Ed., Academic Press, New York, 1986, 317.
34. **Murphy, F. M.,** Togavirus morphology and morphogenesis, in *The Togaviruses: Biology, Structure, and Replication*, Schlesinger, W., Ed., Academic Press, New York, 1986, 241.
35. **Strauss, E. G., Rice, C. M., and Strauss, J. H.,** Complete nucleotide sequence of the genomic RNA of Sindbis virus, *Virology*, 133, 92, 1984.
36. **Faragher, S. G., Marshall, J. D., and Dalgarno, L.,** Ross River virus genetic variants in Australia and the Pacific Islands, *Aust. J. Exp. Biol. Med.*, 63,473,1985.
37. **Trent, D. W. and Grant, J. A.,** A comparison of new world alphaviruses in the western equine encephalitis virus complex by immunochemical and oligonucleotide fingerprint techniques, *J. Gen. Virol.*, 47, 261, 1980.
38. **Trent, D. W., Clewley, J. P., France, J. K., and Bishop, D. H. L.,** Immunochemical and oligonucleotide fingerprint analysis of Venezuelan equine encephalomyelitis complex viruses, *J. Gen. Virol.*, 43, 365, 1979.
39. **Kinney, R. M. and Trent, D. W.,** Comparative immunological and biochemical analysis of viruses in the Venezuelan equine encephalitis complex, *J. Gen. Virol.*, 64, 1111, 1983.
40. **Olson, K. and Trent, D. W.,** Genetic and antigenic variations among geographical isolates of Sindbis virus, *J. Gen. Virol.*, 66, 797, 1985.
41. **Rentier-Delrue, F. and Young, N.,** Genomic divergence among sindbis virus strains, *Virology*, 104, 59, 1980.
42. **Garoff, H., Frischauff, A.-M., Simons, K., Lehrach, H., and Delius, H.,** Nucleotide sequence of cDNA coding for Semliki forest virus membrane glycoproteins, *Nature (London)*, 288, 236, 1980.
43. **Dalgarno, L., Rice, C. M., and Strauss, J. H.,** Ross River virus 24S RNA: complete nucleotide sequence and deduced sequence of the encoded structural proteins, *Virology*, 129, 170, 1983
44. **Kinney, R. M., Johnson, B. B. J., Brown, V. L., and Trent, D. W.,** Nucleotide sequence of the 26S mRNA and deduced sequence of the encoded structural proteins of the virulent Trinidad donkey strain of Venezuelan equine encephalitis virus, *Virology*, 152, 400, 1986
45. **Baltimore, D.,** Evolution of RNA viruses, *Ann. N. Y. Acad. Sci.*, 354, 492, 1980.
46. **Grantham, R., Gautier, C., Gouy, M., Jacobzone, M., and Mercier, R.,** Codon catalog usage is a genome strategy modulated for gene expressivity, *Nucleic Acids Res.*, 9, 43, 1981.
47. **Rice, C. M., Lenches, E. M., Eddy, S. R., Shin, S. J., Sheets, R. L., and Strauss, J. M.,** Nucleotide sequence of yellow fever virus: implications for flavivirus gene expression and evolution, *Science*, 229, 726, 1985.
48. **Rice, C. M., Aebersoid, R., Teplow, D. B., Pata, J., Bell, J. R., Vorndam, A. V., Trent, D. W., Brandriss, M. W., Schlesinger, J. J., and Strauss, J. H.,** Partial N-terminal amino acid sequences of three nonstructural proteins of two flaviviruses, *Virology*, 151, 1, 1986.
49. **Dalgarno, L., Trent, D. W., Strauss, J. H., and Rice, C. M.,** Partial nucleotide sequence of Murray Valley encephalitis virus genome: comparison of the encoded polypeptides with yellow fever virus structural and nonstructural proteins, *J. Mol. Biol.*, 187, 309, 1986.
50. **Castie, E., Nowak, T. H., Leidner, U., Wengler, G., and Wengler, G.,** Sequence analysis of the viral core protein and the membrane associated proteins V1 and NV2 of the flavivirus West Nile virus and of the genome sequence for these proteins, *Virology*, 145, 227, 1985.
51. **Wengler, G., Castle, E., Leidner, U., Nowak, T., and Wengler, G.,** Sequence analysis of the membrane protein V3 of the flavivirus virus and of its gene, *Virology*, 147, 264, 1985.
52. **Trent, D. W., Kinney, R. M., Johnson, B. B. J., Grant, A., Hahn, C., and Rice, C. M.,** Nucleotide sequence of St. Louis encephalitis virus genome encoding the structural proteins and nonstructural protein NS1, *J. Virol.*, 156, 243, 1987.
53. **Trent, D. W.,** Antigenic characterization of flavivirus structural proteins separated by isoelectric focusing, *J. Virol.*, 22, 408, 1977.
54. **Roehrig, J. T., Mathews, J. H., and Trent, D. W.** Identification of epitopes on the E glycoprotein of St. Louis encephalitis virus using monoclonal antibodies, *Virology*, 128, 118, 1983.
55. **Trent, D., Monath, T., Bowen, S., Vorndam, A., Cropp, B., and Kemp, G.,** Variation among strains of St. Louis encephalitis virus: basis for a genetic, pathogenic, and epidemiologic classification, *Ann. N. Y. Acad. Sci.*, 354, 219, 1980.
56. **Trent, D. W., Grant, J. A., Vorndam, A. V., and Monath, T. P.,** Biochemical heterogeneity among St. Louis encephalitis virus isolates of different geographic origin, *Virology*, 114, 319, 1981.
57. **Mitchell, C. J., Gubler, D. J., and Monath, T. P.,** Variation in infectivity of Saint Louis encephalitis viral strains for *Culex pipiens quinquefasciatus* (Diptera:Culicidae), *J. Med. Entomol.*, 20, 526, 1983.
58. **Deubel, V., Digoutte, J.-P., Monath, T. P., and Grant, M.,** Genetic heterogeneity of yellow fever virus strains from Africa and the Americas, *J. Gen. Virol.*, 67, 209, 1986.

59. **Deubel, V., Pailliez, J. P., Cornet, M., Schlesinger, J. J., Diop, M., Diop, A., Digoutte, J. P., and Girard, M.,** Homogeneity among Senegalese strains of yellow fever virus, *Am. J. Trop. Med. Hyg.*, 34, 976, 1985.
60. **Monath, T. P., Kinney, R. M., Schlesinger, J. J., Brandriss, M. W., and Brès, P.,** Ontogeny of yellow fever 17D vaccine: RNA oligonucleotide fingerprint and monoclonal antibody analysis of vaccines produced world wide, *J. Gen. Virol.*, 64, 627, 1983.
61. **Vezza, A. C., Rosen, L., Repik, P., Dalrymple, J., and Bishop, D. H. L.,** Characterization of the viral RNA species of prototype dengue viruses, *Am. J. Trop. Med, Hyg.*, 29, 643, 1980.
62. **Repik, P., Dalrymple, J. M., Brandt, W. E., McCown, J. M., and Russell, P. K.,** RNA fingerprinting as a method for distinguishing dengue 1 virus strains, *Am. J. Trop. Med. Hyg.*, 32, 577, 1983.
63. **Trent, D. W., Grant, J. A., Rosen, L., and Monath, T. P.,** Genetic variation among dengue 2 viruses of different geographic origin, *Virology*. 128, 271, 1983.
64. **Monath, T. P., Wands, J. R., Hill, L. J., Brown, N. V., Marciniak, R. A., Wong, M. A., Gentry, M. K., Burke, D. S., Grant, J. A., and Trent, D. W.,** Geographic classification of dengue 2 virus strains by antigen signature analysis, *Virology*, 154, 313, 1986.
65. **Blok, J., Henchal, E. A., and Gorman, B. M.,** Comparison of dengue viruses and some other flaviviruses by cDNA-RNA hybridization analysis and detection of a close relationship between dengue virus serotype 2 and Edge Hill virus, *J. Gen. Virol.*, 65, 2173, 1984.
66. **Hori, H.,** Oligonucleotide fingerprint analysis of Japanese encephalitis (JE) virus strains of different geographic origins, *Trop. Med.*, 28, 179, 1986.
67. **Burke, D. S., Schmaljohn, C. S., and Dalrymple, M.,** Strains of Japanese encephalitis virus isolated from human bains have a higly conserved genotype compared to strains isolated from other natural hosts, personal communication.
68. **Bell, J. R., Kinney, R. M., Trent, D. W., Lenches, E. M., Dalgarno, L., Strauss, J. H.,** Amino-terminal amino acid sequences of structural proteins of three flaviviruses, *Virology*, 143, 224, 1985.
69. **Bell, J. R., Kinney, R. M., Trent, D. W., Strauss, E. G., and Strauss, J. H.,** An evolutionary tree relating eight alphaviruses, based on amino-terminal sequences of their glycoproteins, *Proc. Natl. Acad. Sci. U.S.A.*, 81, 4702, 1984.
70. **Said, L. H. E., Vorndam, V., Gentsch, J. R., Clewley, J. P., Calisher, C. H., Klimas, R. A., Thompson, W. H., Grayson, M., Trent, D. W., and Bishop, D. H. L.,** A comparison of LaCrosse virus isolates obtained from different ecological niches and an analysis of the structural components of California encephalitis serogroup virus and other bunyaviruses, *Am. J. Trop. Med. Hyg.*, 28, 364, 1979.
71. **Klimas, R. A., Thompson, W. H., Calisher, C. H., Clark, G. G., Grimstad, P. R., and Bishop, D. H. L.,** Genotypic varieties of LaCrosse virus isolated from different geographic regions of the continental United States and evidence for a naturally occurring intertypic recombinant LaCrosse virus, *Am. J. Epidemiol.*, 114, 112, 1981.
72. **Clerx-van Haaster, C., Akashi, H., Auperin, D., and Bishop, D.,** Nucleotide sequence analyses and predicted coding of bunyavirus genome RNA species, *J. Virol.*, 41, 119, 1982.
73. **Palese, P.,** Reassortment continuum, in *Concepts in Viral Pathogenesis*, Notkins, A. L. and Oldstone, M. B. A., Eds., Springer-Verlag, New York, 1984, 137.
74. **Elliott, R. M., Lees, J. F., Watret, G. E., Clark, W., and Pringle, C. R.,** Genetic diversity of bunyaviruses and mechanisms of genetic variation, in *Mechanisms of Viral Pathogenesis: From Gene to Pathogen*, Kohn, A. and Fuchs, P., Eds., Martinus Nijhoff, Boston, 1983, 61.
75. **Pringle, C. R., Lees, J. F., Clark, W., and Elliott, R. M.,** Genome subunit reassortment among bunyaviruses analyzed by dot hybridization using molecularly cloned complementary DNA probes, *Virology*, 135, 244, 1984.
76. **Gorman, B. M., Taylor, J., Walker, P. J., and Young, P. R.,** The isolation of recombinants between related orbiviruses, *J. Gen. Virol.*, 41, 333, 1978.
77. **Casals, J. and Whitman, L.,** Group C, a new serological group of hitherto undescribed arthropod-borne viruses. Immunological studies, *Am. J. Trop. Med. Hyg.*, 10, 250, 1960.
78. **Shope, R. and Causey, O.,** Further studies on the serological relationships of group C arthropod-borne viruses and the application of these relationships to rapid identification of types, *Am. J. Trop. Med. Hyg.*, 11, 283, 1962.
79. **Ushijima, H., Clerx-van Haaster, C., and Bishop, D.,** Analyises of Patois serogroup viruses: evidence for naturally occurring recombinant bunyaviruses and existence of viral coded nonstructural proteins induced in bunyavirus-infected cells, *Virology*, 110, 318, 1981.
80. **Sugiyama, K., Bishop, D., and Roy, P.,** Analysis of the genomes of bluetongue viruses recovered in the United States. I. Oligonucleotide fingerprint studies that indicate the existence of naturally-occurring reassortant BTV isolates, *Virology*, 114, 210, 1981.
81. **Collisson, E. and Roy, P.,** Analysis of the genomes of bluetongue virus serotype 10 vaccines and recent BTV-10 isolate from Washington, *Am J. Vet. Res.*, 44, 235, 1983.

82. **Sugiyama, K., Bishop, D. H. L., and Roy, P.,** Analysis of the genomes of bluetongue viruses recovered from different states of the United States and at different times, *Am. J. Epidemiol.*, 115, 332, 1982.
83. **Kew, O., Nottay, B., Hatch, M., Nakano, J., and Obijeski, J.,** Multiple genetic changes can occur in the oral polio vaccines upon replication in humans. *J. Gen. Virol.*, 56, 337, 1981.
84. **Nottay, B. K., Kew, O., Hatch, M. H., Heyword, J. T., and Obijeski, J. F.,** Molecular variation of type 1 vaccine related and wild poliovirus during replication in humans, *Virology*, 106, 405, 1981.
85. **King, A. M. Q., Underwood, B. O., McCahon, D., Newman, J. W. I., and Brown, F.,** Biochemical identification of virus causing the 1981 outbreaks of foot-and-mouth disease in U.K., *Nature (London)*, 293, 479, 1981.
86. **Sobrino, F., Davila, M., Ortin, J., and Domingo, E.,** Multiple genetic variants arise in the course of replication of foot-and-mouth disease virus in cell culture, *Virology*, 123, 310, 1983.
87. **Domingo, E., Davila, M., and Ortin, J.,** Nucleotide sequence heterogeneity of the RNA from a natural population of foot-and-mouth disease virus, *Gene*, 11, 333, 1980.
88. **Gubler, D. J., Kuno, G., Sather, E., and Waterman, H.,** A case of natural concurrent human infection with two dengue viruses, *Am. J. Trop. Med. Hyg.*, 34, 170, 1985.
89. **Darnell, M. and Plaguemann, P.,** Physical properties of lactic dehydrogenase elevating virus and its ribonucleic acid, *J. Virol.*, 10, 1082, 1972.
90. **Michaelides, M. C. and Schlesinger, S.,** Structural proteins of lactic dehydrogenase virus, *Virology*, 55, 211, 1973.
91. **Notkins, A. L., Mahar, S., Scheele, C., and Goffman, J.,** Infectious virus-antibody complex in the blood of chronically infected mice, *J. Exp. Med.*, 124, 81, 1966.
92. **Notkins, A. L., Mage, M., Ashe, W. K., and Mahar, S.,** Neutralization of sensitized lactic dehydrogenase virus by anti-gamma-globulin, *J. Immunol.*, 100, 314, 1968.
93. **Monath, T. P.,** Neutralizing antibody responses in the major immunoglobulin classes to yellow fever 17D of humans, *Am. J. Epidemiol.*, 93, 122, 1971.
94. **Stott, J. L., Osburn, B. I., and Barber, T. L.,** Recovery of dual serotypes of bluetongue virus from infected sheep and cattle, *Vet. Microbiol.*, 7, 197, 1982.
95. **Watts, D., Pantuwatana, S., Defoliart, G., Yuill, T. M., and Thompson, W. H.,** Transovarial transmission of La Crosse virus (California encephalitis group) in the mosquito *Aedes triseriatus*, *Science*, 182, 1140, 1973.
96. **Edman, J. D. and Downe, A. E. R.,** Host-blood sources and multiple feeding habits of mosquitoes in Kansas, *Mosq. News*, 24, 154, 1964.
97. **Thompson, W. and Beaty, B.,** Venereal transmission of La Crosse (California encephalitis) arbovirus in *Aedes triseriatus* mosquitoes, *Science*, 196, 530, 1977.
98. **Beaty, B. J., Rozhon, E. J., Gensemer, P., and Bishop, D. H. L.,** Formation of reassortant bunyaviruses in dually-infected mosquitoes, *Virology*, 111, 662, 1981.
99. **Beaty, B. J., Sundin, D. R., Chandler, L., Bishop, D. H. L.,** Evolution of bunyaviruses via genome segment reassortment in dually-infected (per os) *Aedes triseriatus* mosquitoes, *Science*, 230, 548, 1985.
100. **Beaty, B. J., Bishop, D. H. L., Gay, M., and Fuller, F.,** Interference between bunyaviruses in *Aedes triseriatus* mosquitoes, *Virology*, 127, 83, 1983.
101. **Florkiewicz, R. Z. and Hewlett, M. J.,** Persistent infections of bunyaviruses in *Aedes albopictus* (mosquito) cells, in *Animal Virus Genetics*, Fields, B. N. and Jaenesch, R., Eds., Academic Press, New York, 1980, 749.
102. **Beaty, B. J., Holterman, M., Tabachnick, W., Shope, R. E., Rozhon, E. J., and Bishop, D. H. L.,** Molecular basis of bunyavirus transmission by mosquitoes: the role of the mRNA segment, *Science*, 211, 1433, 1981.
103. **Beaty, B. J., Miller, B. R., Shope, R. E., Rozhon, E. J., and Bishop, D. H. L.,** Molecular basis of bunyavirus per os infection of mosquitoes: the role of the mRNA segment, *Proc. Natl. Acad. Sci. U. S. A.*, 79, 1295, 1982.
104. **Roehrig, J. T., Corser, J. A., and Schlesinger, M. J.,** Isolation and characterization of hybridoma cell lines producing monoclonal antibodies against the structural proteins of Sindbis virus, *Virology*, 101, 41, 1980.
105. **Boere, W. A. M., Harmsen, M., Benaissa-Trouw, B. J., Kraaijevald, C. A., and Snippe, H.,** Identification of distinct determinants on Semliki Forest virus by using monoclonal antibodies with different antiviral activities, *J. Virol.*, 52, 575, 1982.
106. **Schmaljohn, A. L., Kokubun, J. M., and Cole, G. A.,** Protective monoclonal antibodies define maturational and pH-dependent antigenic changes in Sindbis virus E1 glycoprotein, *Virology*, 130, 144, 1983.
107. **Hunt, A. R. and Roehrig, J. T.,** Identification and characterization of epitopes on the E1 glycoprotein of Western equine encephalitis virus using monoclonal antibodies, *Virology*, 142, 334, 1985.
108. **Roehrig, J. T. and Mathews, J. H.,** The neutralization site on the E2 glycoprotein of Venezuelan equine encephalomyelitis virus is composed of multiple conformationally stable epitopes, *Virology*, 142, 347, 1985.

109. **Peiris, J. S. M., Porterfield, J. S., and Roehrig, J. T.,** Monoclonal antibodies against the flavivirus West Nile, *J. Gen. Virol.*, 58, 283, 1982.
110. **Henchal, E. A., Gentry, M. K., McCown, J. M. and Brandt, W. E.,** Dengue virus-specific and flavivirus group determinants identified with monoclonal antibodies by indirect immunofluorescence, *Am. J. Trop. Med. Hyg.*, 31, 830, 1982.
111. **Kimura-Kuroda, J. and Yasui, K.,** Topographical analysis of antigenetic determinants on the envelope glycoprotein V3 (E) of Japanese encephalitis virus, using monoclonal antibodies, *J. Virol.*, 45, 124, 1983.
112. **Heinz, F. X., Berger, R., Tuma, W., and Kunz, C.,** A topological and functional model of epitopes on the structural glycoprotein of tick-borne encephalitis virus defined by monoclonal antibodies, *Virology*, 126, 525, 1983.
113. **Mathews, J. H., Roehrig, J. T., and Trent, D. W.,** Role of complement and Fc protein of immunoglobulin G in immunity to Venezuelan equine encephalomyelitis virus infection with glycoprotein-specific monoclonal antibodies, *J. Virol.*, 55, 594, 1985.
114. **Hirsch, R. L., Griffin, D. E., and Winkelstein, J. A.,** Role of complement in viral infections: participation of terminal complement components (C5 to C9) in recovery of mice from Sindbis virus infection, *Infect. Immun.*, 30, 899, 1980.
115. **Schmaljohn, A. L., Johnson, E. D., Dalrymple, J. M., and Cole, G. A.,** Non-neutralizing monoclonal antibodies can prevent lethal alphavirus encephalitis, *(Nature London,)* 297, 70, 1982.
116. **Schlesinger, J. J., Walsh, E. E., and Brandriss, M. W.,** Analysis of 17D yellow fever virus envelope protein epitopes using monoclonal antibodies, *J. Gen. Virol.*, 65, 1637, 1984.
117. **Rabinowitz, S. G. and Alder, W. H.,** Host defenses during primary Venezuelan equine encephalomyelitis virus infection in mice. I. Passive transfer of protection with immune serum and immune cells, *J. Immunol.*, 110, 1345, 1973.
118. **Rabinowitz, S. G. and Proctor, R. A.,** In vitro study of antiviral activity of immune spleen cells in experimental Venezuelan equine encephalomyelitis infection in mice, *J. Immunol.*, 112, 1070, 1974.
119. **Wolcott, J. A., Wust, C. J., and Brown, A.,** Immunization with one alphavirus cross-primes cellular and humoral immune responses to a second alphavirus, *J. Immunol.*, 129, 1267, 1982.
120. **Wolcott, J. A., Gates, D. W., Wust, C. J., and Brown, A.,** Cross-reactive cell associated antigen on L929 cells infected with temperature-sensitive mutants of Sindbis virus, *Infect. Immun.*, 36, 704, 1982.
121. **Wolcott, J. A., Wust, C. J., and Brown, A.,** Cross-reactive antigen in cell-mediated cytolysis of cells infected with a temperature-sensitive mutant of Sindbis virus, *J. Gen. Virol.*, 66, 1167, 1985.
122. **Halstead, S. B., Porterfield, J. S., and O'Rourke, E. J.,** Enhancement of dengue virus infection in monocytes by flavivirus antisera, *Am. J. Trop. Hyg.*, 29, 638, 1980.
123. **Schlesinger, J. J. and Brandriss, M. W.,** 17D yellow fever virus infection of P388D1 cells mediated by monoclonal antibodies: properties of the macrophage Fc receptor, *J. Gen. Virol.*, 64, 1255, 1983.
124. **Brandt, W. E., McCown, J. M., Gentry, M. K., and Russell, P. K.,** Infection enhancement of dengue type 2 virus in the U-937 human monocyte cell line by antibodies to flavivirus cross-reactive determinants, *Infect. Immunol.*, 36,1036, 1982.
125. **Halstead, S. B., Venkateshan, C. N., Gentry, M. K., and Larsen, L. K.,** Heterogeneity of infection enhancement of dengue 2 strains by monoclonal antibodies, *J. Immunol.*, 132, 1529, 1984.
126. **Shope, R. E.,** The alphaviruses, in *Virology*, Fields, B., Ed., Raven Press, New York, 1985, 93, 1.
127. **Wengler, G., Wengler, G., and Filipe, A. R.,** A study of nucleotide sequence homology between nucleic acids of different alphaviruses, *Virology*, 78, 124, 1977.
128. **Evans, D., Dunn, G., Minor, P., Schild, G., Cann, A., Stanway, G., Almond, J., Currey, K., and Maizel, J.,** Increased neurovirulence associated with a single nucleotide change in a noncoding region of the Sabin type 3 poliovaccine genome, *Nature (London)*, 314, 548, 1985.
129. **Chang, C. and Trent, D. W.,** unpublished data.
130. **Trent, D. W.,** unpublished data.
131. **Beaty, B. J.,** unpublished data.
132. **Nuttall, P. and Bishop, D. H. L.,** personal communication.
133. **Miller, B. R. and Knudson, D.,** personal communication.
134. **Sundin, D. and Beaty, B. J.,** unpublished data.
135. **Roehrig, J. T.,** unpublished data.

Chapter 4

SUSCEPTIBILITY AND RESISTANCE OF VECTOR MOSQUITOES

James L. Hardy

TABLE OF CONTENTS

I. INTRODUCTION

Most arboviruses are unique among the animal viruses since their survival depends on an ability to infect both vertebrate and invertebrate hosts and to multiply over a broad temperature range. Therefore, a knowledge of virus-mosquito interactions from the genetic to the organismal levels, and the effect of temperature on these interactions, is important to an understanding of the epidemiology of mosquito-borne arbovirus diseases. Of particular concern is the elucidation of factors and mechanisms that contribute to the quantitative differences observed in tissue susceptibilities, extrinsic incubation periods, and transmission efficiencies between females of different mosquito species, and within a single mosquito species, that are exposed to or infected with various arboviruses. It is this aspect of vector competence that is emphasized in this chapter. It must be pointed out, however, that there are other important factors of vector competence which determine if the female mosquito will be exposed to and ingest an infective blood meal from a vertebrate host.[1-10] These include ecological and environmental factors, vertebrate host genetics, and other genetically controlled traits in mosquitoes, such as host preference, autogeny, and diapause. The importance of these factors to the vector competence of mosquitoes for arboviruses and the epidemiology of arbovirus diseases is presented elsewhere in this book.

II. HISTORICAL PERSPECTIVE

The term "vector competence" has been widely used during the last decade to denote, in part, the ability of hematophagous arthropods to become infected with an arbovirus after ingestion of an infective blood meal, and to transmit virus subsequently while feeding on a vertebrate host. As pointed out by Mitchell,[9] the concept of vector competence of mosquitoes for arboviruses is not new. It dates back at least to 1881 when a Cuban physician, Carlos Finlay, hypothesized that the agent of yellow fever (YF) was transmitted by mosquitoes.[11] Finlay singled out *Aedes aegypti* as the most likely mosquito vector of YF and conducted experimental transmission studies with humans and this mosquito. However, it was Major Walter Reed[12] and other members of the American Yellow Fever Commission, who in 1900, reported the preliminary results of experiments which proved that *Ae. aegypti* could transmit the agent of YF from human to human. Critical to this discovery was the recognition that mosquitoes had to be incubated for about 12 days after ingestion of an infective blood meal from a YF patient before they became infective (i.e., could transmit virus perorally). This concept was suggested by Carter in 1898 after he had observed that secondary cases of YF usually occurred 15 to 23 days after a primary case. According to Bauer,[13] Carter referred to "the time from the development of the infecting case to the time the environment is capable of developing infection in other men" as the "extrinsic incubation period". Of interest is that the successful transmission of the YF agent by mosquitoes was reported 1 year before Reed and Carroll[14] demonstrated that this agent was filterable, a property later ascribed to viruses.

Similar studies with dengue soon followed since it was recognized that the epidemiology of dengue was quite similar to that of YF. According to Simmons,[15] Graham reported the successful transmission of the dengue agent from human to human by *Culex quinquefasciatus* in 1903, as did Ashburn and Craig in 1907, but later investigators felt that this was mechanical, rather than biological transmission. Experimental biological transmission of dengue virus from human to human by *Ae. aegypti* was first demonstrated in 1918 by Cleland and associates, and was later confirmed by three other research groups from 1923 to 1928.[15] An extrinsic incubation period of 8 to 14 days was required before *Ae. aegypti* could transmit dengue virus. Simmons et al.[16] reported on the biological transmission of dengue virus by *Ae. albopictus* in 1930.

Early attempts to further characterize YF virus infections in *Ae. aegypti* were hampered by the need to use humans as the vertebrate host. This obstacle was removed in 1928 when Stokes et al.[17] demonstrated that YF virus could be transmitted from human to monkey by injection of infected human blood and then from monkey to monkey, either by injection of infected monkey blood or by the bite of infected *Ae. aegypti*. Monkeys developed a clinical disease following infection with YF virus and consequently provided the laboratory model for the pioneering studies reported by Bauer,[13] Bauer and Hudson,[18] Hindle,[19,20] Philip,[21-23] Davis and Shannon,[24] Davis,[25,26] and Davis et al.[27] from 1928 through 1934. Besides defining more precisely the length of the extrinsic incubation period for YF virus in *Ae. aegypti,* these studies provided the initial evidence for some of our current concepts on the vector competence of mosquitoes for arboviruses. For example, these concepts include: once infected with an arbovirus, the females of the vector mosquito species remain infective for the rest of their lives; mosquitoes belonging to species other than the primary vector species are susceptible to peroral infection and can transmit virus while other mosquito species are susceptible, but cannot transmit virus, and still other species are completely refractory to peroral infection; an arbovirus produces a generalized infection in female mosquitoes; and the length of the extrinsic incubation period of arboviruses in mosquitoes is influenced by environmental temperature. One has to marvel that such far-reaching concepts were derived from experiments in which, by necessity, relatively few animals could be used.

The early observations on the vector competence of mosquitoes for YF and dengue viruses were extended over the next 30 years to include most of the mosquito-borne arboviruses that are of public health and veterinary importance today. These later studies employed more quantitative techniques, as they became available, to determine virus infection and transmission rates, and to titrate or visualize virus in whole mosquitoes, tissues and organs, or hemolymph samples. Consequently, the older concepts have been refined and expanded upon while new concepts have arisen.[1-5,9,10,28-35]

An important concept to emerge recently, or at least become widely recognized, relates to the role of genetics in the vector competence of mosquitoes for arboviruses.[33-35] It is now quite evident that females within a single mosquito species can vary significantly in their peroral susceptibility and/or their ability to transmit an arbovirus that is normally vectored by that mosquito species. These genetic traits are controlled by multiple genes, and their expression can be influenced by nongenetic factors such as virus dose and strain, temperature of extrinsic incubation, larval nutrition, etc. Vector incompetence appears to be associated with several ill-defined infection and dissemination thresholds or barriers in the mosquito midgut (i.e., mesenteron) and salivary glands.[35] Consequently, the proportion of competent and incompetent females for a particular arbovirus in a mosquito vector population can vary intraseasonally, yearly, and geographically, thus potentially influencing the epidemiology and distribution of mosquito-borne arbovirus diseases.

II. BIOLOGICAL TRANSMISSION OF ARBOVIRUSES BY COMPETENT MOSQUITOES

Mechanical transmission of arboviruses, via contaminated mosquito mouthparts, has been demonstrated to occur experimentally within a short time after female mosquitoes probe or feed on vertebrate hosts with high-titered viremias.[29] Although this mode of transmission may explain the explosive nature of virus amplification during epidemics or epizootics of some arboviruses [e.g., Rift Valley fever (RVF)[36]], it is not considered important to the long-term maintenance of arbovirus transmission cycles. Rather, there is overwhelming evidence that the maintenance of arboviruses by mosquitoes in vertebrate populations depends on biological transmission, where the virus has to multiply in the mosquito before it can be transmitted.[29] There is no evidence that mosquitoes can transmit arboviruses by the circulative, nonpropagative type of virus transmission which occurs with some nonenveloped plant viruses in their aphid or leafhopper vectors.[37] In this case, virus passes through the mesenteronal epithelium into the hemocele and from the hemolymph through the cells of the salivary glands into the secretory ducts without multiplication. However, infectious Uganda S[38] and Whataroa (WHA)[39] viruses were found in mosquito hemolymph samples taken within a few hours after ingestion of an infectious meal and before virus multiplication would be expected. It still needs to be determined if the "leaky gut" phenomenon occurs frequently for other arbovirus-mosquito vector systems, and if it has any epidemiological significance.

The discussion of biological transmission in this chapter is confined to peroral transmission of arboviruses to susceptible vertebrate hosts by perorally infected female mosquitoes. Other important mechanisms of biological transmission of arboviruses within a mosquito population (i.e., transovarial and venereal transmission) are presented in Chapter 5.

A. Multiplication of Arboviruses in Mosquitoes

In 1932, Davis[26] found that the length of the extrinsic incubation (EI) period for YF virus in *Ae. aegypti* varied inversely with the incubation temperature, and he believed that this provided evidence for virus multiplication in mosquitoes. However, in subsequent experiments, Davis et al.[27] found that while YF virus titers in whole mosquitoes increased slightly after an initial decrease, they never exceeded the initial quantity of virus in the mosquito. Consequently, they concluded that YF virus did not multiply in *Ae. aegypti*. Unfortunately, the mosquitoes ingested a much higher titer of virus than one would expect to find after YF virus had multiplied to maximal titers in infected females. Studies reported in 1934 by Merrill and TenBroeck[40] provided the first clear-cut evidence for multiplication of an arbovirus in mosquitoes. They infected a group of *Ae. aegypti* females with western equine encephalomyelitis (WEE) virus by feeding them on an infected brain suspension. Then they serially passed the virus at 6 to 7 day intervals from one lot of mosquitoes to another for ten generations by allowing uninfected females to feed on a suspension of infected females from the previous generation. Since each passage represented a 1:100 dilution of the virus and similar virus titers were sustained in mosquitoes during ten passages, they concluded that the virus had multiplied in the mosquitoes. It was not until 1937 that Whitman[41] showed that YF virus multiplied in *Ae. aegypti* when the females were infected on lower concentrations of virus than used by Davis et al.[27]

Subsequent studies done from 1945 to 1962, in which virus titers in whole mosquitoes were determined at various times after ingestion of virus, left no doubt that several alphaviruses, bunyaviruses, flaviviruses, and rhabdoviruses could multiply in vector and some nonvector mosquito species. Virus-mosquito systems included YF virus in *Haemagogus* sp.,[42] Murray Valley encephalitis (MVE) virus in *Ae. vigilax, Ae. vittiger, Culex annulirostris,* and *Cx. fatigans;*[43] eastern equine encephalomyelitis (EEE) virus in *Ae. aegypti*[44] and *Ae. triseriatus;*[44,45] WEE virus in *Ae. aegypti*[44] and *Cx. tarsalis;*[46] West Nile (WN) virus in *Ae. aegypti*[47] and *Cx. molestus;*[48] Semliki Forest (SF) virus in *Ae. aegypti;*[49] Japanese encephalitis (JE) virus in *Cx. tritaeniorhynchus*[50-52] and *Cx. pipiens;*[53] St. Louis encephalitis (SLE) virus in *Cx. pipiens, Cx. quinquefasciatus,* and *Cx. tarsalis;*[54] Tahyna virus in *Ae. vexans;*[55] and vesicular stomatitis (VS) virus in *Ae. aegypti*[56] Although experimental conditions and procedures were quite variable in these studies, a concept emerged for the multiplication cycle of arboviruses in mosquitoes. After ingestion of an infective blood meal, virus titers in whole mosquitoes usually decreased tenfold or more during an "eclipse" phase. The length of the eclipse phase varied from 1 to 4 days and depended on the quantity of virus ingested, the mosquito species, and the EI temperature. The decrease in virus titer during the eclipse phase was most likely related to a loss of virus infectivity through a

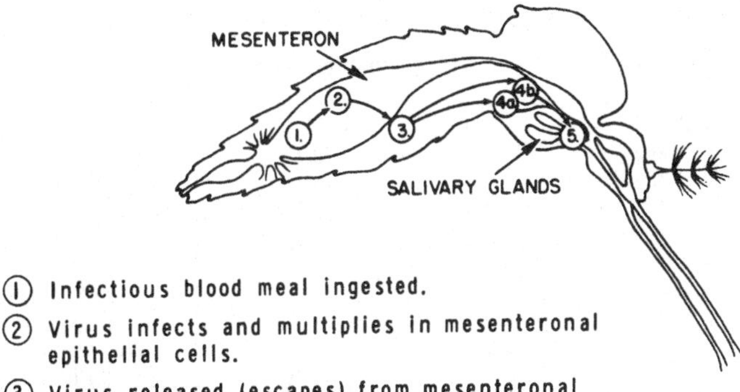

FIGURE 1. Sequential steps required for a competent female mosquito to transmit an arbovirus after ingestion of an infective blood meal.

combination of virus being uncoated during the infection of mesenteronal epithelial cells, inactivation of virus by a "hostile" environment in the lumen of the mesenteron, and anal excretion of virus. The eclipse phase was followed by a phase of rapid virus multiplication during which virus titers in the mosquito increased 1000-fold or more within a few days to 1 week, again depending on virus dose, mosquito species, and EI temperature. A leveling off of virus titer in mosquitoes signified the initiation of a maintenance phase. The early interpretation of the latter phase suggested that there was a finite number of susceptible cells, and that each cell produced only a certain amount of virus. Based on more recent evidence, Murphy,[30] Murphy et al.,[31] and Hardy et al.[35] have suggested that leveling off of arbovirus titers in mosquitoes is related to an innate ability of mosquitoes to control or modulate virus titers and thus avert the potential pathological effects of arboviruses. Evidence to support this concept is presented in Section IV.

B. Sequential Movement of an Arbovirus through a Vector Mosquito

The ability of an arbovirus to produce a generalized infection in mosquitoes was first demonstrated in the 1930s when YF,[24] dengue,[57] and WEE[41] viruses were detected in anatomical units (i.e., head, thorax, and abdomen), or dissected tissues and organs of perorally infected mosquitoes. However, these studies failed to reveal the sequential movement of the virus within the mosquito from the time it was ingested until it was transmitted perorally to a vertebrate host. This was accomplished in subsequent studies where virus titers in dissected organs and tissues or hemolymph samples of mosquitoes were determined at various times between ingestion and transmission of virus.[35,53,58-67] Studies using virus infectivity assays were subsequently complemented by studies using immunofluorescent assays (IFA)[67-72] or immunoperoxidase assays[73] to detect virus antigens in mosquito tissues or organs and electron microscopy (EM)[61,65,73-82] to visualize virus infection and maturation at the cellular level, primarily in epithelial cells of the mesenteron and salivary glands. Unfortunately, most studies employed only one of the three assay systems and many studies examined arbovirus infections in nonvector mosquito species. Nonetheless, it is quite evident that the movement of arboviruses through a mosquito occurs in a step-wise fashion (Figure 1), the rapidity of which depends on the virus, infecting dose of virus, mosquito species/strain, and EI temperature.

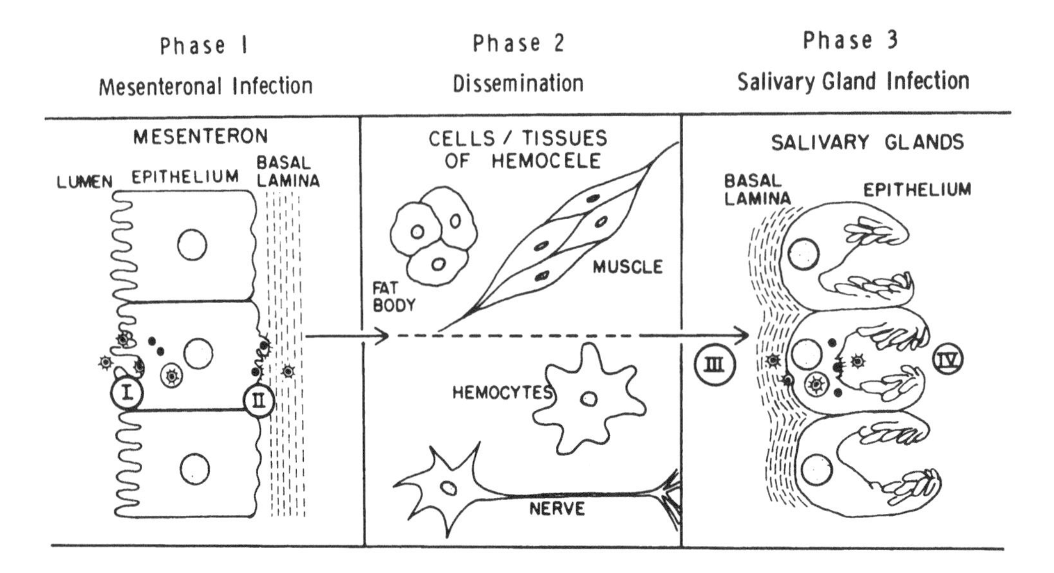

FIGURE 2. Schematic diagram depicting the sequential movement of an arbovirus through various mosquito cells/tissues and the (I) mesenteron infection barrier; (II) mesenteron escape barrier; (III) salivary gland infection barrier; (IV) salivary gland escape barrier that contribute to vector incompetence.

Sequential movement of virus in infected female mosquitoes can be divided into three phases (Figure 2): (1) virus infection and multiplication in mesenteronal epithelial cells and escape into the hemocele; (2) virus dissemination via hemolymph with or without secondary amplification in cells, tissues, and/or organs bathed by hemolymph other than the mesenteron and salivary gland; and finally, (3) virus infection and multiplication in salivary gland cells and subsequent transmission during the feeding process. Each phase is discussed separately.

1. Mesenteronal Infection

The mosquito mesenteron consists of a single layer of columnar epithelial cells surrounded by a porous basal lamina[82] on the abluminal or hemocele side (Figure 3). The structure of the epithelium is relatively simple with a microvillar membrane on the luminal margin, a continuous lateral junction between cells, and a highly involuted basolateral plasma membrane on the abluminal margin. Epithelial cells undergo significant morphological and physiological changes immediately after the deposition of the blood meal in the bulbous, distensible posterior portion of the mesenteron.[83] The cells become flattened into squamous forms and begin the process of blood meal digestion which involves osmoregulation, induction and synthesis of secretory products (e.g., peritropic membrane and digestive enzymes), and digestion and transport of blood meal nutrients across the mesenteronal epithelium into the hemocele. Arbovirus infection and multiplication are initiated simultaneously with blood meal digestion but apparently are not dependent on this process since arboviruses can infect and multiply in the resting mesenteron of unfed female mosquitoes after intrathoracic inoculation.[39,58,61,62,64-66,84]

Initiation of virus infection in the mosquito mesenteron requires that the female ingest a sufficient concentration of virus to overcome a ''threshold of infection''.[29,32,35] This concept was suggested by results reported in 1935 by Merrill and TenBroeck[85] on the susceptibility of *Ae. aegypti* to peroral infection with WEE virus. Also, Bates and Roca-Garcia[42] recognized that the percentage of *Haemagogus* female mosquitoes that became infected was influenced by the amount of virus ingested in the blood meal. However, it was Chamberlain et al.[86] who first presented the concept and developed the methods for establishing mesenteronal

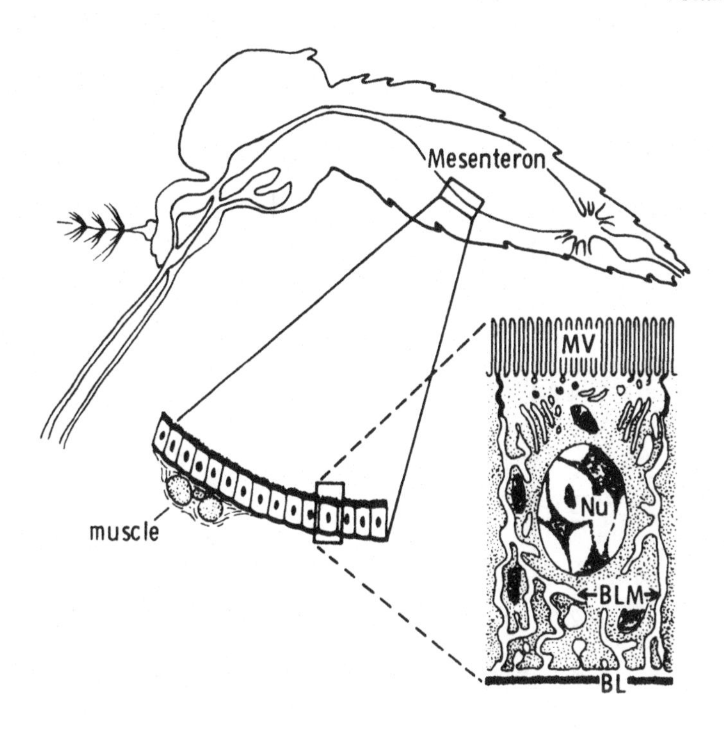

FIGURE 3. Gross and microanatomy of the mosquito mesenteron, mesenteronal epithelium, and mesenteronal epithelial cell. MV = microvilli, Nu = nucleus, BLM = basolateral membrane, and BL = basal lamina. (From Houk, E. J., et al., *Virus Res.*, 2, 123, 1985. With permission.)

infection (MI) thresholds to EEE virus in 21 mosquito species. The existence of an MI threshold has now been demonstrated for numerous arboviruses with many mosquito species or strains.[52,87-100]

After initiation of infection, some arboviruses multiply rapidly in the mosquito mesenterons and reach peak titers within 1 to 3 days. Houk et al.[82] observed significant increases in WEE virus titers in dissected and rinsed mesenterons of *Cx. tarsalis* and *Ae. dorsalis* within 8 to 12 hr after ingestion of 10^4 to 10^6 pfu of virus and incubation at 26°C; peak virus titers of 10^5 to 10^6 pfu per mesenteron occurred at 24 to 30 hr after infection (Figure 4). Thomas[60] also noted peak titers of WEE virus in pooled intestine, mesenteron, and Malpighian tubules of *Cx. tarsalis* at 24 to 48 hr after feeding. Scott et al.[67] reported that EEE virus titers in mesenterons of *Culiseta melanura* peaked at 48 to 72 hr after peroral infection. Japanese encephalitis (JE) virus titers in mesenterons of *Cx. pipiens* were quite variable, but peak titers were obviously reached in some mesenterons by 24 or 48 hr after infection.[53] Thus, the eclipse phase for these viruses in the mosquito species tested was very short. In contrast, McLean[58] did not observe an increase in Murray Valley encephalitis (MVE) virus titers in mesenterons of *Cx. annuliostris* until 3 days after feeding, and peak titers were not reached until 5 days. Similarly, Chernesky[62] reported that an increase in titers of California encephalitis (CE) virus in mesenterons of *Ae. vexans* did not occur until 4 days after infection, but *Ae. vexans* is not a natural vector of this virus.

The IFA (immunofluorescent antibody) technique has been used very successfully to reveal the distribution of arbovirus antigens in the mesenteronal epitheliem at various times after infection. At 4 days after infection, Doi et al.[68] and Doi[69] observed the initiation of JE virus multiplication in isolated foci of epithelial cells located in the posterior portion of mesenterons of *Cx. tritaeniorhynchus and Cx. pipiens* females. By 9 days after infection, JE virus antigen was seen throughout the mesenteronal epithelium. Similarly, Kuberski[72] reported that small

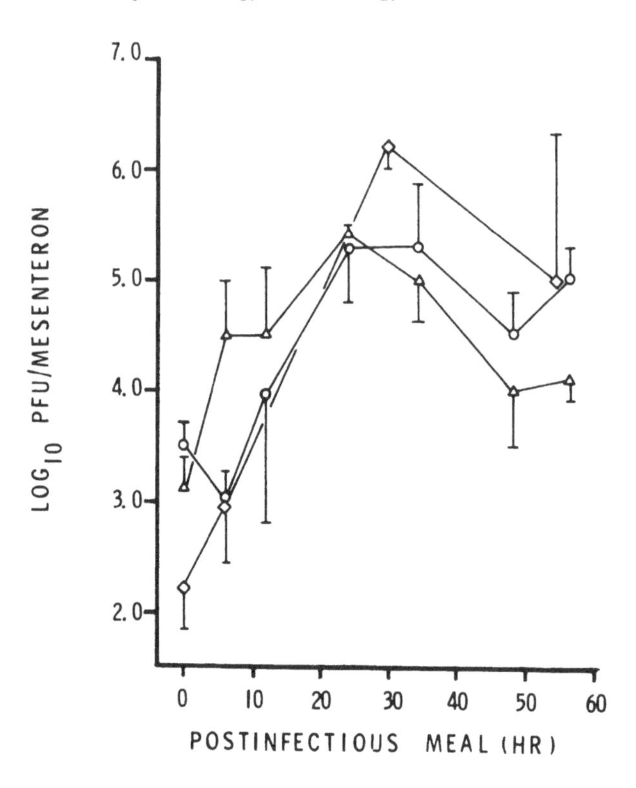

FIGURE 4. Growth of WEE virus in mesenterons of *Ae. dorsalis* (△), *Cx. tarsalis* Knights Landing (◇), and *Cx. tarsalis* WS-3 (○) after ingestion of an infectious blood meal. (Modified from Houk et al.[82])

groups of epithelial cells in the posterior portion of the mesenteron of *Ae. albopictus* were positive by IFA for dengue-2 virus antigen by 2 days postinfection. At 6 days postinfection, dengue-2 virus had infected most of the epithelial cells in the posterior portion of the mesenteron but did not spread to cells in the anterior portion of the mesenteron, as was observed with JE virus in *Cx. tritaeniorhynchus* and *Cx. pipiens*.[68,69] These observations suggest that cell-to-cell spread of virus can occur in the mosquito mesenteron, probably via adsorption and penetration through basolateral plasma membranes since, as stated previously, arboviruses can infect and multiply in the mosquito mesenteron after parenteral infection.[39,58,61,62,64-66,84]

Surprisingly few studies have taken advantage of EM to characterize arbovirus infections in mosquito mesenterons. Whitfield el al.[78] reported on an ultrastructural study of St. Louis encephalitis (SLE) virus infections in one of its vector mosquito species, *Cx. pipiens*. SLE virions were observed in small numbers within the infectious blood meal at 1 to 8 hr after ingestion, but attachment to microvillar membranes was not observed. At 8 hr after feeding, virions were observed within cytoplasmic vacuoles of one cell which might suggest that SLE virus entered the mesenteronal cell by receptor-mediated endocytosis, as proposed by Helenius et al.[101,102] for entry of alphaviruses into vertebrate cells. Untrastructual evidence of virus multiplication was found at 6 days postinfection when mature virions were seen in the cisternae of the endoplasmic reticulum and between the basal membrane of the epithelial cells and the basal lamina. Intracytoplasmic masses of convoluted filaments or tubules, presumably formed by proliferation of endoplasmic reticular membranes, were observed in association with SLE virion maturation. SLE virions were also observed within the basal lamina, ostensibly on their way to the hemocele. Only one in five mesenteronal epithelial cells of *Cx. pipiens* became infected with the SLE virus.

Similar results were reported by Murphy et al.[31] for maturation of EEE virus in mesenteronal epithelial cells of *Ae. triseriatus* females except that an estimated one in three epithelial cells became infected.

Houk et al.[82] recently reported the results of an ultrastructural study with WEE virus infections in the mesenterons of two vector mosquito species, *Cx. tarsalis* and *Ae. dorsalis*. Ultrastructural evidence was presented for the entry of WEE virus into mesenteronal epithelial cells via fusion of the virion envelope with the microvillar membrane rather than receptor-mediated endocytosis (Figure 5A). Naked nucleocapsids were observed only in the cytosol immediately subjacent to the microvillar membrane and not in the endosomal or lysosomal vesicles (Figure 5B). Further, maturation of virions in one strain of *Cx. tarsalis* occurred by budding of nucleocapsids only through the basal membrane of the epithelial cells (Figure 5C); this was first observed at 22 to 24 hr postinfection. However, with *Ae. dorsalis* and a strain of *Cx. tarsalis* selected for high susceptibility to WEE virus, maturation took place in association with endoplasmic reticular membranes, and nucleocapsids were frequently embedded in an amorphous matrix. There was extensive cytoplasmic vacuolization in mesenteronal epithelial cells of *Ae. dorsalis* infected with WEE virus.

Overall, these studies provide substantial evidence to indicate that arboviruses multiply in the mosquito mesenteron after ingestion of a viremic blood meal. The requirement for initiation of infection in the mesenteron is further indicated by the fact that EI periods of arboviruses in mosquitoes are significantly decreased when the mesenteron is by-passed by intrathoracic inoculation of virus. However, there is little direct evidence to explain how the virus is transported through the basal lamina of the mesenteron into the hemocele. The pore size in the basal lamina of *Ae. dorsalis* is only 10 nm in diameter.[103] Furthermore, only particles with a diameter of 5 to 8 nm were able to permeate the basal lamina of mesenterons in *Cx. tarsalis* from the hemocele side.[104] Thus, the mechanism by which virions with a diameter of 50 to 60 nm can traverse the basal lamina of the mosquito mesenteron remains an enigma. The fact that they can go through this structure is suggested by the infection of the mesenteron in parenterally infected females.

2. Dissemination and Secondary Amplification of Virus in Hemocele-Associated Cells, Tissues, and Organs

After escaping from the infected mosquito mesenterons into the hemocele, arboviruses may infect and multiply in numerous tissues and organs, including salivary glands, other tissues of the alimentary tract, reproductive organs, ganglia and brain, fat body, pericardium, Malpighian tubules, or muscle. This has been suggested by the demonstration of increased virus titers in dissected tissues and organs,[53] but the best evidence was provided by IFA studies done with EEE virus in *Cx. melanura*,[67] JE virus in *Cx. tritaeniorhynchus*[68,69] and *Cx. pipiens*,[69] La Crosse (LAC) virus in *Ae. triseriatus*,[71] and dengue-2 virus in *Ae. albopictus*.[72] Additional evidence of multiplication has been demonstrated by EM with SLE virus in cells of abdominal muscles, Malpighian tubules, and ovarian sheath of perorally infected *Cx. pipiens*[78] and for dengue-2 virus in cells of the brain, thoracic ganglion, fat body, body wall lining, anterior intestine, and hemocytes, but not striated muscles, diverticula, posterior intestine, Malpighian tubules, or ovarian follicles of perorally infected *Ae. albopictus*.[79] EM studies also revealed that vesicular stomatitus virus (VSV) replicated in salivary glands, mesenterons, and thoracic and abdominal ganglia of *Ae. aegypti* after intrathoracic inoculation.[61]

At present, it is not clear if arboviruses differ significantly in their tropisms for various tissues and organs of their vector mosquito species because too few studies have been done. Further, it is not evident if secondary amplification of arboviruses in other hemocele-associated cells, tissues, or organs is required for infection of the salivary glands. LaMotte[53] demonstrated increases in titers of JE virus in ganglia, anterior intestine and diverticula,

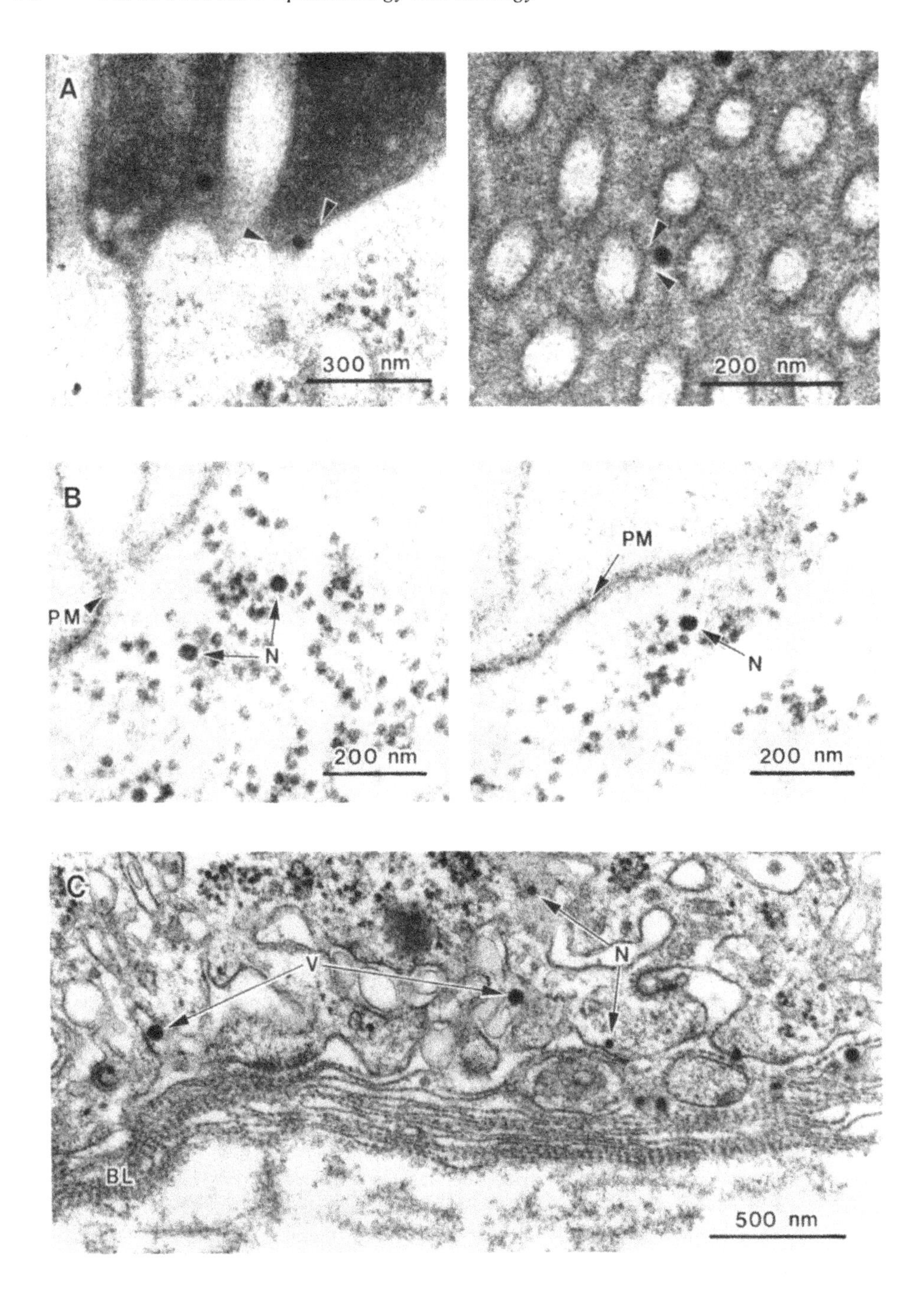

FIGURE 5. Electron micrographs showing the attachment, penetration, and maturation phases of the WEE virus multiplication cycle in mesenteronal epithelial cells of *Cx. tarsalis.* Apparent fusion of WEE virion envelopes and mesenteronal microvillar membranes in *Cx. tarsalis.* (Left) Knights Landing (KL) strain at 5 hr postblood meal (PBM); (right) WR-2 strain at 35 hr PBM. Two examples of WEE nucleocapsids (N), ca. 40 nm particles with electron-lucent centers, immediately subjacent to the mesenteronal lumenal plasma membrane (PM) at 0 hr PBM in *Cx. tarsalis* (WS-3 strain). (C) Nucleocapsids (N) immediately adjacent to the basal plasma membrane of mesenteron of *Cx. tarsalis* (KL strain) at 30 hr PBM. Extracellular virions (V) are apparent. BL = basal lamina. (A and B from Houk, E. J., et al., *Virus Res.*, 2, 123, 1985. With permission.)

ovaries, and hemolymph of *Cx. pipiens* at approximately the same time (i.e., 2 to 3 days after ingestion of viremic blood) that increases in virus titers were observed in the salivary glands. Similar results were reported by Scott et al.[67] for EEE virus in *Cs. melanura* and by Kuberski for dengue-2 virus in *Ae. albopictus*.[72] However, IFA studies reported by Doi et al.[68] and Doi[69] clearly indicated that JE virus multiplied in fat body cells before infection of the salivary glands occurred. Certainly, more studies are required to delineate the sequential infection of hemocele-associated tissues and organs after the escape of arboviruses from the mesenteron. This can be accomplished by use of sensitive immunohistological staining of virus antigens in tissues of microtome-sectioned whole mosquitoes at frequent intervals after infection.

3. Salivary Gland Infection

A pair of salivary glands is located in the anterior portion of the thoracic hemocele of a female mosquito, each consisting of three cylindrical lobes (two lateral, and one median).[105,106] The distal and proximal portions of each lobe are frequently termed acinus and neck, respectively. Each lobe consists of a single layer of cuboidal epithelial cells surrounded by a basal lamina and has a fenestrated, chitinous central duct which is continuous with the salivary gland duct that connects to the hypopharynx. Cells in the lateral acinus are larger than those in the lateral neck; no cells are found in the median neck. Apical plasma membranes have some microvillar projections and are highly convoluted to increase the surface area for discharge of secretions into the apical cavity (diverticulum).

It has been suggested that arboviruses can infect the salivary glands of mosquitoes by either a hemolymph or neural pathway.[29,31] Evidence to support the hemolymph pathway derives from experiments with JE virus in *Cx. pipiens*[53,69] and *Cx. tritaeniorhynchus*[68,69] and EEE virus in *Cx. melanura*,[67] where it was noted that peak virus titers or specific immunofluorescence occurred in salivary glands at the same time or before they occurred in neural tissues. Maguire (unpublished data referenced by Murphy et al.[31]) has suggested that WHA virus may enter the salivary glands of *Ae. australis* by a neural pathway. This was based on IFA studies in which he observed a sequential spread of virus from the mesenteron to the abdominal fat body, thoracic ganglia, cerebral ganglia, and finally, salivary glands. Virus titrations on dissected tissues also suggested that WHA virus infected neural tissue before it infected salivary glands of *Ae. australis*.[66] Similarly, Kuberski[72] noted that specific fluorescence for dengue-2 virus appeared in neural tissues before it was observed in salivary glands of *Ae. albopictus*. However, this was not confirmed by EM studies reported subsequently by Sriurairatna and Bhamarapravati[79] who observed maturation of dengue-2 virions at least 2 days earlier in the salivary glands of *Ae. albopictus* than in the brain. Thus, it remains to be determined if arboviruses other than WHA virus enter the salivary glands of their vector mosquitoes by a neural pathway.

Numerous studies have shown that arboviruses accumulate in the salivary glands of mosquitoes.[29-32,35] This is suggested by finding increased virus titers or antigens in salivary glands over time after infection and by the detection of arbovirus-like particles in salivary glands of infected, and not in uninfected, female mosquitoes. However, these observations do not necessarily indicate that virus multiplied in the salivary glands because similar results would be obtained if the virus in the surrounding hemolymph was merely concentrated by the salivary glands as occurs for some plant viruses in their insect vectors.[37] Therefore, the definitive demonstration of arbovirus multiplication in mosquito salivary glands depended on EM ultrastructural studies in which maturation of virus in infected cells was observed. Such studies have been done with EEE virus in *Ae. triseriatus*,[77] SLE virus in *Cx. pipiens*,[78] dengue-2 virus in *Ae. albopictus*,[79] and JE virus in *Cx. tritaeniorhynchus* and *Cx. pipiens*.[80] In general, these studies revealed that the morphogenesis of arboviruses in acinar cells was similar to that observed in infected mammalian cells, but without any evidence of cytopathology. Nucleocapsids of the alphavirus, EEE, were formed in the cytoplasm of acinar cells, and virion maturation occurred as nucleocapsids acquired an envelope while budding through endoplasmic reticular membranes or the apical plasma membrane. Mature EEE virions within cytoplasmic vacuoles and cisternae of endoplasmic reticulum were most likely released into the apical cavity via a fusion process with apical plasma membranes. Large concentrations of mature virions were seen in the apical cavity, often in crystalline arrays. Interestingly, only 1 in about 20 acinar cells in the salivary glands of *Ae. triseriatus* was infected with EEE. Similar observations were made with the three flaviviruses (dengue-2, JE, and SLE) in the infected salivary glands of their vector mosquito species, except that virus morphogenesis occurred frequently in association with intracytoplasmic matrices. Further, the formation of flavivirus nucleocapsids and early stages of the budding process were not seen.

The tropism of arboviruses for different lobes of the salivary glands is interesting, especially since median and lateral lobes apparently secrete different materials.[107] In IFA studies with VEE and Sindbis (SIN) viruses in *Ae. aegypti,* Gaidamovich et al.[70] observed that virus antigens first appeared in cells of the lateral acinus and then spread to the cells of the lateral neck, but cells in the median lobe apparently did not become infected. Similar observations were reported by Gubler and Rosen[108] for dengue-2 virus in *Ae. albopictus,* except cells of the median lobe apparently became infected when mosquitoes were incubated for long periods. They also found no correlation between virus transmission and the site of virus replication in the lateral lobes (i.e., acinus or neck). Using EM and IFA, Takahashi and Suzuki[80] detected JE virus only in the lateral lobes of salivary glands from *Cx. tritaeniorhynchus* females. However, in more recent IFA studies, Takahashi[109] reported that the median lobe became infected in strains of *Cx. tritaeniorhynchus* that were efficient transmitters of JE virus.

The amount of virus transmitted by a mosquito during the feeding process has been estimated with dengue-2 virus in *Ae. albopictus,*[108] EEE,[110] and VEE[111] viruses in *Ae. aegypti,* and SLE virus in *Cx. pipiens*[112] by various direct and indirect methods. Estimates of virus concentrations ejected ranged from a few infectious units of virus to as high as 100,000 infectious units, depending on the method used and the length of EI.

C. Extrinsic Incubation Period

The EI period, as defined by Bauer and Hudson in 1928,[18] is the "period between the initial infecting feeding and the time at which the mosquitoes are capable of inducing infections" while feeding on a vertebrate host. This definition does not include the quantitative parameters that are required for comparisons of EI periods for different mosquito species or strains infected with various arboviruses. For example, should the end of the EI period be based on first transmission of virus or when a certain percentage of females transmit virus? The potential importance of this is illustrated by studies reported in 1954 by Chamberlain et al.[44] They measured transmission rates over time after infection of *Ae. aegypti* and *Ae. triseriatus* with EEE virus; the EI temperature was 27°C and nearly 100% of females in each mosquito species were infected. At 3 days postinfection, 11% of *Ae. aegypti* and 12% of *Ae. triseriatus* transmitted virus. However, the 50% transmission rate was not reached until 17 days after infection with *Ae. triseriatus* as compared to 9 days with *Ae. aegypti.* Thus, it is evident that an EI period based on the 50% transmission endpoint is epidemiologically more relevant than one based on first transmission.

It is abundantly clear that EI periods for arboviruses in mosquitoes are quite variable no matter what method is used to measure them. This variability in EI periods derives from the effects that genetic and nongenetic factors have on arbovirus infections, multiplication, and dissemination in mosquitoes.[35] These factors include environmental temperature, infecting dose of virus, genetic ability of mosquitoes to become infected with and transmit arboviruses, and virus genetics. Most notable among these is environmental temperature, which has been studied extensively[20,26,29,35,45,113-126] because it is an obvious variable and also is known to influence the epidemiology of arbovirus diseases.[127-129]

1. Effect of Temperature

Most studies have shown that EI periods for arboviruses in mosquitoes are inversely related to the incubation temperature within those temperature ranges which allow virus replication to occur. Bates and Roca-Garcia[42] reported that the EI period for YF virus in *Haemagogus capricornii* was 28 days at a constant 25°C and 12 days at a constant 30°C. Similar observations on the relationship between holding temperature and EI period were made by Chamberlain and Sudia[45] with *Ae. triseriatus* infected with EEE virus and held at 23, 27, or 30°C; by Takahashi[122] with *Cx. tritaeniorhynchus* infected with JE virus and incubated at 20 or 28°C; by Kramer et al.[125] with *Cx. tarsalis* infected with WEE virus and held at 26 or 32°C; and by Turell et al.[126] with *Cx. pipiens* and *Ae. taeniorhynchus* infected with RVF virus and held at 27 or 33°C. However, EI periods at 25 to 27°C were significantly longer for YF virus in *Hg. capricornii* (28 days) than for EEE virus in *Ae. triseriatus* (<6 days), JE virus in *Cx. tritaeniorhynchus* (9 days), WEE virus in *Cx. tarsalis* (<6 days), and RVF virus in *Cx. pipiens* (5 days) and *Ae. taeniorhynchus* (10 days).

Since mosquitoes in nature are exposed to daily fluctuating temperatures rather than constant temperatures, Bates and Roca-Garcia[42] also evaluated the effect of an alternating daily temperature (20 hr at 25°C and 4 hr at 35°C/day) on the EI period of YF virus in *Hg. capricornii* and found that the EI period was similar to that obtained when mosquitoes were held at a constant 30°C. Using a similar protocol, Chamberlain and Sudia[45] also observed that a daily alternating temperature had the effect of an intermediate constant temperature on the EI period for EEE virus in *Cx. pipiens*. Shichijo et al.[116] conducted studies with JE virus in *Cx. tritaeniorhynchus* using simulated ambient air temperatures that occur throughout most of the year in Japan; unfortunately, the effect of temperature on EI period was not determined.

Presumably, there is a "zero development" temperature for each arbovirus-mosquito system, at and below which viruses fail to multiply to detectable levels in the mosquito after ingestion of virus. This has been estimated to be 17.52°C for SLE virus with a strain of *Cx. quinquefasciatus*.[117] However, other studies have shown that even though virus multiplication was minimal or not demonstrable at low EI temperatures, virus titers increased rapidly when mosquitoes were transferred to a higher EI temperature.[42,115,116,126] In addition to a decrease in virus titer, Turell et al.[126] observed that infection rates also were reduced for RVF virus in perorally infected *Cx. pipiens* and *Ae. taeniorhynchus* held at 13°C. However, when females were transferred to 26°C, infection rates quickly increased to the level observed for females held at a constant 26°C. Shichijo et al.[116] noted that JE virus multiplication, as detected by IFA, was restricted to a few cells in the posterior portion of the mesenteron of *Cx. tritaeniorhynchus* incubated in a biotron under simulated early winter temperatures in Japan (i.e., decreasing from about 14 to 8°C over 60 days). When mosquitoes were then transferred to late winter temperatures that increased from about 12 to 18°C over 40 or more days, virus spread throughout the mesenteron and eventually escaped from the mesenteron to infect fat body cells, thoracic ganglia, and salivary glands. These studies illustrate that arboviruses can persist for prolonged periods in mosquito mesenterons at low EI temperatures, presumably in a slowly replicating state. This is epidemiologically important if mosquitoes become infected with arboviruses in the late fall because the virus could persist in the mesenteron of overwintering females and then complete its EI period in the mosquito when temperatures increased in the late winter and early spring.

Thus, it is quite clear that environmental temperature has a dramatic effect on EI periods for arboviruses in competent mosquito species. One has to assume that temperature exerts this influence either directly by its effect on virus replication, or indirectly by its effect on mosquito physiology, or perhaps both. This is indicated by the direct relationship that exists between EI temperature and rate of arbovirus multiplication in mosquitoes. However, higher EI temperature may evoke genetically controlled factors in mosquitoes that limit virus multiplication.[35] This is discussed in Section IV.

2. Effect of Infecting Virus Dose

It is reasonable to expect that the infecting virus dose would have an effect on the EI period of arboviruses in mosquitoes since it would determine, along with the MI threshold, the number of mesenteronal epithelial cells that initially become infected with virus. This, in turn, would influence the dissemination of virus to the salivary glands. If true, then virus dose should be inversely related to the length of the EI period.

This concept was first proposed by Bates and Roca-Garcia[42] when they observed that the minimal EI period for YF virus in *Hg. capricornii* at 30°C was 13 days after females ingested a "moderate amount of virus" and 10 days after females ingested a "large amount of virus". Similar observations have been reported for JE virus in *Cx. tritaeniorhynchus*[52] and West Nile (WN) virus in *Culex gelidis*[130] after ingestion of $10^{0.4}$ to $10^{3.3}$ weanling mouse (WM)-LD_{50}/0.04 mℓ of viremic pig blood and $10^{1.5}$ to $10^{4.8}$ WM-LD_{50}/0.03 mℓ of viremic chick blood, respectively. Other vector competence studies did not measure virus transmission by mosquitoes at frequent enough intervals to determine the effect of virus dose on the EI period. The genetics of mosquitoes and arboviruses obviously influence the EI period, and is discussed along with vector incompetence in Section IV.

IV. FACTORS AND BARRIERS AFFECTING ARBOVIRUS INFECTION AND TRANSMISSION IN MOSQUITOES

As pointed out previously, early investigators documented the existence of an infection threshold in mosquitoes that governs the concentration of an arbovirus required in the viremic blood meal to initiate infection in the mesenteron.[29,32,35] This threshold has been termed "gut barrier", "midgut barrier", "intestinal barrier", and more recently, "mesenteronal infection (MI) barrier" by various investigators (Figure 2). In addition, early researchers recognized that some mosquitoes can "become infected for life without being able to transmit (virus) by bite".[23] For example, only the abdomens were infected in *Ae. (Aedimorphus) irritans* that failed to transmit YF virus 51 days postinfection. With *Ae. (Neomelanoconion) punctocostalis,* YF virus was present in the abdomens, thoraxes, and heads but females still failed to transmit virus to monkeys. These observations can now be explained on the basis of recent studies which indicate that dissemination barriers exist in virus-susceptible female mosquitoes.[35] These barriers influence the escape of virus from the infected mesenteron and the infection of the salivary glands, as well as per os transmission of virus by females with infected salivary glands (Figure 2). Therefore, populations of a vector mosquito species can be composed of females that are completely refractory to peroral infection, those that are susceptible to peroral infection but are unable to transmit virus, and those that are fully competent to vector a particular arbovirus. Thus, an understanding of the various factors that affect the expression of these barriers to arbovirus infection, dissemination and transmission in incompetent mosquitoes, and the mechanisms involved, is important to our overall knowledge of arbovirus transmission cycles in vertebrate host and human populations.

A. Factors Influencing the Vector Competence of Mosquitoes for Arboviruses

Various extrinsic and intrinsic factors have been identified that can decrease or enhance the vectorial capacity of mosquitoes for arboviruses by influencing virus infection, multiplication, dissemination, or transmission, and have been reviewed by Mitchell,[9] Chamberlain

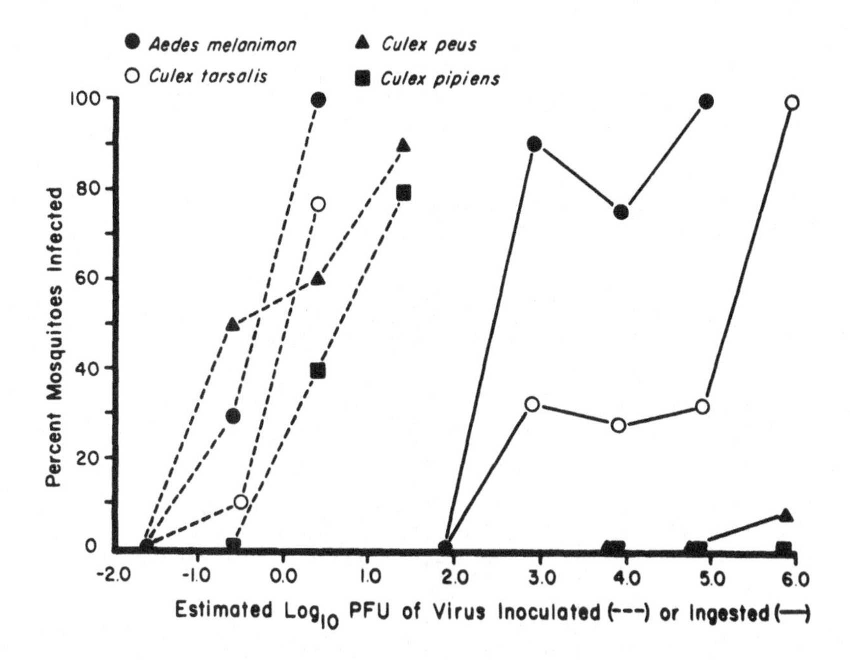

FIGURE 6. Parenteral and peroral suceptibility profiles of WEE virus in females of 4 mosquito species collected at field sites in California during 1972 and evaluated simultaneously for virus suceptibility. For peroral infection, females were allowed to feed on gauze pledgets soaked with a mixture of virus, defibrinated rabbit blood, and sucrose (2.5%). Infection status of each female was determined after incubation of fed or inoculated females for 12 to 14 days at 24°C. (From Hardy, J. L., et al., *Artic and Tropical Arboviruses*, Kurstak, E., Ed., Academic Press, New York, 1979, 157. With permission.)

and Sudia,[29] Murphy,[30] Murphy et al.,[31] Grimstad,[10,33] Gubler et al.,[34] and Hardy et al.,[35] Identifiable factors include mosquito and virus genetics, environmental temperature, nutrition, coinfection with exogenous or endogenous microorganisms, and virus-induced pathology in mosquitoes.

1. Mosquito Species or Strain

Variations in the ability of female mosquitoes in different mosquito species to become infected after ingestion of selected arboviruses, and to subsequently transmit virus have been well documented.[10,29] This is amply illustrated by recent studies done with WEE and SLE viruses and mosquito species occurring in California. By feeding on gauze pledgets soaked with a virus-defibrinated blood-sucrose mixture, *Ae. melanimon* was found to be the most susceptible of the mosquito species tested to peroral infection with WEE virus followed by *Cx. tarsalis, Cx. peus,* and *Cx. pipiens,* in that order (Figure 6).[131] In fact, only an occasional *Cx. pipiens* female became infected after ingestion of $10^{6.2\text{-}7.1}$ pfu of WEE virus from viremic chickens.[132] Differences noted in the susceptibilities of these four mosquito species to per os infection with WEE virus appeared to be associated with an MI barrier since all four species are equally susceptible to parenteral infection (Figure 6).[131,132] In contrast, *Cx. peus* was the most susceptible *Culex* mosquito to per os infection with SLE virus, as well as the most efficient transmitter of SLE virus, followed by *Cx. tarsalis, Cx. quinquefasciatus* (Figure 7), and *Cx. erythrothorax,* in that order.[133,134] Thus, one mosquito species may be a competent vector of one virus and an incompetent vector for another virus (e.g., *Cx. peus* for SLE and WEE viruses, respectively) or one mosquito species may be a competent vector of more than one virus in the same geographic locality (e.g., *Cx. tarsalis* vectors both WEE and SLE viruses). In fact, *Cx. tarsalis* is a natural vector of arboviruses belonging to five

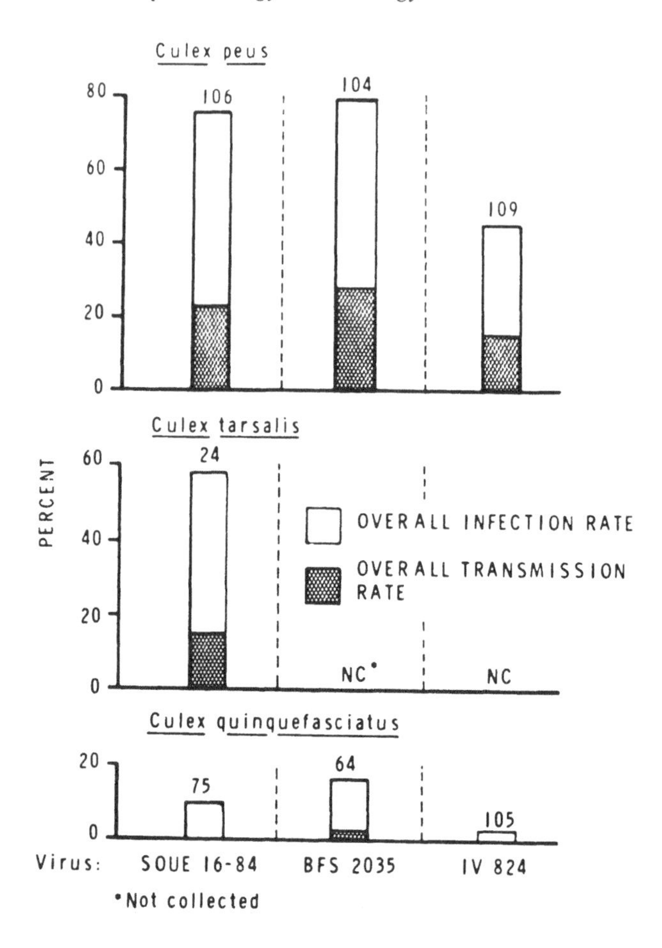

FIGURE 7. Infection and transmission rates obtained with 3 California strains of SLE virus and 3 *Culex* species collected in the Greater Los Angeles area in October and November 1984. Females fed on pledgets soaked with sweetened, defibrinated viremic chicken blood containing $10^{3.6}$, $10^{3.4}$, and $10^{2.0}$ pfu/0.1 mℓ of BFS 2035, SOUE 16-84, and IV 824 viruses, respectively, and were incubated at 25° for 14 to 21 days before transmission, and infection rates were measured.[133]

virus families: Togaviridae (WEE), Flaviviridae (SLE), Bunyaviridae (Turlock), Rhabdoviridae (Hart Park), and Reoviridae (Llano Seco).[135] It is assumed that interspecific variations in the ability of mosquitoes to vector arboviruses are controlled, in part, by mosquito genetics.

The possibility of intraspecific variation in the vector competence of mosquitoes for arboviruses was apparently first suggested by Legendre (cited by Hindle[19]) who postulated that the absence of YF in the Orient might be related to the resistance of Asian strains of *Ae. aegypti* to YF virus. After conducting rather limited transmission experiments with African and Indian strains of *Ae. aegypti* infected with YF virus, Hindle[19] concluded that "there is no foundation for the suggestion made by Legendre". Dudley[136] came to the same conclusion after reviewing Hindle's data. However, even though the Indian strain of *Ae. aegypti* transmitted YF virus, it appeared to be a less efficient transmitter than was the African strain of *Ae. aegypti*. Thus, Hindle[19] did demonstrate polymorphism in the ability of different geographic strains of *Ae. aegypti* to transmit YF virus.

Studies reported in the last 10 years clearly indicate that the vector competence of mosquitoes for arboviruses is controlled, in part, by mosquito genetics.[33-35] This has been shown for three alphaviruses,[131,137-141] one bunyavirus,[142] and eight flaviviruses[34,143-155] in their vector mosquito species (Table 1). Most of the evidence comes from studies in which genetic polymorphism in virus susceptibility, modulation or transmission was observed with geographic strains of vector mosquitoes. This is exemplified in Figure 8, which presents the susceptibility profiles obtained with WEE virus on four field populations of *Cx. tarsalis* collected in the Central Valley of California during 1972. The ID_{50} for different populations varied by more than 200-fold by peroral infection and only 25-fold by parenteral infection. Thus, like interspecific variation (Figure 6), the intraspecific variation in the susceptibility of *Cx. tarsalis* to infection with WEE virus appeared to be associated with an MI barrier (Figure 8).

Genetic selection of mosquito lines with increased or decreased vector competence has been attempted, with varying degrees of success, for a few alpha- and flaviviruses (Table 1). Tesh et al.[137] failed to select lines of *Ae. albopictus* with increased or decreased susceptibility to peroral infection with chikungunya (CHIK) virus. However, Hardy et al.[131,140,141] were able to select lines of *Cx. tarsalis* that were either highly susceptible (WS) or highly resistant (WR) to per os infection with WEE virus (Figure 9). There was a greater than 100,000-fold difference in susceptibility between WS and WR females by the peroral route of infection. Susceptibility was dominant and appeared to be polyfactorial. Interestingly, WS and WR females were equally susceptible to peroral infection with SLE and Turlock viruses.[134]

In addition to peroral susceptibility, the ability of WEE virus to multiply in cells of parenterally infected *Cx. tarsalis* males and females is genetically controlled.[35,134] Genetic lines of high virus producer (HVP) and low virus producer (LVP) *Cx. tarsalis* were selected that allowed WEE virus to multiply to mean titers of 10^7 pfu and less than 10^1 pfu per mosquito, respectively, after parenteral infection (Figure 10). Reciprocal crossmatings and backcrosses with HVP and LVP parents indicated that the HVP trait was incompletely dominant and possibly controlled by a single autosomal gene.[134] The expression of the LVP trait (i.e., virus modulation) was influenced greatly by virus dose, length of EI, and temperature. Modulation of CHIK virus in *Ae. albopictus* was also observed by Tesh et al.,[137] but genetic selection was not attempted.

Genetic selection for mosquito lines with increased or reduced vector competence has been attempted with three flaviviruses. Wallis et al.[155] were unable to select lines of *Ae. aegypti* that were significantly more susceptible or resistant to per os infection with YF virus. However, Gubler and Rosen[144] were able to decrease the peroral susceptibility of a geographic strain of *Ae. albopictus* for dengue-2 virus from 73 to 13% after two generations of selection, but a completely refractory line could not be selected. The susceptibility of F_1 progeny, derived from reciprocal matings between naturally resistant and susceptible strains of *Ae. albopictus,* had nearly the same per os susceptibility to dengue-2 virus as the susceptible strain which suggested that susceptibility was dominant. Of interest was that variations observed in the susceptibility of different geographic strains of *Ae. albopictus* to dengue-2 virus correlated with those observed with other dengue serotypes (i.e., 1, 3, and 4),[144] but not with CHIK virus.[137] Similar results were obtained with dengue viruses and geographic strains of *Ae. aegypti,*[143] except that resistance rather than susceptibility appeared to be dominant. Also, peroral ID_{50} obtained for dengue serotypes 1 to 4 with different strains of *Ae. aegypti* was not always the same. The investigators concluded that peroral susceptibility for dengue viruses in both *Ae. aegypti*[143] and *Ae. albopictus*[144] was most likely polyfactorial.

More conclusive studies were reported recently by Hayes et al.[150] on the role of genetics in the vector competence of *Cx. tritaeniorhynchus* for WN virus. They were able to select females with increased but not decreased susceptibility to peroral infection with WN virus. They also demonstrated that WN virus multiplied to significantly higher titers in males and

Table 1
EVIDENCE FOR GENETIC CONTROL OF THE VECTOR COMPETENCE OF MOSQUITOES FOR ARBOVIRUSES

Virus	Mosquito species	Trait examined	Genetic polymorphism observed	Genetic selection	Mode of inheritance of trait	Ref.
Alphavirus						
Chikungunya	*Ae. albopictus*	Susceptibility	Yes	No	—	137
		Modulation[a]	Yes	Not done	—	
Ross River	*Ae. polynesiensis*	Susceptibility	?	Not done	—	138
		Transmission	Yes	Not done	—	138
WEE	*Ae. trivittatus*	Susceptibility	?	Not done	—	139
	Cx. tarsalis	Susceptibility	Yes	Yes	Dominant, autosomal, polyfactorial?	131, 134, 140, 141
		Modulation[a]	Yes	Yes	Recessive, autosomal, monofactorial?	35, 134
Bunyavirus						
La Crosse	*Ae. triseariatus*	Susceptibility	Yes	Not done	—	142
		Transmission	Yes	Not done	—	142
Flavivirus						
Dengue	*Ae. aegypti*	Susceptibility	Yes	Not done	Recessive, polyfactorial?	34, 143
	Ae. albopictus	Susceptibility	Yes	Partial	Dominant, polyfactorial?	144
JE	*Cx. tritaeniorhynchus*	Susceptibility	Yes	Not done	—	109, 145
		Transmission	Yes	Not done	—	109
Kunjin	*Cx. annulirostris*	Susceptibility	Yes	Not done	—	146
		Transmission	Yes	Not done	—	146
MVE	*Cx. annulirostris*	Susceptibility	Yes	Not done	—	146
		Transmission	Yes	Not done	—	146
	Cx. pipiens complex	Susceptibility	Yes	Not done	—	147
Rocio	*Cx. pipiens* complex	Susceptibility	?		—	148
		Transmission	Yes	Not done	—	148
SLE	*Cx. pipiens* complex	Susceptibility	Yes	Not done	—	34
West Nile	*Cx. tritaeniorhynchus*	Susceptibility	Yes	Yes	Dominant, polyfactorial?	149, 150
		Modulation[a]	Yes	Yes	Recessive, polyfactorial?	150
YF	*Ae. aegypti*	Susceptibility	Yes	No	—	34,151,153—155
		Transmission	Yes	Not done	—	19, 34, 151, 152

[a] Modulation is ability of mosquito to control virus multiplication.

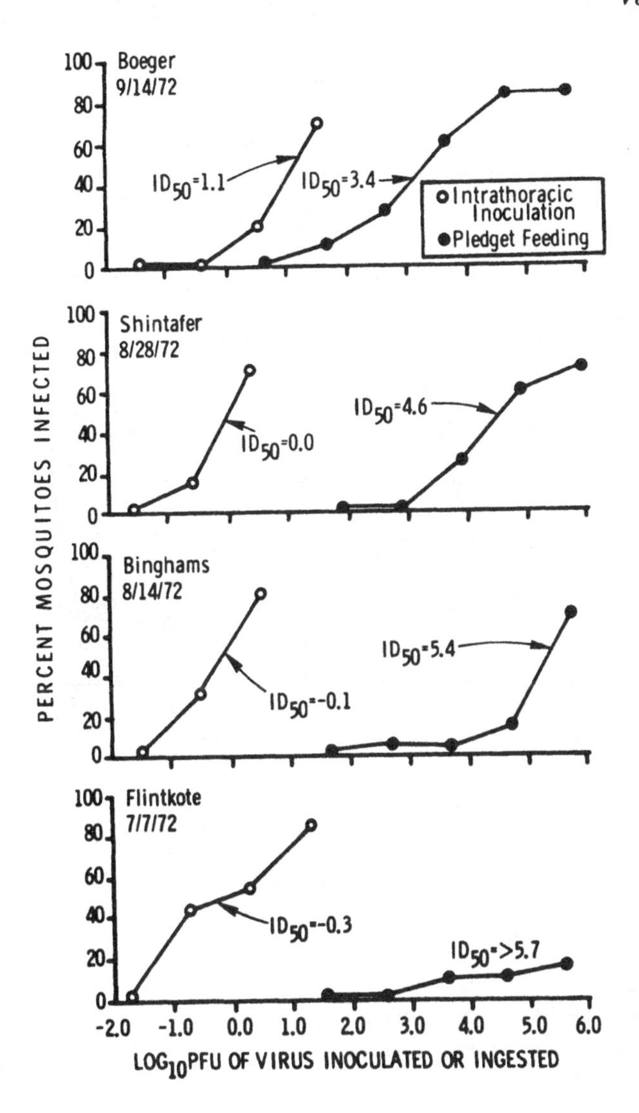

FIGURE 8. Parenteral and peroral suceptibility profiles of WEE virus in *Cx. tarsalis* collected at four field sites in California during 1972. For peroral infection, females were allowed to feed on gauze pledgets soaked with a mixture of virus, defibrinated rabbit blood, and sucrose (2.5%). Infection status of each female was determined after 12 to 14 days of incubation at 24°C. (Modified from Hardy et al.[131])

females from susceptible strains of *Cx. tritaeniorhynchus* than from less susceptible strains after either peroral or parenteral infection. Genetic mating experiments revealed that the trait for increased susceptibility was dominant and the trait for enhanced virus multiplication was incompletely dominant. Both traits were apparently polyfactorial and associated with multiple chromosomes of *Cx. tritaeniorhynchus*.

Evidence that mosquito strains can vary in their vectorial capacity for arboviruses is further indicated by changes observed in peroral susceptibility or ability to transmit virus during colonization of field strains. This is not surprising since other genetically controlled traits are known to change when field populations of mosquitoes are colonized.[156] Meegan et al.[157] observed a decrease in the ability of *Cx. pipiens* to transmit RVF virus during several generations of colonization in the laboratory. This observation was confirmed in a more

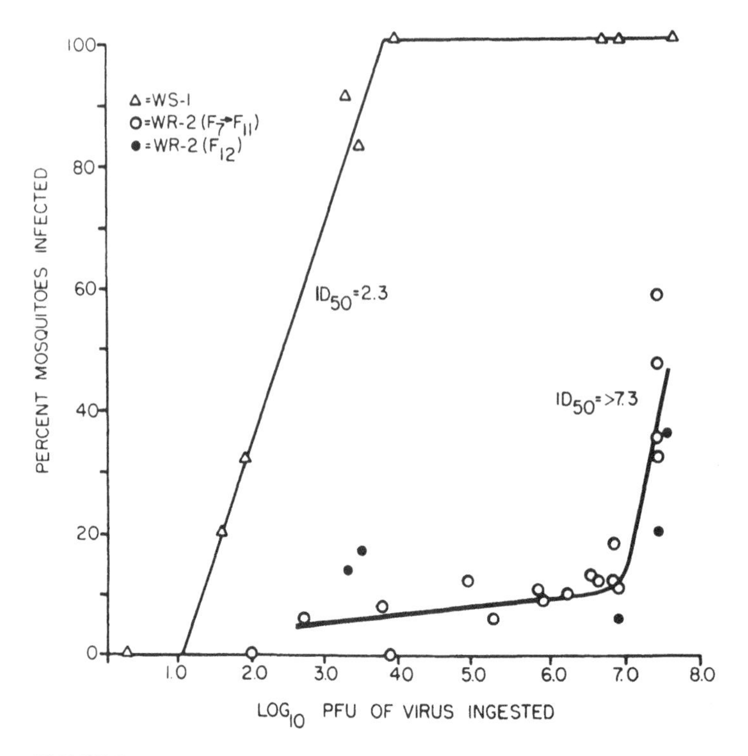

FIGURE 9. Peroral susceptibility profiles for WEE virus in genetic lines of *Cx. tarsalis* selected for high susceptibility (WS-1) and resistance (WR-2) to peroral infection with virus. Females were fed on viremic chickens and then incubated at 24°C for 12 to 14 days before testing each female for virus. (Modified from Hardy et al.[131])

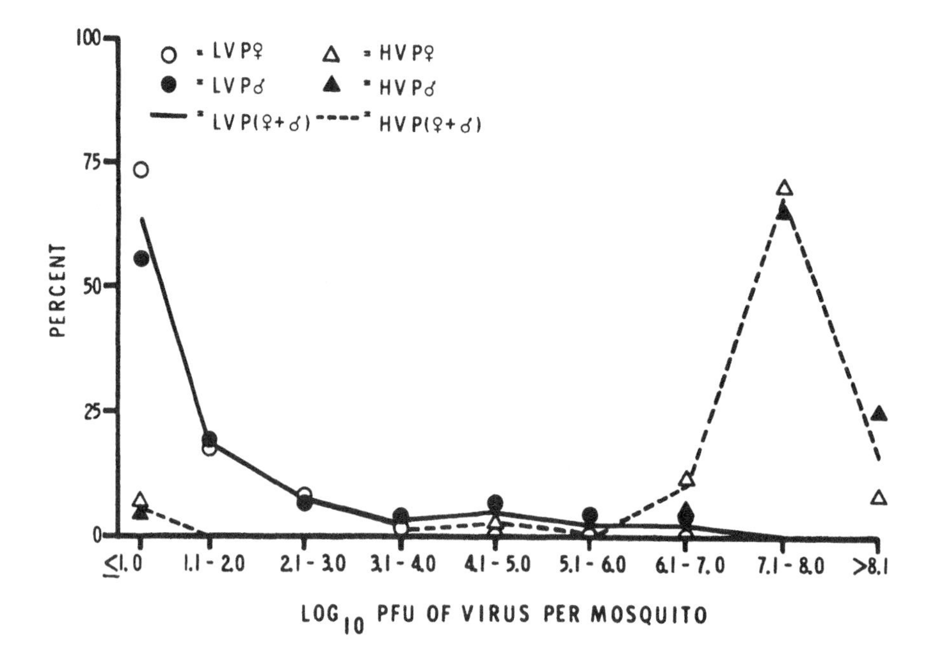

FIGURE 10. WEE virus titers in males and females of genetic lines of *Cx. tarsalis* selected for ability to allow WEE virus to multiply to high titers or to modulate WEE virus to low titers after parenteral infection. Mosquitoes were inoculated intrathoracically with 100 pfu of virus and incubated at 32°C for 72 hr. Virus in each female was titered by plaque assay in primary chicken embryonic cell culture.

comprehensive study conducted by Gargan et al.[158] They found that the susceptibility of *Cx. pipiens* females to peroral infection with a high dose of RVF virus increased over 16 generations of colonization while ability to develop a disseminated infection and transmit virus decreased. Lorenz et al.[159] also observed significant variations in the peroral susceptibility of different generations of *Ae. aegypti* to YF virus during ten generations of colonization. Of interest was that differences in peroral susceptibility of *Ae. aegypti* to YF virus correlated with genetic variations in the locus for malate dehydrogenase.

Thus, there is increasing evidence that mosquitoes can genetically control arbovirus infections, which suggests a long-time association between mosquitoes and arboviruses. Genetic control can be exerted at the level of mesenteronal infection or virus replication in cells, or both. Similar observations were made earlier with bluetongue virus in *Culicoides variipennis,*[160] plant viruses in their insect vectors,[161] and filarial and malarial parasites in mosquitoes.[34]

2. Virus Strains and Mutants

Based on studies in vertebrate hosts, one would expect that different strains of an arbovirus might vary in their ability to infect and/or be transmitted by different mosquito species or strains. This concept is supported by several vector competence studies that have been done with multiple strains of the same arbovirus.

While evaluating the vector competence of *Anopheles quadrimaculatus* for two isolates of Cache Valley-like virus, Saliba et al.[162] observed that both isolates were equally infectious for females, but only one of them could be transmitted by infected females. Kramer and Scherer[163] used vector competence of mosquitoes as a marker to differentiate enzootic and epizootic strains of VEE virus. They found that epizootic strains of VEE virus were transmitted more efficiently by perorally infected *Ae. taeniorhynchus* and multiplied to higher titers after parenteral infection than enzootic strains. In later studies, Scherer et al.[98] demonstrated that enzootic strains of VEE virus were more infectious than epizootic strains for *Cx. (Melanoconion) taeniopus* by the peroral route of infection. Resistance of this mosquito to an epizootic strain of VEE virus was associated with an MI barrier.[164] Deubel et al.[165] noted that African strains of YF virus developed faster in *Ae. aegypti* and were transmitted more efficiently than a South American strain of YF virus. Further evidence for virus strain variation was reported by Mitchell et al.,[166] who demonstrated that strains of SLE virus isolated from mosquitoes and rodents in Argentina differed significantly in their per os infectivity for an Argentina strain of *Cx. quinquefasciatus.* The rodent isolates of SLE virus, which were the least infectious for mosquitoes, also multiplied poorly in the mesenterons of infected females and failed to escape from the mesenteron into the hemocele. Strains of dengue-1 and -3 viruses varied significantly in their ability to infect *Ae. aegypti, Ae. albopictus,* or *Ae. polynesiensis* by the peroral but not by the parenteral route of infection.[167] Overall, these studies illustrate that different strains of arbovirus can vary in their infectivity for and/or transmissibility by mosquitoes. This has obvious epidemiological significance since the genetic characteristics of the virus strain may determine, in part, which mosquito species or perhaps strain can serve as a vector.

Arbovirus mutants produced in the laboratory can also differ from parental or wild type *(wt)* viruses in their ability to infect or be transmitted by mosquitoes. Of course, the trait for decreased infectivity or transmissibility is very desirable for attenuated arbovirus vaccines as was demonstrated for JE vaccine strains in *Cx. tritaeniorhynchus*[168,169] and dengue-2 vaccine in *Ae. aegypti.*[170] Also, it is fortunate that phenotypic markers, including attenuation, seemed to be quite stable upon growth and passage of these vaccine strains in mosquitoes. *Aedes aegypti* also failed to become infected and transmit YF (17D) vaccine virus after feeding on viremic monkeys and humans.[171] However, Whitman[171] concluded that this was most likely due to the low viremia titers produced by this attenuated virus in monkeys and humans and not necessarily to an inability of the virus to infect *Ae. aegypti* females.

Other arbovirus mutants, particularly small plaque (sp) mutants, are frequently infectious when ingested by mosquitoes but are not transmissible. These include sp mutants of Middelburg,[172] VEE,[173] and SIN[174] viruses in *Ae. aegypti*. With VEE virus,[173] it was shown that sp mutants failed to infect the salivary glands. In another study, a mutant cloned from prototype LAC virus infected the mesenteron of *Ae. triseriatus*, but failed to escape into the hemocele and infect other cells or tissues.[175] This mutant was similar to the prototype virus in all other properties examined, including plaque size and mouse virulence. Molecular genetic studies done with reassortants of LAC and snowshoe hare (SSH) viruses[176] suggested that the failure of the LAC mutant to escape from the infected mesenterons of *Ae. triseriatus* females may be associated with changes in one or both of the two envelope glycoproteins, which are coded by the M RNA segment of bunyaviruses.[177] Both LAC and SSH viruses readily infected the mesenteron of *Ae. triseriatus*;[172] however, SSH virus escaped from the mesenteron in only 17% of infected females as compared to 100% for LAC virus. Genetic reassortants with the LAC M RNA segment disseminated efficiently from the mesenteron whereas those with the SSH M RNA segment disseminated poorly. Therefore, the M RNA segment appears to be a major determinant of the dissemination of LAC virus, and possibly other bunyaviruses from the infected mesenteronal cells of mosquitoes.

Well-characterized mutants of arboviruses produced in the laboratory are clearly important tools for the elucidation of mechanisms associated with the various barriers to arbovirus infection, dissemination, and transmission found in incompetent mosquitoes. It is important to remember, however, that mutant arboviruses arising in infected mosquito vectors or vertebrate hosts in nature would not be selected out unless they were able to infect and be transmitted efficiently by mosquitoes. As an example, sp variants of WEE virus arise frequently during passage in cell cultures, but only large plaque variants are isolated from naturally infected *Cx. tarsalis*.[178]

3. Environmental Factors

a. Temperature

The enhancing effect that increasing environmental temperatures have on the vector competence of mosquitoes for arboviruses was presented previously in relation to the EI period (see Section III.C.1). However, high EI temperatures may decrease vector efficiency as was shown for *Cx. tarsalis* infected with WEE virus.[35,125] It was found that prolonged incubation of *Cx. tarsalis* at a constant elevated EI temperature of 32°C caused the infection rate for WEE virus to decrease from 71% on day 6 to 22% on day 12 after feeding on viremic chickens (Figure 11). This phenomenon was not associated with a selective mortality of females infected with WEE virus at the high temperature. Rather it appeared that the ability of *Cx. tarsalis* to modulate WEE virus titers to low or undetectable levels was enhanced at 32°C. This observation seems to be unique since elevated EI temperatures of 30 to 33°C had a positive effect on infection rates of YF virus in *Haemagogus* species[41] and RVF virus in *Cx. pipiens*,[126] and no effect on infection rates of EEE virus in *Ae. triseriatus*[45] and RVF virus in *Ae. taeniorhynchus*.[126]

One must question the epidemiological relevance of vector competence data derived from laboratory experiments in which adult female mosquitoes infected with arboviruses are incubated at constant temperatures of 30 to 33°C since this would not occur in nature. Even with cyclic temperatures found in hotter climates, it is unlikely that resting mosquitoes would experience such high temperatures for more than a few hours each day, if at all. However, since immature mosquito stages are captives of their aquatic environments, they could be exposed to unusually high temperatures during the middle of each day which might effect the ability of the adult females to vector arboviruses. This possibility was examined with *Cx. tritaeniorhynchus* and WN virus, but no significant differences were noted in susceptibility of adult females derived from larvae reared at 22, 28, or 32°C.[179] In another study,

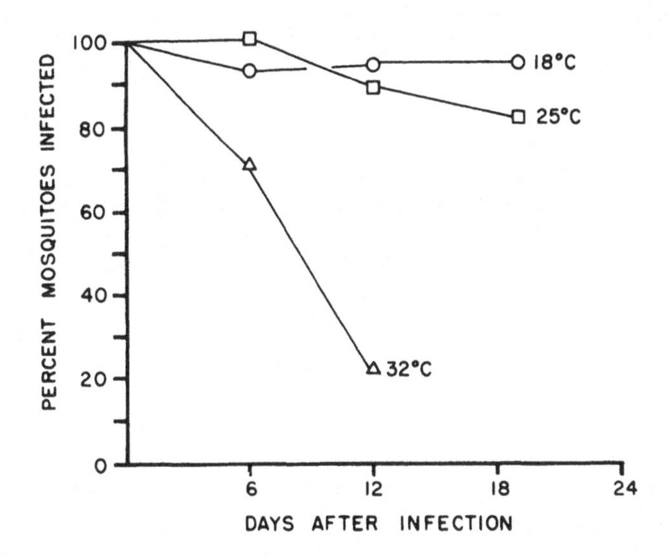

FIGURE 11. Effect of a high extrinsic incubation temperature on WEE virus infection rates in *Cx. tarsalis* [F_1 (WR females × WS males)]. Each female ingested $10^{3.8}$ to $10^{5.6}$ pfu of virus.[125]

cohorts of F_1 progeny derived from matings between WR female and WS male *Cx. tarsalis* were reared at a constant 27 or 32°C for 21 continuous generations and evaluated periodically for per os susceptibility with low and high doses of WEE virus.[125] The trait for decreased susceptibility, after ingestion of low virus doses and incubation at 32°C, was retained or selected for only in adult females that were reared at 32°C as larvae. Obviously, additional studies need to be done using cyclic water temperatures that simulate those found in natural breeding sites before it can be determined whether the vector competence of adult female mosquitoes to arboviruses is affected in any way by temperatures experienced during the larval stage.

b. Nutrition

Nutrients available during larval development in the aquatic environment could potentially affect the vector competence of adult female mosquitoes for arboviruses. This has been evaluated by using different levels of nutrition and/or crowding during larval development. With *Cx. tritaeniorhynchus,* Takahashi[122] found that a significantly higher proportion of adult females secreted JE virus by the serum-agar method from 6 to 14 days postinfection when they were reared as larvae on a ''low'' dietary regimen of powdered yeast rather than on a ''high'' dietary regimen of powdered yeast plus lactalbumin hydrolysate. Similar results were obtained with LAC virus in *Ae. triseriatus* by Grimstad and Haramis,[180] using a dietary regimen that produced small-, normal-, and large-sized adults; small females that were nutritionally deprived as larvae were the most efficient transmitters of LAC virus. The enhanced transmission efficiency of LAC virus by small adult female *Ae. triseriatus* correlated with increased virus titers and dissemination rates in these females as compared to those in the other size classes of adult female. In contrast, Baqar et al.[179] failed to show any significant differences in WN virus infection rates for *Cx. tritaeniorhynchus* females reared as larvae on decreased or increased levels of food; however, these diet regimens did result in significant differences in median times to pupation, pupation rates, and wing lengths of males. Larval crowding may have affected the peroral susceptibility of *Cx. tritaeniorhynchus* to infection with WN virus since adult females reared as larvae at a density of 2 larvae per milliliter were significantly less susceptible to infection than those reared at densities of 0.5 or 4 larvae per milliliter.

It is difficult to explain why nutritional deprivation of mosquito larvae produces adult females that are more competent vectors of an arbovirus, but the data indicate that it does occur. Grimstad and Haramis[180] found that small *Ae. triseriatus* females ingested a larger amount of blood, and thus a larger dose of virus, than did normal- and large-sized females in proportion to body weight, and suggested that this may explain their increased vector efficiency. Whatever the mechanism, it is clear that nutrition during larval development is a potentially important determinant of the vector competence of mosquitoes for arboviruses and could influence arbovirus transmission in nature. As pointed out by Grimstad and Haramis,[180] vector competence data obtained with mosquitoes reared in the laboratory on optimal diets probably do not reflect the true vector competence of field mosquito populations for arboviruses.

4. Coinfection of Mosquitoes with Exogenous and Endogenous Infectious Agents

Mosquito cell cultures persistently infected with one arbovirus may be resistant to superinfection with the homologous virus and sometimes heterologous viruses.[181,182] Likewise, interference of arbovirus infection, multiplication, or transmission has been observed when mosquitoes are coinfected with some, but not other, combinations of arboviruses. In general, homologous interference seems to occur more frequently than heterologous interference. Further, infection with the one virus has to be established in the mosquito before it can interfere with the second virus.

Sabin[183] suggested that the capacity of *Ae. aegypti* to transmit YF virus was reduced when females were previously infected with dengue virus. Similarly, interference of peroral infection or transmission was also demonstrated with mosquitoes coinfected with MVE and JE viruses[89] and with mutant and wild type strains of WN virus.[184] However, Lam and Marshall[185] found no evidence of interference when *Ae. aegypti* females were infected perorally or parenterally with various combinations of flaviviruses (dengue, WN, and MVE) and alphaviruses (SIN and SF). Similarly, Chamberlain and Sudia[186] observed that virus titers and transmission rates of EEE and WEE viruses were identical for singly and dually infected *Cx. tarsalis*.

Using only the parenteral route of infection, Davey et al.[187] demonstrated interference of *wt* SF virus in *Cx. annulirostris* infected 7 days previously with a temperature sensitive *(ts)* sp mutant isolated from persistently infected *Ae. albopictus* cell cultures; no interference was noted if mosquitoes were infected simultaneously with both viruses. Similar results were reported by Beaty et al.[188] with *Ae. triseriatus* infected parenterally with an LAC *ts* mutant followed 7 days later with *wt* LAC virus. In addition, they noted that prior infection with the LAC *ts* mutant interfered with the multiplication of other California serogroup bunyaviruses, but not with the multiplication of a Bunyamwera serogroup bunyavirus, flavivirus, or rhabdovirus. To the contrary, Turell et al.[189] found that *Ae. dorsalis* infected transovarially with CE virus was resistant to parenteral challenge with the homologous CE virus but not with another California serogroup bunyavirus (Jamestown Canyon) as determined by CO_2 sensitivity.

Thus, evidence for heterologous interference of arboviruses in dually infected mosquitoes is mixed. However, even if arboviruses interfere with each other in dually infected mosquitoes in the laboratory, it is questionable whether this has any epidemiological significance since dual infection of mosquitoes with two arboviruses in nature would be a rare event, given the usual arbovirus infection rates in field mosquito populations.

Interference of arbovirus infections in mosquitoes that are infected with endogenous viruses may be of more importance epidemiologically if a significant proportion of a vector mosquito population is infected with such viruses. Unfortunately, the impact of this potentially important variable on the vector competence of mosquitoes for arboviruses cannot be assessed at present because interference, if any, between arboviruses and endogenous viruses in dually infected mosquitoes has yet to be examined.

Several studies addressed the question of whether concurrent protozoan and filarial infections in mosquitoes affect their ability to vector arboviruses. Results reported by Rozeboom et al.[190] suggested that prior infection of *Ae. aegypti* with *Plasmodium gallinaceum* decreased their susceptibility to peroral infection with WN virus. However, earlier studies by Barnett[191] indicated that *P. relictum* infection in *Cx. tarsalis* had no effect on their ability to transmit WEE virus. Likewise, *Ascocystis barretti* infections in *Ae. triseriatus* larvae did not affect their ability to become infected after ingestion of LAC virus and to transmit virus as adults.[192] In contrast, concurrent infection of *Ae. taeniorhynchus* females with RVF virus and the microfilaria, *Brugia malayi,* resulted in enhanced infection, dissemination, and transmission rates of RVF virus.[193] Similar results were obtained when *Culicoides nubeculosus* were concurrently infected with *Onchocerca cervicalis* and bluetongue virus.[194]

5. Pathological Effects Associated with Arbovirus Infections in Mosquitoes

It is generally assumed that arboviruses do not produce cytopathology in cells and tissues of infected mosquitoes. This concept derives from early studies which demonstrated that there was no difference in the survival of virus-infected vs. uninfected females.[29] However, Mims et al.[195] observed that transmission rates of SF virus for *Ae. aegypti* declined to zero by 3 weeks after peroral or parenteral infection and that this decrease was correlated with the degree of cytological changes observed in the salivary glands. This observation was confirmed by Lam and Marshall.[196] They further showed that SF virus infections in *Ae. aegypti* reduced transmission rates of a second noncytopathic virus (i.e., WN or SIN) introduced after the SF virus had replicated in the mosquito. Interestingly, when females were first infected with the noncytopathic virus and then SF virus, they continued to transmit the noncytopathic virus even though SF virus produced cytopathic effects in the salivary glands and transmission rates of SF virus declined to zero. However, the epidemiological significance of these observations is not clear since *Ae. aegypti* is apparently not a natural vector of SF virus.[135]

However, signs of virus-induced morbidity have been observed with other arboviruses in their natural or epidemic mosquito vectors. Grimstad et al.[197] noted a reduced ability of *Ae. triseriatus* females that were infected perorally with LAC virus to refeed on vertebrate hosts. Turell et al.[198] made a similar observation with *Cx. pipiens* infected with RVF virus. In addition, they found that the number of eggs oviposited by RVF virus-infected *Cx. pipiens* females was significantly reduced as compared to uninfected females. No attempt was made in either of these studies to correlate morbidity with cytopathology. Nonetheless, arbovirus-induced cytological or behavioral changes in mosquitoes might well affect their vector competence. Obviously, there is a need for further research in this area with other arboviruses and their natural mosquito vectors.

B. Barriers to Arbovirus Infection, Dissemination, and Transmission in Incompetent Mosquitoes

1. Mesenteronal Infection Barrier

The existence of an MI barrier to an arbovirus in mosquitoes was first demonstrated by Merrill and TenBroeck[85] using *Ae. aegypti* that were completely refractory to peroral infection with EEE virus. They demonstrated that *Ae. aegypti* females could be induced to transmit EEE virus if the virus was inoculated directly into the hemocele or the abdomen was punctured with a sterile needle immediately after ingestion of a blood meal. This phenomenon had been demonstrated earlier by Storey[199] using a strain of leafhoppers, *Cicadulina mbila,* that was genetically resistant to per os infection with maize-streak virus. Similar observations were subsequently reported for other arbovirus-mosquito combinations. McLean[58] found that MVE virus multiplied after peroral and parenteral infection of *Cx. annulirostris,* but only after parenteral infection of *Anopheles annulipes.* Hardy et al.[131] demonstrated that field strains of *Ae. melanimom, Cx. peus, Cx. pipiens,* and *Cx. tarsalis* were equally susceptible to parenteral infection with WEE virus, whereas only *Ae. melanimom* and *Cx. tarsalis* were readily infected after ingestion of virus (Figure 6). The results with *Cx. peus, Cx. pipiens,* and *Cx. tarsalis* were confirmed using colonized mosquito strains.[132]

Resistance of geographical strains of a single mosquito species to per os infection with an arbovirus also appears to be associated with an MI barrier. This was demonstrated for dengue 2 virus in naturally refractory strains of *Ae. albopictus*[144] and *Ae. aegypti*[143] and for WEE virus in field-collected (Figure 8) and colonized strains of *Cx. tarsalis.*[131,140] Further, the WR line of *Cx. tarsalis* that was genetically selected for resistance to per os infection with WEE virus had the same susceptibility as the WS strain of *Cx. tarsalis* by parenteral infection.[141]

Various hypotheses have been proposed to explain the MI barrier to infection of mosquitoes with arboviruses.[5,29,31,32,35,199] These include diversion of the blood meal into diverticula, inactivation of virus by digestive enzymes in the lumen of the mesenteron, occlusion of virus by the peritrophic membrane, absence or reduced number of cellular receptor sites for virus attachment on the microvillar membrane, and abortive replication of virus in mesenteronal epithelial cells. Attempts to experimentally verify or disprove these hypotheses have been done primarily with WEE virus using an interspecific mosquito model with *Cx. tarsalis* Knights Landing (KL) and *Cx. pipiens* as the susceptible and resistant species, respectively (Figure 12A), and an intraspecific model with WS *Cx. tarsalis* and WR *Cx. tarsalis* as the susceptible and refractory genetic lines, respectively (Figure 12B).[35,132]

Current evidence indicates that diversion of the blood meal into diverticula, peritrophic membrane occlusion, and degradation of ingested virions by digestive enzymes are not significant components of the MI barrier to infection of *Culex* species with WEE virus.[35,83,200] Instead, the resistance of *Cx. pipiens* females to per os infection with WEE virus appears to be related to a failure of virus to bind to the mesenteronal microvillar membrane. This latter conclusion derives from several lines of evidence. Pattyn and de Vleesschauwer[201] showed that incorporation of DEAE-dextran, a polycationic substance, into the blood meal enhanced the per os infectivity of several arboviruses for *Ae. aegypti,* presumably by dampening the electrostatic repulsion between the highly anionic virus envelope and microvillar membrane surfaces. When this experimental approach was tried with *Cx. pipiens,* it was found that WEE virus infected and multiplied in their mesenterons.[132] Further, multiplication of WEE virus occurred in the mesenterons of *Cx. pipiens* when they were infected by intrathoracic inoculation.[132] The results obtained by these two experimental approaches indicated that the MI barrier to infection of *Cx. pipiens* females with WEE virus was associated with the attachment, penetration, or uncoating step of the virus multiplication cycle and not with an inability of virus to replicate in the mesenteronal epithelial cells.

Since pH influences the surface charge along membrane surfaces, it was thought that lowering the pH of the blood meal would increase the positive charge on the microvillar membrane and allow WEE virus to infect *Cx. pipiens.* However, the occasional infection of a *Cx. pipiens* female with WEE virus seemed to be a random event since alteration of the blood meal pH over a range from 6.0 to 8.5 had no effect on susceptibility.[132] In contrast, a definite optimal pH of 8.0 was observed for infection of KL *Cx. tarsalis* with WEE virus. Preliminary experiments[202] using isolated brush border fragments from mosquito mesenterons demonstrated that ^{3}H-labeled WEE virions had a significantly higher binding affinity for brush border fragments isolated from WS *Cx. tarsalis* than from *Cx. pipiens.*[202] Recent competitive binding experiments with unlabeled and ^{32}P-labeled WEE virions indicated that binding to brush border fragments from *Cx. tarsalis* was specific, whereas that to brush border fragments from *Cx. pipiens* was nonspecific.[203] These data suggest that mesenteronal microvillar membranes of *Cx. pipiens* lack specific receptor sites for WEE virus

Similar studies done with WS *Cx. tarsalis* — WR *Cx. tarsalis* intraspecific mosquito model clearly indicate that refractoriness to peroral infection of WS *Cx. tarsalis* is associated

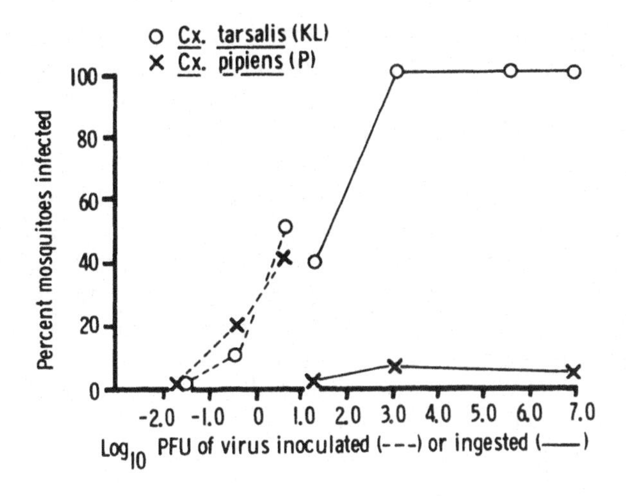

A

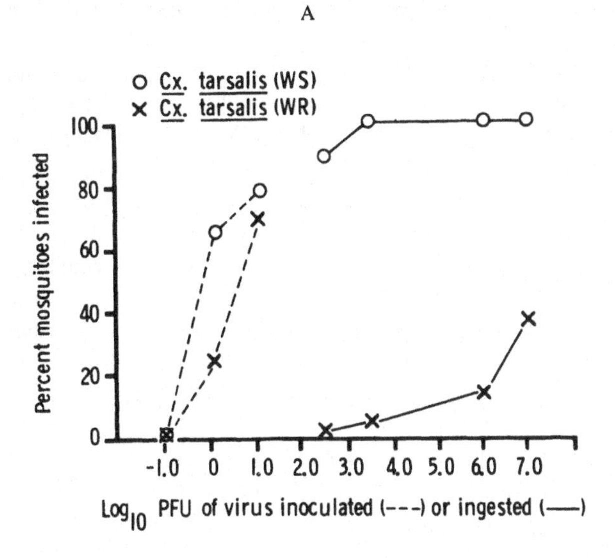

B

FIGURE 12. (A) Inter- and (B) intraspecific mosquito models used to study the mesenteronal infection barrier to peroral infection with WEE virus. The interspecific mosquito model includes unselected strains of *Cx. tarsalis* (KL) and *Cx. pipiens* as the perorally susceptible and refractory species, respectively, and the intraspecific model is comprised of the genetic lines of WS and WR *Cx. tarsalis* selected for high susceptibility and resistance to peroral infection, respectively. Females of all species and genetic lines have nearly equal susceptibility by the parenteral route of infection.

with an MI barrier,[141] but it is not clear if resistance occurs at the level of virus attachment-penetration or intracellular replication.[134] The susceptibility of WR *Cx. tarsalis* to WEE virus was not influenced by blood meal pH, whereas it was for WS *Cx. tarsalis* (i.e., optimal pH of 8.0). Also, brush border fragments from mesenterons of WR *Cx. tarsalis* had a lower binding affinity for WEE virus than did those from WS *Cx. tarsalis*.[202] These observations suggest that different mechanisms are involved in mesenteronal infection of these two genetic strains of *Cx. tarsalis*. However, the addition of DEAE-dextran to the infectious blood meal failed to increase WEE virus infection rates in WR *Cx. tarsalis* females. Further, mesenterons of WR *Cx. tarsalis* rarely became infected after intrathoracic inoculation of WEE virus. Differences observed between WEE virus refractory *Cx. pipiens* and WR *Cx. tarsalis* females undoubtedly are related to an ability of WR *Cx. tarsalis* females, and not *Cx. pipiens* females, to modulate WEE virus titers after parenteral infection.[134] Thus, current evidence suggests that the MI barrier to peroral infection of WR *Cx. tarsalis* with WEE virus may be associated with at least two genetically controlled mechanisms: susceptibility of mesenteronal epithelial cells to infection and modulation of virus multiplication in cells that become infected.

Refractoriness of some geographic strains of *Ae. aegypti* and *Ae. albopictus* to peroral infection with dengue-2 virus appears to be associated with only a MI barrier since there do not seem to be any subsequent barriers to virus dissemination and multiplication in females that become infected.[143,144] As yet, no studies on mechanisms involved in the MI barrier to DEN virus infection in these mosquito species have been reported.

2. Dissemination Barriers

Numerous investigators have observed that some females of various mosquito species/strains can become infected with certain arboviruses, but are unable to transmit virus perorally after a suitable EI period.[23,29,48,52,86,95,109,118,131,148,158,204,216] In some cases, inability of infected female mosquitoes to transmit virus was correlated with a low infecting dose of virus[48,118,214] or low virus titers in nontransmitting females.[131,148,158,208,210,214-216] Collins et al.[205] noted that transmission thresholds of EEE virus for *Anopheles albimanus* and *An. quadrimaculatus* were significantly higher than MI thresholds. However, Kramer et al.[214] were the first to demonstrate that the inability of infected *Cx. tarsalis* to transmit WEE virus was associated with two dose-dependent dissemination barriers which have been termed mesenteronal escape (ME) and salivary gland infection (SGI) barriers. Similar barriers have now been demonstrated for RVF virus in *Cx. pipiens*[158,216] and VEE virus in *Cx. taeniopus*.[164]

Virus in female mosquitoes exhibiting an ME barrier was restricted to the mesenteron and virus titers in the mesenteron were usually lower than those found in mesenterons of the SGI barrier and fully competent females (Figure 13). With WEE virus in *Cx. tarsalis*, the MI barrier was not time-dependent since MI barrier rates did not increase significantly with time of EI.[214] However, Sudia[217] found that the MI barrier to SLE virus in *Cx. quinquefasciatus* was time-dependent; this explained why the EI period for SLE virus in *Cx. quinquefasciatus* was significantly longer than in *Cx. pipiens*.[54] EI temperatures inversely affected ME barrier rates for WEE virus in *Cx. tarsalis* during the first 7 days of incubation after feeding on viremic chicks; however, ME barrier rates were nearly the same in females incubated at 18, 25, or 32°C for 12 to 19 days.[125] Similar observations were made by Turell et al.[126] for RVF virus in *Cx. pipiens* incubated at 13, 26, or 33°C. However, dissemination rates of RVF virus were significantly lower for *Ae. taeniorhynchus* females incubated at 13°C as compared to those incubated at 26 or 33°C. Thus, both time and temperature of EI may have different effects on the expression of ME barriers in different arbovirus-mosquito systems.

WEE virus titers in female mosquitoes expressing the SGI barrier were normal in mesenterons, reduced in hemolymph and remnants, and not detectable in thoracic/abdominal ganglia and salivary glands (Figure 13).[35,125] Thus, virus had escaped from the infected mesenteron but did not multiply normally in other cells and tissues bathed by the hemolymph. SGI barrier rates decreased with time of incubation at 26°C, which suggested that a critical level of virus was probably required in the hemolymph to infect the salivary glands.[125,214] However, when *Cx. tarsalis* infected with WEE virus were incubated at 32°C, the SG infection rate appeared to increase rather than decrease from day 6 to 12 postinfection (i.e., mosquitoes were less efficient transmitters of virus after prolonged incubation at 32°C). In contrast, Turell et al.[126] observed that transmission rates of RVF virus for both *Cx. pipiens* and *Ae. taeniorhynchus* were higher at an EI temperature of 33°C than at 26°C.

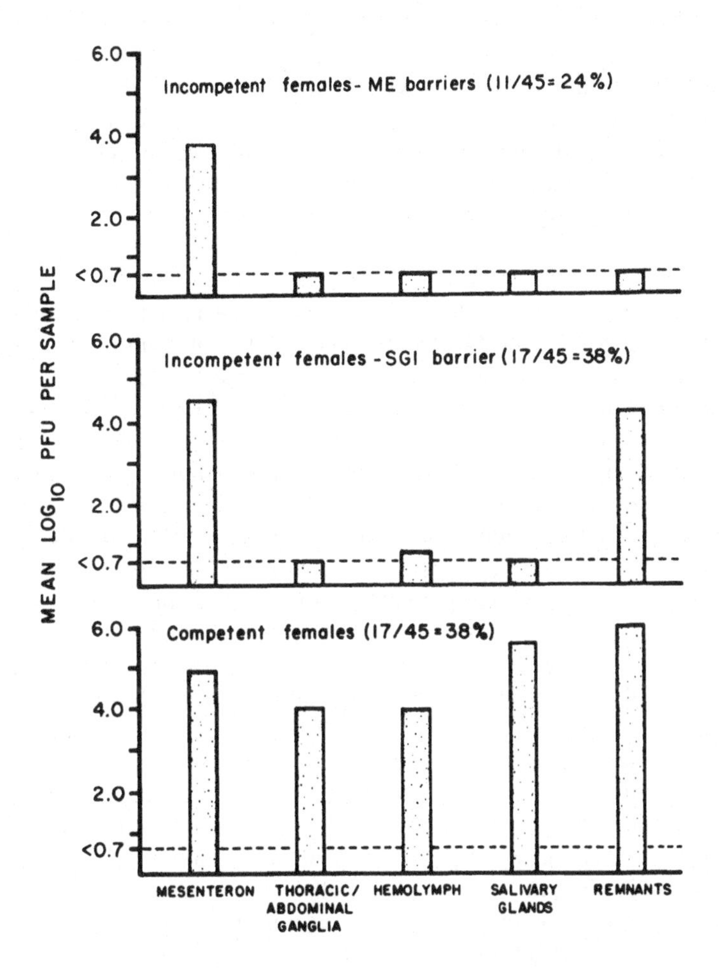

FIGURE 13. Mean WEE virus titers in selected tissues/fluids of competent and incompetent *Cx. tarsalis* (WS) females infected per os on a low dose of WEE virus ($10^{1.8}$ to $10^{2.8}$ pfu per female) and incubated for 20 to 21 days at 26°C. (From Hardy, J. L., et al., *Ann. Rev. Entomol.*, 28, 229, 1983. With permission.)

A mechanism to explain the barriers to arbovirus dissemination in susceptible mosquitoes has yet to be elucidated. However, the association of dissemination barriers with reduced or nil multiplication of virus in some tissues of susceptible females suggests that the genetic trait associated with modulation of arbovirus replication in mosquitoes may be involved. Modulation of alphavirus titers in mosquito cell cultures has been well documented.[131,182] Further, a small molecular weight host protein has been recovered from *Ae. albopictus* cells persistently infected with SIN or SF virus that specifically interferes with the multiplication of the inducer virus in *Ae. albopictus* cells.[182] However, if a host protein controls WEE virus multiplication in LVP *Cx. tarsalis,* then it must be different than inhibitors of SIN and SF viruses produced by *Ae. albopictus* cells since modulation in LVP *Cx. tarsalis* is alphavirus-specific but not WEE virus-specific.[134] A further indication that the mosquito can control arbovirus infections is suggested by the observation that the expression of dissemination barriers is dose dependent. If the female mosquito is not initially overwhelmed with a high dose of virus, then she seems to be able to bring the infection under control at various phases of tissue infection. It is tempting to speculate that modulation of arbovirus replication in mosquitoes is a form of mosquito immunity.

Alternatively, arbovirus mutants (e.g., sp, *ts,* and deletion) or gene products could limit virus multiplication and dissemination in mosquitoes.[30,31,35] Various types of alphavirus mutants have been recovered from persistently infected mosquito cell cultures that specifically interfere with superinfection of the cells with *wt* virus.[181,182] Also, the matrix or membrane polypeptide of the flavivirus, Banzi, was produced in high concentrations in persistently infected *Ae. albopictus* cells and controlled Banzi virus replication, possibly by regulating virus RNA synthesis. Whether similar mechanisms are operative in intact mosquitoes infected with arboviruses has not been reported.

3. Salivary Gland Escape (SGE) Barrier

While reviewing this subject in 1981, Hardy et al.[35] speculated that an arbovirus transmission barrier, analogous to the ME barrier, may exist in mosquito salivary glands since there was evidence that some mosquitoes with infected salivary glands were unable to transmit virus.[29,119] The existence of such a barrier has now been definitively demonstrated for JE virus in *Cx. tritaeniorhynchus,*[109] SSH virus in *Ae. triseriatus,*[218] LAC virus in *A. hendersoni,*[219] and SIN virus in *Cx. theileri,*[220] In *Ae. hendersoni,*[219] the SGE barrier was a major contributor to vector incompetence since 65% of one infected population (Palmetto strain) had salivary gland infections by infectious virus assay, and only 5% of the infected females transmitted virus at 21 days postinfection. IFA studies indicated that high concentrations of LAC virus antigens were present in salivary glands of the eight *Ae. hendersoni* females examined.

Although definitive studies have yet to be done, one can speculate on several possible mechanisms that might explain the SGE barrier to transmission of arboviruses by mosquitoes. One mechanism might be that too little virus is being produced in the salivary glands or secreted during the feeding process to infect vertebrate hosts. The former did not seem to be the case with LAC virus in *Ae. hendersoni* since virus titers in salivary glands of this species were identical to those produced in salivary glands of *Ae. triseriatus* which transmitted LAC virus much more efficiently than *Ae. hendersoni.* Another possibility is that the virus titers are modulated in mosquito salivary gland cells to low or undetectable levels. This may explain the observations of McIntosh and Jupp[221] who noted that transmission rates of CHIK virus for *Ae. aegypti* decreased from 78% at 21 days to 12% at 49 days postinfection. Similar observations were made by Mangiafico[206] for CHIK virus with both *Ae. aegypti* and *Ae. albopictus.* Also, Kramer et al.[125] observed that transmission rates of WEE virus for *Cx. tarsalis* decreased concurrently with infection rates from 6 to 12 days postinfection and incubation at 32°C. These results with alphaviruses may also correlate with observations made on the multiplication of SIN virus in *Ae. albopictus* cells.[182] Apparently each virion that enters the cell is compartmentalized into a cytoplasmic "virus factory" where progeny virus are produced and then released from the cell by membrane fusion. Thus, continued infection of the cell depends on reinfection by released virus. If a similar phenomenon exists in mosquito salivary glands infected with alphaviruses in which the flow of virus is unidirectional, then a decrease in viremic titers in the hemolymph below the SGI threshold level would interrupt infection of the salivary gland cells. However, a decrease in the ability of mosquitoes with infected salivary glands to transmit alphaviruses may simply be due to cytopathology as shown for SF virus in *Ae. aegypti.*[195,196]

Takahashi[109] found that JE virus infections in the salivary glands of non- or low-transmitting *Cx. tritaeniorhynchus* females were frequently confined to cells in the lateral neck, and that infection of lateral acinar cells was usually required for virus transmission. This may also explain failure of mosquitoes to transmit other arboviruses when the salivary glands are infected.

Finally, the failure of an arbovirus to be transmitted from infected mosquito salivary glands may be determined by virus genetics and not necessarily host genetics. This is indicated by the studies of Beaty et al.[218] in which genetic reassortants of LAC and SSH viruses were inoculated intrathoracically into *Ae. triseriatus.* Reassortants with the M RNA segment of LAC virus were transmitted efficiently by *Ae. triseriatus* whereas reassortants with the M RNA segment of SSH virus were not, even though the salivary glands were infected as evidenced by IFA. Thus, the envelope glycoproteins coded by the M RNA segment of bunyaviruses may be important determinants of peroral transmission of these viruses, as they are for dissemination of virus from the infected mesenteron.[176]

V. SUMMARY

Research done since 1928 has provided a broad base of knowledge on the susceptibility and resistance of numerous mosquito species and strains to various arboviruses. The biological transmission of arboviruses by fully competent vector mosquitoes requires that the virus initiate infection in the mesenteron following the ingestion of an infective blood meal, escape from the mesenteron and disseminate via the hemolymph throughout the hemocele, establish infection in the salivary glands with or without secondary amplification in extra mesenteronal tissues, and finally be released into the salivary gland secretions where it can be transmitted to a vertebrate host during the feeding process. Infection of the mesenteron, and possibly the salivary glands, is virus-dose dependent since an infection threshold must be overcome by the virus. The concentration of an arbovirus required to overcome the mesenteronal infection threshold varies significantly from one arbovirus-mosquito system to another. The time interval between ingestion of an infective blood meal and peroral transmission of an arbovirus (i.e., extrinsic incubation period) is also quite variable among different arbovirus-mosquito systems and is influenced greatly by incubation temperature and possibly infecting dose of virus. The most competent vector mosquito species/strains for an arbovirus are obviously those that have a low MI threshold and allow the virus to complete its EI incubation period in a relatively short time. However, this is not always necessary for the maintenance of arbovirus transmission cycles in nature if vertebrate hosts develop high levels of viremias and/or the mosquito vector is relatively long-lived.

Interspecific variations in the ability of different mosquito species to vector a particular arbovirus and the association of resistance to an MI barrier were established by early investigators. However, the most important concept to be recognized in the last decade is that individual mosquitoes within a proven vector mosquito species can vary significantly in their ability to vector an arbovirus since susceptibility/resistance and/or ability of arboviruses to multiply are apparently controlled by mosquito genetic factors. Incompetence is expressed as a series of barriers that affect the infection of the mesenteron (i.e., MI barrier), dissemination of virus from the mesenteron (i.e., ME barrier), infection of the salivary glands (i.e., SGI barrier), or escape of virus into the salivary gland secretions (i.e., SGE barrier). The expression of these barriers is influenced by numerous factors, such as mosquito strain, virus strain, extrinsic incubation temperature, nutrition during larval development, rearing temperature during larval development, and endogenous or exogenous infection of mosquitoes with other infectious agents. Consequently, a vector mosquito population can be composed of varying proportions of incompetent and competent females for a particular arbovirus, and these proportions can change intraseasonally, annually, or geographically, thus potentially affecting the occurrence and distribution of arbovirus diseases.

However, there are still many unanswered questions regarding the vector competence of mosquitoes for arboviruses. For example, how do arboviruses attach to, and penetrate into, mesenteronal and salivary gland epithelial cells? Is secondary amplification of virus in extramesenteronal cells/tissues required for infection of the salivary gland? What are the genetic and molecular bases of the various barriers to arbovirus infection, dissemination, multiplication, and transmission in mosquitoes? What temperatures do immature and adult mosquitoes experience in nature, and how do these temperatures affect arbovirus EI periods in mosquitoes and the vector competence of mosquitoes? Do variations that occur in the vector competence of mosquitoes for arboviruses in the field actually influence the occurrence and distribution of arbovirus diseases? Thus, future research should be directed at answering these and other important questions on the nature of arbovirus-mosquito interactions, and the effect of the environment on these interactions.

ACKNOWLEDGMENTS

The author is indebted to Dr. E.J. Houk, Dr. L.D. Kramer, and Ms. Sharon Lynn for assistance in preparing the figures; Dr. W.C. Reeves for his critical review of the manuscript; and Ms. Miyuki Takahashi for assistance in preparing the typescript.

REFERENCES

1. **Smith, C. E. G.,** Factors influencing the behavior of viruses in their arthropodan hosts, in *Host-Parasite Relationships in Invertebrate Hosts,* Blackwell Scientific, Oxford, 1964, 1.
2. **Smith, C. E. G.,** Factors in the transmission of virus infections from animals to man, *Sci. Basis Med. Annu. Rev.,* 125, 1964.
3. **Hurlbut, H. S.,** Arthropod transmission of animal viruses, *Adv. Virus Res.,* 11, 277, 1965.
4. **Reeves, W. C.,** Factors that influence the probability of epidemics of Western equine, St. Louis, and California encephalitis in California, *Calif. Vector Views,* 14, 13, 1967.
5. **Chamberlain, R. W.,** Arboviruses, the arthropod-borne animal viruses, *Curr. Top. Microbiol. Immunol.,* 42, 38, 1968.
6. **Smith, C. E. G.,** The spread and maintenance of infections in vertebrates and arthropods, *J. Invert. Pathol.,* 18, i, 1971.
7. **Hardy, J. L.,** Arbovirus research program at the University of California, Berkeley, *Proc. Calif. Mosq. Control Assoc.,* 43, 15, 1975.
8. **Sellers, R. F.,** Weather, host and vector — their interplay in the spread of insect-borne animal virus diseases, *J. Hyg. (Cambridge),* 85, 65, 1980.
9. **Mitchell, C. J.,** Mosquito vector competence and arboviruses, *Curr. Top. Vector Res.,* 1, 62, 1983.
10. **Grimstad, P. R.,** Mosquitoes and the incidence of encephalitis. *Adv. Virus Res.,* 28, 357, 1983.
11. **Finlay, C. E.,** in *Carlos Finlay and Yellow Fever,* Kahn, M. C., Ed., Oxford University Press, New York, 1940.
12. **Reed, W., Carroll, J., Agramonte, A., and Lazear, J. W.,** The etiology of yellow fever — a preliminary note, in *Yellow Fever: A Compilation of Various Publications,* Owen, Ed., U.S. Government Printing Office, Washington, D.C., 1911, 56.
13. **Bauer, J. H.,** The transmission of yellow fever by mosquitoes other than *Aedes aegypti, Am. J. Trop. Med.,* 8, 261, 1928.
14. **Reed, W. and Carroll, J.,** The etiology of yellow fever — a supplemental note, in *Yellow Fever: A Compilation of Various Publications,* Owen, Ed., U.S. Government Printing Office, Washington, D.C., 1911, 149.
15. **Simmons, J. S.,** Dengue fever, *Am. J. Trop. Med.,* 11, 77, 1931.
16. **Simmons, J. S., St. John, J. H., and Reynolds, F. H. K.,** Dengue fever transmitted by *Aedes albopictus* Skuse, *Am. J. Trop. Med.,* 10, 17, 1930.
17. **Stokes, A., Bauer, J. H., and Hudson, N. P.,** Experimental transmission of yellow fever to laboratory animals, *Am. J. Trop. Med.,* 8, 103, 1928.
18. **Bauer, J. H. and Hudson, N. P.,** The incubation period of yellow fever in the mosquito, *J. Exp. Med.,* 48, 147, 1928.
19. **Hindle, E.,** An experimental study of yellow fever, *Trans. R. Soc. Trop. R. Med. Hyg.,* 22, 405, 1929.
20. **Hindle, E.,** The transmission of yellow fever, *Lancet,* 219, 835, 1930.
21. **Philip, C. B.,** Preliminary report of further tests with yellow fever transmission by mosquitoes other than *Aedes aegypti, Am. J. Trop. Med.,* 9, 267, 1929.
22. **Philip, C. B.,** Studies on transmission of experimental yellow fever by mosquitoes other than *Aedes, Am. J. Trop. Med.,* 10, 1, 1930.
23. **Philip, C. B.,** Transmission of yellow fever virus by aged *Aedes aegypti* and comments on some other mosquito-virus relationships, *Am. J. Trop. Med. Hyg.,* 11, 697, 1962.

24. **Davis, N. C. and Shannon, R. C.,** The location of yellow fever virus in infected mosquitoes and the possibility of hereditary transmission, *Am. J. Hyg.*, 11, 335, 1930.
25. **Davis, N. C.,** The transmission of yellow fever on the possibility of immunity in *Stegomyia* mosquitoes, *Am. J. Trop. Med.*, 11, 31, 1931.
26. **Davis, N. C.,** The effect of various temperatures in modifying the extrinsic incubation period of the yellow fever virus in *Aedes aegypti*, *Am. J. Hyg.*, 16, 163, 1932.
27. **Davis, N. C., Frobisher, M., and Lloyd, W.,** The titration of yellow fever virus in *Stegomyia* mosquitoes, *J. Exp. Med.*, 58, 211, 1933.
28. **Chamberlain, R. W.,** Virus-vector-host relationships of the American arthropod-borne encephalitides, *Proc. 10th Int. Congr. Entomol.*, 3, 567, 1956 (1958).
29. **Chamberlain, R. W. and Sudia, W. D.,** Mechanism of transmission of viruses by mosquitoes, *Ann. Rev. Entomol.*, 6, 371, 1961.
30. **Murphy, F. A.,** Cellular resistance to arbovirus infection, *Ann. N. Y. Acad. Sci.*, 266, 197, 1975.
31. **Murphy, F. A., Whitfield, S. G., Sudia, W. D., and Chamberlain, R. W.,** Interactions of vector with vertebrate pathogenic viruses, in *Invertebrate Immunity*, Maramorosch, K. and Shope, R. E., Eds., Academic Press, New York, 1975, 25.
32. **McLintock, J.,** Mosquito-virus relationships of American encephalitides, *Ann. Rev. Entomol.*, 23, 17, 1978.
33. **Grimstad, P. R.,** Genetics of vector competence, *Misc. Publ. Entomol. Soc. Am.*, 11, 29, 1980.
34. **Gubler, D. J., Novak, R., and Mitchell, C. J.,** Arthropod vector competence — epidemiological, genetic, and biological considerations, in *Recent Developments in the Genetics of Insect Disease Vectors*, Steiner, W. W. M., Tabachnick, W. J., Rai, K. S., and Narang, S., Eds., Stipes Publishing, Champaign, Ill., 1982, 343.
35. **Hardy, J. L., Houk, E. J., Kramer, L. D., and Reeves, W. C.,** Intrinsic factors affecting vector competence of mosquitoes for arboviruses, *Ann. Rev. Entomol.*, 28, 229, 1983.
36. **Hoch, A. L., Gargan, T. P., II, and Bailey, C. L.,** Mechanical transmission of Rift Valley fever virus by hematophagous *Diptera*, *Am. J. Trop. Med. Hyg.*, 34, 188, 1985.
37. **Harrison, B. D. and Murant, A. F.,** Involvement of virus-coded proteins in transmission of plant viruses by vectors, in *Vectors in Virus Biology*, Mays, M. A. and Harrap, K. A., Eds., Academic Press, London, 1984, 1.
38. **Boorman, J.,** Observations on the amount of virus present in the haemolymph of *Aedes aegypti* infected with Uganda S, yellow fever and Semliki Forest Viruses, *Trans. R. Soc. Trop. Med. Hyg.*, 54, 362, 1960.
39. **Miles, J. A. R., Pillai, J. S., and Maguire, T.,** Multiplication of Whataroa virus in mosquitoes, *J. Med. Entomol.*, 10, 176, 1973.
40. **Merrill, M. H. and TenBroeck, C.,** Multiplication of equine encephalomyelitis virus in mosquitoes, *Proc. Soc. Exp. Biol.*, 32, 421, 1934.
41. **Whitman, L.,** The multiplication of the virus of yellow fever in *Aedes aegypti*, *J. Exp. Med.*, 66, 133, 1937.
42. **Bates, M. and Roca-Garcia, M.,** The development of the virus of yellow fever in *Haemagogus* mosquitoes, *Am. J. Trop. Med.*, 26, 585, 1946.
43. **McLean, D. M.,** Transmission of Murray Valley encephalitis virus by mosquitoes, *Aust. J. Exp. Biol. Med.*, 31, 148, 1953.
44. **Chamberlain, R. W., Corristan, E. C., and Sikes, R. K.,** Studies on the North American arthropod-borne encephalitides. V. The extrinsic incubation of eastern and western equine encephalitis in mosquitoes, *Am. J. Hyg.*, 60, 269, 1954.
45. **Chamberlain, R. W. and Sudia, W. D.,** The effects of temperature upon the extrinsic incubation of eastern equine encephalitis in mosquitoes, *Am. J. Hyg.*, 62, 295, 1955.
46. **Chamberlain, R. W. and Sudia, W. D.,** The North American arthropod-borne encephalitis viruses in *Culex tarsalis* Conquillett, *Am. J. Hyg.*, 66, 151, 1957.
47. **Davies, A. M. and Yoshpe-Purer, Y.,** Observations on the biology of West Nile virus, with special reference to its behavior in the mosquito *Aedes aegypti*, *Am. Trop. Med. Parasit.*, 48, 46, 1954.
48. **Tahori, A. S., Sterk, V. V., and Goldblum, N.,** Studies on the dynamics of experimental transmission of West Nile virus by *Culex molestus*, *Am. J. Trop. Med. Hyg.*, 4, 1015, 1955.
49. **Davies, A. M. and Yoshpe-Purer, Y.,** The transmission of Semliki Forest virus by *Aedes aegypti*, *J. Trop. Med. Hyg.*, 57, 273, 1954.
50. **Hale, J. H., Colless, D. H., and Lim, K. A.,** Investigation of the Malaysian form of *Culex tritaeniorhynchus* as a potential vector of Japanese B encephalitis virus on Singapore Island, *Ann. Trop. Med. Parasit.*, 51, 17, 1957.
51. **Gresser, I., Hardy, J. L., Hu, S. M. K., and Scherer, W. F.,** The growth curve of Japanese encephalitis virus in the vector mosquito of Japan, *Culex tritaeniorhynchus*, *Jpn. J. Exp. Med.*, 28, 243, 1958.

52. **Gresser, I., Hardy, J. L., Hu, S. M. K., and Scherer, W. F.,** Factors influencing transmission of Japanese B encephalitis virus by a colonized strain of *Culex tritaeniorhynchus* Giles, from infected pigs and chicks to susceptible pigs and birds, *Am. J. Trop. Med. Hyg.*, 7, 365, 1958.
53. **LaMotte, L. C.,** Japanese B encephalitis virus in the organs of infected mosquitoes, *Am. J. Hyg.*, 72, 73, 1960.
54. **Chamberlain, R. W., Sudia, W. D., and Gillett, J. D.,** St. Louis encephalitis virus in mosquitoes, *Am. J. Hyg.*, 70, 221, 1959.
55. **Simkova, A., Danielova, V., and Bardos, V.,** Experimental transmission of the Tahyna virus by *Aedes vexans* mosquitoes, *Acta Virol.*, 4, 341, 1960.
56. **Mussgay, M. and Suarez, O.,** Multiplication of vesicular stomatitis virus in *Aedes aegypti* (L.) mosquitoes, *Virology*, 17, 202, 1962.
57. **Holt, R. L. and Kintner, J. H.,** Location of dengue virus in the body of mosquitoes, *Am. J. Trop. Med.*, 11, 103, 1931.
58. **McLean, D. M.,** Multiplication of viruses in mosquitoes following feeding and injection into the body cavity, *Aust. J. Exp. Biol. Med.*, 33, 53, 1955.
59. **Danielova, V.,** Multiplication dynamics of Tahyna virus in different body parts of *Aedes vexans* mosquito, *Acta Virol.*, 6, 227, 1962.
60. **Thomas, L. A.,** Distribution of the virus of western equine encephalomyelitis in the mosquito vector, *Culex tarsalis*, *Am. J. Hyg.*, 78, 150, 1963.
61. **Bergold, G. H., Suarez, O. M., and Munz, K.,** Multiplication in and transmission by *Aedes aegypti* of vesicular stomatitis virus, *J. Invert. Pathol.*, 11, 406, 1968.
62. **Chernesky, M. A.,** Transmission of California encephalitis virus by mosquitoes, *Can. J. Microbiol.*, 14, 19, 1968.
63. **Danielova, V.,** Penetration of the Tahyna virus to various organs of the *Aedes vexans* mosquito, *Folia Parasit.*, 15, 87, 1968.
64. **Ogunbi, O.,** Ukauwa virus proliferation in mosquitoes, *Can J. Microbiol.*, 14, 125, 1968.
65. **Peers, R. R.,** Bunyamwera virus replication in mosquitoes, *Can. J. Microbiol.*, 18, 741, 1972.
66. **Miles, J. A. R., Pillai, J. S., and Maguire, T.,** Multiplication of Whataroa virus in mosquitoes, *J. Med. Entomol.*, 10, 176, 1973.
67. **Scott, T. W., Hildreth, S. W., and Beaty, B. J.,** The distribution and development of eastern equine encephalitis virus in its enzootic mosquito vector, *Culiseta melanura*, *Am. J. Trop. Med. Hyg.*, 33, 300, 1984.
68. **Doi, R., Shirasaka, A., and Sasa, M.,** The mode of development of Japanese encephalitis virus in the mosquito *Culex tritaeniorhynchus summorosus* as observed by the fluorescent antibody technique, *Jpn. J. Exp. Med.*, 37, 227, 1967.
69. **Doi, R.,** Studies on the mode of development of Japanese encephalitis virus in some groups of mosquitoes by the fluorescent antibody technique, *Jpn. J. Exp. Med.*, 40, 101, 1970.
70. **Gaidamovich, S. Ya., Khutoretskaya, N. V., Lvova, A. I., and Sveshnikova, N. A.,** Immunofluorescent staining study of the salivary glands of mosquitoes infected with Group A arboviruses, *Intervirology*, 1, 193, 1973.
71. **Beaty, B. J. and Thompson, W. H.,** Tropisms of La Crosse virus in *Aedes triseriatus* (Diptera:Culicidae) following infective blood meals, *J. Med. Entomol.*, 14, 499, 1978.
72. **Kuberski, T.,** Fluorescent antibody studies in the development of dengue-2 virus in *Aedes albopictus* (Diptera:Culicidae), *J. Med. Entomol.*, 16, 343, 1979.
73. **McLean, D. M., Gubash, S. M., Grass, P. N., Miller, M. A., Petric, M., and Walters, T. E.,** California encephalitis virus development in mosquitoes as revealed by transmission studies, immunoperoxidase staining, and electron microscopy, *Can. J. Microbiol.*, 21, 453, 1975.
74. **Bergold, G. H. and Weibel, J.,** Demonstration of yellow fever virus with electron microscopy, *Virology*, 17, 554, 1962.
75. **Janzen, H. G., Rhodes, A. J., and Doane, F. W.,** Chikungunya virus in salivary glands of *Aedes aegypti* (L.): an electron microscopic study, *Can. J. Microbiol.*, 16, 581, 1970.
76. **Larsen, J. R. and Ashley, R. F.,** Demonstration of Venezuelan equine encephalomyelitis virus in tissues of *Aedes aegypti*, *Am. J. Trop. Med Hyg.*, 20, 754, 1971.
77. **Whitfield, S. G., Murphy, F. A., and Sudia, W. D.,** Eastern equine encephalomyelitis virus: an electron microscopic study of *Aedes triseriatus* (Say) salivary gland infection, *Virology*, 43, 110, 1971.
78. **Whitfield, S. G., Murphy, F. A., and Sudia, W. D.,** St. Louis encephalitis virus: an ultrastructural study of infection in a mosquito vector, *Virology*, 56, 70, 1973.
79. **Sriurairatna, S. and Bhamarapravati, N.,** Replication of dengue-2 virus in *Aedes albopictus* mosquitoes. An electron microscopic study, *Am. J. Trop. Med. Hyg.*, 26, 1199, 1977.
80. **Takahashi, M. and Suzuki, K.,** Japanese encephalitis virus in mosquito salivary glands, *Am. J. Trop. Med. Hyg.*, 28, 122, 1979.

81. **Scott, T. W. and Burrage, T. G.,** Rapid infection of salivary glands in *Culiseta melanura* with eastern equine encephalitis virus: an electron microscopic study, *Am. J. Trop. Med. Hyg.*, 33, 961, 1984.
82. **Houk, E. J., Kramer, L. D., Hardy, J. L., and Chiles, R. E.,** Western equine encephalomyelitis virus: *in vivo* infection and morphogenesis in mosquito mesenteronal epithelial cells, *Virus Res.*, 2, 123, 1985.
83. **Houk, E. J. and Hardy, J. L.,** Midgut cellular responses to bloodmeal digestion in the mosquito, *Culex tarsalis* Coquillett (Diptera:Culicidae), *Int. J. Insect Morphol. Embryol.*, 11, 109, 1982.
84. **Liu, I. K. M. and Zee, Y. C.,** The pathogenesis of vesicular stomatitis virus, serotype Indiana, in *Aedes aegypti* mosquitoes. I. Intrathoracic injection, *Am. J. Trop. Med. Hyg.*, 25, 177, 1976.
85. **Merrill, M. H. and TenBroeck, C.,** The transmission of equine encephalomyelitis virus by *Aedes aegypti*, *J. Exp. Med.*, 62, 687, 1935.
86. **Chamberlain, R. W., Sikes, R. K., Nelson, D. B., and Sudia, W. D.,** Studies on the North American arthropod-borne encephalitides. VI. Quantitative determinations of virus-vector relationships, *Am. J. Hyg.*, 60, 278, 1954.
87. **Hurlbut, H. S.,** West Nile virus infection in arthropods, *Am. J. Trop. Med. Hyg.*, 5, 76, 1956.
88. **Barnett, H. C.,** The transmission of western equine encephalitis virus by the mosquito *Culex tarsalis* Coq., *Am. J. Trop. Med. Hyg.*, 5, 86, 1956.
89. **Altman, R. M.,** The behavior of Murray Valley encephalitis virus in *Culex tritaeniorhynchus* Giles and *Culex pipiens quinquefasciatus* Say, *Am. J. Trop. Med. Hyg.*, 12, 425, 1963.
90. **Collins, W. E., Harrison, A. J., and Jumper, J. R.,** Infection and transmission thresholds of eastern encephalitis virus to *Aedes aegypti* as determined by a membrane feeding technique, *Mosq. News*. 25, 293, 1965.
91. **Jupp, P. G. and McIntosh, B. M.,** Quantitative experiments on the vector capability of *Culex (Culex) pipiens fatigans* Wiedemann with West Nile and Sindbis viruses, *J. Med. Entomol.*, 7, 353, 1970.
92. **Jupp, P. G. and McIntosh, B. M.,** Quantitation experiments on the vector capability of *Culex (Culex) univittatus* Theobald with West Nile and Sindbis viruses, *J. Med. Entomol.*, 7, 371, 1970.
93. **Hayles, L. B., McLintock, J., and Saunders, J. R.,** Laboratory studies on the transmission of western equine encephalitis virus by Saskatchewan mosquitoes. I. *Culex tarsalis*, *Can, J. Comp. Med.*, 36, 83, 1972.
94. **Jupp, P. G., McIntosh, B. M., and Dickinson, D. B.,** Quantitative experiments on the vector capability of *Culex (Culex) theileri* Theobald with West Nile and Sindbis viruses, *J. Med. Entomol.*, 9, 393, 1972.
95. **Schiefer, B. A. and Smith, J. R.,** Comparative susceptibility of eight mosquito species to Sindbis virus, *Am. J. Trop. Med. Hyg.*, 23, 131, 1974.
96. **Howard, J. J. and Wallis, R. C.,** Infection and transmission of eastern equine encephalomyelitis virus with colonized *Culiseta melanura* (Coquillett), *Am. J. Trop. Med. Hyg.*, 23, 522, 1974.
97. **Scherer, W. F., Cupp, E. W., Lok, J. B., Brenner, R. J., and Ordonez, J. V.,** Intestinal threshold of an enzootic strain of Venezuelan encephalitis virus in *Culex (Melanoconion) taeniopus* mosquitoes and its implications to vector competency and vertebrate amplifying hosts, *Am. J. Trop. Med. Hyg.*, 30, 862, 1981.
98. **Scherer, W. F., Cupp, E. W., Dziem, G. M., Brenner, R. J., and Ordonez, J. V.,** Mesenteronal infection threshold of an epizootic strain of Venezuelan encephalitis virus in *Culex (Melanoconian) taeniopus* mosquitoes and its implication to the apparent disappearance of this virus strain from enzootic habitat in Guatemala, *Am. J. Trop. Med. Hyg.*, 31, 1030, 1982.
99. **Freier, J. E. and Beier, J. C.,** Oral and transovarial transmission of La Crosse virus by *Aedes atropalpus*, *Am. J. Trop. Med Hyg.*, 33, 708, 1984.
100. **Patrican, L. A., DeFoliart, G. R., and Yuill, T. M.,** Oral infection and transmission of La Crosse virus by an enzootic strain of *Aedes triseriatus* feeding on chipmunks with a range of viremia levels, *Am. J. Trop. Med. Hyg.*, 34, 992, 1985.
101. **Helenius, A., Kartenbeck, J., Simons, K., and Fries, E.,** On the entry of Semliki Forest virus into BHK-21 cells, *J. Cell. Biol.*, 84, 404, 1980.
102. **Helenius, A., Marsh, M., and White, J.,** The entry of viruses into animal cells, *Trends Biochem. Sci.*, 5, 104, 1980.
103. **Houk, E. J., Chiles, R. E., and Hardy, J. L.,** Unique midgut basal lamina in the mosquito, *Aedes dorsalis* (Meigen) (Insecta:Diptera), *Int. J. Insect Morphol. Embryol.*, 9, 161, 1980.
104. **Houk, E. J., Hardy, J. L., and Chiles, R. E.,** Permeability of the midgut basal lamina in the mosquito, *Culex tarsalis* Coquillett (Insecta, Diptera), *Acta Trop.*, 38, 163, 1981.
105. **Wright, K. A.,** The anatomy of salivary glands of *Anopheles stephensi* Liston, *Can. J. Zool.*, 47, 579, 1969.
106. **Lanzen, H. G. and Wright, K. A.,** The salivary glands of *Aedes aegypti* (L.): an electron microscopic study, *Can. J. Zool.*, 49, 1343, 1971.
107. **Orr, C. W. M., Hudson, A., and West, A. S.,** The salivary glands of *Aedes aegypti*. Histological-histochemical studies, *Can. J. Zool.*, 39, 265, 1961.

108. **Gubler, D. J. and Rosen, L.,** A simple technique for demonstrating transmission of dengue virus by mosquitoes without use of vertebrate hosts, *Am. J. Trop. Med. Hyg.*, 25, 146, 1976.
109. **Takahashi, M.,** Differential transmission efficiency for Japanese encephalitis virus among colonized strains of *Culex tritaeniorhynchus, Jpn. J. Sanit. Zool.*, 33, 325, 1982.
110. **Chamberlain, R. W., Kissling, R. E., and Sikes, R. K.,** Studies on the North American arthropod-borne encephalitides. VII. Estimation of the amount of eastern equine encephalitis virus inoculated by infective *Aedes aegypti, Am. J. Hyg.*, 60, 286, 1954.
111. **Mellink, J. J.,** Estimation of the amount of Venezuelan equine encephalomyelitis virus transmitted by a single infected *Aedes aegypti (Diptera:Culicidae), J. Med. Entomol.*, 19, 275, 1982.
112. **Hurlbut, H. S.,** Mosquito salivation and virus transmission, *Am. J. Trop. Med. Hyg.*, 15, 989, 1966.
113. **Bates, M. and Roca-Garcia, M.,** Laboratory studies of the *Saimiri-Haemagogus* cycle of jungle yellow fever, *Am. J. Trop. Med.*, 25, 203, 1945.
114. **Huang, C. H.,** Studies on virus factors as causes of inapparent infection in Japanese B encephalitis: virus strains, viremia, stability to heat and infective dosage, *Acta Virol.*, 1, 36, 1957.
115. **LaMotte, L. C., Jr.,** Effect of low environmental temperature upon Japanese B encephalitis virus multiplication in the mosquito, *Mosq. News*. 23, 330, 1963.
116. **Shichijo, A., Mifune, K., Hayashi, K., Wada, Y., Oda, T., and Omori, N.,** Experimental infection of *Culex tritaeniorhynchus summorosus* mosquitoes reared in biotron with Japanese encephalitis virus, *Trop. Med.*, 14, 218, 1972.
117. **Hurlbut, H. S.,** The effect of environmental temperature upon the transmission of St. Louis encephalitis virus by *Culex pipiens quinquefasciatus, J. Med. Entomol.*, 10, 1, 1973.
118. **Jupp, P. G.,** Laboratory studies on the transmission of West Nile virus by *Culex (Culex) univittatus* Theobald; factors influencing the transmission rate, *J. Med. Entomol.*, 11, 455, 1974.
119. **McLean, D. M., Clarke, A. M., Coleman, J. C., Montalbelli, C. A., Skidmore, A. G., Walters, T. E., and Wise, R.,** Vector capability of *Aedes aegypti* mosquitoes for California encephalitis and dengue viruses at various temperatures, *Can. J. Microbiol.*, 20, 225, 1974.
120. **McLean, D. M., Miller, M. A., and Grass, P. N.,** Dengue virus transmission by mosquitoes incubated at low temperatures, *Mosq. News*, 35, 322, 1975.
121. **McLean, D. M., Grass, P. N., Miller, M. A., and Wong, K. S. K.,** Arbovirus growth in *Aedes aegypti* mosquitoes throughout their viable temperature range, *Arch. Virol.*, 49, 49, 1975.
122. **Takahashi, M.,** The effects of environmental and physiological conditions of *Culex tritaeniorhynchus* on the pattern of transmission of Japanese encephalitis virus, *J. Med. Entomol.*, 13, 275, 1976.
123. **McLean, D. M., Grass, P. N., and Judd, B. D.,** California encephalitis virus transmission by Arctic and domestic mosquitoes, *Arch. Virol.*, 55, 39, 1977.
124. **McLean, D. M., Grass, P. N., Judd, B. D., Stolz, K. J., and Wong, K. K.,** Transmission of Northway and St. Louis encephalitis virus by Arctic mosquitoes, *Arch. Virol.*, 57, 315, 1978.
125. **Kramer, L. D., Hardy, J. L., and Presser, S. B.,** Effect of temperature of extrinsic incubation on the vector competence of *Culex tarsalis* for western equine encephalomyelitis virus, *Am. J. Trop. Med. Hyg.*, 32, 1130, 1983.
126. **Turell, M. J., Rossi, C. A., and Bailey, C. L.,** Effect of extrinsic incubation temperature on the ability of *Aedes taeniorhynchus* and *Culex pipiens* to transmit Rift Valley fever virus, *Am. J. Trop. Med. Hyg.*, 34, 1211, 1985.
127. **Reeves, W. C. and Hammon, W. McD.,** Epidemiology of the Arthropod-Borne Viral Encephalitides in Kern County, California 1943-1952, University of California Publications in Public Health, Vol.4, University of California Press, Berkeley, 1962, 1.
128. **Hess, A. D., Cherubin, C. E., and LaMotte, L. C.,** Relation of temperature to activity of western and St. Louis encephalitis viruses, *Am. J. Trop. Med. Hyg.*, 12, 657, 1963.
129. **Mogi, M.,** Relationship between number of human Japanese encephalitis cases and summer meteorological conditions in Nagasaki, Japan, *Am. J. Trop. Med. Hyg.*, 32, 170, 1983.
130. **Gould, D. J., Barnett, H. C., and Suyemoto, W.,** Transmission of Japanese encephalitis virus by *Culex gelidus, Trans. R. Soc. Trop. Med. Hyg.*, 56, 429, 1962.
131. **Hardy, J. L., Reeves, W. C., Bruen, J. P., and Presser, S. B.,** Vector competence of *Culex tarsalis* and other mosquito species for western equine encephalomyelitis virus, in *Arctic and Tropical Arboviruses*, Kurstak, E., Ed., Academic Press, New York, 1979, 157.
132. **Houk, E. J., Kramer, L. D., Hardy, J.L.,and Presser, S. B.,** An intraspecific mosquito model for the mesenteronal infection barrier to western equine encephalomyelitis virus *(Culex tarsalis* and *Culex pipiens)*, *Am. J. Trop. Med. Hyg.*, 35, 632, 1986.
133. **Hardy, J. L., Presser, S. B., Meyer, R. P., Reisen, W. K., Kramer, L. D., and Vorndam, A. V.,** Comparison of a 1984 Los Angeles strain of SLE virus with earlier California strains of SLE virus: mouse virulence, chicken viremogenic, RNA oligonucleotide and vector competence characteristics, *Proc. Calif. Mosq.Vector Control Assoc.*, 53, 10, 1985.
134. **Hardy, J. L.,** unpublished data, 1985.

135. **Karabatsos, N., Ed.,** *International Catalogue of Arboviruses*, 3rd ed., American Society of Tropical Medicine and Hygiene, San Antonio, Tex., 1985, 1147.
136. **Dudley, S. F.,** Can yellow fever spread into Asia? An essay with ecology of mosquito-borne disease, *J. Trop. Med. Hyg.*, 37, 273, 1934.
137. **Tesh, R. B., Gubler, D. J., and Rosen, L.,** Variation among geographic strains of *Aedes albopictus* in susceptibility to infection with chikungunya virus, *Am. J. Trop. Med. Hyg.*, 25, 326, 1976.
138. **Gubler, D. J.,** Transmission of Ross River virus by *Aedes polynesiensis* and *Aedes aegypti*, *Am. J. Trop. Med. Hyg.*, 30, 1303, 1981.
139. **Green, D. W., Rowley, W. A., Wong, Y. W., Brinker, J. P., Dorsey, D. C., and Hausler, W. J., Jr.,** The significance of western equine encephalomyelitis viral infections in *Aedes trivittatus* (Diptera:Culicidae) in Iowa. I. Variation in susceptibility of *Aedes trivittatus* to experimental infection with three strains of western equine encephalomyelitis virus, *Am. J. Trop. Med. Hyg.*, 29, 118, 1980.
140. **Hardy, J. L., Reeves, W. C., and Sjogren, R. D.,** Variations in the susceptibility of field and laboratory populations of *Culex tarsalis* to experimental infection with western equine encephalomyelitis virus, *Am. J. Epidemiol.*, 103, 498, 1976.
141. **Hardy, J. L., Apperson, G., Asman, S. M., and Reeves, W. C.,** Selection of a strain of *Culex tarsalis* highly resistant to infection following ingestion of western equine encephalomyelitis virus, *Am. J. Trop. Med. Hyg.*, 27, 313, 1978.
142. **Grimstad, P. R., Craig, G. B., Jr., Ross, Q. E., and Yuill, T. M.,** *Aedes triseriatus* and La Crosse virus: geographic variation in vector susceptibility and ability to transmit, *Am. J. Trop. Med. Hyg.*, 26, 990, 1977.
143. **Gubler, D. J., Tan, N. R., Saipan, H., and Sulianti Saroso, J.,** Variation in susceptibility to oral infection with dengue viruses among geographic strains of *Aedes aegypti*, *Am. J. Trop. Med. Hyg.*, 28, 1045, 1979.
144. **Gubler, D. J. and Rosen, L.,** Variation among geographic strains of *Aedes albopictus* in susceptibility to infection with dengue viruses, *Am. J. Trop. Med. Hyg.*, 25, 318, 1976.
145. **Takahashi, M.,** Variation in susceptibility among colony strains of *Culex tritaeniorhynchus* to Japanese encephalitis virus infection, *Jpn. J. Med. Sci. Biol.*, 33, 321, 1980.
146. **Kay, B. H., Fanning, I. D., and Carley, J. G.,** The vector competence of Australian *Culex annulirostris* with Murray Valley encephalitis and Kunjin viruses, *Aust. J. Biol. Med. Sci.*, 62, 641, 1984.
147. **Kay, B. H., Fanning, I. D., and Carley, J. G.,** Vector competence of *Culex pipiens quinquefasciatus* for Murray Valley encephalitis, Kunjin and Ross River viruses from Australia, *Am. J. Trop. Med. Hyg.*, 31, 844, 1982.
148. **Mitchell, C. J., Monath, T. P., and Cropp, C. B.,** Experimental transmission of Rocio virus by mosquitoes, *Am. J. Trop. Med. Hyg.*, 30, 465, 1981.
149. **Ahmed, T., Hayes, C. G., and Baqar, S.,** Comparison of vector competence for West Nile virus of colonized populations of *Culex tritaeniorhynchus* from southern Asia and the Far East, *Southeast Asian J. Trop. Med. Public Health*, 10, 498, 1979.
150. **Hayes, C. G., Baker, R. H., Baqar, S., and Ahmed, T.,** Genetic variation for West Nile virus susceptibility in *Culex tritaeniorhynchus*, *Am. J. Trop. Med. Hyg.*, 33, 715, 1984.
151. **Aitken, T. H. G., Downs, W. G., and Shope, R. E.,** *Aedes aegypti* strain fitness for yellow fever virus transmission, *Am. J. Trop. Med. Hyg.*, 26, 985, 1977.
152. **Beaty, B. J. and Aitken, T. H. G.,** In vitro transmission of yellow fever virus by geographic strains of *Aedes aegypti*, *Mosq. News*, 39, 232, 1979.
153. **Tabachnick, W. J., Aitken, T. H. G., Beaty, B. J., Miller, B. R., Powell, J. R., and Wallis, G. P.,** Genetic approaches to the study of vector competency of *Aedes aegypti*, in *Recent Developments in the Genetics of Insect Disease Vectors*, Steiner, W., Tabachnick, W., Rai, K., and Narang, S., Eds., Stipes Publishing, Champaign, Ill., 1982, 413.
154. **Tabachnick, W. J., Wallis, G. P., Aitken, T. H.G., Miller, B. R., Amato, G. D., Lorenz, L., Powell, J. R., and Beaty, B. J.,** Oral infection of *Aedes aegypti* with yellow fever virus: geographic variation and genetic considerations, *Am. J. Trop. Med. Hyg.*, 34, 1219, 1985.
155. **Wallis, G. P., Aitken, T. H. G., Beaty, B. J., Lorenz, L., Amato, G. D., and Tabachnick, W. J.,** Selection for susceptibility and refractoriness of *Aedes aegypti* to oral infection with yellow fever virus, *Am. J. Trop. Med. Hyg.*, 34, 1225, 1985.
156. **Craig, G. B., Jr.,** Application of genetic technology of mosquito rearing, *Bull. WHO*, 31, 469, 1964.
157. **Meegan, J. M., Khalil, G. M., Hoogstraal, H., and Adhan, F. K.,** Experimental transmission and field isolation studies implicating *Culex pipiens* as a vector of Rift Valley fever virus in Egypt, *Am. J. Trop. Med. Hyg.*, 29, 1405, 1980.
158. **Gargan, T. P., II, Bailey, C. L., Higbee, G. A., Gad, A., and Said, S. E.,** The effect of laboratory colonization on the vector-pathogen interactions of Egyptian *Culex pipiens* and Rift Valley fever virus, *Am. J. Trop. Med. Hyg.*, 32, 1154, 1983.

159. **Lorenz, L., Beaty, B. J., Aitken, T. H. G., Wallis, G. P., and Tabachnick, W. J.,** The effect of colonization upon *Aedes aegypti* susceptibility to oral infection with yellow fever virus, *Am. J. Trop. Med. Hyg.*, 33, 690, 1984.
160. **Jones, R. H. and Foster, N. M.,** Oral infection of *Culicoides variipennis* with bluetongue virus: development of susceptible and resistant lines from a colony population, *J. Med. Entomol.*, 11. 316, 1974.
161. **Briese, D. T.,** Resistance of insect species to microbial pathogens, in *Pathogenesis of Invertebrate Microbial Diseases*, Davidson, E. W., Ed., Allanheld-Osman, Montclair, N. J., 1981, chap. 18.
162. **Saliba, E. K., DeFoliart, G. R., Yuill, T. M., and Hanson, R. P.,** Laboratory transmission of Wisconsin isolates of a Cache Valley-like virus by mosquitoes, *J. Med. Entomol.*, 10, 470, 1973.
163. **Kramer, L. D. and Scherer, W. F.,** Vector competence of mosquitoes as a marker to distinguish Central American and Mexican epizootic from enzootic strains of Venezuelan encephalitis virus, *Am. J. Trop. Med. Hyg.*, 25, 336,1976.
164. **Weaver, S. C., Scherer, W. F., Cupp, E. W., and Castello, D. A.,** Barriers to dissemination of Venezuelan encephalitis viruses in the Middle American enzootic vector mosquito, *Culex (Melanoconian) taeniopus*, *Am. J. Trop. Med. Hyg.*, 33, 953, 1984.
165. **Deubel, V., Camicas, J. L., Pandare, D., Robert, V., Digoutte, J. P., and Germain, M.,** Developpement de souches sauvages et vaccinales du virus de la fievre jaune dans les cellules de *Aedes aegypti* et transmission au souriceau, *Ann. Virol. (Inst. Pasteur)*, 132, 41, 1981.
166. **Mitchell, C. J., Gubler, D. J., and Monath, T. P.,** Variation in infectivity of Saint Louis encephalitis viral strains for *Culex pipiens quinquefasciatus* (Diptera:Culicidae), *J. Med. Entomol.*, 5, 526, 1983.
167. **Rosen, L., Roseboom, L. E., Gubler, D. J., Lien, J. C., and Chaniotis, B. N.,** Comparative susceptibility of mosquito species and strains to oral and parenteral infection with dengue and Japanese encephalitis viruses, *Am. J. Trop. Med. Hyg.*, 34, 603, 1985.
168. **Takahashi, M., Yabe, S., and Okada, T.,** Effects of various passages on some properties of an attenuated strain of Japanese encephalitis virus with special regard to mosquito infectivity, *Jpn. J. Med. Sci. Biol.*, 22, 163, 1969.
169. **Chen, B. Q. and Beaty, B. J.,** Japanese encephalitis vaccine (2-8 strain) and parent (SA 14 strain) viruses in *Culex tritaeniorhynchus* mosquitoes, *Am. J. Trop. Med. Hyg.*, 31, 403, 1982.
170. **Miller, B. R., Beaty, B. J., Aitken, T. H. G., Eckels, K. H., and Russell, P. K.,** Dengue-2 vaccine: oral infection, transmission, and lack of evidence for reversion in the mosquito, *Aedes aegypti*, *Am. J. Trop. Med. Hyg.*, 31, 1232, 1982.
171. **Whitman, L.,** Failure of *Aedes aegypti* to transmit yellow fever cultured virus (17D), *Am. J. Trop. Med.*, 19, 19, 1939.
172. **Pattyn, S. R. and De Vleesschauwer, L.,** The multiplication of Middelburg s and l plaque viruses in *Aedes aegypti* mosquitoes, *Acta Virol.*, 12, 347, 1968.
173. **Gaidomovich, S. Ya., Tsilinsky, Y. Y., Lvova, A. I., and Khutoretskaya, N. V.,** *Aedes aegypti* mosquitoes as an experimental model for studies on the ecology and genetics of Venezuelan equine encephalomyelitis virus, *Acta Virol.*, 15, 301, 1971.
174. **Peleg, J.,** *In vivo* behavior of a Sindbis virus mutant isolated from persistently infected *Aedes aegypti* cell cultures, *Ann. N. Y. Acad. Sci.*, 266, 204, 1975.
175. **Miller, B. R.,** A variant of La Crosse virus attenuated for *Aedes triseriatus* mosquitoes, *Am. J. Trop. Med. Hyg.*, 32, 1422, 1983.
176. **Beaty, B. J., Miller, B. R., Shope, R. E., Rozhon, E. J., and Bishop, D. H. L.,** Molecular basis of bunyavirus per os infection of mosquitoes: role of the middle-sized RNA segment, *Proc. Natl. Acad. Sci. U.S.A.*, 79, 1295, 1982.
177. **Gentsch, J. R. and Bishop, D. H. L.,** M viral RNA segment of bunyaviruses codes for two glycoproteins, G1 and G2, *J.Virol.*, 30, 767, 1979.
178. **Marshall, I. P., Scrivani, R. P., and Reeves, W. C.,** Variation in the size of plaques produced in tissue culture by strains of western equine encephalitis virus, *Am. J. Hyg.*, 76, 216, 1962.
179. **Baqar, S., Hayes, C. G., and Ahmed, T.,** The effect of larval rearing conditions and adult age on the susceptibility of *Culex tritaeniorhynchus* to infection with West Nile virus, *Mosq. News*, 40, 165, 1980.
180. **Grimstad, P. R. and Haramis, L. D.,** *Aedes triseriatus* (Diptera: Culicidae) and La Crosse virus. III. Enhanced oral transmission by nutrition-deprived mosquitoes, *J. Med. Entomol.*, 21, 249, 1984.
181. **Stollar, V.,** Togaviruses in cultured arthropod cells, in *The Togaviruses*, Schlesinger, R. W., Ed., Academic Press, New York, 1980, 585.
182. **Brown, D. T. and Condreay, L. D.,** Replication of alphaviruses in mosquito cells, in *The Togaviridae and Flaviviridae*, Schlesinger, S., Ed., Plenum Press, New York, 1986, 171.
183. **Sabin, A. B.,** Research on dengue during World War II, *Am. J. Trop. Med. Hyg.*, 1, 30, 1952.
184. **Rozeboom, L. and Kassira, E. J.,** Dual infection of mosquitoes with strains of West Nile virus, *J. Med. Entomol.*, 6, 407, 1969,
185. **Lam, K. S. K. and Marshall, I. D.,** Dual infections of *Aedes aegypti* with arboviruses. I. Arboviruses that have no apparent cytopathic effect in the mosquito, *Am. J. Trop. Med. Hyg.*, 17, 625, 1968.

186. **Chamberlain, R. W. and Sudia, W. O.**, Dual infections of eastern and western equine encephalitis virus in *Culex tarsalis*, *J. Infect. Dis.*, 101, 233, 1957.
187. **Davey, M. W., Mahon, R. J., and Gibbs, A. J.**, Togavirus interference in *Culex annulirostris* mosquitoes, *J. Gen. Virol.*, 42, 641, 1979.
188. **Beaty, B. J., Bishop, D. H. L., Gay, M., and Fuller, F.**, Interference between bunyaviruses in *Aedes triseriatus* mosquitoes, *Virology*, 127, 83, 1983.
189. **Turell, M. J., Hardy, J. L., and Reeves, W. C.**, Sensitivity to carbon dioxide in mosquitoes infected with California serogroup arboviruses, *Am. J. Trop. Med. Hyg.*, 31, 389, 1982.
190. **Rozeboom, L. E., Behin, R., and Kassira, E. N.**, Dual infections of *Aedes aegypti* L. with *Plasmodium gallinaceum brumpt* and West Nile virus, *J. Parasitol.*, 52, 579, 1966.
191. **Barnett, H. C.**, Experimental studies of concurrent infection of canaries and of the mosquito *Culex tarsalis* with *Plasmodium relictum* and western encephalitis virus, *Am. J. Trop. Med. Hyg.*, 5, 99, 1956.
192. **Miller, B. R. and DeFoliart, G. R.**, Infection rates of Ascocystis-infected *Aedes triseriatus* following ingestion of La Crosse virus by larvae, *Am. J. Trop. Med. Hyg.*, 28, 1064, 1979.
193. **Turell, M. J., Rossignol, P. A., Spielman, A., Rossi, C. A., and Bailey, C. L.**, Enhanced arboviral transmission by mosquitoes that concurrently ingested microfilariae, *Science*, 225, 1039, 1984.
194. **Mellor, P. S. and Boorman, J.**, Multiplication of bluetongue virus in *Culicoides nubeculosus* (Meigen) simultaneously infected with the virus and the microfilarie of Onchocera cervicalis (Railliet & Henry), *Ann. Trop. Med. Parasitol.*, 74, 463, 1980.
195. **Mims, C. A., Day, M. F., and Marshall, I. D.**, Cytopathic effect of Semliki Forest virus in the mosquito *Aedes aegypti*, *Am. J. Trop. Med. Hyg.*, 15, 775, 1966.
196. **Lam, K. S. K. and Marshall, I. D.**, Dual infections in *Aedes aegypti* with arboviruses. II. Salivary-gland damage by Semliki Forest virus in relation to dual infections, *Am. J. Trop. Med. Hyg.*, 17, 637, 1968.
197. **Grimstad, P. R., Ross, Q. E., and Craig, G. B., Jr.**, *Aedes triseriatus* (Diptera:Culicidae) and La Crosse virus. II. Modification of mosquito feeding behavior by virus infection, *J. Med. Entomol.*, 17, 1, 1980.
198. **Turell, M. J., Gargan, T. P., II, and Bailey, C. L.**, *Culex pipiens* (Diptera:Culicidae) morbidity and mortality associated with Rift Valley fever virus infection, *J. Med. Entomol.*, 22, 332, 1985.
199. **Tinsley, T. W.**, Factors affecting infection of insect gut tissue, in *Invertebrate Immunity*, Maramorosch, K. and Shope, R. E., Eds., Academic Press, New York, 1975, 55.
200. **Houk, E. J., Obie, F., and Hardy, J. L.**, Peritrophic membrane formation in the mosquito, *Culex tarsalis* (Insecta, Diptera), *Acta Trop.*, 36, 39, 1979.
201. **Pattyn, S. R. and de Vleesschauwer, L.**, The enhancing affect of diethylamino-ethyl dextran on the infectivity of arboviruses for *Aedes aegypti*, *Arch. Ges. Virusforsch.*, 31, 175, 1970.
202. **Arcus, Y. M., Houk, E. J., and Hardy, J. L.**, Comparative *in vitro* binding of an arbovirus to midgut microvillar membranes from susceptible and refractory *Culex* mosquitoes, *Fed. Proc.*, 42, 2141, 1983.
203. **Houk, E. J.**, personal communication, 1986.
204. **Collins, W. E.**, Studies on the transmission of Semliki Forest virus by anopheline mosquitoes, *Am. J. Hyg.*, 77, 109, 1963.
205. **Collins, W. E., Harrison, A. J., and Jumper, J. R.**, Infection and transmission studies with eastern encephalitis virus and *Anopheles albimanus* and *A. quadrimaculatus*, *Mosq. News*, 25, 296, 1965.
206. **Mangiafico, J. A.**, Chikungunya virus infection and transmission in five species of mosquito, *Am. J. Trop. Med. Hyg.*, 20, 642, 1971.
207. **Pantuwatana, S., Thompson, W. H., Watts, D. M., and Hanson, R. P.**, Experimental infection of chipmunks and squirrels with La Crosse and trivittatus viruses and biological transmission of La Crosse virus by *Aedes triseriatus*, *Am. J. Trop. Med. Hyg.*, 21, 476, 1972.
208. **Watts, D. M., Grimstad, P. R., DeFoliart, G. R., Yuill, T. M., and Hanson, R. P.**, Laboratory transmission of La Crosse encephalitis virus by several species of mosquitoes, *J. Med. Entomol.*, 6, 583, 1973.
209. **McIntosh, B. M., Jupp, P. G., Anderson, D., and Dickinson, D. B.**, Rift Valley fever. II. Attempts to transmit virus with seven species of mosquito, *J. S. Afr. Vet. Assoc.*, 44, 57, 1973.
210. **Simasathien, P. and Olson, L. C.**, Factors influencing the vector potential of *Aedes aegypti* and *Culex quinquefasciatus* for Wesselsbron virus, *J. Med. Entomol.*, 10, 587, 1973.
211. **Watts, D. M., Grimstad, P. R., DeFoliart, G. R., and Yuill, T. M.**, *Aedes hendersoni:* failure of laboratory-infected mosquitoes to transmit La Crosse virus (California encephalitis group), *J. Med. Entomol.*, 12, 451, 1975.
212. **Watts, D. M., DeFoliart, G. R., and Yuill, T. M.**, Experimental transmission of trivittatus virus (California virus group) by *Aedes trivittatus*, *Am. J. Trop. Med. Hyg.*, 25, 173, 1976.
213. **Kay, B. H., Carley, J. G., Fanning, I. D., and Filippich, C.**, Quantitative studies of the vector competence of *Aedes aegypti*, *Culex annulirostris* and other mosquitoes (Diptera:Culicidae) with Murray Valley encephalitis and other Queensland arboviruses, *J. Med. Entomol.*, 16, 59, 1979.

214. **Kramer, L. D., Hardy, J. L., Presser, S. B., and Houk, E. J.,** Dissemination barriers for western equine encephalomyelitis virus in *Culex tarsalis* infected after ingestion of low viral doses, *Am. J. Trop. Med. Hyg.*, 30, 190, 1981.
215. **Scott, T. W., Cord, C. S., Francy, D. B., Mitchell, C. J., and McLean, R. G.,** Turlock virus infection and transmission by *Culex* mosquitoes (Diptera:Culicidae), *J. Med. Entomol.*, 20, 682, 1983.
216. **Turell, M. J., Gargan, T. P., II, and Bailey, C. L.,** Replication and dissemination of Rift Valley fever virus in *Culex pipiens*, *Am. J. Trop. Med. Hyg.*, 33, 176, 1984.
217. **Sudia, W. D.,** The multiplication of St. Louis encephalitis virus in two species of mosquitoes: *Culex quinquefasciatus* Say and *Culex pipiens* Linnaeus, *Am. J. Hyg.*, 70, 237, 1959.
218. **Beaty, B. J., Holterman, M., Tabachnick, W., Shope, R. E., Rozhon, E. J., and Bishop, D. H. L.,** Molecular basis of bunyavirus transmission by mosquitoes: role of the middle-sized RNA segment, *Science*, 211, 1433, 1981.
219. **Grimstad, P. R., Paulson, S. L., and Craig, G. B., Jr.,** Vector competence of *Aedes hendersoni* (Diptera:Culicidae) for La Crosse virus and evidence of a salivary-gland escape barrier, *J. Med. Entomol.*, 22, 447, 1985.
220. **Jupp, P. G.,** *Culex theileri* and Sindbis virus; salivary glands infection in relation to transmission, *J. Am. Mosq. Control Assoc.*, 1, 374, 1985.
221. **McIntosh, B. M. and Jupp, P. G.,** Attempts to transmit chikungunya virus with six species of mosquito, *J. Med. Entomol.*, 7, 615, 1970.

Chapter 5

HORIZONTAL AND VERTICAL TRANSMISSION OF VIRUSES BY INSECT AND TICK VECTORS

Michael J. Turell

TABLE OF CONTENTS

I. INTRODUCTION

When an arthropod feeds on a viremic host, there is potential for the arthropod to transmit the virus either mechanically or biologically. In the former case, the vector need not be susceptible to viral infection; while in the latter, virus must replicate in the vector before the virus can be transmitted. Once it has replicated in the vector, the virus may be transmitted either horizontally to a vertebrate by bite, or vertically to the vector's progeny by transovarial transmission. Transovarially infected individuals then have the potential to retransmit the virus vertically to their progeny, or to transmit it horizontally either by bite to a vertebrate or venereally during mating. This chapter discusses various factors that have been shown to affect the efficiency with which arboviruses are transmitted by their potential vectors.

II. TRANSMISSION CYCLES

Traditionally, arboviruses are considered to be viruses that are capable of replication in both vertebrate and arthropod hosts and are transmitted by arthropods to vertebrates. Transmission cycles range from the relatively simple, involving only a single vector and vertebrate, to the highly complex, involving numerous vertebrate and vector species. As detailed natural histories are given for the various viruses in their respective chapters, only a few representative cycles are discussed here to illustrate general principles. Also, in this section, we consider only the horizontal or amplification cycle. The mechanism(s) by which arboviruses persist from year to year are discussed in the section on the survival of virus during periods of vector inactivity.

Examples illustrating various degrees of complexity of transmission cycles are shown in Figure 1. The transmission cycle for urban yellow fever (YF) virus is relatively straightforward, involving a single vertebrate host, man, and a single vector, *Aedes aegypti*. In a slightly more complicated cycle, as typified by Colorado tick fever (CTF), the virus is maintained enzootically in a population of vertebrates (in this case rodents, such as ground squirrels and chipmunks) by the tick *Dermacentor andersoni*.[1] However, CTF virus is occasionally transmitted to other hosts (e.g., man) when they enter an area where infected ticks are found. The natural history of eastern equine encephalitis (EEE) virus represents an even more complicated cycle. In its basic enzootic amplification cycle, EEE virus is maintained in wild passerine bird populations by *Culiseta melanura*[2] However, depending on environmental conditions and density of specific vector and vertebrate species, this virus may be picked up by other vector species, such as *Ae. sollicitans,* and then transmitted to mammalian hosts, such as horses and man.[3] In addition to the traditional vector-vertebrate cycles, many arboviruses, such as La Crosse (LAC) virus, are now recognized to be transmitted in a vector-vector cycle by either transovarial or venereal transmission.

The ability of an arthropod to serve as a vector for a certain arbovirus is dependent on many factors.[4] These include vertebrate host characteristics (e.g., susceptibility to infection, duration and titer of viremia produced, population size, attractiveness to vectors, behavior, etc.), vector characteristics (e.g., relative ability to transmit virus, longevity, feeding behavior, population density, etc.), and environmental factors (e.g., temperature, humidity, rainfall, etc.). An understanding of the natural history of an arbovirus, including the factors that affect the relative efficiency of this system, is a prerequisite to the development of a control program. Thus, it is necessary to determine both the potential arthropod vectors and the vertebrate hosts involved in the transmission cycle.

While many of the principles of arbovirus transmission apply equally to tick or insect transmission cycles, there are some inherent differences. One of the most important of these is that in general, while ticks obtain blood meals in all developmental stages (i.e., as larvae, nymphs, and adults), insect vectors take blood meals only in the adult stage. Because most

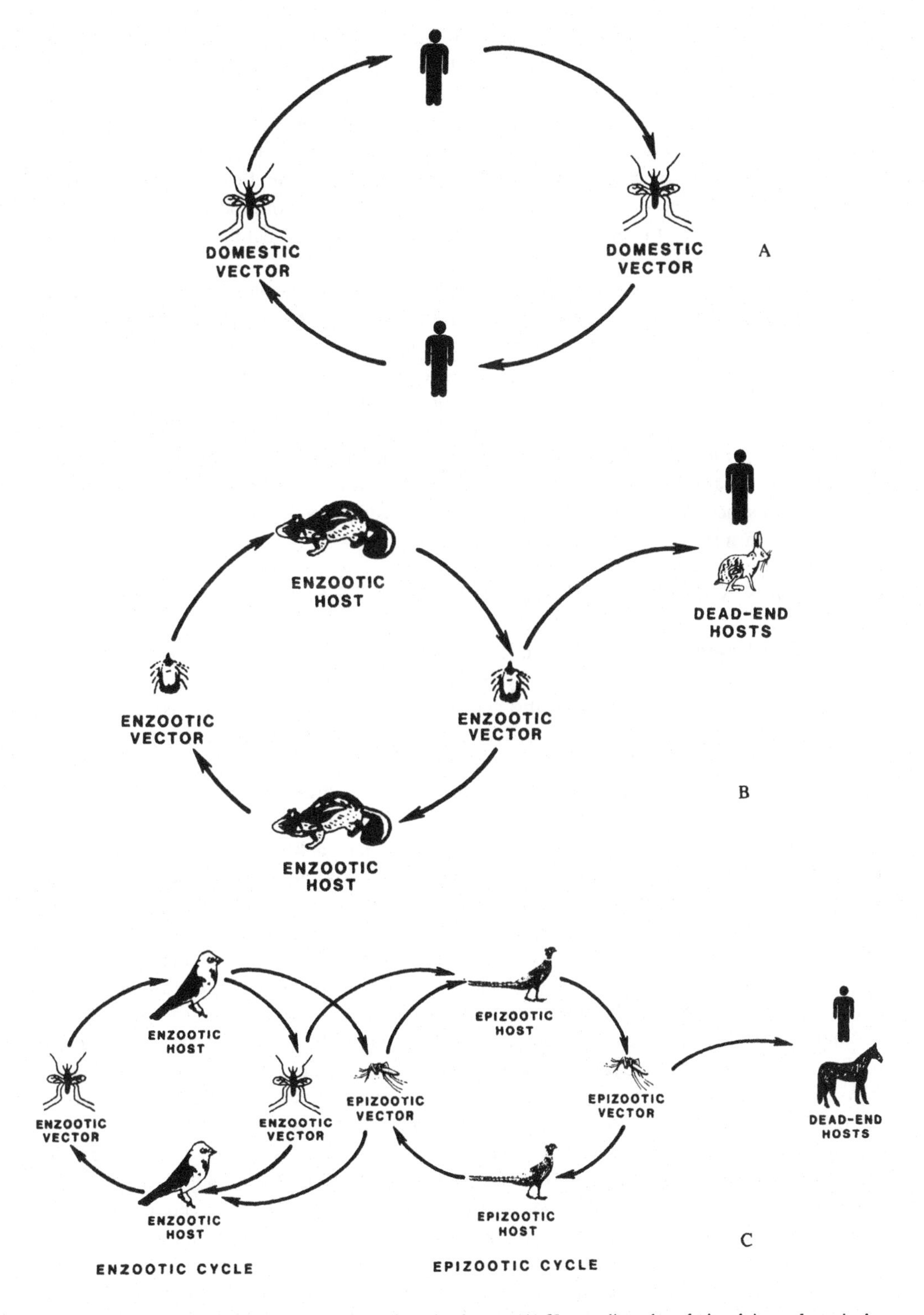

FIGURE 1. Generalized transmission cycles for arboviruses. (A) Uncomplicated cycle involving only a single vertebrate host and a single primary vector species. (B) Intermediate cycle where virus is maintained in an enzootic cycle and man or other vertebrates are involved only tangentially, usually as dead-end hosts. (C) Complex cycle involving both en- and epizootic subcycles with different vectors and vertebrate hosts in each subcycle, with man and domestic animals becoming involved tangentially from the epizootic portion of the cycle.

insect vectors have a relatively short adult life stage, one of the critical factors for virus transmission by an insect vector is whether it lives long enough to transmit the virus. As a corollary to the need for rapid transmission, insects that feed on a wide variety of vertebrates, including ones that do not produce a viremia, are less efficient vectors than ones that feed preferentially on a vertebrate that does produce a viremia. This is because the vector must feed at least twice to transmit a virus, once to become infected and once to transmit it. If either the first or second feeding was on a vertebrate that could not produce a viremia, then the vector would have to survive to at least a third blood meal before it would have the opportunity to transmit the virus.

In contrast, because ticks may take up to several years to develop to adults and many arboviruses can be transmitted transstadially (i.e., from larva to nymph or from nymph to adult) by their tick vectors, rapid ability to transmit a virus is not as important in ticks as it is in mosquitoes. In fact, because many ticks feed on both small and large vertebrates at different stages in their life cycle, the critical component in a transmission system could be a vertebrate that is essential for tick survival, but that is not susceptible to infection with that particular virus. In addition, avian and large mammalian hosts, even if they are not susceptible to viral infection themselves, may transport infected ticks over great distances. Thus, the presence or absence of such a vertebrate would determine the range of the tick species, and in turn, the distribution of the disease.

III. INCRIMINATION OF AN ARBOVIRUS TRANSMISSION CYCLE

Reeves[5] and Barnett[6] reviewed the criteria needed to implicate both the vector and vertebrates in a transmission cycle. Briefly, to incriminate a particular species, or group of species, there must be evidence that this species is involved in nature. This is provided by the repeated isolation of the virus from field-collected specimens. The species must also be a competent vector of that virus (i.e., individuals must become infected and be able to transmit virus by bite after ingestion of a viremic blood meal). This is normally confirmed for the suspected species by laboratory studies. Finally, the suspected vector must be associated with the suspected vertebrate hosts in nature. This is accomplished by identifying blood meals from field-collected, engorged specimens. Vertebrate hosts are incriminated similarly. They must be infected in nature. This is demonstrated by either the recovery of virus from infected individuals, or by detection of virus-specific antibodies in these hosts. When uninfected individuals are exposed to virus, they must produce a viremia of sufficient titer to infect potential vectors, and as mentioned above, there must be a close relationship between the vertebrate and the suspected vector.

Numerous studies have been conducted in an attempt to determine the vectors involved in the natural transmission cycle. While isolations of virus from field-collected individuals may give an indication that a particular species is involved in nature, sometimes this information can be misleading, as virus has been recovered from species not involved in the natural transmission cycle. For example, arthropods that have recently fed on a viremic host may contain virus, even if they are not susceptible to infection. Also, a leg may break off one specimen and be placed in a pool of a different species during the handling of field-collected specimens. A single mosquito leg may contain $>10^{4.7}$ plaque forming units (pfu) of Rift Valley fever (RVF) virus.[7] When large numbers of specimens are handled, there is always the possibility that a few of the specimens may have been misclassified or added to the wrong pool. In addition to these problems that may allow for recovery of virus from a species that is not even susceptible to viral infection, some species are susceptible to viral infection, but are incapable of transmitting virus. Thus, laboratory studies are needed to confirm the ability of a potential vector to transmit an arbovirus.

IV. MECHANICAL TRANSMISSION

Arthropods have long been known to be able to serve as "flying contaminated needles" that are capable of transmitting a virus. While these arthropods do not fit the classical model of an arboviral vector (i.e., the virus does not need to replicate in this "vector" in order for mechanical transmission to occur), mechanical transmission does play a major role in the epidemiology of certain viruses. For example, even though myxoma virus has not been shown to replicate in any arthropod, it is maintained by mechanical transmission by various species of mosquitoes.[8] Similarly, mechanical transmission has been demonstrated for the virus causing rinderpest by the tsetse fly, *Glossina morsitans*,[9] and for the virus of swine pox by the hog louse, *Haematopinus suis*.[10] The potential for mechanical transmission, in addition to biological transmission, has also been shown for numerous arboviruses. These include the demonstration in the laboratory of mechanical transmission of YF virus by *Stomoxys calcitrans*;[11] vesicular stomatitis (VS) virus (New Jersey) by numerous species of *Tabanus*, *Chrysops*, and four species of mosquitoes;[12] western equine encephalitis (WEE) virus by *Culex tarsalis*;[13] EEE virus by *S. calcitrans* and *Ae. triseriatus*;[14] Ross River virus by *Ae. vigilax*;[15] and RVF virus by *S. calcitrans*, *G. morsitans*, *Lutzomyia longipalpis*, *Culicoides variipennis*, and three species of mosquitoes.[16]

The importance of mechanical transmission in the epidemiology of any given arbovirus depends on several factors. Mechanical transmission would be more efficient if the virus induced a high-titered viremia and was stable under various environmental conditions. This stability, in turn, is influenced by the temperature and humidity during the time interval between the two feedings. The length of this interval depends on the arthropod's feeding behavior (i.e., frequency of partial blood meals and multiple feedings). Also, the surface area capable of carrying virus, protection of the virus from exposure to the environment, and the likelihood of virus being removed from the mouthparts are influenced by both the vector's type of mouthparts and their cleaning behavior. Finally, the potential for mechanical transmission is enhanced if the vertebrate hosts are highly susceptible to viral infection. In general, because viral replication is not necessary for mechanical transmission, there is no incubation period required before transmission can occur, and there is relatively little vector specificity. While arboviruses may be maintained primarily by biological transmission, mechanical transmission may be important in explaining explosive outbreaks of these diseases if there are large numbers of potential vectors and if the factors described above are favorable for mechanical transmission.

V. BIOLOGICAL TRANSMISSION

In general, before an uninfected arthropod can transmit an arbovirus biologically, the following steps must occur: (1) ingestion of a viremic blood meal, (2) infection of the cells of the midgut, (3) escape of virus into the hemocele, (4) infection of the salivary glands, (5) secretion of virus into the saliva, and (6) refeeding on a susceptible vertebrate. As an alternative to ingestion of a viremic blood meal, arthropods could also be infected venereally or transovarially. These two options are discussed in the section devoted to vertical transmission.

A. Infection of the Vector

While a wide range of hematophagus arthropods have the potential to mechanically transmit a given virus, there tends to be a relatively high degree of vector specificity for viruses transmitted biologically because of the need for that virus to replicate in the vector. This specificity is often determined by the susceptibility of the midgut to infection following ingestion of a viremic blood meal. Many studies have been conducted to determine the range of species capable of becoming infected following ingestion of a particular arbovirus. In addition to finding that various arthropods differ greatly in their susceptibility to a particular arbovirus, these studies have also found that strains of a given species collected at different locations or time periods may vary considerably in their susceptibility to a specific virus. This difference in susceptibility to viral infection of geographical or temporal strains of vector populations has been shown for the four serotypes of dengue virus both in *Ae. aegypti*[17] and in *Ae. albopictus,*[18] for chikungunya (CHIK) virus in these same populations of *Ae. albopictus,*[19] for WEE virus in *Cx. tarsalis,*[20] for LAC virus in *Ae. triseriatus,*[21] and for YF virus in *Ae. aegypti.*[22]

This variability in susceptibility to viral infection within a species is indicative of potential genetic control of the suceptibility of the midgut to viral infection. Hardy et al.[23] showed that both susceptible [infectious dose — 50% (ID_{50}) $\leq 10^{2.3}$ pfu] and resistant ($ID_{50} > 10^{6.7}$ pfu) strains of *Cx. tarsalis* could be selected for WEE virus. They concluded that the two lines differed in susceptibility of the midgut to infection, rather than a general refractoriness to viral replication in the resistant line, because both lines remained equally susceptible to WEE virus by intrathoracic inoculation (i.e., by-passing the need for infecting the midgut eliminated the difference in susceptibility between the two lines). Hayes et al.[24] demonstrated genetic control of susceptibility to West Nile (WN) virus in *Cx. tritaeniorhynchus.* The genetic susceptibility to viral infection within a species also appears to be virus specific, as the susceptibility of the strains of *Ae. albopictus* for the dengue viruses were not correlated with their susceptibility to CHIK virus.[18,19] Similarly, susceptibility of a vector to one virus does not imply susceptibility to others. In fact, strains of the same virus may differ greatly in their ability to infect the midgut of a given strain of vector. The importance of the susceptibility of the midgut to viral infection has been known for many years, and the concept of a ''gut barrier'' to viral infection was first proposed by Storey.[25]

In summary, the susceptibility of the gut to viral infection is the primary determinant as to whether an arthropod feeding on a viremic host will become infected. Susceptibility to viral infection appears to be under genetic control and varies not only between species, but also among various geographical populations of the same species for many viruses. Also, the virus-vector relationship appears to be relatively specific, with susceptibility to one virus not being correlated with susceptibility to other viruses.

B. Dissemination to the Hemocele

For many years, the gut barrier was believed to be the principal determinant of the ability of a species to transmit a virus. In other words, if the gut was susceptible, then it was merely a matter of time until virus would infect the salivary glands, and the vector would be able to transmit the virus. However, Kramer et al.[26] recently described both a mesenteronal or midgut escape (ME) and a salivary gland infection (SGI) barrier for WEE virus in *Cx. tarsalis.* While the ME barrier occurred in *Cx. tarsalis* only when low does of virus were ingested, a ME barrier was the principal determinant of the ability of *Cx. pipiens* to transmit RVF virus[7] and of *Cx. taeniopis* to transmit an epizootic strain of Venezuelan encephalitis virus (VE), regardless of the dose ingested.[28]

Studies with YF,[29] SLE (St. Louis encephalitis),[30] WEE,[31] WN,[32] and LAC[33] viruses have shown that ingestion of larger infectious doses not only increased infection rates, but also were associated with higher transmission rates. This increase in transmission rates was beyond that expected from the increase in infection rates alone. While these studies did not examine viral dissemination directly, apparently the larger infectious doses, enabled virus to disseminate more efficiently, resulting in a higher transmission rate.

C. Infection of the Salivary Glands and Transmission

Once virus escapes from the midgut, there are two distinct routes it could follow to the salivary glands. First, virus could be transported by the hemolymph to various target organs (e.g., salivary glands, fat body, ovaries, brain, etc.). Following replication, if necessary, in any of these organs, virus could then be carried directly to the salivary glands. Second, it is possible that some of the neurotropic viruses are transported along nerves to the salivary glands in much the same way that rabies virus is transported along the nervous system in vertebrates. While many arboviruses readily infect invertebrate nervous tissue, and Maguire in Murphy et al.[34] found some evidence that Whataroa (WHA) virus may reach the salivary glands by this route, most studies indicate that the salivary glands become infected with virus directly from the hemolymph.[35]

Because of the inability to detect very low levels of virus and the potential for contamination of the salivary glands with virus from the hemolymph, it is difficult to differentiate mosquitoes in which the salivary glands are refractory to infection from those in which the salivary glands become infected, but which do not pass infectious virus in the saliva. Perhaps electron microscopy or immunohistochemical studies will help elucidate the role of the salivary glands in biological transmission.

As an alternative to salivary gland infection and transmission by bite, virus may be secreted in the coxal fluid of certain soft ticks. This virus may then be introduced into the vertebrate host when it scratches the bite site, thus by-passing the need for salivary gland infection.

D. Extrinsic Incubation Period

The time interval between ingestion of a viremic blood meal and the ability of a vector to transmit the virus by bite is known as the extrinsic incubation period. The length of this interval has major implications for the capacity of an arthropod to serve as a vector. Because of the high daily mortality rate associated with many species, a long extrinsic incubation period may preclude most, if not all, of the infected individuals from surviving long enough to maintain the transmission cycle. Thus, any phenomenon that shortens the extrinsic incubation period may enhance the ultimate capacity of a particular species to transmit a specific virus. Both environmental temperature (see section on temperature of extrinsic incubation) and the presence of other agents (see section on the presence of other agents) have been shown to have a major influence on the extrinsic incubation period. Also, a ''leaky midgut'' phenomenon in which a small portion of the viremic blood meal leaks directly into the hemocele, as described by Hardy et al.,[23] may greatly affect the vectorial capacity of a species. Not only may a leaky midgut increase the infection rate, but more importantly, it can reduce the extrinsic incubation period by the time interval needed for virus to replicate in and be disseminated from the midgut. This time interval is not insignificant and may be 25 or more days for certain virus and vector pairs. Even though a specific extrinsic incubation period may be applied to a particular virus and arthropod pair, one should remember that this term is applied to a population, and not to the individual vectors. Thus, while the majority of a certain vector population may not be able to transmit a virus until at least 21 days extrinsic incubation, a few members of that population may be competent vectors as early as 4 days after the infectious blood meal.

VI. FACTORS AFFECTING VECTOR COMPETENCE

The ability of virus ingested in a viremic blood meal to infect the vector and ultimately be transmitted by bite to another vertebrate host is affected by a number of factors. Hardy et al.[35] recently reviewed factors intrinsic to the vector that affect viral transmission by mosquitoes. Intrinsic factors are also discussed in the chapter on susceptibility and resistance of vector mosquitoes in the text. Other reviews by Hoogstraal,[36] Reeves,[37,38] Downs,[39] and Nuttall[40] have examined other factors that affect transmission of arboviruses in both insect and tick vectors.

These include environmental parameters such as rainfall, temperature, and humidity. These will not only influence the number of potential vectors available and their relative survival, but also may affect the arthropod's ability to become infected or transmit a virus. In addition, man-made interventions in the environment such as draining swamps, damming streams, or flooding pastures for irrigation may greatly increase or decrease vector populations.

Vertebrate host factors affecting transmission include susceptibility to viral infection; viremia titer produced; population number, density, and turnover rate; attractiveness to vectors; and behavior as it relates to vector feeding. Greater vertebrate susceptibility and/or a higher viremia produced may allow less efficient vectors to be able to transmit a virus. Greater population size and density allow for an increased chance of a vector finding a vertebrate host. The more rapid the turnover in a vertebrate population, the lower the percentage of that population that will be immune; thus, the greater the chances are that a vector will feed on a susceptible host. An animal's behavior can affect both the likelihood of encountering a vector and the vector's chances of obtaining a blood meal once contact is established.

In addition to the ability of an arthropod to transmit a specific virus, other factors affect potential transmission. These include host preference, vector population number and density, autogeny, rate of dispersal, infection threshold, and relationship between life span and the extrinsic incubation period. For example, vectors that feed preferentially on a vertebrate population that is both susceptible to viral infection and produces a viremia are more likely to become infected and transmit a virus than are arthropods that feed on a wide variety of hosts, including ones that are not susceptible to viral infection. While larger numbers (and/or density) of a vector population may increase the likelihood that a vector will feed on a viremic host and thus become infected and potentially transmit virus, this may not always be the case. For example, Olson et al.[41] found that transmission of both WEE and SLE virus by *Cx. tarsalis* was highest when moderate, rather than high, numbers of this species were present in California. One possible explanation for the decrease in transmission efficiency with very large populations of *Cx. tarsalis* may be a switch in feeding behavior associated with large numbers of mosquitoes being present. Nelson et al.[42] reported that when *Cx. tarsalis* were given a choice between an avian host, either a chicken or a pheasant, and a rabbit in a stable trap, feeding success was less and feeding rates on the rabbit were greater when larger numbers of mosquitoes were present. This switch in blood hosts associated with high mosquito density could reduce transmission efficiency by causing potential vectors to feed on nonproductive hosts. The presence of autogeny in an arthropod population can also reduce the potential for viral transmission. Because autogenous females do not require a blood meal to produce their first ovarian cycle eggs, they must survive that many days longer before their first blood meal (i.e., first potential viral exposure). Thus, given the same extrinsic incubation period, these autogenous vectors would have to survive longer than anautogenous ones in order to transmit an arbovirus. While it would appear that the more sensitive a vector was to viral infection, the more important the vector would be, this may not always be true. Increased virulence is often associated with an increased viremia. Because the less susceptible vectors are likely to become infected only when they ingest blood from hosts with higher viremias, they may select for a more virulent virus, and thus a more severe disease. As discussed earlier, strains of a given arbovirus differ with respect to their ability to infect and be transmitted by potential vectors. Thus, environmental, vertebrate, vector, and viral factors influence the potential for arboviral transmission. A few of these factors are discussed in more detail in the following sections.

A. Effect of Extrinsic Incubation Temperature

Numerous studies have shown that the ability of a potential vector to transmit an arbovirus is affected by the extrinsic incubation temperature. These studies have consistently demonstrated that the extrinsic incubation period is inversely related to the temperature at which the vectors are held.[29,43-47] In contrast, however, environmental temperature appears to affect both infection and transmission rates differently, depending on the virus and vector combination. For example, while reduced infection rates with YF virus in *Haemagogus* mosquitoes were associated with low environmental temperatures (20°C),[29] in *Cx. tarsalis* reduced infection rates with WEE virus were associated with high (32°C) environmental temperatures.[46] No effect of environmental temperature on infection rates was found for EEE virus in *Ae. triseriatus*[44] or for WN virus in *Cx. univittatus*.[32] The importance of indicating the specific virus and vector under consideration is further demonstrated by an examination of the role of environmental temperature on the vector competence of *Cx. pipiens* and *Ae. taeniorhynchus* for RVF virus.[47] While infection rates in *Ae. taeniorhynchus* were not affected by incubation temperatures, virus was recovered from 75% of *Cx. pipiens* maintained at 26°C as compared to only 38% of those held at 13°C, even though both groups had fed concurrently on the same viremic hamster. In contrast, while the rate of viral dissemination in infected *Cx. pipiens* was not affected by incubation temperature, only 7% of infected *Ae. taeniorhynchus* developed a disseminated infection when held at 13°C as compared to 57% of those held at 26°C. For both species, the time interval between the viremic blood meal and either viral dissemination or transmission was inversely related to incubation temperature. Transmission rates, even after adjusting for differences in extrinsic incubation periods, were directly related to incubation temperature. Thus, in this study, even though environmental temperature affected the interaction of RVF virus with each of the species differently (i.e., infection rate in *Cx. pipiens* and viral dissemination in *Ae. taeniorhynchus*), the end results were similar: faster, more efficient transmission occurred at higher environmental temperatures. However, while Kramer et al.[46] found more rapid transmission at higher incubation temperatures, they also found that *Cx. tarsalis* held at elevated temperatures were less efficient vectors of WEE virus than those held at lower temperatures (18 or 25°C). This may help to explain a study by Hess et al.,[48] who found that WEE outbreaks usually were preceded by unusually cool springtime temperatures, while SLE outbreaks followed warmer springs. Thus, even though environmental temperature may play a critical role in vector competence, few generalizations can be made other than that viral transmission normally occurs sooner at higher environmental temperatures.

B. Effect of Viral Infection on the Vector

It has long been thought that an infection with an arbovirus has no detrimental effects on its vector.[49] In part, this is due to the belief that arboviruses and their vectors have evolved to the degree that these viruses no longer cause morbidity or mortality in their natural vectors. While Mims et al.[50] found that infection with Semliki Forest (SF) virus was associated with degeneration of the salivary glands in *Ae. aegypti*, this effect was attributed to an unnatural virus/vector relationship. SF virus has never been isolated from *Ae. aegypti* in nature.[51] Several recent studies, however, have indicated that infection in the vector may not be benign. Effects associated with arboviral infection include reduced fecundity,[52] increased developmental time from egg to pupae,[53-56] and reduced efficiency in obtaining a blood meal.[52,57] In contrast to these detrimental effects, Patrican et al.[58] reported that feeding ability was similar for uninfected-control and transovarially infected *Ae. triseriatus*.

Whether these effects are (1) due to an actual pathology caused by viral infection, (2) related to an increased energy requirement needed to support both the vector's needs (e.g., egg production and larval development) and viral replication, or (3) due to a combination of the above, have not yet been determined. Several hypotheses may explain the reduction in feeding efficiency. These include pathology in the salivary glands that affects either the quantity or the quality of saliva produced. An enzyme, apyrase, has recently been shown to be present in mosquito saliva and to be used in blood vessel location.[59] Alternatively, viral infection may result in neural damage and affect either the receptors that indicate the presence of blood, or their interpretation by the central nevous system. Additional studies are needed to determine the actual mechanism(s) by which these effects occur.

Epidemiologically, the most important of these effects is the reduced efficiency in obtaining a blood meal. While it would seem to be to the virus' advantage to cause as little damage as possible to the vector, this reduction in refeeding efficiency may actually enhance the vectorial capacity of the infected vector. If an infected vector feeds to repletion at its first feeding attempt, then it would not normally be available for a second feeding (i.e., transmission attempt) until after it had completed its gonotrophic cycle. Because of the relatively high daily mortality for many vector species, few of these vectors would live to have a second or third transmission attempt. In contrast, if at the first refeeding attempt, the vector failed to obtain a blood meal, it would be available, almost immediately, for a second transmission attempt. Mosquitoes that probe but fail to obtain blood, are capable of transmitting virus.[57,60] In a study with RVF virus, transmission rates were similar for mosquitoes that fed to repletion and for those that probed but did not obtain blood during the transmission attempt.[61] Also, in this same study, transmission rates were not affected by the order of feeding (i.e., transmission rates were similar at the first, second, and third refeeding attempt). Thus, because probing vectors are potentially as efficient vectors as those that feed to repletion, the failure to obtain a blood meal at a given feeding attempt should greatly increase the vector potential for that individual vector. This increased potential is demonstrated by the transmission of RVF virus to five consecutive hamsters within a total of 10 min by a single female *Cx. pipiens*.[61]

C. Effect of Viral Infection on the Vertebrate Host

Several factors affect the likelihood of a vertebrate providing a blood meal for a potential vector. These include the vertebrate's attractiveness to host-seeking arthropods, its behavior once arthropods try to feed, and its physiological state affecting blood-finding success (e.g., dilation of surface blood vessels or impaired coagulation mechanism). Body heat[62-64] and CO_2[65-68] given off by vertebrates have long been known to attract host-seeking arthropods. Female *Ae. aegypti* were able to detect differences as small as 1°C, and probed more frequently at the warmer temperature when given a choice between tubes of circulating water at two different temperatures.[69] Many vertebrates have behavioral displays that are directed against host-seeking mosquitoes. This type of behavior has been termed anti-mosquito behavior[70] and has been shown to limit the ability of host-seeking mosquitoes to obtain a blood meal.[70-73] Viral infections in vertebrates are usually associated with a febrile response. This elevated body temperature can then lead to both increased CO_2 output and a dilation of surface blood vessels. Also, anti-mosquito behavior may be diminished in infected animals.[74] Thus, not only are host-seeking arthropods more likely to be attracted to viremic than to well animals, but also, once they arrive, they are more likely to obtain a complete blood meal from an infected vertebrate as compared to an uninfected one.

Various studies have attempted to quantify this potential for increased feeding on viremic hosts. Traps baited with a chicken infected with Sindbis (SIN) virus attracted more mosquitoes than did traps baited with an uninfected chicken.[75] As part of this study, *Cx. annulirostris* were allowed to feed on a pair (one infected with SIN virus and one uninfected control) of unrestrained chickens in a 40 × 40 × 40 cm cage. Overall, 68% (96/141) of the engorged mosquitoes fed on the viremic chicken. In a similar study conducted with RVF virus in lambs, 65% (1264/1943) of *Cx. pipiens* fed on the viremic lamb.[76] In order to examine the role of viral infection and host physiology on feeding success, *Ae. aegypti* were allowed to feed on a series of anesthetized hamsters. These hamsters were either uninfected controls or viremic with RVF virus. Mosquitoes probing on the viremic hamsters required a median of 95 sec to locate blood as compared to 155 sec for those probing on the control hamsters.[77]

This difference of 60 sec could have a major effect in nature, where mosquitoes must locate blood rapidly to feed successfully. Thus, viremic vertebrate hosts may provide a greater than expected percentage of blood meals to host-seeking arthropods. This excess of feeding on viremic animals may need to be incorporated into models of the natural history of arboviruses.

D. Presence of Other Agents

As was discussed in the sections on infection of the vector and dissemination into the hemocele, virus must infect the midgut and disseminate into the hemocele before it can be biologicallly transmitted. Merrill and TenBroeck[78] found that *Ae. aegypti* that ingested EEE virus normally failed to transmit this virus by bite. However, if the midgut of the freshly engorged mosquitoes was punctured with a needle, they were able to transmit EEE virus. The puncture of the midgut apparently allowed virus to by-pass both the midgut infection and dissemination phases. Because certain microfilariae also puncture the midgut shortly after ingestion, it has been hypothesized that concurrent ingestion of an arbovirus and microfilariae may enhance the arthropod's capacity to transmit that virus.[79,80] This hypothesis is supported by a study in which *Culicoides nuberculosus* became infected with bluetongue virus if the virus was ingested concurrently with microfilariae of *Onchocerca cervicalis*, but not if they ingested virus alone.[79] While this study did not examine virus transmission, a study by Turell et al.[80] showed that *Ae. taeniorhynchus* that fed on a gerbil with both a RVF viremia and a *Brugia malayi* microfilaremia had a fourfold higher viral dissemination rate and an over sixfold higher viral transmission rate than did *Ae. taeniorhynchus* that ingested the same amount of RVF virus from a gerbil infected with virus alone. Moreover, the microfilarie enabled the virus to disseminate into the hemocele more rapidly than would have occurred normally. This, in turn, shortened the extrinsic incubation period.

Filarial infections produce microfilaremias of long duration, often lasting several years. Because of the generally high prevalence rate of these infections in both domestic and feral animals, as well as in man in the tropics, it is likely that many arboviral infections are, in fact, dual infections with microfilariae. Thus, the increased vectorial capacity for an arbovirus associated with concurrent ingestion of microfilariae may have to be considered in control programs aimed at either the arbovirus or the filarial worm.

In contrast to the situation with microfilariae, Barnett[81] found that concurrent ingestion of an avian malarian parasite, *Plasmodium relictum*, did not affect the ability of *Cx. tarsalis* to transmit WEE virus. He also found that viremia levels were significantly lower in canaries infected with *P. relictum* as compared to those infected with WEE virus alone. Thus, if mosquitoes were not matched by infectious dose, as was done in Barnett's experiments, infection and transmission rates might have been lower in mosquitoes feeding on the dually infected canaries. Infection with the gregarine parasite, *Ascocystis barretti*, did not affect infection rates in *Ae. triseriatus* larvae orally exposed to LAC virus.[82] The key difference between these two studies and the ones involving microfilariae appears to be the rapid penetration of the midgut by the microfilariae, allowing virus to escape into the hemocele. Thus, concurrent ingestion of an arbovirus and an agent that allows the virus direct access to the hemocele may enhance vectorial capacity for the virus by shortening its extrinsic incubation period and by enabling the virus to by-pass both the midgut infection and dissemination to the hemocele phases of the replication cycle.

VII. SURVIVAL OF VIRUS DURING PERIODS OF VECTOR INACTIVITY

In most parts of the world there is either a dry or a cold season during which vectors are not active. Because arboviral infections in vertebrates generally produce a viremia of short duration followed by a life-long immunity, the method(s) by which arboviruses maintain themselves in nature during this period of vector inactivity is (are) of considerable importance. The long-term persistence of an arbovirus may be particularly susceptible to intervention and control at this time because of the reduction in both vector and virus activity. Reeves[83] reviewed various hypotheses to explain virus persistence through these periods. These included: (1) maintenance in the vector, either in long-lived individuals and/or by transovarian transmission; (2) a vertebrate host with either a chronic or recurring viremia; (3) maintenance within an internal parasite, such as a *Plasmodium* species, where it would be protected against the host's immune system; (4) in a food source such as a plant; or (5) reintroduction each season by migrating animals such as birds.

A. Survival in Long-Lived Vectors

While vector feeding activity may cease during periods that are unfavorable to vector activity, many vector species survive this period in the adult stage. These "overwintering" vectors may serve as the reservoir to reintroduce virus during the next season of vector activity. For example, Lewis[84] demonstrated that Nairobi sheep disease (NSD) virus persisted in the tick, *Rhipicephalus appendiculatus,* for at least 871 days. Similarly, *Culiseta annulata,* when held at 6 to 11°C, remained infected with Tahyna (TAH) virus for at least 181 days;[85] snowshoe hare (SSH) virus persisted in *Culiseta inornata* for 329 days when held at 4°C;[86] and Japanese encephalitis (JE) virus survived for at least 82 days in *Cx. quinquefasciatus* held at 8 to 13°C.[87] These *Cx. quinquefasciatus* successfully transmitted JE virus after 3 days subsequent incubation at 30°C. *Cx. tarsalis* and *Cx. quinquefasciatus* remained infected with WEE and SLE viruses, respectively, for over 110 days when held under natural overwintering conditions in an unheated cellar, and the *Cx. tarsalis* successfully transmitted WEE virus after 109 days.[88]

In support of this mechanism for viral persistence, various arboviruses have been isolated from vectors during periods of relative vector inactivity or in early spring before the vectors became active. These include isolations of Fort Morgan virus from the cliff swallow bug, *Oeciacus vicarius;*[89] WEE virus from *Cx. tarsalis;*[90,91] SLE virus from *Cx. tarsalis*[91] and *Cx. pipiens;*[92] and JE virus from *Cx. pipiens* by Wang as cited by Bailey et al.[92] However, several studies have indicated that mosquitoes that obtain a blood meal may not survive the winter. For example, Slaff and Crans[93] found that parous females are seldom found among overwintering *Cx. pipiens.* In fact, they found that only 7/493 (1%) of the females caught from November to January were parous, and none of 280 females caught in February or March was parous. Thus, they felt that "some parous mosquitoes enter hibernation, but few, if any, survive the winter". However, when *Cx. pipiens* were preconditioned for diapause by exposure to short daylengths and reduced temperatures, their ovaries did not always develop following ingestion of a blood meal.[94] In these mosquitoes, protein from the blood meal went to develop fat body instead of eggs. Thus, the lack of ovarian development in overwintering females reported by Slaff and Crans[93] does not eliminate the possibility that some of these nonparous mosquitoes had obtained a blood meal prior to diapause. The potential for engorged mosquitoes to overwinter was also supported by a study that showed that *Cx. pipiens* preconditioned for diapause in the laboratory survived natural winter conditions following engorgement and gonotrophic dissociation (i.e., blood feeding without ovarian development) as well as preconditioned mosquitoes that had not fed, or wild-caught, hibernating mosquitoes.[95] While the data supporting overwintering of mosquito-borne viruses in their mosquito vectors are inconclusive, there is evidence that some of these viruses may survive periods of vector inactivity in a long-lived, overwintering vector.

In contrast to the situation with mosquitoes, numerous tick-borne viruses have been shown to survive periods of vector inactivity in their tick vectors.[36,83] Ticks are relatively long-lived and most ticks require at least two active seasons to complete their life cycle. Thus, ticks that can transmit an arbovirus transstadially (i.e., larva to nymph or nymph to adult) can serve as an "overwintering" reservoir for that virus. Transstadial transmission has been demonstrated for numerous tick-borne viruses, including Russian spring-summer encephalitis (RSSE) virus in *Ixodes persulcatus*,[36] louping ill virus in *Ixodes ricinus*,[36] CTF virus in *Dermacentor andersoni*,[96] African swine fever (ASF) virus in *Ornithodoros moubata*,[97] and Crimean-Congo hemorrhagic fever (CCHF) virus in *Hyalomma marginatum* by Chumakov as reported by Hoogstraal.[36]

B. Transovarial Transmission

While transovarial transmission (i.e., transmission of virus from an infected female to her progeny) of tick-borne viruses has been accepted for many years,[36,83,98] it was not until fairly recently that transovarial transmission was conclusively demonstrated for a mosquito-borne virus.[99] Following the recovery of LAC virus from progeny of infected *Ae. triseriatus* by Watts et al.,[99] transovarial transmission has since been shown in the laboratory for a wide variety of mosquito-borne viruses (Table 1), including viruses in the Togaviridae, Rhabdoviridae, Bunyaviridae, and Reoviridae. Transovarial transmission has also been shown for several reo-, phlebo-, and rhabdoviruses in sand flies.[100-105] Evidence of transovarial transmission under natural conditions has also been demonstrated for numerous diptera-borne viruses by recovery of virus either from field-collected males or from adults reared from field-collected larvae (Table 1). Similarity of infection rates in progeny derived from untreated or surface-sterilized eggs indicates that these progeny were infected transovarially, and not by contamination of the egg surface with virus.[55,99,106,107] Also, various studies[54,56,108] have shown that vertical transmission rates for female mosquitoes infected by intrathoracic inoculation are similar to those for females infected orally to their second or later ovarian cycles.

As was described for horizontal transmission, a virus must infect and pass through a series of membranes before it can be transovarially transmitted. After infecting the midgut and disseminating into the hemocele, the virus must still cross the ovarian and the ovariole sheaths before it can infect the follicular epithelium and thus have access to the developing oocytes. If infection of the oocyte did not occur early in oogenesis, the developing chorion would probably protect the egg from infection. Because oogenesis begins shortly after the blood meal is obtained, it is unlikely that virus ingested in a particular blood meal would have sufficient time to escape from the midgut, reach the ovaries, and infect the developing oocyte during that ovarian cycle. Thus, transovarial transmission would not normally occur until the second or even later ovarian cycle after the infectious blood meal.

Several studies have followed the course of viral infection in potential insect[109-112] and tick[96,113] vectors. The importance of localization of viral infection in the ovaries was shown by a study of EEE virus in *Culiseta melanura*.[112] Even though virus was detected in both the common and lateral oviducts, it was not recovered from within the ovarioles themselves, and attempts to demonstrate transovarial transmission with this species have been unsuccessful.[114-116] Similarly, RVF virus was detected in the ovarian and ovariole sheaths as well as in the calyx and the common and lateral oviducts of *Cx. pipiens*, but not in the follicular epithelium.[117] As expected from the lack of infection of the follicular epithelium, none of the over 7000 progeny of RVF virus-infected *Cx. pipiens* contained this virus.[118] In contrast, LAC virus has been detected in the follicles of *Ae. triseriatus*,[110] a species known to transovarially transmit this virus.[99] Thus, detection of virus in the ovaries of a vector is not sufficient evidence that it can transmit that virus transovarially.

Both mosquito and viral genetics have been shown to affect vertical transmission. For example, various geographic strains of *Ae. albopictus* differed significantly in their ability to transovarially transmit both San Angelo (SA) and Kunjin viruses.[54] Furthermore, because the strain of *Ae. albopictus* that was most efficient for SA virus was one of the least efficient

Table 1
ARBOVIRUSES FOR WHICH TRANSOVARIAL TRANSMISSION HAS BEEN SHOWN

Virus	Arthropod	Source	Ref.
Togaviridae			
Alphavirus			
Eastern equine encephalitis	*Culiseta melanura*	F-lar	14
Ross River	*Aedes vigilax*	L	15
Western equine encephalitis	*Dermacentor andersoni*	L	140
Flavivirus			
Absettarov	*Ixodes ricinus*	L	51
	Ix. persulcatus	L	51
Bussuquara	*Ae. albopictus*	L	142
Dengue-1	*Ae. aegypti*	L	56
	Ae. albopictus	L	56
Dengue-2	*Ae. aegypti*	L	143
	Ae. aegypti	F-lar, F-ad	144
	Ae. albopictus	L	56
	Ae. fulcifer/taylori	Males	145
Dengue-3	*Ae. albopictus*	L	56
Dengue-4	*Ae. albopictus*	L	56
	Ae. aegypti	F-ad	146
Hypr	*Ix. ricinus*	L	147
Ilheus	*Ae. albopictus*	L	142
JE	*Ae. albopictus*	L	128
	Ae. togoi	L	128
	Cx. tritaeniorhynchus	L	148
Kokobera	*Ae. albopictus*	L	142
Koutango	*Ae. aegypti*	L	149
Kunjin	*Ae. albopictus*	L	54
Kyasanur Forest disease	*Haemaphysalis spinigera*	L	150
Murray Valley encephalitis	*Ae. aegypti*	L	151
Russian spring-summer encephalitis	*Ix. persulcatus*	L	36
St. Louis encephalitis	*Ae. epactius*	L	125, 127
	Ae. atropalpis	L	125, 127
	Cx. pipiens	L	126, 127
	Cx. quinquefasciatus	L	127
	Cx. tarsalis	L	127
	Dermacentor variabilis	L	139
Uganda-S	*Ae. albopictus*	L	142
Yellow fever	*Ae. aegypti*	L	53, 107
	Ae. fulcifer/taylori	Males	152
	Ae. mascarensis	L	53
	Haemagogus equinus	L	153
	Amblyomma variegatum	L	141
Bunyaviridae			
Bunyavirus			
Cache Valley	*Culiseta inornata*	L	154
California encephalitis	*Ae. dorsalis*	L	119, 122
	Ae. dorsalis	F-ad	155
	Ae. melanimon	L	119, 156
	Ae. melanimon	Males	157
Gamboa	*Aedeomyia squamipennis*	F-ad	51
Jamestown Canyon	*Ae. triseriatus*	F-ad	158
Keystone	*Ae. atlanticus*	F-lar, F-ad	159
La Crosse	*Ae. triseriatus*	L	99,121
	Ae. triseriatus	F-lar, F-ad	160
San Angelo	*Ae. albopictus*	L	54
Snowshoe hare	*Ae. implicatus*	F-ad	161
	Ae. communis	F-ad	162

Table 1 (continued)
ARBOVIRUSES FOR WHICH TRANSOVARIAL TRANSMISSION HAS BEEN SHOWN

Virus	Arthropod	Source	Ref.
Tahyna	*Ae. aegypti*	L	163
	Ae. vexans	L	124
	Culiseta annulata	F-lar	164
Trivittatus	*Ae. trivittatus*	F-lar-ad	165
	Ae. trivittatus	L	106
Nairovirus			
Crimean-Congo hemorrhagic fever	*Hyalomma marginatum rufipes*	L	166
	Hyalomma marginatum marginatum	L	166
	Rhipicephalus rossicus	L	166
	Dermacentor marginatus	L	166
Dugbe	*Amblyomma variegatum*	L	167
Nairobi sheep disease	*Rhipicephalus appendiculatus*	L	168
Phlebovirus			
Aguacate	*Lutzomyia trapidoi*	Males	138
Arbia	*Phlebotomus pernicious*	L	105
	Phlebotomus sp.	Males	169
Cacao	*Lutzomyia trapidoi*	Males	138
Chagres	*Lutzomyia* sp.	Males	101
Chilibre	*Lutzomyia* sp.	Males	101
Karimabad	*Phlebotomus papatasi*	L	105
	Phlebotomus sp.	Males	170
Pacui	*Lutzomyia flaviscutellata*	L	171
	Lutzomyia flaviscutellata	Males	172
	Lutzomyia longipalpis	L	105
	Phlebotomus papatasi	L	105
Punta Toro	*Lutzomyia trapidoi*	Males	101
Rift Valley fever	*Ae. lineatopennis*	F-ad	173
Rio Grande	*Lutzomyia anthophora*	L	102
Saint-Floris	*Phlebotomus papatasi*	L	105
Sandfly fever (Sicilian)	*Phlebotomus papatasi*	L	105
	Phlebotomus papatasi	Males	174
Toscana	*Phlebotomus perniciosus*	L	105
	Phlebotomus sp.	Males	169
Uukuvirus			
Uukuniemi	*Ix. ricinus*	L	142
Bunyavirus-like			
Kaisodi	*Haemaphysalis spinigera*	L	51
Tamdy	*Hyalomma asiaticum*	F-lar	175
Rhabdoviridae			
Vesiculovirus			
Carajas	*Lutzomyia longipalpis*	L	176
Chandipura	*Phlebotomus papatasi*	L	104
Maraba	*Lutzomyia* sp.	Males	176
	Lutzomyia longipalpis	L	176
Vesicular stomatitis (Indiana)	*Lutzomyia trapidoi*	L	100
	Lutzomyia ylephilator	L	100
Iridoviridae			
African swine fever	*Ornithodoros moubata*	L	97
	Ornithodoros puertoicensis	L	177
Reoviridae			
Orbivirus			
Changuinola	*Lutzomyia* sp.	Males	101
Orungo	*Aedes* sp.	Males	178

Note: L = Laboratory study, males = virus isolated from field-collected males, F-lar = virus isolated from field-collected larvae, and F-ad = virus isolated from adults reared from field-collected larvae.

for Kunjin virus, this relationship appears to be virus-specific. Thus, as for oral exposure, increased efficiency (susceptibility) for (to) one virus does not imply increased efficiency for others. Likewise, the efficiency with which California encephalitis (CE) was transovarially transmitted by *Ae. dorsalis* was dependent on the viral strain utilized.[119] Similar variations in vertical transmission rates by both viral and mosquito strains have been reported for the various serotypes of dengue virus in *Ae. albopictus*.[56]

Several measures of the efficiency of transovarial transmission are used.[120] These include the transovarial transmission rate, which is defined as the percentage of infected females that transmit virus to their progeny, regardless of the infection rate in these progeny. The filial infection rate refers to the percentage of the progeny of an infected female that are transovarially infected, and when this rate is generalized to the percentage of progeny infected from a population of infected females, it called the vertical transmission rate. Because the transovarial transmission rate refers only to the parental females, it may give a misleading impression on the efficiency of transovarial transmission. For example, Miller et al.[121] found that 53/54 (98%) of transovarially infected *Ae. triseriatus* transmitted LAC virus, and Christensen et al.[106] found that 4/6 (67%) of *Ae. trivittatus* with 10 or more progeny transmitted trivittatus (TVT) virus; however, only 14/126 (11%) of the progeny were infected with TVT virus. Similarly, while Turell et al.[122] found that all *Ae. dorsalis* (8) and *Ae. melanimon* (16) transmitted CE virus, only about 17% of the progeny of either species were infected with CE virus. Thus, vertical transmission rates are probably a better measure of the potential of transovarial transmission to serve as a long-term maintenance mechanism for the viruses.

Studies of arboviruses tested for their ability to be vertically transmitted by their respective vectors have indicated that efficiency of vertical transmission appears to be related to the virus' taxonomic classification. For example, most of the Bunyaviridae tested, particularly members of the California serogroup and the phleboviruses, tend to be vertically transmitted relatively efficiently, often with vertical transmission rates in excess of 20%. While vertical transmission has also been demonstrated for most flaviviruses tested, this group tends to be transmitted inefficiently, with vertical transmission rates often $<1\%$. In contrast, despite extensive studies with EEE,[114-116] CHIK, Gateh, SIN, SF,[103] and WEE[123] viruses, the only demonstrations of vertical transmission of an alphavirus by an insect vector remains a single laboratory study with Ross River virus and *Ae. vigilax*[15] and an isolation of EEE virus from field-collected larval *Culiseta melanura*.[14]

In most studies on transovarial transmission, eggs are hatched shortly after embryogenesis is complete. However, in order for transovarial transmission to serve as an overwintering mechanism, not only must the eggs become infected, but they must also remain infected until the next period of vector activity. It is possible that virus is lost with time or that virus-infected eggs have a lower survival (hatch) rate than uninfected eggs. This may be further complicated by the repeated freeze-thaws to which the eggs are exposed in a temperature climate. Beaty et al.[53] and Tesh[54] showed that YF and SA viruses, respectively, remained infectious in transovarially infected eggs for at least 3 months, and Danielova and Ryba[124] found that TAH virus remained infectious in overwintering *Ae. vexans* eggs. Similarly, Turell et al.[55] found that CE virus remained infectious in *Ae. dorsalis* eggs for at least 19 months, even when these eggs were exposed to repeated freeze-thaws. In this study, vertical transmission rates were not affected by the length of time the eggs were held prior to hatching, and hatch rates remained high, even after 19 months. Thus, the presence of virus in the eggs does not appear to affect egg survival, and virus may remain infectious in these eggs for a considerable period of time, even when exposed to extremes of temperature. While the storage conditions of the eggs did not affect infection rates in the resulting progeny, the temperature at which mosquito larvae are reared has been shown to influence the ultimate infection rate in the adult progeny. For example, infection rates in adult progeny of SLE-inoculated *Ae. epactius* were over 24-fold higher in progeny reared at 18°C (2.0%, 30/1529) than in those reared at 27°C (0.08%, 2/2463). However, larval infection rates were similar in larvae reared at either temperature.[125] A similar increase in adult SLE infection rates was also observed for *Cx. pipiens* reared at 18°C as compared to 25°C,[126] and in a second study by Hardy et al.,[127] where infection rates were higher in *Culex (Cx. pipiens, Cx. quinquefasciatus,* and *Cx. tarsalis)* larvae reared at 18°C (0.5%, 9/1850) than in those reared at 27°C (0/1850). Thus, it appears that either virus is being lost from most of the infected individuals during larval development and pupation, or infected individuals have a significantly higher, temperature-dependent mortality rate during this time period.

In addition, if transovarial transmission is to provide a means of overwintering, transovarially infected individuals must be able to transmit the virus by bite to susceptible vertebrates in order to reinitiate the summer amplification cycle. The ability of transovarially infected females to transmit virus by bite has been shown for VS (Indiana),[100] LAC,[99] JE,[128] and SLE[126] viruses. More importantly, *Ae. triseriatus* was able to transmit LAC virus by bite after six consecutive transovarial passages.[121] Similarly, *Ae. albopictus* transmitted SA virus [129] and *Ae. dorsalis* transmitted CE virus [122] after 14 and 5 consecutive transovarial passages, respectively. Thus, these viruses may be maintained in their vectors by transovarial transmission for at least several generations and still be transmitted to vertebrates.

While virus may be transovarially transmitted for several consecutive generations, most studies have found relatively low rates of vertical transmission. In fact, with the exception of members of the California serogroup in mosquitoes and some of the phleboviruses in sand flies, vertical transmission rates have generally been <1%, and often <0.1% in the progeny of the infected parental females. In addition, several studies have shown that transovarially infected larvae take longer to develop to the pupal stage than do their uninfected siblings reared in the same water container.[53-56] This delay in larval development may result in an increased mortality of infected larvae, particularly with some of the flood water species, where the larval habitat may dry before all of the larvae complete pupation. Thus, it appears that unless there is some sort of amplification cycle to increase the prevalence of infection, virus would disappear within a few generations. Various models have been proposed to quantify these cycles.[120,130,131]

Basically, if an infected female transmits virus to 20% of her progeny, then there must be a fivefold increase in the rate of infections due to the amplification cycle. It should be pointed out that the amplification cycle referred to here relates to an increase in the prevalence rate of infection in the vector, and not to an increased number of infected vectors or to an increase in infected vertebrate hosts. Also, this example assumes that there are no negative effects on the vector due to viral infection and that vertical transmission rates are not affected by the number of ovarian cycles completed. While the delayed development may have only a slight negative effect on survival, transovarial transmission to the first ovarian cycle following an infectious blood meal is significantly less than to later ovarian cycles.[126,132] This low vertical transmission rate to the first ovarian cycle is compounded with the relatively small percentage of mosquitoes that survive to complete two or more ovarian cycles and the failure of several arboviruses to escape from the midgut of infected mosquitoes. For example, if a particular species had an 80% daily survival rate with an ovarian cycle taking 7 days, then only about 20% of the females that fed on a viremic animal and oviposited would survive to oviposit a second time. If this species transmitted virus to 25% of its progeny once virus had disseminated to the ovaries, then because virtually all eggs laid in the first ovarian cycle would not be infected, and only 20% of these females would have 25% of their eggs infected, the realistic vertical transmission rate for this species would be about 8%. This, in turn, would require a 12-fold increase in the prevalence rate of viral infection in the vector between the time when the overwintering eggs hatched and when the next overwintering eggs were produced. For vector systems where the realistic vertical transmission rate is less than 1%, an over 100-fold increase in prevalence rate must occur

Table 2
FIELD INFECTION RATES BY GENERATION FOR A VIRUS MAINTAINED SOLELY BY TRANSOVARIAL TRANSMISSION FOR VECTORS WITH NONSTABILIZED AND STABILIZED INFECTIONS

No. of generations	Nonstabilized infection[a]	Stabilized infection[b]
0	1/1,000	1/1,000
1	1/12,500	1/1,080
2	1/156,000	1/1,160
3	1/1,950,000	1/1,240
4	1/24,300,000	1/1,340
5	1/304,000,000	1/1,440
6	1/3,800,000,000	1/1,550

[a] Effective vertical transmission rate of 8%, based on a vertical transmission rate of 25% to the second ovarian cycle eggs, 80% daily survival, and a 7-day gonotrophic cycle.
[b] Effective vertical transmission rate of 93%.

during the active season. Increases of this magnitude have never been reported. In fact, in a study conducted in California, no consistent changes in the prevalence rates of CE virus infections in *Ae. melanimon* were observed from May through October.[122] If, as in this study, there is no major amplification cycle, then virus should disappear within several generations (Table 2). Given the lack of evidence for an amplification cycle of the magnitude needed to maintain virus in an area with the vertical transmission rates observed in the laboratory, it would appear that vertical transmission could not maintain an arbovirus in nature.

However, recent studies have shown that arboviruses may exist in a stabilized state in their vectors, as described for sigma virus in *Drosophila melanogaster*.[133] When a vector is first infected, either orally or by intrathoracic inoculation, virus replicates to a relatively high titer. Some individuals may transmit virus transovarially to a percentage of their progeny by direct infection of the oocytes. Such individuals with high-titered infections not involving the germinal tissue are referred to as having a nonstabilized infection. However, some transovarially infected individuals may have their germinal tissue infected. In general, these individuals contain less virus than their nonstabilized mothers, but because their germinal tissue is infected, virtually all of their progeny, and likewise, their progeny's progeny, will be infected. A stabilized state has been shown to exist for *Ae. albopictus* and SA virus[129] and for *Ae. dorsalis* and CE virus.[122] In both of these models, vertical transmission rates were in excess of 90%. Also, because these mosquitoes emerge with their germinal tissue already infected, they transmit virus at this high rate to their first ovarian cycle progeny. Thus, even without an amplification cycle in vertebrates, these viruses could remain enzootic for several generations (Table 2).

C. Venereal Transmission

While male insects do not take a blood meal, and thus cannot be infected horizontally, they can be infected transovarially. In addition to this route of infection, male ticks can also be infected transstadially from larvae or nymphs that fed on a viremic host. These infected males may then transfer virus venereally to females as has been demonstrated for LAC virus by transovarially infected male *Ae. triseriatus*[134] and for ASF virus by orally infected male *Ornithodorus moubata.*[113] Similarly, SIN virus-inoculated male *Ae. australis* transmitted this virus during copulation.[135] The venereally infected *Ae. triseriatus* were able to transmit LAC virus both horizontallly to vertebrates by bite and transovarially to their progeny. Because infection rates in male and female progeny of infected mosquitoes are similar, venereal transmission may account for part of the amplification cycle needed to bring the effective vertical transmission rate back to 100%. While venereal transmission rates were relatively low if mating took place prior to the female's first blood meal, 49% of the female *Ae. triseriatus* became infected with LAC virus if they had obtained a blood meal prior to copulation.[136] Horizontal and vertical transmission rates by these females were correspondingly higher if they had fed prior to copulation. However, if the blood meal prior to copulation was obtained from a LAC-immune chipmunk, the horizontal transmission rate was significantly reduced as compared to those obtaining this meal from a LAC-susceptible chipmunk.[137] Thus, the potential for ingested vertebrate antibody to regulate viral development in the vector requires further study.

D. Summary

The demonstration of stabilized infections for several members of the California serogroup brings up the possibility that these viruses may have evolved as mosquito viruses that are maintained in a mosquito-mosquito cycle by transovarial and venereal transmission. The infection of vertebrates with these viruses by transovarially infected mosquitoes may have evolved later, and may play little, if any, role in the long-term maintenance of these viruses in nature. While stabilized infections have not been shown for vesiculoviruses or phleboviruses in sand flies, the difficulty in orally infecting sand flies and the lack of vertebrates capable of producing a viremia sufficient to infect a sand fly,[138] along with the relatively high rates of vertical transmission of these viruses in sand flies, indicate that these viruses may also be maintained primarily, if not entirely, by transovarial/venereal transmission in nature. The hypothesis that certain vector-borne agents may be maintained primarily by transovarial transmission has been reviewed by Tesh and Shroyer.[129]

In contrast, vertical transmission appears to be much less efficient in the insect-borne flaviviruses and nearly nonexistent in the alphaviruses. Thus, other methods may be used by these viruses for long-term persistence. For example, migrating birds may reintroduce viruses each year. Also, viruses may survive in vectors not normally associated with them. For example, SLE,[139] WEE,[140] and YF[141] viruses have all been shown to be transovarially transmitted by ticks, and Fort Morgan virus can overwinter in the cliff swallow bug, *Oeciacus vicarius.*[89] Because viral persistance may be more susceptible to intervention during periods unsuitable for vector activity than when vectors are actively transmitting these agents, additional studies are needed to elucidate these methods for long-term viral persistence in nature.

VIII. SUMMARY

Numerous factors affected the ability of arthropods to transmit viruses. These include both biotic and abiotic factors inherent to the ecosystem, vector, vertebrate host, and virus, as well as the interactions between them. Transmission cycles range in complexity from the relatively simple, involving only a single species of vertebrate host and a single primary vector, to the highly complex, which may involve several completely different sets of vertebrate hosts and arthropod vectors in separate or overlapping en- and epizootic cycles before man or domestic animals become involved. Mechanisms for viral persistence from year to year include survival in long-lived vectors, migration of infected vertebrate hosts, and transstadial and transovarial transmission in appropriate vectors.

REFERENCES

1. **Burgdorfer, W. and Eklund, C. M.,** Studies on the ecology of Colorado tick fever virus in western Montana, *Am. J. Hyg.*, 69, 127, 1959.
2. **Williams, J. E., Young, O. P., and Watts, D. M.,** Relationship of density of *Culiseta melanura* mosquitoes to infection of wild birds with eastern and western equine encephalitis viruses, *J. Med. Entomol.*, 11, 352, 1974.
3. **Crans, W. J.,** The status of *Aedes sollicitans* as an epidemic vector of eastern equine encephalitis in New Jersey, *Mosq. News*, 37, 85, 1977.
4. **Reeves, W. C.,** Factors that influence the probability of epidemics of western equine, St. Louis, and Califiornia encephalitis in California, *Vector Views*, 14, 13, 1967.
5. **Reeves, W. C.,** Arthopods as vectors and resevoirs of animal pathogenic viruses, in *Handbuch der Virus Forschung*, Vol. 4, Supplement 3, Hallauer, C. and Meyer, K. F., Eds., Springer-Verlag, Vienna, 1957, 117.
6. **Barnett, H. C.,** The incrimination of arthropods as vectors of disease, 11th Int. Cong. Entomol., 1962, 341.
7. **Turell, M. J., Gargan, T. P., II, and Bailey, C. L.,** Replication and dissemination of Rift Valley fever virus in *Culex pipiens*, *Am. J. Trop. Med. Hyg.*, 33, 176, 1984.
8. **Day, M. F., Fenner, F., Woodroofe, G. M., and McIntyre, G. A.,** Further studies on the mechanism of mosquito transmission of myxomatosis in the European rabbit, *J. Hyg.*, 54, 258, 1956.
9. **Hornby, H. E.,** Studies on rinderpest immunity: two methods of infection, *Vet. J.* 82, 348, 1926.
10. **Shope, R. E.,** Swine pox, *Arch. Ges. Virusforsh.*, 1, 457, 1940.
11. **Hoskins, M.,** An attempt to transmit yellow fever virus by dog fleas *(Ctenocephalides canis* Curt) and flies *(Stomoxys calcitrans* Linn.), *J. Parasitol.*, 19, 299, 1933.
12. **Ferris, D. H., Hanson, R. P., Dicke, R. J., and Roberts, R. H.,** Experimental transmission of vesicular stomatitis virus by diptera, *J. Infect. Dis.*, 96, 184, 1955.
13. **Barnett, H. C.,** The transmission of western equine encephalitis virus by the mosquito *Culex tarsalis* Coq., *Am. J. Trop. Med. Hyg.*, 5, 86, 1956.
14. **Chamberain, R. W. and Sudia, W. D.,** Mechanism of tranmission of viruses by mosquitoes, *Ann. Rev. Entomol.*, 6, 371, 1961.
15. **Kay, B. H.,** Three modes of transmission of Ross River virus by *Aedes vigilax*, *Aust. J. Exp. Biol. Med. Sci.*, 60, 339, 1982.
16. **Hoch, A. L., Gargan, T. P., II, and Bailey, C. L.,** Mechanical transmission of Rift Valley fever virus by hematophagus diptera, *Am. J. Trop. Med. Hyg.*, 34, 188, 1985.
17. **Gubler, D. J., Nalim, S., Tan, R., Saipan, H., and Sulianti Saroso, J.,** Variation in susceptibility to oral infection with dengue viruses among geographic strains of *Aedes aegypti*, *Am. J. Trop. Med. Hyg.*, 28, 1045, 1979
18. **Gubler, D. J. and Rosen, L.,** Variation among geographic strains of *Aedes albopictus* in susceptibility to infection with dengue viruses, *Am. J. Trop. Med. Hyg.*, 25, 318, 1976.
19. **Tesh, R. B., Gubler, D. J., and Rosen, L.,** Variation among geographic strains of *Aedes albopictus* in susceptibility to infection with chikungunya virus, *Am. J. Trop. Med. Hyg.*, 25, 326, 1976.
20. **Hardy, J. L., Reeves, W. C., and Sjogren, R. D.,** Variations in the susceptibility of field and laboratory populations of *Culex tarsalis* to experimental infection with western equine encephalomyelitis virus, *Am. J. Epidemiol.*, 103, 498, 1976.
21. **Grimstad, P. R., Craig, G. B., Jr., Ross, Q. E., and Yuill, T. M.,** *Aedes triseriatus* and La Crosse virus: geographic variation in vector susceptibility and ability to transmit, *Am. J. Trop. Med. Hyg.*, 26, 990, 1977.
22. **Tabachnick, W. J., Wallis, G. P., Jr., Aitken, T. H. G., Miller, B. R., Amato., G. D., Lorenz, L., Powell, J. R., and Beaty, B. J.,** Oral infection of *Aedes aegypti* with yellow fever virus: geographic variation and genetic considerations, *Am. J. Trop. Med. Hyg.*, 34, 1219, 1985.
23. **Hardy, J. L., Apperson, G., Asman, S. M., and Reeves, W. C.,** Selection of a strain of *Culex tarsalis* highly resistant to infection following ingestion of western equine encephalomyelitis virus, *Am. J. Trop. Med. Hyg.*, 27, 313, 1978.
24. **Hayes, C. G., Baker, R. H., Baqar, S., and Ahmed, T.,** Genetic variation for West Nile virus susceptibility in *Culex tritaeniorhynchus*, *Am. J. Trop. Med. Hyg.*, 33, 715, 1984.

25. **Storey, H. H.,** Investigations of the mechanisms of the transmission of plant viruses by insect vectors, *Proc. R. Soc. London Ser. B.*, 113, 463, 1933.
26. **Kramer, L. D., Hardy, J. L., Presser, S. B., and Houk, E. J.,** Dissemination barriers for western equine encephalomyelitis virus in *Culex tarsalis* infected after ingestion of low viral doses, *Am. J. Trop. Med. Hyg.*, 30, 190, 1981.
28. **Weaver, S. C., Scherer, W. F., Cupp, E. W., and Castello, D. A.,** Barriers to dissemination of Venezuelan encephalitis virus in the middle American enzootic vector mosquito, *Culex (Melanoconion) taeniopis, Am. J. Trop. Med. Hyg.*, 33, 953, 1984.
29. **Bates, M. and Roca-Garcia, M.,** The development of the virus of yellow fever in *Haemagogus* mosquitoes, *Am. J. Trop. Med.*, 26, 585, 1946.
30. **Chamberlain, R. W., Sudia, W. D., and Gillett, J. D.,** St. Louis encephalitis virus in mosquitoes, *Am. J. Hyg.*, 70, 221, 1959.
31. **Thomas, L. A.,** Distribution of the virus of western equine encephalomyelitis in the mosquito vector *Culex tarsalis, Am. J. Hyg.*, 78, 150, 1963.
32. **Jupp, P. G.,** Laboratory studies on the transmission of West Nile virus by *Culex (Culex) univittatus* Theobald; factors influencing the transmission rate, *J. Med. Entomol.*, 11, 455, 1974.
33. **Watts, D. M., DeFoliart, G. R., and Yuill, T. M.,** Experimental transmssion of trivittatus virus (California virus group) by *Aedes trivittatus, Am. J. Trop. Med. Hyg.*, 25, 173, 1976.
34. **Murphy, F. A., Whitfield, S. G., Sudia, W. D., and Chamberlain, R. W.,** Interactions of vector with vertebrate pathogenic viruses, in *Invertebrate Immunity*, Maramorosch, K. and Shope, R. E., Eds., Academic Press, New York, 1975, 25.
35. **Hardy, J. L., Houk, E. J., Kramer, L. D., and Reeves, W. C.,** Intrinsic factors affecting vector
36. **Hoogstraal, H.,** Ticks in relation to human diseases caused by viruses, *Ann. Rev. Entomol.*, 11, 261, 1966.
37. **Reeves, W. C.,** Ecology of mosquitoes in relation to arboviruses, *Ann. Rev. Entomol.*, 10, 25, 1965.
38. **Reeves, W. C.** Factors that influence the probability of epidemics of western equine, St. Louis, and California encephalitis in California, *Vector Views*, 14, 13, 1967.
39. **Downs, W. G.,** Arboviruses, in *Viral Infections of Humans*, Evens, A. S., Ed., Plenum Press, New York, 1983, 95.
40. **Nuttall, P. A.,** Transmission of viruses to wild life by ticks, in *Vectors in Virus Biology*, Mayo, M. A. and Harrap, K. A., Eds., Academic Press, New York, 1984, 135.
41. **Olson, J. G., Reeves, W. C., Emmons, R. W., and Milby, M. M.,** Correlation of *Culex tarsalis* population indices with the incidence of St. Louis encephalitis and western equine encephalomyelitis in California, *Am. J. Trop. Med. Hyg.*, 28, 335, 1979.
42. **Nelson, R. L., Tempelis, C. H., Reeves, W. C., And Milby, M. M.,** Relation of mosquito density to bird-mammal feeding ratios of *Culex tarsalis* in stable traps, *Am. J. Trop. Med. Hyg.*, 25, 664, 1976.
43. **Davis, N. C.,** The effect of various temperatures in modifying the extrinsic incubation period of yellow fever virus in *Aedes aegypti, Am. J. Hyg.*, 16, 163, 1932.
44. **Chamberlain, R. W. and Sudia, W. D.,** The effects of temperature upon the extrinsic incubation of eastern equine encephalitis in mosquitoes, *Am. J. Hyg.*, 62, 295, 1955.
45. **Hurlbut, H. S.,** The effects of environmental temperature upon the transmission of St. Louis encephalitis virus by *Culex pipiens quinquefasciatus, J. Med. Entomol.*, 10, 1, 1973.
46. **Kramer, L. D., Hardy, J. L., and Presser, S. B.,** Effect of temperature of extrinsic incubation on the vector competence of *Culex tarsalis* for western equine encephalomyelitis virus, *Am. J. Trop. Med. Hyg.*, 32, 1130, 1983.
47. **Turell, M. J., Rossi, C. A., and Bailey, C. L.,** Effect of extrinsic incubation temperature on the ability of *Aedes taeniorhynchus* and *Culex pipiens* to transmit Rift Valley fever virus, *Am. J. Trop. Med. Hyg.*, 34, 1211, 1985.
48. **Hess, A. D., Cherubin, C. E., and LaMotte, L. C.,** Relation of temperature to activity of western and St. Louis encephalitis viruses, *Am. Trop. Med. Hyg.*, 12, 657, 1963.
49. **Manson-Bahr, P. F. C. and Apted, F. I. C.,** *Manson's Tropical Diseases*, Balliere Tindall, London, 1982, 259.
50. **Mims, C. A., Day, M. F., and Marshall, I. D.,** Cytopathic effect of Semliki Forest virus in the mosquito
51. **Karabatsos, N., Ed.,** *International Catalogue of Arboviruses Including Certain Other Viruses of Vertebrates*, 3rd ed., American Society of Tropical Medicine and Hygiene, Washington, D. C., 1985.
52. **Turell, M. J., Gargan, T. P., II, and Bailey, C. L.,** *Culex pipiens* (Diptera: Culicidae) morbidity and mortality associated with Rift Valley fever virus infection, *J. Med. Entomol.*, 22, 332, 1985.
53. **Beaty, B. J., Tesh, R. B., and Aitken, T. H. G.,** Transovarial transmission of yellow fever virus in *Stegomyia* mosquitoes, *Am. J. Trop. Med. Hyg.*, 29, 125, 1980.
54. **Tesh, R. B.,** Experimental studies on the transovarial transmission of Kunjin and San Angelo viruses in mosquitoes, *Am. J. Trop. Med. Hyg.*, 29, 657, 1980.

55. **Turell, M. J., Reeves, W. C., and Hardy, J. L.,** Transovarial and transstadial transmission of California encephalitis virus in *Aedes dorsalis* and *Aedes melanimon*, *Am. J. Trop. Med. Hyg.*, 31, 1021, 1982.
56. **Rosen, L., Shroyer, D. A., Tesh, R. B., Freier, J. E., and Len, J. C.,** Transovarial transmission of dengue viruses by mosquitoes: *Aedes albopictus and Aedes aegypti*, *Am. J. Trop. Med. Hyg.*, 32, 1108, 1983.
57. **Grimstad, P. R., Ross, Q. E., and Craig, G. B., Jr.,** *Aedes triseriatus* (Diptera: Culicidae) and La Crosse virus. II. Modification of mosquito feeding behavior by virus infection, *J. Med. Entomol.*, 17, 1, 1980.
58. **Patrican, L. A., DeFoliart, G. R., and Yuill, T. M.,** La Crosse viremias in juvenile, subadult and adult chipmunks *(Tamias striatus)* following feeding by transovarially-infected *Aedes triseriatus*, *Am. J. Trop. Med. Hyg.*, 34, 596, 1985.
59. **Ribeiro, J. M. C., Rossignol, P. A., and Spielman, A.,** Role of mosquito saliva in blood vessel location, *J. Exp. Biol.*, 108, 1, 1984.
60. **Gargan, T. P., II, Bailey, C. L., Higbee, G. A., Gad, A., and El Said, S.,** The effect of laboratory colonization on the vector-pathogen interactions of Egyptian *Culex pipiens* and Rift Valley fever virus, *Am. J. Trop. Med. Hyg.*, 32, 1154, 1983.
61. **Turell, M. J. and Bailey, C. L.,** Transmission studies in mosquitoes (Diptera: Culicidae) with disseminated Rift Valley fever virus infections, *J. Med. Entomol.*, 24, 11, 1987.
62. **Howlett, F. M.,** The influence of temperature on the biting of mosquitoes, *Parasitology*, 3, 479, 1910.
63. **Peterson, D. G. and Brown, A. W. A.,** Studies on the responses of the female *Aedes* mosquito. III. The response of *Aedes aegypti* (L.) to a warm body and its radiation, *Bull. Entomol. Res.*, 42, 535, 1951.
64. **Wright, R. H.,** Tunes to which mosquitoes dance, *New Sci.*, 37, 694, 1968.
65. **Brown, A. W. A.,** Factors in the attractiveness of bodies for mosquitoes, *Nature (London)*, 167, 202, 1951.
66. **Reeves, W. C.,** Field studies on carbon dioxide as a possible stimulant to mosquitoes, *Proc. Soc. Exp. Biol. Med.*, 77, 64, 1951.
67. **Garcia, R.,** Carbon dioxide as an attractant for certain ticks (Acarina: Argasidae and Ixodidae), *Ann. Entomol. Soc. Am.*, 55, 505, 1962.
68. **Wilson, J. G., Kinzer, D. R., Sauer, J. R., and Hair, J. A.,** Chemoattraction in the lone star tick. I. Response of different developmental stages to carbon dioxide administered via traps, *J. Med. Entomol.*, 9, 245, 1972.
69. **Gillett, J. D. and Connor, J.,** Host temperature and the transmission of arboviruses by mosquitoes, *Mosq. News*, 36, 472, 1976.
70. **Edman, J. D. and Kale, H. W.,** Host behavior: its influence on the feeding success of mosquitoes, *Ann. Entomol. Soc. Am.*, 64, 513, 1971.
71. **Webber, L. A. and Edman, J. D.,** Anti-mosquito behavior of ciconiiform birds, *Anim. Behav.*, 20, 228, 1972.
72. **Edman, J. D., Webber, L. A., and Schmid, A. A.,** Effect of host defenses on the feeding pattern of *Culex nigripalpis* when offered a choice of blood sources, *J. Parasitol.*, 60, 874, 1974.
73. **Klowden, M. J. and Lea, A. O.,** Effect of defensive host behavior on the blood meal size and feeding
74. **Day, J. F., Ebert, K. M., and Edman, J. D.,** Feeding patterns of mosquitoes (Diptera: Culicidae) simultaneously exposed to malarious and healthy mice, including a method for separating blood meals from conspecific hosts, *J, Med. Entomol.*, 20, 120, 1983.
75. **Mahon, R. and Gibbs, A.,** Arbovirus-infected hens attract more mosquitoes, in *Viral Diseases in S. E. Asia and Western Pacific*, Mackensie, J. S., Ed., Academic Press, Sydney, 1982, 502.
76. **Turell, M. J., Bailey, C. L., and Rossi, C. A.,** Increased mosquito feeding on Rift Valley fever virus&infected lambs, *Am. J. Trop. Med. Hyg.*, 33, 1232, 1984.
77. **Rossignol, P. A., Ribeiro, J. M. C., Jungery, M., Turell, M. J., and Bailey, C. L.,** Enhanced mosquito blood-finding success on parasitemic hosts: evidence for vector-parasite mutualism, *Proc. Natl. Acad. Sci. U. S. A.*, 82, 7725, 1985.
78. **Merrill, M. H. and TenBroeck, C.,** The transmission of equine encephalitis virus by *Aedes aegypti*, *J. Exp. Med.*, 62, 687, 1935.
79. **Mellor, P. S. and Boorman, J.,** Multiplication of blue tongue virus in *Cuclicoides nubeculosus* (Meigen) simultaneously infected with the virus and the microfilariae, of *Onchocerca cervicalis* (Railliet and Henry), *Ann. Trop. Med. Parasitol.*, 74, 463, 1980.
80. **Turell, M. J., Rossignol, P. A., Spielman, A., Rossi, C. A., and Bailey, C. L.,** Enhanced arboviral transmission by mosquitoes that concurrently ingested microfilariae, *Science*, 225, 1039, 1984.
81. **Barnett, H. C.,** Experimental studies of concurrent infection of canaries and of the mosquito *Culex tarsalis* with *Plasmodium relictum* and western equine encephalitis virus, *Am. J. Trop. Med. Hyg.*, 5, 99, 1956.
82. **Miller, B. R. and DeFoliart, G. R.,** Infection rates of Ascocystis&infected *Aedes triseriatus* following ingestion of La Crosse virus by the larvae, *Am. J. Trop. Med. Hyg.*, 28, 1064, 1979.

83. **Reeves, W. C.,** Overwintering of arboviruses, *Prog. Med. Virol,* 17, 93, 1974.
84. **Lewis, E. A.,** Nairobi sheep disease: the survival of the virus in the tick *Rhipicephalus appendiculatus, Parasitology,* 37, 55, 1946.
85. **Danielova, V. and Minar, J.,** Experimental overwintering of the virus Tahyna in mosuitoes *Culiseta annulata* (Schik.) *(Diptera, Culicidae), Folia Parasit.,* 16, 285, 1969.
86. **McLean, D. M., Grass, P. N., Judd, B. D., and Stolz, K. J.,** Bunyavirus development in arctic and *Aedes aegypti* mosquitoes as revealed by glucose oxidase staining and immunofluorescence, *Arch. Virol.,* 62, 313, 1979.
87. **Hurlbut, H. S.,** The transmission of Japanese B encephalitis by mosquitoes after experimental hibernation, *Am. J. Hyg.,* 51, 265, 1950.
88. **Bellamy, R. E., Reeves, W. C., and Scrivani, R. P.,** Relationships of mosquito vectors to winter survival of encephalitis viruses. II. Under experimental conditions, *Am. J. Hyg.,* 67, 90, 1958.
89. **Hayes, R. O., Francy, D. B., Lazuick, J. S., Smith, G. C.,. and Gibbs, E. P. J.,** Role of the cliff swallow bug *(Oeciacus vicarius)* in the natural cycle of a western equine encephalitis-related alphavirus, *J. Med. Entomol.,* 14, 257, 1977.
90. **Blackmore, J.S. and Winn, J. F.,** A winter isolation of western equine encephalitis virus from hibernating *Culex tarsalis* Coquillett, *Proc. Soc. Exp. Biol. Med.,* 91, 146, 1956.
91. **Reeves, W. C., Bellamy, R. E., and Scrivani, R. P.,** Relationships of mosquito vectors to winter survival of encephalitis viruses. I. Under natural conditions, *Am. J. Hyg.,* 67, 78, 1958.
92. **Bailey, C. L., Eldridge, B. F., Hayes, D. E., Watts, D. M., Tammariello, R. F., and Dalrymple, J. M.,** Isolation of St. Louis encephalitis virus from overwintering *Culex pipiens* mosquitoes, *Science,* 199, 1346, 1978.
93. **Slaff, M. E. and Crans, W. J.,** Parous rates of overwintering *Culex pipiens pipiens* in New Jersey, *Mosq. News,* 37, 11, 1977.
94. **Eldridge, B. F.,** The effect of temperature and photoperiod on blood feeding and ovarian development in mosquitoes of the *Culex pipiens* complex, *Am. J. Trop. Med. Hyg.,* 17, 133, 1968.
95. **Bailey, C. L., Faran, M. E., Gargan, T. P., II, and Hayes, D. E.,** Winter survival of blood-fed and nonblood-fed *Culex pipiens* L., *Am. J. Trop. Med. Hyg.,* 31, 1054, 1982.
96. **Rozeboom, L. E. and Burgdorfer, W.,** Development of Colorado tick virus in the Rocky Mountain wood tick, *Dermacentor andersoni, Am. J. Hyg.,* 69, 138, 1959.
97. **Plowright, W., Perry, C. T., and Peirce, M. A.,** Transovarial infection with African swine fever virus in the argasid tick, *Ornithodoros moubata porcinus* Walton, *Res. Vet. Sci.,* 11, 582, 1970.
98. **Burgdorfer, W. and Varma, M. G. R.,** Trans-stadial and transovarial development of disease agents in arthropods, *Ann. Rev. Entomol.,* 12, 347, 1967.
99. **Watts, D. M., Pantuwatana, S., DeFoliart, G. R., Yuill, T. M., and Thompson, W. H.,** Transovarial transmission of La Crosse virus (California encephalitis group) in the mosquito, *Aedes triseriatus, Science,* 180, 1140, 1973.
100. **Tesh, R. B., Chaniotis, B. N., and Johnson, K. M.,** Vesicular stomatitis virus (Indiana serotype): transovarial transmission by phlebotomine sandflies, *Science,* 175, 1477, 1972.
101. **Tesh, R. B. and Chaniotis, B. N., Peralta, P. H., and Johnson, K. M.,** Ecology of viruses isolated from Panamanian phlebotomine sandflies, *Am. J. Trop. Med. Hyg.,* 23, 258, 1974.
102. **Endris, R. G., Tesh, R. B., and Young, D. G.,** Transovarial transmission of Rio Grande virus *(Bunyaviridae: Phlebovirus)* by the sand fly, *Lutzomyia anthophora, Am. J. Trop. Med. Hyg.,* 32, 862, 1983.
103. **Tesh, R. B.,** Transovarial transmission of arboviruses in their invertebrate vectors, *Curr. Tropics Vector Res.,* 2, 57, 1984.
104. **Tesh, R. B. and Modi, G. B.,** Growth and transovarial transmission of Chandipura virus (Rhabdoviridae: *Vesiculovirus*)in *Phlebotomus papatasi, Am. J. Trop. Med. Hyg.,* 32, 621, 1983.
105. **Tesh, R. B. and Modi, G. B.,** Studies on the biology of phleboviruses in sand flies (Diptera:psychodidae). I. Experimental infection of the vector, *Am. J. Trop. Med. Hyg.,* 33, 1007, 1984.
106. **Christensen, B. M., Rowley, W. A., Wong, Y. W., Dorsey, D. C., and Hausler, W. J., Jr.,** Laboratory studies of transovarial transmission of trivittatus virus by *Aedes trivittatus, Am. J. Trop. Med. Hyg.,* 27, 184, 1978.
107. **Aitken, T. H. G., Tesh, R. B., Beaty, B. J., and Rosen, L.,** Transovarial transmissiom of yellow fever virus by mosquitoes *(Aedes aegypti), Am. J. Trop. Med. Hyg.,* 28, 119, 1979.
108. **Turell, M. J.,** The Role of Transovarial Transmission in the Natural History of California Encephalitis Virus in California, Ph. D. dissertation, University of California, Berkeley, 1981.
109. **Thomas, L. A.,** Distribution of the virus of western equine encephalomyelitis in the mosquito vector, *Culex tarsalis, Am. J. Hyg.,* 78, 150, 1963.
110. **Beaty, B. J. and Thompson, W. H.,** Tropisms of La Crosse virus in *Aedes triseriatus* (Diptera:Culicidae) following infective blood meals, *J. Med. Entomol.,* 14, 499, 1978.

111. **Kuberski, T.,** Fluorescent antibody studies on the development of dengue-2 virus in *Aedes albopictus* (Diptera:Culicidae), *J. Med. Entomol.*, 16, 343, 1979.
112. **Scott, T. W., Hildreth, S. W. and Beaty, B. J.,** The distribution and development of eastern equine encephalitis virus in its enzootic mosquito vector, *Culiseta melanura, Am. J. Trop. Med. Hyg.*, 33, 300, 1984.
113. **Plowright, W., Perry, C. T., and Greig, A.,** Sexual transmission of African swine fever virus in the tick *Ornithodoros moubata porcinus* Walton, *Res. Vet. Sci.*, 17, 106, 1974.
114. **Morris, C. D. and Srihongse, S.,** An evaluation of the hypothesis of transovarial transmission of eastern equine encephalomyelitis virus by *Culiseta melanura, Am. J. Trop. Med. Hyg.*, 27, 1246, 1978.
115. **Sprance, H. E.,** Experimental evidence against the transovarial transmission of eastern equine encephalitis virus in *Culiseta melanura, Mosq. News*, 41, 168, 1981.
116. **Clark, G. G., Crans, W. J., and Crabbs, C. L.,** Absence of eastern equine encephalitis (EEE) virus in immature *Coquillettidia perturbans* associated with equine cases of EEE, *J. Am. Mosq. Control Assoc.*, 1, 540, 1985.
117. **Romoser, W. S., Faran, M. E., and Bailey, C. L.,** unpublished data, 1986.
118. **Turell, M. J.,** unpublished data, 1986.
119. **Turell, M. J., Reeves, W. C., and Hardy, J. L.,** Evaluation of the efficiency of transovarial transmission of California encephalitis viral strains in *Aedes dorsalis* and *Aedes melanimon, Am. J. Trop. Med. Hyg.*, 31, 382, 1982.
120. **Fine, P. E. M.,** Vectors and vertical transmision: an epidemiologic perspective, *Ann. N. Y. Acad. Sci.*, 266, 173, 1975.
121. **Miller, R. B., DeFoliart, G. R., and Yuill, T. M.,** Vertical transmission of La Crosse virus (California encephalitis group): transovarial and filial infection rates in *Aedes triseriatus* (Diptera:Culicidae), *J. Med. Entomol.*, 14, 437, 1977.
122. **Turell, M. J., Hardy, J. L., and Reeves, W. C.,** Stabilized infection of California encephalitis virus in *Aedes dorsalis*, and its implications for viral maintenance in nature, *Am. Trop. Med. Hyg.*, 31, 1252, 1982.
123. **Hardy, J. L. and Turell, M. J.,** unpublished data, 1986.
124. **Danielova, V. and Ryba, J.,** Laboratory demonstration of transovarial transmission of Tahyna virus in *Aedes vexans* and the role of this mechanism in the overwintering of this arbovirus, *Folia Parasitol.*, 26, 361, 1979.
125. **Hardy, J. L., Rosen, L., Kramer, L. D., Presser, S. B., Shroyer, D. A., and Turell, M. J.,** Effect of rearing temperature on transovarial transmission of St. Louis encephalitis virus in mosquitoes, *Am. J. Trop. Med. Hyg.*, 29, 963, 1980.
126. **Francy, D. B., Rush, W. A., Montoya, M., Inglish, D. S. and Bolin, R. A.,** Transovarial transmission of St. Louis encephalitis virus by *Culex pipiens* complex mosquitoes, *Am. J. Trop. Med. Hyg.*, 30, 699, 1981.
127. **Hardy, J. L., Rosen, L., Reeves, W. C., Scrivani, R. P., and Presser, S. B.,** Experimental transovarial transmission of St. Louis encephalitis virus by *Culex* and *Aedes* mosquitoes, *Am. J. Trop. Med. Hyg.*, 33, 166, 1984.
128. **Rosen, L., Tesh, R. B., Lien, J. C., and Cross, J. H.,** Transovarial transmission of Japanese encephalitis virus by mosquitoes, *Science*, 199, 909, 1978.
129. **Tesh, R. B. and Shroyer, D. A.,** The mechanism of arbovirus transovarial transmission in mosquitoes: San Angelo virus in *Aedes albopictus, Am. J. Trop. Med. Hyg.*, 29, 1394, 1980.
130. **Fine, P. E. M. and LeDuc, J. W.,** Towards a quantitative understanding of the epidemiology of Keystone virus in the eastern United States, *Am. J. Trop. Med. Hyg.*, 27, 322, 1978.
131. **DeFoliart, G. R.,** *Aedes triseriatus:* vector biology in relationship to the persistence of La Crosse virus in endemic foci, in *California Serogroup Viruses*, Calisher, C. H. and Thompson, W. H., Eds., Alan R. Liss, New York, 1983, 89.
132. **Miller, B. R., DeFoliart, G.R., and Yuill, T. M.,** *Aedes triseratius* and La Crosse virus: lack of infection in eggs of the first ovarian cycle following oral infection of females, *Am. J. Trop. Med. Hyg.*, 28, 897, 1979.
133. **Seecof, R. L.,** The sigma virus infection of *Drosophila melanogaster, Curr. Topics Microbiol. Immunol.*, 42, 59, 1968.
134. **Thompson, W. H. and Beaty, B. J.,** Venereal transmission of La Crosse virus from male to female *Aedes triseriatus, Am. J. Trop. Med. Hyg.*, 27, 187, 1978.
135. **Ovenden, J. R. and Mahon, R. J.,** Venereal transmission of Sindbis virus between individuals of *Aedes australis* (Diptera:Culicidae), *J. Med. Entomol.*, 21, 292, 1984.
136. **Thompson, W. H.,** Higher venereal infection and transmission rates with La Crosse virus in *Aedes triseriatus* engorged before mating, *Am. J. Trop. Med. Hyg.*, 28, 890, 1979.
137. **Thompson, W. H.,** Lower rates of oral transmission of La Crosse virus by *Aedes triseriatus* venereally exposed after engorgement on immune chipmunks, *Am. J. Trop. Med. Hyg.*, 32, 1416, 1983.

138. **Tesh, R. B. and Chaniotis, B. N.,** Transovarial transmission of viruses by phlebotomine sandflies, *Ann. N. Y. Acad. Sci.*, 266, 125, 1975.
139. **Blattner, R. J. and Heys, F. M.,** Blood-sucking vectors of encephalitis: experimental transmission of St. Louis encephalitis (Hubbard strain) to white Swiss mice by the American dog tick, *Dermacentor variabilis* Say, *J. Exp. Med.*, 79, 439, 1944.
140. **Syverton, J. T. and Berry, G. P.,** Hereditary transmission of the western type of equine encephalomyelitis virus in the wood tick, *Dermacentor andersoni* Stiles, *J. Exp. Med.*, 73, 507, 1941.
141. **Saluzzo, J. F., Herve, J. P., Salaun, J. J., Germain, M., Cornet, J. P., Camicas, J. L., Heme, G., and Robin, Y.,** Caracteristiques des souches du virus de la fièvre jaune isolées des oeufs et des larves d'une tique *Amblyomma variegatum*, recoltée sur le betail a Bangui (Centrafrique), *Ann. Virol. (Inst. Pasteur)*, 131, 155, 1980.
142. **Tesh, R. B.,** Transovarial transmission of arboviruses in their invertebrate vectors, *Curr. Topics Vector Res.*, 2, 57, 1984.
143. **Jousset, F. X.,** Geographic *Aedes aegypti* strains and dengue 2 virus susceptibility ability to transmit to vertebrate and transovarial transmission, *Ann. Virol.*, 132, 357, 1982.
144. **Khin, M. M. and Than, K. A.,** Transovarial transmission of dengue 2 virus by *Aedes aegypti* in nature, *Am. Trop. Med. Hyg.*, 32, 590, 1983.
145. **Cordellier, R., Bouchite, B., Roche, J. C., Monteny, N., Diaco, B., and Akoliba, P.,** The sylvatic distribution of dengue 2 virus in the sub-Sudanese savanna areas of the Ivory Coast in 1980. Entomological data and epidemiological study, *Cah. Orstom Ser. Entomol. Med. Parasitol.*, 21, 165, 1983.
146. **Hull, B., Tikasingh, E., De-Souza, M., and Martinez, R.,** Natural transovarial transmission of dengue 4 virus in *Aedes aegypti* in Trinidad, *Am. J. Trop. Med. Hyg.*, 33, 1248, 1984.
147. **Rehacek, J.,** Transovarial transmission of tick-borne encephalitis virus by ticks, *Acta Virol.*, 6, 220, 1962.
148. **Rosen, L., Shroyer, D. A., and Lien, J. C.,** Transovarial transmission of Japanese encephalitis virus by *Culex tritaeniorhynchus* mosquitoes, *Am. J. Trop. Med. Hyg.*, 29, 711, 1980.
149. **Coz, J., Valade, M., Cornet, M., and Robin, Y.,** Transmission transovarienne d'un *Flavivirus*, le virus Koutango chez *Aedes aegypti* L., *C. R. Acad. Sci. Paris*, 283, 109, 1976.
150. **Singh, K. R. P., Pavri, K., and Anderson, C. R.,** Experimental transovarial transmission of Kyasanur Forest disease virus in *Haemaphysalis spinigera*, *Nature (London)*, 199, 513, 1963.
151. **Kay, B. H. and Carley, J. G.,** Transovarial transmission of Murray Valley encephalitis virus by *Aedes aegypti*, *Aust. J. Exp. Biol. Med. Sci.*, 58, 501, 1980.
152. **Cornet, M., Robin, Y., Heme, G., Adam C., Renaudet, J., Valade, M., and Eyraud, M.,** Une poussée epizootique de fievre jaune selvatique au Senegal oriental. Isolement du virus de lots de moustiques adultes males et femelles, *Med. Mal. Infect.*, 9, 63, 1979.
153. **Dutary, B. E. and LeDuc, J. W.,** Transovarial transmission of yellow fever virus by a sylvatic vector, *Haemagogus equinus*, *Trans. R. Soc. Trop. Med. Hyg.*, 75, 128, 1981.
154. **Corner, L. C., Robertson, A. K., Hayles, L. B., and Iverson, J. O.,** Cache Valley virus: experimental infection in *Culiseta inornata*, *Can. J. Microbiol.*, 26, 287, 1980.
155. **Crane, G. T. and Elbel, R. E.,** Transovarial transmission of California encephalitis virus in the mosquito *Aedes dorsalis* at Blue Lake, Utah, *Mosq. News*, 37, 479, 1977.
156. **Turell, M. J., Hardy, J. L., and Reeves, W. C.,** Demonstration of transovarial transmission of California encephalitis virus in experimentally infected *Aedes melanimon*, *Proc. Calif. Mosq. Vector Control Assoc.*, 48, 15, 1980.
157. **Reeves, W. C. and Hardy, J. L.,** unpublished data, 1986.
158. **Berry, R. L., Lalonde-Weigert, B. J., Calisher, C. H., Parsons, M. A., and Bear, G. T.,** Evidence for transovarial transmission of Jamestown Canyon virus in Ohio, *Mosq. News*, 37, 494, 1977.
159. **LeDuc, J. W., Suyemoto, W., Eldridge, B. F., Russell, P. K., and Barr, A. R.,** Ecology of California encephalitis viruses on the Delmarva Peninsula. II. Demonstration of transovarial transmission, *Am. J. Trop. Med. Hyg.*, 24, 124, 1975.
160. **Pantuwatana, S., Thompson, W. H., Watts, D. M., Yuill, T. M., and Hanson, R. P.,** Isolation of La Crosse virus from field collected *Aedes triseriatus* larvae, *Am. J. Trop. Med. Hyg.*, 23, 246, 1974.
161. **McLintock, J., Carry, P. S., Wagner, R. J., Leung, M. K. and Iverson, J. O.,** Isolation of snowshoe hare virus from *Aedes implicatus* larvae in Saskatchewan, *Mosq. News*, 36, 233, 1976.
162. **Belloncik, S., Poulin, L., Maire, A., Aubin, A., Fauvel, M., and Jousset, F. X.,** Activity of California encephalitis group viruses in Entrelacs (Province of Quebec, Canada), *Can. J. Microbiol.*, 28, 572, 1982.
163. **Labuda, M., Ciampor, F., and Kozuch, O.,** Experimental model of transovarial transmission of Tahyna virus in *Aedes aegypti* mosquitoes, *Acta Virol.*, 27, 245, 1983.
164. **Bardos, V., Ryba, J., and Hubalek, Z.,** Isolation of Tahyna virus from field collected *Culiseta annulata* (Schik.) larvae, *Acta Virol.*, 19, 446, 1975.
165. **Andrews, W. N., Rowley, W. A., Wong, Y. W., Dorsey, D. C., and Hausler, W. J., Jr.,** Isolation of trivittatus virus from larvae and adults reared from field-collected larvae of *Aedes trivittatus* (Diptera: Culicidae), *J. Med. Entomol.*, 13, 699, 1977.

166. **Hoogstraal, H.,** The epidemiology of tick-borne Crimean-Congo hemorrhagic fever in Asia, Europe, and Africa, *J. Med. Entomol.*, 15, 307, 1979.
167. **Huard, M., Cornet, J. P., Germain, M., and Camicas, J. L.,** Passage transovarienne du virus Dugbe chez la tique *Amblyomma variegatum* (Fabricius), *Bull. Soc. Pathol. Exot.*, 71, 19, 1978.
168. **Davies, F. G. and Mwakima, F.,** Qualitative studies of the transmission of Nairobi sheep disease virus by *Rhipicephalus appendiculatus* (Ixodoidea: Ixodidae), *J. Comp. Pathol.*, 92, 15, 1982.
169. **Ciufolini, M. G., Maroli, M., and Verani, P.,** Growth of two phleboviruses after experimental infection of their suspected sand fly vector, *Phlebotomus perniciosus* (Diptera: Psychodidae), *Am. J. Trop. Med. Hyg.*, 34, 174, 1985.
170. **Tesh, R., Saidi, S., Javadian, E., and Nadim, A.,** Studies on the epidemiology of sandfly fever in Iran. I. Virus isolates obtained from *Phlebotomus*, *Am. J. Trop. Med. Hyg.*, 26, 282, 1977.
171. **Herve, J. P., Travassas da Rosa, A. P. A., Sa-Filho, G. C., Travassos da Rosa, J. F., and Pinheiro, F. P.,** Demonstration de la transmission transovarienne due virus Pacui chez *Lutzomyia flaviscutellata* (Phlebotomine). Importance epidemiologique, *Cah. ORSTOM Ser. Entomol. Med. Parasitol.*, 22, 207, 1984.
172. **Aitken, T. H. G., Woodall, J. P., DeAndrade, A. H. P., Bensabath, G., and Shope, R. E.,** Pacui virus, phlebotomine flies, and small mammals in Brazil: an epidemiological study, *Am. J. Trop. Med. Hyg.*, 24, 358, 1975.
173. **Linthicum, K. J., Davies, F. G., Kaird, A., and Bailey, C. L.,** Rift Valley fever virus (family Bunyaviridae, genus *Phlebovirus*). Isolations from diptera collected during an inter-epizootic period in Kenya, *J. Hyg.*, 95, 197, 1985.
174. **Schmidt, J. R., Schmidt, M. L., and Said, M. I.,** Phlebotomus fever in Egypt. Isolation of phlebotomus fever viruses from *Phlebotomus papatasi*, *Am. J. Trop. Med. Hyg.*, 20, 483, 1971.
175. **Lvov, D. K., Sidorova, G. A., Gromashevskii, V. L., Skvortsova, T. M., Aristova, V. A., Ipatov, V. P., Neronov, V. M., Zakaryan, V. A., and Karimov, S. K.,** Isolation of tamdy virus (Bunyaviridae) pathogenic for man from natural sources in Central Asia Kazakh SSR and transcaucasus USSR, *Vopr. Virusol.*, 29, 487, 1984.
176. **Travassos da Rosa, A. P. A., Tesh, R. B., Travassos da Rosa, J. F., Herve, J. P., and Main, A. J., Jr.,** Carajas and Maraba viruses, two new vesiculoviruses isolated from *Phlebotomine* sand flies in Brazil, *Am. J. Trop. Med. Hyg.*, 33, 999, 1984.
177. **Endris, R. and Hess, A. D.,** unpublished data, 1986.
178. **Cordellier, R., Chippaux, A., Monteny, N., Heme, G., Courtois, B., Germain, M., and Digoutte, J.- P.,** Isolements du virus Orungo a partir de femelles et de males d' *Aedes* selvatiques captures en Cote d'Ivoire, *Cah. ORSTOM Ser. Entomol. Med. Parasitol.*, 20, 265, 1982.

Chapter 6

BLOOD-FEEDING BY VECTORS: PHYSIOLOGY, ECOLOGY, BEHAVIOR, AND VERTEBRATE DEFENSE

John D. Edman and Andrew Spielman

TABLE OF CONTENTS

I. INTRODUCTION

The basis for transmission of arboviral agents from one host to another derives from the blood-feeding requirement of certain arthropods, mainly mosquitoes and ticks. Except for vertically and venereally transmitted agents, this act of feeding provides the circumstances for transmitting such pathogens and determines their degree of communicability.

Successful hematophagy culminates in an intricate cascade of events. As a consequence of their interdependency, certain ixodid ticks insure perpetuation by ingesting massive quantities of blood, thereby producing extraordinary numbers of offspring; certain hematophagic flies exploit the reciprocal strategy by developing some eggs without taking any blood at all (autogeny). Vector species, however, tend to depend upon a close association with particular hosts, thereby optimizing feeding success and survival. Longevity and narrowness of host-range dominate as crucial attributes of an effective vector. Human transmission of arboviruses, however, often contradicts this logic because the majority of these pathogens are zoonoses in which human infections represent nonperpetuating aberrations of normal maintenance cycles.[1,2]

The more "primitive" nematocerous Diptera take blood solely as females and in the adult stage. In contrast, ticks normally feed on vertebrate blood in each developmental stage. Ixodid (hard) ticks attach for several days, argasids (soft) ticks for less than 1 hr, and

Nematocera only a few minutes. Because host-contact is fraught with risk, the slow feeding ticks seem to specialize in avoiding their host's inflammatory response, and flies in rapidity of feeding.[3-8]

A complex array of factors, intrinsic as well as extrinsic to the vector, contribute to blood-feeding success. These interrelationships determine the manner in which pathogens are perpetuated and influence epidemiological risk. Accordingly, we shall describe the principal factors that influence the outcome of a vector's attempt to feed, arranging these topics in the order in which they naturally occur. Emphasis will be placed on mosquitoes, the biological vectors of most arboviruses and, to a lesser extent, on ixodid ticks. Biting midges, sandflies, and argasid ticks, which transmit certain arboviral agents, will be mentioned in passing because information on the feeding of these insects is limited. Our objective is to demonstrate both the fragility and resiliency of the host-arbovirus-vector relationship.

II. INITIATION OF QUESTING BEHAVIOR

A. Hormonal Regulation

Juvenile hormone (JH) appears to conserve maturation in mosquitoes, as in the classical model. Thus, mosquitoes enter their "pupal" pharate adult stage with JH essentially absent.[9] Apolysis between the adult and pupal cuticles occurs some hours before adult ecdysis (shedding of the exocuticle), and JH may be secreted during this interval, at least in certain mosquitoes that produce eggs without blood feeding. Other mosquitoes, however, enter the adult stage with their ovaries in a stage of immaturity, i.e., Christophers' Stage I,[10] suggesting that this hormone has not yet been released. Maturation of the teneral (newly emerged) adult to a condition capable of reproductive activity then awaits release of JH from its site of synthesis in the corpora allata.

Secretion of JH appears to prepare: (1) the ovarian follicle to progress to Christophers' stage IIb (resting stage), a vitellogenic-competent stage of development;[11] (2) the capacity of females to accept a mate;[12] and (3) the competence of the fat body to produce vitellogenin.[13] In *Culex,*[14] but not *Aedes,* the allata may regulate host questing behavior. Allatum activity, thus, seems to signal full maturity of the adult, and this generally occurs during the second day after adult emergence.

The actual timing of JH-induced adult maturation depends on a variety of circumstances. Adults deriving from larvae that developed under deprived conditions may not mature until they have fed either on sugar or blood, although better favored larvae develop into adults that mature without additional feeding. Both conditions seem to describe certain *Anopheles* and *Ae. aegypti* found in nature.[15] Temperate *Culex* mosquitoes fail to undergo JH-mediated ovarian development when their pupae are subjected to winter light and temperature conditions.[16] Although they do not seek vertebrate hosts, mating behavior occurs normally. *Anopheles,* on the other hand, may enter a winter condition that has been termed "gonotrophic dissociation".[17] Here, ovaries are in diapause, but questing and mating occur normally. These separations of effects are paradoxical because both functions share common hormonal stimuli. Blood feeding and ovarian maturation resume when summer conditions ensue.

Hormonal regulation of oogenesis in mosquitoes has been a subject for intense and continuing debate. Physical as well as chemical properties of the blood meal[18] appear to stimulate certain endocrine cells contained in the midgut wall[19] to produce a hormone that causes the ovary to release a hormone[20-21] that causes the neurohemal organs in the cerebral neurosecretory system[22] to release brain hormone[23] that causes the ovary to secrete ecdysone[24] that simulates the fat body to produce 20-hydroxyecdysone that stimulates the oocyte to incorporate vitellogenin that is also produced by the fat body.[25] The sequence is further complicated by an apparent postblood-feeding role for JH and the still-unexplained process of maturation of the egg and its enveloping membranes.

Host-seeking behavior is suppressed by the physical distension of the abdomen after mosquitoes ingest blood.[26] Following diuresis and consequent reduction of this distension, a humoral factor serves to maintain this behavioral inhibition,[27] acting to suppress receptor (e.g., antennal) sensitivity to host stimuli such as lactic acid. The fat body may be the source of this inhibitory hormone,[28] but its nature has not been established. Questing behavior resumes following oviposition.

The manner in which ixodid ticks regulate development is poorly understood. Fusion of body segments and consequent compression of the various organs renders it difficult to design experiments requiring surgical ablation. Indeed, we lack clear evidence identifying the source of any humoral regulatory factor in ticks. Neurosecretory vesicles have been demonstrated in various tissues, particularly the synganglion, and ecdysteroids have been extracted and identified.[29] Experiments employing chemical ablation or hormone antagonists followed by hormonal rescue procedures, confirm that 20-hydroxyecdysone and JH-III influence pheromone production, spermatogenesis, and oogenesis; however, the details of these putative regulatory roles have yet to be established. Increased temperature resulting from host contact, pheromonal and mechanical stimuli associated with mating, and distension and nutritional factors that follow feeding combine to provide the signals that disrupt developmental diapause in these organisms. To render this matter even more ambiguous, related argasid ticks express various forms of autogeny.[30] Thus, the requirement of ticks for feeding may be facultative.

The sequence of these hormonal interactions determines the pace and rhythm of the development of vector arthropods and thereby their abundance and frequency of contact with the reservoir hosts of any arboviral pathogens that they may transmit.

B. Age and Gonotrophic Status

Although adult mosquitoes generally emerge with their ovaries in stage I, they do not quest until the ovaries have become vitellogenic-competent (resting stage). Ovarian stage, however, does not appear to regulate questing. Instead, blood-seeking behavior and ovarian development proceed independently, and both mature at about 2 days after emergence. Clear evidence that undernourished mosquitoes will quest before their ovaries reach the resting stage was provided by Feinsod and Spielman.[15] The occasional nutritional requirement of two blood meals for the initial gonotropic cycle, along with partial feeding and transovarial or venereal transmission, provides another mechanism whereby nulliparous females may become infective with virus.

Large synchronous broods of univoltine spring *Aedes* emerge several days prior to the first evidence of biting activity in the field. Nonfed *Culiseta morsitans* were first captured in resting boxes 3 to 5 weeks before blood-engorged individuals were found.[31] It is not clear what controls such prolonged postemergence delays in questing or what the biological advantages might be. Insemination also was delayed about 3 weeks in *Cs. morsitans* and biting inhibition among virgin females has been reported for some species. Nulliparous females also persisted for several days among marked and recaptured *Ae. triseriatus*,[32] and nonfertilized females of this species commonly quest for blood. Male questing and mating often occurs near the host in this and several other species.[33-36] In the lab, *Ae. triseriatus* begin questing 2 days after emergence and peak feeding occurs on days 3 and 4,[37] so age does not appear to be the factor delaying questing in the field. Of course, nullipars may persist simply because suitable hosts are scarce.

Argasid ticks, like hematophagic hemipterans, are essentially nest parasites living in close association with their host; however, ixodid ticks usually quest repeatedly for free-ranging hosts. Although discontinuous, host-seeking behavior extends for several weeks when hosts are not contacted;[38] lab-reared ticks begin to quest within a few days after hatching or molting. Tick questing activity may be accelerated or inhibited by conditions of daylength or temperature, regardless of age.[39]

C. Autogeny

Although mosquitoes mainly require a meal of blood in order to initiate oogenesis, many diverse species include individuals that produce eggs entirely on nutrient stored from the larval stage. The following discussion develops from a previous review.[40] The term "autogeny" designates this departure from the general parasitic mode of life.[41] Some genera, such as *Toxorhynchites* and *Opifex*, and such species as *Wyeomyia smithii* and *Ae. atropalpis* are entirely autogenous. Other species contain populations that are largely autogenous. *Cx. pipiens* provides the best-known example of such genetic polymorphism.

Populations of yet other species contain varied numbers of autogenous individuals. The list of such latently autogenous populations is so long that the trait appears virtually to be universal. Perhaps autogeny provides a conservative resource enabling mosquitoes to survive periods or situations in which hosts are scarce.

In general, autogenous mosquitoes produce their first clutch of eggs without having fed on vertebrate blood. The ovaries of such females seem to be more highly developed at the time of emergence than are those of anautogenous mosquitoes. They are at a stage of development that suggests prior stimulation by JH, an observation that suggests that autogenous ovarian development results from compression into the pupal stage of hormonal stimuli otherwise expressed by adults. Perhaps the hormonal milieu at the time of ecdysis resembles that following blood feeding in that oogenesis results when the system has been potentiated by prior secretion of JH. Larvae that develop into autogenous adults generally extend their last larval instar by a day or so, thereby permitting their body mass to increase in order to store nutrient that ultimately will be converted into eggs. Thus, autogenous mosquitoes develop somewhat more slowly than do their anautogenous siblings, and they produce fewer eggs. In hematophagic species, only the first clutch of eggs is autogenous. Autogeny, then, delays blood feeding.

The manifestations of autogeny are diverse. Certain arctic *Aedes* are facultatively autogenous. In the event that they fail to feed on blood within 1 week or so after emergence, a few eggs develop without blood feeding. Multiple autogeny seems to characterize *Wy. smithii*;[41] successive clutches of eggs are produced without blood nutriment. In contrast, undernourished individuals of some species are multiply anautogenous.[15] Thus, autogenous ovarian development may affect blood feeding in various ways and both insemination and sugar-feeding can potentiate the expression of autogeny.[41] In general, however, autogeny serves to reduce vectorial capacity by delaying hematophagy and advancing the age at which a vector becomes capable of transmitting infection. Indeed, no autogenous population has been implicated as a vector.

Although the term is most often applied to mosquitoes, "autogeny" has been used to describe a variety of conditions. In this sense, any form of development can be regarded as autogenous if normally mediated by hematophagy. Thus, larval *Ornithodorus moubata* would be autogenous because they moult without nutriment. Similarly, spinous ear ticks *(Otobius megnini)* blood-feed as larvae and nymphs, but adults reproduce without the benefit of a blood meal.

It seems axiomatic that autogeny would invariably detract from the vectorial capacity of a vector population. Although this property helps the population to perpetuate in the absence of vertebrate hosts, autogenous individuals invariably experience fewer host-contacts during the course of their life. In mosquitoes, the lost contact generally occurs during the first oogenic cycle, thereby extending the extrinsic incubation period of any pathogen into the oldest age cohort of the vector population. The depressing effect upon the force of transmission would exceed the linear.

D. Mating

Female mosquitoes generally mate only once.[42,43] Although females in such exceptional genera as *Opifex* and *Deinocerites* become inseminated at the moment of adult ecdysis (the male assists the female to emerge and mates as her abdomen clears the exuvia),[44] most become sexually receptive at about 2 days of age,[45] just before they seek vertebrate hosts. The copulatory apparatus of males is directed dorsally at the time of ecdysis; complete ventrad rotation is required before mating can be consummated, and this occupies about 1 $^1/_2$ days. Copulatory behavior of males seems to mature without hormonal stimulus. Sperm transfer in culicine mosquitoes is a surprisingly complex process, requiring an intricate fusion of structures everted from the genital openings of both sexual partners.[46]

At least in *Ae. aegypti*, mating seems to occur in association with the vertebrate hosts on which the females feed.[47] Indeed, males of this and certain other species are frequently taken by collectors attempting to sample mosquitoes that feed on people.[33-36] This host-associated mating behavior is reminiscent of the "following-swarm" of tsetse and may characterize the strategy of other species with low, asynchronous populations. Because vertebrate hosts serve to concentrate females, their presence might equally well serve as a marker for questing males. Still other mosquitoes initiate mating contact when the female enters a coordinated swarm of dancing males. These prominent aggregations form over some visible marker as a prominent tree or a pool of water. Mating commences once the male's antennal Johnston's organs sense the peculiar point-source of sound produced by the female's wings. Contact pheromones then guide consummation of the mating act.[48] Any interruption in the continuity of this sequence serves to abort insemination.

Several kinds of ticks are parthenogenetic, but the vector ixodid species mate in the course of reproduction. Various prostriate species in the genus *Ixodes* may mate even before attaining host contact. Male *Ix. muris*, for example, do not attach to hosts, however, metastriate ticks and even some *Ixodes* species mate on the host and delay oogenesis until mating has occurred. Females of most species become attached before producing the pheromone that attracts mates.[29] Nonmated female ticks feed slowly during this initial period of attachment, and engorge only to a certain point. A sex attractant, 2,6-dichlorophenol (2,6-DCP) is then produced from the foveal glands of various female metastriates.

This pheromone has been most thoroughly studied in American *Dermacentor* species.[29] The pheromone is synthesized and stored in droplets of oil contained within the foveal glands beginning at approximately 1 week after adult ecdysis. Host attachment stimulates renewed synthesis of 2,6-DCP, presumably through the same neurotransmitter and hormone systems that stimulated the initial cycle of synthesis. Synthesis is inhibited by the presence of the oil, by mating, and by ingestion of blood. Males become sensitive to the pheromone only after they have fed on blood to repletion, whereupon sensillae in the Haller's organ become competent to receive the olfactory signal. They are then attracted to slow-feeding females and undergo an intricate 9-step courtship routine that culminates with a venter-to-venter posture and placement of the spermatheca by means of the male's chelicerae. This pheromone may serve to aggregate questing males and females of other species, thereby increasing the mass of ticks that feed on particular hosts. Mating becomes enhanced. Specificity in mate recognition systems is provided by other pheromones, differing concentrations of 2,6-DCP, and avoidance behavior by the attached female.

Questions relating to the mating behavior by vector arthropods impinges on epidemiology and control of pathogenic viruses in several ways:

1. The traditional view focuses on preservation of the integrity of the structure of sexual populations and regards mating selectivity in terms of preventing introgression.
2. Vector populations may one day be replaced by others that are incompetent as hosts for the pathogen in question. Molecular biologists hope to produce such release-strains, but success depends upon the ability of arthropods containing such "designer genes" to mate with the target population in nature.

3. Sexual selectivity, however, may also be relevant to the displacement of parapatric species.[49] Interspecific mating may permit displacement of species by others that have males that mate nonselectively and disperse readily.

E. Nectar Feeding

Ticks feed only on blood (and atmospheric water), but female mosquitoes, biting midges, and phlebotomine sand flies also feed on nectar and other plant juices. Fully autogenous species feed only on plant juices, and in nature, females appear to take a sugar meal before their initial blood meal.[50-53] The diel rhythm of sugar feeding seems to mirror that of blood feeding, but mosquitoes apparently do not seek both foods simultaneously.[54-57] Captive females, deprived of sugar from emergence, seldom survive long enough to blood-feed.[58-60] Presumably the same would happen in the field.[61,62]

Female mosquitoes, held several days on concentrated sugar, become increasingly reluctant to feed on blood.[63] Subsequent sugar deprivation reverses inhibition but overstarved females fail to seek blood. It is a common laboratory practice to withhold sugar for 1 or 2 days before blood feeding. The former, i.e., sugar loading, may be a laboratory artifact, but the latter, i.e., deprivation-induced inhibition, suggests that the choice between seeking nectar or blood may be regulated by the insect's carbohydrate reserves — the survival fuel.[64] The sources and frequency of sugar meals are not well documented. Still, substantial portions of field populations of *Aedes, Anopheles, Culex, Culiseta, Coquillettidia, Psorophora, and Wyeomyia* contain plant sugar in their crops, as recorded by the cold anthrone test.[65-68] Seasonal and habitat differences in sugar feeding have been found,[65-69] but the significance of these differences in vector survival and in virus maintenance and transmission in unclear. Nor do we understand how or how difficult it is for mosquitoes to find natural sugar sources. Based on fructose-positive rates and digestion times, sugar feeding is more frequent than blood feeding. The importance of sugar for survival (a critical component of vectorial capacity) and its interrelationship with blood feeding[57,63] dictate the need for more information on this aspect of vector biology.

F. Daily Activity Rhythms

Soft ticks are nocturnal[70] but hard ticks may quest at any time of day. Depending on species, stage, and environmental conditions, questing may occur throughout the diel or ascent to questing sites and subsequent descent to resting sites may follow a daily periodicity.[71] Species with set patterns tend to quest mainly during daylight but activity is often reduced during midday when temperatures are highest and relative humidity lowest.[72]

Blood-feeding Nematocera fall into four basic activity patterns that are cued by daily light changes: diurnal, nocturnal, diurnal-crepuscular, and nocturnal-crepuscular.[73,74] Diurnal species such as *Ae. aegypti, Ae. triseriatus, Culicoides paraensis,* and species of *Anopheles (Kerteszia), Haemagogus, Sabethes,* and *Wyeomyia* generally are most active during the middle of the day (i.e., 0900 to 1700 hr), but adverse environmental conditions may at times inhibit activity and skew this pattern.[75-79] Mosquito species having strict diurnal feeding patterns generally dwell in the forest and are associated with forest mammals. In direct contrast, nocturnal species generally feed between 2200 and 0400 hr. Some human-feeding species of *Anopheles* are prime examples of this pattern. Diurnal species never seem to feed after dark and nocturnal species do not bite during the day.

Crepuscular species generally describe either of two activity patterns. Temporary-water *Aedes* and *Psorophora* species tend to be diurnal-crepuscular in that they readily bite during daytime if disturbed in their resting habitat but peak flight activity occurs from shortly before sunset until the end of evening twilight and from the beginning of morning twilight until sunrise.[80-81] These species tend to fly in unforested sites and feed mainly on large herbivores and such crepuscular mammals as rabbits.[82] This activity pattern tends to break down in the far north where summer twilight is extended[83] and in the tropics where it is abbreviated.[84]

Nocturnal-crepuscular vectors, e.g., most biting midges, sand flies, and *Anopheles, Coquillettidia, Culex, Culiseta, Deinocerites,* and *Mansonia* mosquitoes, tend to be active throughout the night but most are active just after dark and just before morning light. This group has the greatest feeding diversity since roosting birds and a wide variety of mammals, amphibians, and reptiles are potential hosts.[82,85-96]

Many instances in which the biting periodicity of a species varies between collection habitats, heights, and times of year have been recorded. Such disparities appear to relate either to temporal or spatial differences in illumination, temperature, or moisture levels, all of which can influence the intensity of activity.

Duration of activity of individual insects is far less than for populations as a whole. In some species, subpopulations representing different age or physiological classes may be active at different times.[97] The advantages and mechanisms of such apparently rare intraspecific, temporal partitioning are obscure. *Anopheles gambiae* and *Cx. quinquefasciatus* shift their flight periodicity after insemination, an effect caused by male accessory gland secretions.[98,99] In other instances, parous subpopulations may quest later than nullipars, reflecting oviposition activity during the earlier part of the daily activity period.

The importance of diel activity rhythms relates to their impact on host feeding patterns, sampling strategy,[100] and perhaps survival as well. For example, diurnal species fail to develop feeding associations with most birds. The daily periodicity of each species limits the range of hosts available to it and the types of arboviruses potentially acquired and transmitted.

G. Environmental Influences

Field biologists often observe day-to-day variation in the biting rate of mosquitoes, sand flies, and biting midges and ascribe both high and low biting levels to unusually favorable or unfavorable climatic conditions. Adult *Ix. dammini,* for example, will quest during unusually warm days even in midwinter, if snow cover is absent.

Humidity, temperature, illumination, wind velocity, and atmospheric pressure are most often credited with causing variations in intensity of vector activity. On a daily basis, however such differences probably have little epidemiological significance. Seasonal trends or an accumulation of several days of exceptionally favorable or unfavorable questing conditions can influence virus cycles if they occur during critical periods of transmission. For example, seasonal shifts from bird to mammal feeding by the St. Louis encephalitis (SLE) vector *Cx. nigripalpus* were associated with environmental moisture levels because they influenced the evening dispersal of mosquitoes from prime woodland resting sites to adjacent fields where many mammals graze.[87] Other vectors vary seasonally in their feeding patterns.[85,90-93] Direct or indirect environmental causes should be suspected for these cyclic and potentially epidemiologically significant changes in behavior.[101]

H. Diapause, Hibernation, and Estivation

In temperate regions and sites having long dry seasons, arthropods generally adapt to environmental inconstancy by undergoing a period of developmental arrest. The eggs of *Aedes* mosquitoes, as well as those of certain related genera, fail to hatch until stimulated by certain environmental clues. The univoltine species require a complex series of such stimuli, including a winter-like period of chilling. Reduced oxygen tension, associated with flooding, seems to provide the proximal clue for hatching in all species that have been studied. Hatching is then rapid, often occurring in seconds. Such estivation is oxygen-, rather than water-mediated.

Duration of daylength generally regulates seasonal diapause, of which reproductive diapause in *Culex* mosquitoes has been most thoroughly studied.[40,102] The pupal stage of these mosquitoes is somehow sensitive to duration of lighting; across the northern tier of the U.S.

early August seems to provide the inducing diel clue. This response, however, is conditional upon coolness and is regulated by JH. *Culex* mosquitoes diapause with ovaries in an early stage of development, hypertrophic fat, and spermathecae filled with sperm. This presents a paradox in that this stage of ovarian development would not persist if JH were secreted yet mating could not have occurred unless it were secreted.

Gonotrophic dissociation describes the overwintering state of certain mosquitoes, mainly *Anopheles* in which blood is taken, but the ovaries fail to develop.[103] Such *Anopheles* occasionally quest for blood during the winter, but *Culex* rarely do so, despite their continuing vagility.[104] This feeding pattern seems to have been the source of some midwinter transmission of malaria in the Netherlands but its role in arboviral transmission is less apparent. SLE-infected, hibernating *Cx. pipiens* have been discovered,[105] and probably resulted from transovarial infection. Hibernating mosquitoes presumably would not have fed on a virus-infected host. Virus would then be transmitted the following spring, when daylength becomes sufficiently long to disrupt diapause.[106]

Diapause patterns of ixodid ticks are more complex than those of mosquitoes because the life cycles of different species may span various periods of time.[107] Even within one genus, a particular species may experience photoperiodically regulated diapause in different stages of the life cycle, including egg, larva, nymph, and adult. Diapause may result from the action either of a long or a short diel. *Ix. ricinus,* for example, diapauses in various stages and may survive more than one winter. A three-host tick that diapauses for one or more winters would have greater capacity as a vector of larval-nymphal transmitted infection than one that completes its development within one season. Only then could the appearance of nymphs precede that of larvae.

III. SEARCH FOR VERTEBRATE HOSTS

A. Stimulatory Cues and Sensory Aspects of Host Detection

Host-associated visual and olfactory cues and the receptors used by vectors to detect and process sensory information have been reviewed several times.[108-111] Visual cues appear to be used by most blood-feeding arthropods.[112-115] They tend to be particularly important for diurnal species, less so for crepuscular species, and least for noctural species. Visual aspects of hosts that are most important in discrimination are size, shape, contrast with the background, and motion.[116,117] Hence, dark objects or shadows provide better cues for diurnal species and light objects or shadows provide better cues for those that quest at night. Although mosquitoes fly little in absolute darkness, nocturnal species successfully navigate on moonless nights, and this indicates that their eyes can function at low light intensities. Diurnal species have great difficulty locating hosts unless visual information supplements olfactory cues. Nocturnal-crepuscular species can locate hidden hosts, but questing is more efficient when hosts are visible.[115]

Carbon dioxide is the single most important kairomone used by arthropods in questing for hosts, yet it does not adequately account for some documented cases of differential attraction, and it is never as strong a cue as the intact host.[118] Laboratory efforts to isolate specific chemicals supporting specific vector-host associations[119-121] have not been particularilty satisfying. This should not be surprising since the species tested have been mosquitoes that display remarkable latitude in the hosts attacked in nature. Introduced exotics such as the armadillo in Florida often become quickly incorporated into the diet of indigenous vectors.[82,87,122] Questing vectors with more host-specific tendencies (e.g., *Uranotaenia lateralis* and fish) generally exist in close association with localized and reliable host populations. They may "read" the specific blend of chemical cues produced by their hosts, somewhat in the manner that different blends of the same chemicals are used as sex pheromones for mate recognition by related species of moths.[123] Host chemicals have been shown to elicit an electrophysiological or orientation response in some mosquitoes,[124] but such information has not yielded a unified theory of how different species locate similar and/or dissimilar hosts.

Most active vertebrates produce sounds associated with movement or communication; Nematocera are well equipped to receive and discriminate sound frequencies. *Corethrella* use the mating calls of their frog host[125] to locate their position, and other species may use more general sounds to assist them in locating hosts. *Cx. territans, Uranotaenia lowii,* and *Ficalbia* would be good candidates for sound studies. *Ornithodoros concanensis* reportedly use host vocalizations to orient toward swallows in the nest,[126] and waiting ixodids respond to the sound vibrations of approaching hosts by extending their legs or crawling in the direction of the host.[114]

B. Active vs. Passive Strategies

There are two basic strategies for finding hosts: waiting in particular locations until hosts enter the field of attack (passive search strategy), or embarking on daily host-seeking excursions at times and in places where hosts or signals of their near presence are likely to be encountered (active search strategy). Arbovirus vectors use both strategies. Ticks and diurnal mosquitoes rely heavily on the passive approach, whereas crepuscular and nocturnal Nematocera appear to embark on daily searching (appetitive) flights of unknown duration to increase their chances of encountering a host. Ticks and diurnal mosquitoes orient toward hosts once cues are received, but little evidence of undirected searching behavior exists for either of these groups. Also, as mentioned earlier, diurnal-crepuscular mosquitoes will opportunistically feed on hosts that invade their daytime resting habitat. It is unknown what portion of blood meals are obtained by this supplemental, passive strategy. Among ticks, the passive approach to host finding can be subdivided into ambush and hunter strategies, depending on whether ticks move toward detected hosts or wait for the host to come within their grasp.[114] Some hunter ticks, e.g., lone star ticks, will crawl as far as 21 m toward a CO_2 source.

C. Host Specificity: Generalists vs. Specialists

Because a particular host species may not always be available, most mosquitoes retain the capacity to seek out and feed on a variety of vertebrates.[3] Vectors that specialize in a particular kind of host are best adapted to transmit pathogens associated with that group, but most arboviruses are zoonotic infections, often involving a variety of native and exotic vertebrates.[127] In many cases, it is not clear which vertebrates function as major enzootic (maintenance), amplification, or zoonotic (bridge) hosts. An understanding of the mode of perpetuation of the agent is key to the biology of the virus.

Anthroponotic arboviruses, such as dengue, must be perpetuated by human-associated vectors (e.g., *Ae. aegypti*). Still, the anthropophilic feeding pattern of the urban form of *Ae. aegypti* evolves more from environmental circumstances than from any biological predeliction of the vector for human blood.[128-130] In urban New Orleans, where human hosts are less accessible than in the tropics, dogs are the principal host of urban *Ae. aegypti.*[131] One important arbovirus vector known to have strong feeding specificity is *Culiseta melanura,* the enzootic vector of eastern equine encephalitis (EEE) virus. Its high degree of specificity for passerine birds makes it an efficient enzootic vector, but seems to preclude any involvement in EEE transmission under epidemic conditions.[131-133] Hence, a second, less host-specific vector is required for epidemic cycles of this disease and may help to explain the rarity of human and horse cases, despite frequent enzootic transmission near populated regions.

Ticks tend to fall into three categories: (1) species that are specific to one kind of host or nearly so, (2) species specific to a broad group of related hosts, and (3) nonspecific feeders. One-host ambush species are the most specific and three-host hunter-species tend to be the least specific. Host-specific ticks generally fail to attach to nonpreferred hosts or, if attached, they may fail to engorge or develop eggs.[114]

D. Orientation to the Search Habitat

As with temporal activity patterns, the place where vectors quest strongly influences feeding relationships. Bidlingmayer and Hem[134-136] grouped mosquitoes into three general types based on their habitat preferences: woodland, field, and commuter species. Woodland species seldom leave the forest biome except to disperse and are, therefore, limited to feeding on forest-dwelling vertebrates. Nocturnal and crepuscular-nocturnal forms often feed on roosting avians or reptiles and amphibians, whereas diurnal species, e.g., *Ae. triseriatus* and *Ps. ferox,* are mammal-associated. Field species, e.g., *Ae. sollicitans* and *Ps. colombiae,* rest, feed, and oviposit in campestral settings and quest for grassland mammals. Commuter species are more diverse, resting in forest by day but flying to the forest edge or into open fields to search for hosts.[137] Mosquitoes use visual cues in orienting to their respective flight and resting habitats.[138]

The greatest diversity and abundance of vertebrate species characterize edge or transition zones between forest and field, many of which are physically active and more visible during crespuscular periods. This zone also forms the prime habitat for many ixodid ticks, but these vectors tend to occupy the same habitat, i.e., field, forest, or edge, throughout their life unless they attach to aberrant hosts or inadvertantly drop off in aberrant habitats.

A vertical dimension may also characterize search patterns, particularly those of forest-inhabiting vectors. Several studies have documented the flight altitude preferences of different mosquito species; of special interest are those of Nasci[139] and Novak and Rohrer[140] who studied the flight elevation and feeding patterns of two sibling treehole mosquitoes in wood-lots enzootic for California group viruses. *Haemaphysalis leporis-palustris* were observed to quest on vegetation at a height closely corresponding to rabbits, the main host for all stages.[141]

Despite the importance of questing behavior in limiting the variety of hosts that a vector is likely to encounter, this aspect of vector biology remains poorly known.

E. Searching Behavior

Mosquitoes and related Diptera engage in upwind or partly crosswind flight until wind-borne odors are detected.[142,143] The height of this flight varies depending on wind speed and perhaps other factors.[144-146] This search strategy optimizes the probability of encountering scattered hosts that are producing downwind odor plumes. Where wind patterns are more erratic and turbulent, as in dense forest, search strategies involving flight above the canopy or along natural breaks may serve to optimize plume detection. Diptera with passive search strategies may "perch" in elevated border locations to enhance their ability to detect hosts passing nearby.[104]

F. Orientation Toward Recognized Hosts

Once kairomones (interspecific chemical signals) are detected, the ensuing consummatory flight toward the source becomes increasingly oriented. Preprogrammed counterturning and casting manuevers serve to bring the insect back into contact with elusive and discontinuous odor trails whenever contact is broken during the zigzag flight toward the host.[147-149] Long-range flight toward odor sources has been studied most intensely with day-active male moths orienting toward females emitting sex pheromones. It is tempting to analogize this behavior with that of blood-feeding Diptera following host-generated odor plumes, especially since night observations are difficult and diurnal blood feeders tend to wait rather than search for host cues.[150,151] The speed of orientation flights toward hosts by West African mosquitoes has been reestimated at 1.4 to 1.8 m/sec.[144] Effectiveness of searching as well as orienting flight is reduced at very low and especially at high (i.e., >2 m/sec) wind speeds and may result in complete cessation of activity. Novel methods for directly observing and recording the host-orientation behavior of individual flying vectors are needed.

G. Pheromone and Kairomone Production by Feeding Vectors

Biting mosquitoes and sand flies may arrive at hosts in waves rather than at random intervals[152] and feed in clusters.[10,117] Because increased densities of biting insects can result in reduced engorgement success (see discussion of host defensive behavior), it would seem maladaptive for vectors to group-feed or to assist in any way questing siblings or nonsiblings in locating a host. However, if host-associated kairomones were associated solely with successful feeding (e.g., a defecation product of the vector), it would assure questing mosquitoes of the presence of receptive hosts. One would expect such a cue to be nonspecific, and in fact, the "invitation effect" observed by Alekseev et al.[153] and by Ahmadi and McClelland[152] seems to act in this way. Similarly, field evidence of screened chicks attracting fewer mosquitoes than exposed chicks suggests a nonspecific cue.[2,154] Schlein et al.[155] observed a species-specific pheromone produced by feeding *Phlebotomus papatasi* and suggested that the palpal glands may be the site of production. Many sand flies mate near the host, and we suggest that this feeding attractant may be a sex pheromone that secondarily serves questing females as a host location kairomone. Ticks produce pheromones as aggregation, mating, and mate finding signals,[29] but apparently not to promote host location or feeding. Of these, 2,6-DCP is best known, and 20-hydroxy ecdysone appears to regulate its production.[156] Mating generally occurs on the host, following attachment and ingestion of some blood. Certain *Ixodes* and other metastriate ticks, however, mate before attaining host contact. The pheromonal signals have not yet been established.

H. Individual Differences in Host Attractiveness

Young, nonimmune vertebrates are critical to the perpetuation of most virus cycles. This has formed the basis for speculation on whether young animals may be particularly prone to attack, presumably due to enhanced attraction. In fact, studies comparing the attractiveness of young and of adult animals of the same species tend to indicate the opposite.[109,157] Young hosts are smaller and therefore seem to provide weaker cues to questing vectors than that of their mature counterparts unless they group together. Blackmore and Dow[158] observed differential feeding success (but not attractiveness) on adult and on nestling birds; they attributed this to the nakedness and quiescence of squabs.

In addition to age, individual humans vary in their attractiveness to mosquitoes depending on sex, skin color, and temperature, and other unknown factors.[109] Some evidence that blood type may influence the selection process was presented by Wood,[159,160] but was not confirmed in subsequent studies.[161]

Mahon and Gibbs[162] and Turell et al.[163] suggested that viremic birds and lambs are more attractive to mosquitoes than are co-exposed healthy animals. This possibility deserves more attention since it would dramatically alter vectorial capacity.[164] Unfortunately, the inherent hazard of exposing viremic hosts to vector populations prevents critical testing in nature. The issue of how viremia might enhance attractiveness is open, but both increased body temperature and carbon dioxide production have been offered as possibilities. The reduced body temperature of malarious mice did not affect their attractiveness at close range.[165] Elevated temperature could have a more distant effect, although heat is usually considered a short-range attractant. Reduced CO_2 production results in fewer mosquitoes arriving at hosts,[166] and dry ice is often added to host-baited traps to increase collections. Whether the increased CO_2 emanations that might accompany viral infection measurably affect attraction remains unanswered. Moreover, arboviral infections can produce a thrombocytopenia which may permit vectors to feed on viremic hosts more effectively than on noninfected hosts, and the resulting preponderance of virus-infected mosquitoes might erroneously be attributed to differential attractiveness.

I. Host Availability

The suitability of a population of vertebrates as hosts for hematophagic arthropods depends upon its abundance, body mass, and pattern of distribution.[167-171] For example, a tendency to gather in herds can greatly reduce questing efficiency since several hosts in one spot do not attract as many vectors as the same number of hosts spaced, so their odor plumes do not overlap.[4,172] These three factors (i.e., host abundance, size, and spatial distribution) mainly determine the feeding patterns of generalist blood feeders. Of course, rare or small vertebrates may fall victim to opportunistic attack by an abundant vector.[173] Normally, however, such animals only serve as hosts for highly specialized or localized ectoparasites occupying undisturbed habitats. Arboviral transmission involving these unusual host associations falls into the ''zoonosis'' category. The classical association involves human invasion of recently disturbed sites, as in the case of *Haemogogus*, jungle yellow fever, and wood cutters.

Large vertebrates attract vectors over greater distance than do smaller vertebrates[169] and they may differentially be selected even when hosts of different size coexist.[170,171] It has been suggested that mosquitoes, which feed on small animals such as passerine birds, may differentially repond to dilute CO_2, and perhaps even be repelled when that attractant is more concentrated; however, this concentration effect has never been well demonstrated in the field.[174]

Immature ixodids, on the other hand, regularly attach to small animals, which may be abundant within the restricted questing habitat of the species. Furthermore, the nonrandom distribution of ixodids both in the habitat[175,176] and on hosts[177,178] suggests that host density and home range play major roles in the attachment success of these ticks. Host density regulates the numbers of triatomid bugs[168] and the same appears to hold for other such nest parasites as soft ticks and swallow bugs.

J. Synchronization of Vector and Host Ecology

The feeding impact of the diel or seasonal activity and habitat or geographic restrictions of each vector and host species has been discussed. It follows that spatial and temporal synchronization between host and vector populations will often result in greater vector-host contact. Ecological overlap, however, often is defined too broadly. Nasci[139] observed that in the same wood lot, two sibling species of *Aedes*, one with greater tendency to remain in the canopy, feed on arboreal and ground mammals at correspondingly different rates. Moreover, Day and Edman[179] found the general activity pattern of small animals to be an important aspect of availability to vectors, even though both remain in the same habitat throughout the diel. Animals in deep nests, subterranean burrows, or grass-covered runs are, in fact, generally inaccessible to free-ranging, hematophagic arthropods. Immature ixodids and certain burrow-occupying sand flies would be better adapted to such a feeding situation than are mosquitoes.[180] In addition, when animal burrows are used as resting or overwintering sites, the proximity of host and vector may promote feeding.[74,89]

K. Geographic and Seasonal Differences in Feeding Patterns

Regional differences in the feeding patterns of the same vector species[92,181] generally relate to corresponding geographic differences in the host populations available to the vector. Similarly, questing activity of most blood-feeding species is seasonally limited to certain months of the year, even in the tropics. Hence, hosts with seasonal migration, reproduction, mortality, or behavior patterns may not equally be available to all blood feeders that occur in a given habitat. For example, bird migration was the suspected cause for seasonal variation in vector feeding in an African study.[182] Similarly, diurnal species questing during the annual harvest may have more human-feeding opportunities than those questing during other seasons. Nasci[183] found a high *Aedes* feeding rate on horses only in early-week collections, resulting from weekend use of the site by an equestrian club.

Seasonal differences in feeding may also result from behavioral changes in the vector, as in the case of *Cx. nigripalpus*.[87] The possible importance of this phenomenon in the cycling of viruses that use a single vector species to bridge the gap between diverse vertebrate groups has been pointed out.[101,181,184] Differences in feeding based solely on population differences in host preference are occasionally found,[185,186] but these populations likely represent genetically isolated sibling species or subspecies.[130,187] Numerous short-term attemps to select for feeding preference in the laboratory have had marginal results.

L. Effect of Habitat Alteration and Human Intervention

Since the feeding and oviposition patterns of most arbovirus vectors are plastic, they are readily modified by human activity. Mass destruction or burning of habitat, critical for vector breeding and survival, may render species rare or nonexistent, thereby eliminating their potential as vectors.[122,188] Conversely, enrichment of surface waters and the introduction of substitute habitats, such as clay pots, automobile tires, plant axils, rice fields, storage reservoirs, irrigation ditches, and borrow pits, frequently promote vector abundance and cause disastrous outbreaks of disease. This is particularly true when the disturbed site lies near human habitations.

Human intervention also eliminates native hosts and introduces exotic species, including domestic animals and the human species itself. By building new settlements (incuding modern subdivisions) in or near known enzootic arbovirus foci, human populations become exposed to increased contact with infected vectors. Examples of this can be drawn from Brazil to central Florida. Disease risk may not be considered by governmental agencies in zoning and development decisions unless impounded water or wetlands are involved, and then only after disaster seems imminent.

Population pressure and poverty in many tropical areas have led to displacement of native mammals, leaving vectors no option but to feed on human or domestic hosts. Consequently, enzootic viruses may emerge rapidly in disturbed environments. At times, abundant livestock may prophylactically serve to shield human hosts from infection,[189-192] but they also may encourage more vectors to share space with human hosts, thereby increasing human-vector contact. Disruption of the natural habitat may also create situations in which accidentally introduced exotic vectors thrive and outcompete native species. Interestingly, introduced pathogens and parasites are transmitted by native vectors more frequently than exotic vectors transmit indigenous disease organisms.

IV. SELECTION OF FEEDING SITES

A. Dermal and Mouthpart Characteristics

The association between the structure of the mouthparts of temporary blood-feeders and the nature of the host's skin tends to be weak. Structural differences relate mainly to the two diverse strategies for taking blood, i.e., pool and vessel feeding, which require somewhat different equipment. Regardless of strategy, probing sites generally are the more exposed skin surfaces. Mosquitoes have long stylets capable of penetrating thick, coarse clothing; erect hair does not appear to provide a sufficient purchase point for penetration.[193] Pool feeders have more latitude in choosing probing sites; black flies are able to crawl between long hairs or feathers provided they are not too dense. The length of the labrum of certain tabanids correlates with the depth of the hair covering their preferred feeding sites on cattle.[194]

Ixodid ticks specialize pharmacologically, perhaps salivating different enzymes or vasoactive substances in response to a particular feeding stimuli.[195]

B. Landing and Foraging Behavior

The host cues used by Nematocera in landing and selecting suitable feeding sites are not well understood but vision plays a critical role in diurnal species. Some black flies attracted to people fail to land and probe for blood; inappropriate body shape seems to be a major cause.[117] In bait-trap studies, some mosquito species are occasionally collected at hosts on which they seem reluctant to feed,[195-198] but the unnatural setting of hosts in such experiments complicates interpretation. It is maladaptive for vectors to expend energy orienting to hosts only to reject their blood.[110] Such abortive feeding appears to occur mainly when exotic hosts become newly introduced into a vector's environment.

After landing, mosquitoes pause before foraging for a probing site — an apparent response to the increased risk of detection when first contacting the host.[199] Foraging in flight would seem inherently safer than more terrestrial modes of locomotion, and mosquitoes use both methods in locating suitable feeding sites.[193] Perhaps this adaptation results from the defensive reactions of hosts to the sound made by the wings of mosquitoes. Foraging for suitable feeding sites is a random process involving frequent labellar tapping by *Ae. triseriatus*[193] but may be more direct in other species.[200] In general, the strategies of abundant mosquitoes (generally pests rather than vectors) seem less well adapted to the responses of their hosts. Similarly, winged vectors that regularly attack intolerant and dexterous hosts (usually small animals or primates) show greater behavioral adaptation.

After grasping a passing host, ticks crawl to their eventual attachment sites. If other ticks are already attached, they may be directed by aggregation or sex pheromones to these same attachment sites.[29] These chemical cues, however, operate only over short distances (i.e., few centimeters). Other factors, therefore, must first direct foraging ticks to the general body region, particularly on the larger animals on which adult ticks often feed. Some species of ticks (generally one-host ticks) and black flies feed almost exclusively inside the ears of their host. The mechanisms used by these species to locate their restricted attachment sites are not known, although unique secretions associated with the ear should be suspect. Body hair and grooming behavior limit the variety of sites suitable for feeding by many species. Moreover, naked skin tends to be more richly supplied with blood (for thermoregulation), which makes these sites particularly suitable for quickly locating blood.

C. Influence of Host Grooming Patterns

Animals normally limit ectoparasite populations through grooming behavior.[7] This is particularly true of such slow feeders as ixodid ticks or of lice and the various fleas that live on the host and feed frequently. A flattened shape, attachment cement, and rapid engorgement of females just prior to drop-off apparently constitute ixodid adaptations to the selective pressure of host grooming. Populations of reduviid bugs are regulated by host influences on the size and frequency of their blood meals.[201] It would not be surprising, therefore, to find the same relationship among the winged vectors that are limited to feeding on hosts (especially small ones) that groom intensively.

Grooming may influence the distribution of attachment or probing sites on the host, because all regions of the body may not be groomed equally. Social animals such as primates[202] and penguins[203] partially solve this problem by grooming each other. Ticks and black flies that feed inside the ears tend to avoid grooming risks. The underbelly region of large mammals provides another preferred feeding site for many Diptera[204] because it lies beyond the range of most grooming activity. Underbelly hair is often white, which may visually divert some vectors away from this unprotected region.[4,117] Many diurnal, man-feeding mosquitoes and black flies feed on the lower legs and ankles, and this appears to be an adaptation to avoid host detection.

D. Partitioning of Hosts and Competition for Feeding Sites

Permanent and semipermanent coectoparasites partition their hosts both temporally and spatially, apparently to reduce competition for living space and/or feeding sites.[205] Sympatric tabanids are the only Diptera that partition their host during brief feeding contact.[2,194] These are large, painful, day biters that evoke extreme reactions from their hosts, even when attacking in small numbers. The observed correlation between reduced mosquito engorgement rates and apparent abundance of these insects was originally attributed to competition for feeding sites. Other more plausible explanations, however, can now be offered.[181]

Ticks often feed on particular body regions and in dense groups, presumably a reflection of some combination of host-grooming, pheromone-mediated behavior and skin characteristics.[206-208] There is a clear trade-off, since group attachment tends to increase pathology while permitting species to exploit the most favored body regions and to aggregate for mating.

V. HOST DEFENSIVE BEHAVIOR AND VECTOR FEEDING SUCCESS

A. Host Age and Size Relationship to Behavior

Implicit in the preceding discussion of grooming is the notion that defensive behavior by vertebrate hosts destroys vectors or deters their feeding. The degree of this impact varies according to various factors, not the least of which are the age and size of the host. In general, small vertebrates possess a more intense and efficient defensive repertoire, but young animals are usually less defensive than adults.[209-211] In compensation, many immature vertebrates are protected in some fashion by their parents or the brood chamber. Age differences are less pronounced among larger animals, and when adults are relatively tolerant of vector attack, their young may be more defensive. Among birds, the young of precocial species have well developed defensive behavior (even more than their parents), whereas altricial species do not.

B. Defensive Behavior and General Activity Patterns

Detailed descriptions and comparisons have been made of the different defensive movements expressed by various herons[212] and rodents[213] when attacked by mosquitoes. The same phenomenon has been observed, but not quantified, for many other birds and mammals. Both the type and frequency of defensive movements influence the efficiency of different species in preventing successful feeding. The effectiveness of such behavior ranges from nil to virtually complete prevention of feeding.

Depending on their nature and timing, defensive movements may cause a significantly higher rate of interrupted and multiple feeding.[214,215] Individual hosts of the same species may behave quite differently. In some cases, this seemed to relate to previous experience or to dominant vs. submissive status within social groups.[216] Experienced and dominant individuals are the most successful in preventing mosquito engorgement. Within herons, differences in defensiveness are associated with the active vs. passive fishing strategy of the species; passive feeders are more tolerant of mosquito annoyance.[212] Chipmunks, important hosts for La Crosse (LAC) virus, and to a lesser extent tree squirrels, are unusual rodents in that they tolerate feeding by modest numbers of mosquitoes. This may relate to day-active status and predator avoidance strategy of these small forest mammals.[213] Nocturnal rodents, e.g., cotton rats *(Sigmodon hispidus)* and cotton mice *(Peromyscus gossypinus)*, though extremely intolerant of mosquitoes in the laboratory, are important hosts of the *Culex (Melanoconion)* spp. which transmit Everglades virus and other enzootic strains of Venezuelan encephalitis virus.[89] It is not clear how these vectors are able to circumvent the efficient anti-mosquito capabilities of their hosts. Similarly, most passerines are highly defensive.[211] Nonetheless, certain mosquitoes (e.g., *Cs. melanura)* regularly feed on roosting passerines and in so doing, effectively transmit arboviruses.

Defensive responses to vectors may not always be in the form of direct movements to prevent feeding attacks. Baboons may leave roosting areas where ixodid ticks are abundant[202] and sea birds abandon nests with eggs and young when argasid ticks become intolerably numerous.[216] Herding behavior among large herbivores[172] and tree climbing among porcupines[2] have been suggested as responses to mosquito attack under certain conditions. Presumably, birds have more opportunity to change habitat than most other vertebrates. Colonial roosting on offshore islands may have evolved partially as a response to blood-feeding pressure, especially in saltmarsh environments. The nest structures of birds also may have been influenced by pressures imposed by parasites, especially in environments where pathogens affecting survival are regularly transmitted.

C. Correlation with Vector Density

The responses of various vertebrate species to vector attack correlates with vector density. The grooming intensity of baboons correlates seasonally with tick densities,[202] and the defensive behavior of several species of birds and mammals increases markedly with increasing mosquito density.[213,217-220] High tsetse densities have been associated with reduced fly/human contact and consequent transmission of trypanosomes.[221] Even hosts that are relatively tolerant of mosquito attack become defensive when mosquitoes are abundant. Most vertebrates are relatively tolerant of sparse mosquito populations (i.e.,<25 per night), although some active rodents consistently react to the approach of even a single mosquito.

It has been suggested that seasonal increases in vector density may alter blood-feeding patterns and even precipitate virus transmission to new kinds of hosts.[219] This could occur only if vector abundance differentially stimulates the defensiveness of different kinds of hosts and therefore on the feeding success of vectors attacking these groups. Two factors complicate our understanding of the relationship of vector density to feeding success: the difficulty of measuring actual natural attack rates and, even when measured, these rates tend to be a composite for all arthropods (vector and nonvector alike) that happen to be attacking the same host during a given time period. For example, the response of a bird to *Cs. melanura* will be influenced by the number of *Cs. morsitans, Cx. pipiens, Cx. restuans,* etc., biting on the same night. Hence, the initial spring generation of *Cs. melanura* may forage on a more tolerant avian population (which also includes nestlings) than subsequent generations feeding in mid- and late summer when other avian-feeding species are most abundant. In fact, the feeding pattern of this species is more diverse later in the season when epizootic and epidemic transmission occurs.[132]

D. Biting Persistence and Feeding Success

Anecdotal evidence suggests that nematoceran species vary in their feeding aggressiveness. Unless hosts are completely tolerant, the success rate of each species should be influenced by their innate biting persistence. Species comparisons are lacking, but a method for measuring persistence was evaluated with *Ae. triseriatus.*[222] Attack duration was the criterion used to measure persistence. Attacks were time limited and influenced by carbohydrate reserves but not by chronological age of the host. Persistence may also be modified by viral infection and by other factors such as ambient temperature and physiological age of the vector. Evidence that some attacking vectors soon voluntarily break contact with defensive hosts points to a mechanism that promotes multiple host contacts within a single gonotrophic cycle. Number of host contacts is critical to any transmission model.

E. Impact of Host Behavior on Viral Transmission

Such mosquito-tolerant species as night herons have, at times, been singled out for suspected greater importance in arbovirus amplification because of their unusual prevalence of seroreactivity.[223,224] Because certain highly defensive vertebrates are similarly frequently infected,[225] that association could be coincidental. As indicated earlier, vector species generally are so scarce as to not invoke as strong a behavioral response from their hosts as more abundant species. Epidemic vectors of zoonotic viruses, on the other hand, often are singularly abundant during outbreaks of human disease. Perhaps epidemic transmission is due partially to the impact of vector density on the behavior of zoonotic hosts. Differential feeding on young, nonimmune animals whose defensive behavior may be imperfect is another important way in which host behavior might affect transmission.

The grooming behavior of rodents, primates, and birds often includes the capture and eating of vectors that are attempting to bite.[202,212,214] If vectors are infected, hosts could conceivably acquire an infection via this oral route. This possibility was demonstrated with both chipmunks and mice after eating LAC virus-infected mosquitoes[214,226] and with mice after eating malaria-infected mosquitoes.[214] Ingestion of ticks by birds and rodents also has been suggested as a possible mode of transmission in the epidemiology of Rocky Mountain spotted fever (RMSF).[227] The importance of this form of transmission is difficult to document in the field, but if hosts regularly eat vectors, it should not be ignored.

F. Disease-Induced Changes in Host Behavior

Normal defensive behavior is compromised by several factors, including nutritional state, pregnancy, competing activities, and disease. Many blood-borne infections result in intense lethargy which reduces an animal's ability to groom ticks and repel Nematocera. Such an effect has been documented in the case of malaria and SLE virus-infected rodents[214,228,229] and with Rift Valley fever (RVF)-infected lambs.[163] The significance of this in the epidemiology of any arbovirus remains to be shown. It stands to reason that selective or enhanced feeding on viremic hosts, whatever the cause, would accelerate transmission. Unknown is whether such disproportionate feeding is necessary for the maintenance, amplification, and/or epidemic transmission of any arbovirus. Sick horses have been suggested as point sources in the rapid spread of epidemic VEE (Venezuelan equine encephalitis) and viremic birds could be a link between the endemic *Cs. melanura* cycle of EEE and the epidemic cycle involving less host-specific mosquitoes.

VI. BLOOD LOCATION AND INGESTION

A. Vessel- vs. Pool-Feeding Strategy

A hematophagic arthropod shares the problem of a blind phlebotomist. Blood comprises only about one tenth of skin volume, and canulatible vessels will seldom be encountered by chance. Pool-feeding arthropods locate the blood that they ingest by disrupting capillaries either by mechanical laceration or an enzymatic process. Vessel-feeding insects scan deeper tissues in a search for venules and arterioles. Both strategies face the platelet-mediated hemostatic mechanisms of their hosts as primary obstacles in their search for blood. Clotting occurs too slowly to affect feeding by most insects. The clot would have solidified long after the mosquito had completed engorgement. Ixodid ticks, on the other hand, which take days to feed, must combat hemostatic mechanisms based on both platelet-aggregation and clotting. Saliva combats hemostasis. The following discussion describes both feeding strategies.

B. Vessel Location by Mosquitoes

Once a mosquito has attained contact with a potential host, the labial sheath draws back to expose the mouthparts, and the bundle of stylets is inserted into the skin. The mandibles seem to serve a piercing function while the toothed maxillae rasp the incision larger.[230] Their track then becomes filled by the apposed food (labral) and salivary (hypopharyngeal) channels. This fascicle penetrates deep into the skin, beyond the capillary loops, where the structure comes to probe horizontally to the skin surface. Capillaries are too small to be cannulated, and larger vessels are most numerous in this deeper plane. With each inward thrust, saliva is distributed along the track of penetration.[231] When blood is tasted as the mouthpart fascicle is withdrawn through the saliva-bathed path of penetration, movements cease.[232] Although vessels are often cannulated by chance, presence of a hematoma greatly facilitates the process, both by enhancing the effective diameter of a vessel and by providing a field of graded viscosity that assists entry into the vessel. Mosquitoes randomly probe in this manner some 15 times before desisting to another site.

The saliva that bathes the track of penetration promotes hematoma formation by preventing platelets from aggregating.[232] ADP is perhaps the most potent of known platelet activators. Together with ATP, ADP is present in interstitial spaces solely when cells are disrupted, thereby serving as a signal of injury to living tissue. An apyrase converts both nucleotides to platelet-inactive AMP. This enzyme is present in all hematophagic arthropods that have been studied, and it appears to be the main antihemostatic salivary component of mosquitoes. Kissing bugs, on the other hand, are armed with a variety of platelet-inactivating salivary components.[233] In *Anopheles,* the quantity of apyrase produced correlates with efficiency of blood finding.[234] The presence of apyrase along the track of mouthpart penetration would then promote hematoma formation, and this would enlarge the phagostimulatory field available for detection by the fascicle as it slides back toward the surface of the skin.

C. Pool-Feeding by Ticks

The manner of feeding of pool-feeding arthropods is not well understood. Most research attention has been directed toward ixodid ticks, but without clear evidence of the manner in which the feeding chamber is created. No lytic enzymes have been reported in saliva, and no clear evidence of regurgitation has yet been presented. Clearly, the mouthparts draw blood from a space hollowed in the dermis of the host. Perhaps ticks take advantage of some host-derived lytic mechanism.

D. Ingestion of Blood

Certain hematophagic arthropods cease probing the skin of their vertebrate hosts and begin to feed in apparent response to chemosensory cues derived from the cells contained in the host's blood.[235] Those culicine mosquitoes that have been studied seem to respond to the taste of cell-derived ATP or perhaps some related adenine nucleotide.[236,237] Anopheline mosquitoes, however, respond to stimuli mainly contained in the plasma.[238] Ticks recognize glutathione, but various cellular and plasma factors synergize this response.[239] Whatever the nature of the blood-derived stimulus, once probing ceases, suction would draw any lacerated vessel toward the functional oral opening at the tip of a mosquito's feeding stylet. The vascular source of a hematoma would thereby cannulate itself.

The presence of salivary apyrase along the track of penetration of the feeding stylets presents an apparent paradox. Indeed, enzymatic degradation of ATP potentially deprives apyrase-secreting insects of that gustatory cue. Perhaps stimulation depends upon some functionally related compound or on a transient condition of concentration.

Arboviruses, as well as other salivarian pathogens, are transmitted to vertebrate hosts mainly during probing.[240] Once feeding has begun, the much greater magnitude of the flow of ingested blood would cause any salivary secretion to be drawn into the feeding stylet. Contained viral particles would be ingested by the vector. The infectious inoculum would not be ''mainlined''. Arboviral particles (or malaria sporozoites) seem mainly to be deposited in interstitial spaces of the skin and only later pass into the circulation.

Arboviruses and their vectors are bound in a mutualistic relationship.[240] Certain arboviral, as well as malarial pathogens, cause lesions in the salivary glands of their vectors that markedly reduce apyrase content, thereby frustrating the vector's attempts to find blood.[241]

Such a prolongation in feeding amounts to functional castration. Infected vectors probe longer than their noninfected siblings,[242,243] they succeed in feeding less often, and tend to probe additional hosts; because they frequently fail to undergo oogenesis, daily questing frequently continues without interruption. Of course, only old mosquitoes are affected. Vectorial capacity is enhanced.

This mutualism may extend in another dimension. A few arboviral, and some other salivarian pathogens produce a thrombocytopenia in their hosts that enhances hematoma formation.[244] Thus, infected hosts may be better sources of blood for mosquitoes than are noninfected vertebrates. Such disproportionate vulnerability to successful feeding would contribute yet another increment to vectorial capacity.

Ingested solutions pass down the alimentary canal of mosquitoes, through the pharynx, and into the crop or the gut. Nectar or sugar solutions are destined for the crop, and later are slowly delivered to the midgut. The wall of the crop is muscled, but the wall is nonsecretory. Nor does saliva contain sugar-digesting enzymes. This structure seems to serve as an inert reservoir. The gut of such sugar-fed mosquitoes never distends. Blood- or ATP-containing solutions, on the other hand, pass directly to the midgut. The nature of this ATP shunt, which has long stimulated experimental investigation, has not yet been defined.

Engorgement on blood results in massive distension of the midgut and of the abdomen, and this seems to play a role in oogenesis,[245] but the quality of ingested material redundantly appears to be important as well. Because the physical stimulus of abdominal distension is mediated via the ventral nerve cord,[18] it parallels the manner in which volume of ingestion is mediated.[246] Three blood-feeding sequelae follow surgical sectioning of the nerve cord at the base of the abdomen: mosquitoes feed until their abdomen bursts, oogenesis fails to commence following ingestion of minimal quantities of blood, immediate host-questing fails to be inhibited.[18] Later, humoral factors promote oogenesis and inhibit questing.[26,27] Blood-feeding is potentiated again only after oviposition.

Ixodid ticks remain attached to their hosts for prolonged periods of time before commencing their characteristic period of rapid engorgement. This preparatory phase of feeding results in a series of hormonally mediated changes that cause the cuticle to become plastic, spermatids to mature, and in the case of metastriate ticks, production of mating pheromones. Ecdysteroids[156] and juvenoids[247] appear to mediate the completion of this stage of developement, mating, and the onset of the "big sip". Piroplasms,[248] as well as other tick-borne pathogens, seem to be transmitted at this time.

E. Blood Volume and the Initiation of Egg Development

If feeding is not interrupted, most vessel feeders rapidly fill with blood until neural cues triggered by abdominal distension signal them to stop. Hence, the replete volume ingested may not vary greatly within the species. Pool feeders fill up more slowly and volume may be more variable, especially in species of tabanids which require several feeds to fully engorge.

Ae. aegypti were observed to take smaller meals when the blood contained *Dirofilaria immitis* microfilaria,[249] but no similar observation has been made with viremic blood. The size of individual females regulates the volume of blood ingested to some extent. Physiologic and gonotrophic age[250] have variously been shown to both increase and decrease meal size. Fecundity generally declines with gonotrophic age, but this may be unrelated to meal size. Different feeding sites on the same host may yield larger or smaller meals, and different blood sources (see later discussion) seem to influence the volume of blood ingested by mosquitoes as well as ticks.

F. Interrupted and Multiple Feeding

Small numbers of both partially fed and gravid female mosquitoes are taken in biting collections.[251,252] Interpretations have varied, but Klowden[18] has documented and reviewed the impact of both meal size and ovarian stage on subsequent questing behavior. Unless mosquitoes ingest the critical amount of blood required to initiate egg development, feeding episodes appear to have little impact on future questing or feeding behavior. Once egg development has been initiated, some mosquitoes may be reluctantly induced to feed if placed directly on the host, but generally without questing. As eggs near maturity, biting inhibition becomes less absolute, which may account for the occasional gravid female observed questing in the field.

Mosquitoes dislodged from their host during blood-feeding can still acquire as well as transmit viral pathogens. Even if infected vectors are dislodged before ingestion begins, virus may be transmitted.[253] This finding has served to justify including nulliparous females in pools scheduled for virus isolation attempts. It is difficult to establish the frequency of such events in nature. The frequency of probing without feeding is presently impossible to record, and small blood meals quickly disappear. Multiple feedings on the same species are difficult to recognize, and one part of a mixed meal may be too small or too digested for identification by serology. In controlled experiments in Africa using haptoglobins as blood markers for different members of certain households, it was estimated that 10% of feedings involved more than one human host.[254] Body movement was cited as the main cause of interrupted feeding.[255]

If for any reason critical volumes of blood are not ingested, then egg development will not result and questing may continue. The critical amount required to stimulate egg developement can be influenced by body size, blood type, and nutritional reserves.[256]

G. Detachment and Departure from the Host

After fully engorging and withdrawing the mouthparts from the skin, flying vectors immediately leave the host. Engorged vectors are particularly vulnerable to host grooming so time on the host is minimized through rapid departure behavior. Engorged females may rest on nearby vegetation, etc. until the blood meal has been concentrated through water excretion, after which they seek out more permanent resting habitats. These resting sites, especially in the case of commuter species, may be remote from the site of feeding, and flights of 2 km or more by recently fed mosquitoes have been recorded.[257] Flight activity during the egg development phase of a vector's life may be reduced, but not curtailed entirely. Nectar feeding may take place and daily movements, presumably to seek more optimum microenvironments as conditions change, are well documented.

Tick detachment from the host is a complex process requiring dissolution of the cement that holds the hypostome in place. Because the rapid phase of feeding is delayed in the absence of mating, the factors that regulate detachment must be independent of those that signal these earlier events. The "big sip" generally commences at the time of day in which the host begins to rest, and the tick detaches late in the resting period.

VII. SALIVATION, DIGESTION, AND POSTENGORGEMENT ACTIVITY

A. Feeding-Associated Pathology in Hosts

Anemia and even death caused by removal of blood by hematophagic arthropods is often reported, mostly for ixodid ticks and livestock.[258] In general, disease vectors are less likely to attain the high population levels required to induce such effects on large animals. Anemia is most apt to occur among small animals with low blood volumes — animals that are observed infrequently in nature. Twenty mosquitoes feeding on a 1-day-old mouse caused immediate death so anemia can be produced by a very small number of blood meals.[259] Many blood-borne infections also produce anemia so the effects of feeding and disease can be compounded. In cattle, anemia due to tick feeding may not relate just to blood loss;

salivary substances may induce anorexia or other toxic effects which antagonize restoration of blood values to normal after feeding has ceased.

B. Pharmacological Effects of Saliva

Salivary products injected into vertebrate hosts during probing include compounds that produce diverse effects that may vary between species. Systemic reactions may occur such as the ascending flaccid paralysis that accompanies the feeding of certain ticks.[260] Localized tissue damage around tick feeding sites can result from both immune complexes and toxins. Either cause can produce appropriate histopathological effects.

Pool-feeding Diptera, particularly black flies, also produce salivary products which may produce toxemia reactions, including a hemorrhagic syndrome, in their hosts.[261] Deaths have been reported, but extraordinary numbers of biting flies are required to produce such extreme toxic manifestations.

Ixodid ticks feed for 1 week or more from the prominent cavity that they produce in the skin of their hosts. Salivary secretions mediate this prolonged and intimate association with vertebrate hosts. Of these salivary moieties, the cement substance has long been associated with salivary secretion, but the four different types of salivary acini, each containing diverse cell types, serve a variety of other functions.[262] Not all salivary products are invoved in blood feeding: the spermatophore becomes coated by a sticky salivary lubricant; a hygroscopic salivary secretion makes it possible for ticks to imbibe atmospheric water.

A copious watery solution, isotonic with vertebrate tissues, is injected into the host during the last day or so of attachment in the course of the rapid-engorgement phase of feeding.[263] This diuresis permits the tick to retain a highly concentrated pack of ingested blood cells while preserving the host from shock and dehydration. In this respect, the salivary glands of ticks share an important function with the malpighian tubules of hematophagic insects.

The prolonged course of feeding of ixodid ticks provides their hosts with sufficient time to mount an immune reponse against their tissues, and various salivary secretions serve to prevent such rejection.[264] Ixodid saliva is particularly rich in prostaglandin E_2(PGE_2), a powerful immunosuppressive that prevents T-cell activation. By acting at the site in which antigen is deposited, PGE_2 would block initiation of the cascade of cellular events that leads to inflammation and production of humoral antibodies. Indeed, this may explain the failure of vaccines to protect against the bites of ticks. Release of pain-potentiating bradykinin is a counterproductive by-product of PGE_2 injection. A salivary kininase, however, elegantly hydrolyzes this substrate produced by another salivary component. A spectrum of these as well as other anti-inflammatory and immunosuppressive constituents appear to match the immune mechanisms of particular hosts, and this may explain the stable associations that characterize the relationships between particular ticks and the hosts to which they are adapted.

C. Immunological Responses of Hosts

Although salivary products typically contain antigenic substances that invoke host immune responses, anaphylactic reactions to the bites of arthropods are rare.[265] At least in the case of the rapid-feeding nematocerans, immediate inflammatory reactions occur at about the time that the mouthparts are withdrawn. Thus, immune responses may constitute a strong selective force favoring rapid blood location and uptake. Hosts generally become desensitized by repeated feeding to particular species, and closely related vector species appear to share some common salivary antigens. Humoral antibodies reacting with salivary antigens appear to have little adverse effect when ingested by blood-feeding arthropods.[266] Interestingly, however, exposure of hosts to antigens of other arthropod tissues may induce antibodies that are highly pathogenic when ingested by that arthropod.

Although host immune responses to slow-feeding ixodids can occur well before engorgement is complete, they are most evident after repeated infestations.[258,265,267] Such host-mounted immunity can influence both feeding success of ticks as well as the volume of blood ingested. This response can thereby serve to both regulate the population density of ticks and to promote seasonal and circadian attachment patterns in order to maximize attachment on nonimmune hosts. Not all hosts mount effective immune responses against all tick species regardless of feeding interval or intensity of exposure. The reasons for this are not clear. Compounding this picture is the observation that hosts, which regularly mount an effective immunity through tick feeding, are difficult to artificially immunize with these same salivary secretions.

Acquired resistance to ticks has long been recognized,[267] including the peculiarly narrow specificity of tolerance that generally characterizes these relationships. Indeed, in 1939, Trager[269] established that certain rodent hosts *(Microtus pennsylvanicus* but not *Peromyscus leucopus),* tolerated repeated feeding by dog ticks *(Dermacentor variabilis).* With the deer tick *(I. dammini),* on the other hand, this stable, chronic association is reversed.[292] In the artificial associations that have been described, immune-mediated, inflammatory skin reactions are responsible for this expression of host resistance to feeding ticks.[265] Ticks feeding on hypersensitive guinea pigs, for example, may die before engorgement or perhaps imbibe a small meal of blood and fail to thrive. Host basophils become recruited to the attachment site where they degranulate and release some (apparently) nonhistaminic mediator.[270] Natural hosts, however, do not reject their tick parasites, and this tolerance presents a fascinating study in coadaptation.

Repeated exposure to biting insects induces hypersensitivity to their antigens.[271] The immediate erythematous, edematous reaction is characterized by polymorphonuclear leukocytes; a delayed reaction, characterized by a lymphocytic, eosinophilic infiltrate, follows 1 day or so later. Continued exposure generally results in desensitization. Immunoglobulin E mediates such reactions. Because mosquitoes and other Nematocera detach from their hosts within a few minutes, such inflammatory reactions have little effect upon them. Edema and vasoconstriction would provide the main barrier to feeding. The effect on pathogens deposited in the skin remains to be determined. Humoral antibody, however, may profoundly affect the insect after it has detached because immunoglobulins penetrate the insect's body.[272] Presumably, salivary secretions contain few antigens in common with other, more exposed tissues. The prolonged period of attachment of ixodid ticks causes them to confront a more critical problem resulting from the inflammatory reactions of their hosts.

D. Enzyme Secretion

Much of the protein contained in ingested blood is converted by hematophagous arthropods into the yolk contained in eggs. Not all the nitrogen contained in a blood meal, however, is transferred to eggs. Synthesis and secretion of trypsin in mosquitoes is stimulated by the presence of soluble protein,[273] but a lag phase of at least 8 hr occurs after the blood meal before tryptic activity rises above the background level. Developing ovaries provide the positive feedback required for the endocrine-dependent continuation of this process. There is good correlation and synchronization between tryptic activity, protein degradation, excretion, vitellogenesis, and egg maturation. The actual rate of digestion, though temperature dependent, does not appear to vary greatly with age or gonotrophic cycle. The duration of the tryptic lag phase is prolonged with age, and this could be the reason why older females are less susceptible to certain infections than are their younger counterparts.

E. Peritrophic Membrane Formation

The timing of peritrophic membrane (PM) formation by mosquitoes following blood feeding has been a frequent subject of study because its appearance suggests that it may serve as a barrier to invasion of pathogens into the body of the vector. Of course, the term "membrane" is something of a misnomer for this noncellular structure. The PM of mosquitoes assumes its discrete, textured form at about 8 hr after blood has been ingested. Reports of precursor material being secreted within approximately 1 hr require critical interpretation — they may represent artifacts of fixation. This delayed deposition of the structured PM corresponds to the lag phase of tryptic activity,[274] and occurs long after arboviruses, ookinetes, or microfilariae have penetrated into the gut wall.[275]

The peritrophic membrane of Nematocera today remains nearly the enigma that it was more than a decade ago.[276] The structure is continuously secreted in larval mosquitoes as a lamellar sleeve originating in specialized cells in the intersuscepted junction of the fore- and hindgut. The feces of these insects comes packaged, sausage-like, in the trailing posterior end of this noncellular membrane. In adult mosquitoes, as in those of other nematocerous flies, secretion is discontinuous. The membrane becomes evident about 12 hr after the blood meal and is ultimately discarded in the feces. In general, however, other adult insects secrete this structure continuously. An analogous peritrophic membrane is present in ticks.[277] Chitin is an important constituent of the peritrophic membrane of various insects and of *Aedes* but not *Anopheles* mosquitoes.[278]

The peritrophic membrane serves a variety of functions. It may protect the gut wall from mechanical damage, as from the hemoglobin of certain vertebrates that forms jagged crystals during digestion. The ectoperitrophic space provides opportunity for countercurrent circulation to distribute digested material across the entire midgut lumen, thereby promoting absorbtion. Otherwise, the absorbtive surface would be limited because the cylindrical mass of ingested food becomes progressively digested as it passes posteriorly. The membrane is permeable to proteolytic enzymes secreted by the gut wall, but retains polypeptides.[279,280] Thus, by packaging ingested material, the membrane enhances digestion because only the surface of the food bolus is presented to the secreted enzyme. Enzyme concentration is thereby maximized.

The role of the peritrophic membrane in vector competence remains unclear. The membrane may limit the passage of microorganisms,[281] but this traditional view requires reexamination. Its formation may serve equally as an opportunity for pathogen attachment as a barrier to their escape. Piroplasm ookinetes attain such close contact with the peritrophic membrane of ixodid ticks that junctions seem to form.[277] Perhaps the forming membrane coincides with a disappearance of receptors for pathogen attachment. In either case, pathogens fail to pass the mature membrane.

F. Digestion of Blood

The enormous quantity of blood generally ingested by vector insects impedes their departure from the immediate vicinity of a vertebrate host and provides an obstacle to subsequent host contact, at least until diuresis and digestion reduces its mass. This feeding event produces dramatic changes. The midgut wall stretches, converting it within 1 min from a pallisade-like epithelium into a squamous membrane about one tenth of its former thickness.[282] The microvillar apex of these cells disappears, intercellular junctions distend, and the overlying basal lamina becomes stressed. About 1 day later, the gut begins to return to its previous form. Blood-borne pathogens penetrate into the body of the vector during this transient period of extraordinary stress and morphogenesis.

The mass of ingested blood undergoes a series of changes soon after it reaches the midgut. The pharyngeal armament of certain species may cause mechanical disruption of some blood cells during ingestion.[283] The blood may clot or ingested erythrocytes may aggregate. In any event, diuresis soon compacts the ingested cells into a tightly joined mass.[282] The effect of these physical changes on the ability of ingested pathogens to attain contact with the midgut wall has not yet been defined. Malaria ookinetes and microfilariae locomote through the gut contents and presumably concentrate against the wall through their own activity. No such arbovirological effect has been demonstrated.

These first 4 hr have been designated as the osmoregulatory phase of digestion in mosquitoes.[282] The second 4 hr after engorgement comprise the presecretory phase of digestion, when the secretory machinery of the midgut matures. The smooth as well as rough endoplasmic reticulum proliferates and becomes reorganized. The third, proteolytic secretory phase extends from 8 to about 30 hr after the blood meal. Trypsin-like enzymes are released into the midgut of mosquitoes at this time. The fourth, absorbtive phase overlaps, beginning at about 16 hr and extending until 2 to 3 days after feeding, when the remnants of the blood meal, fragments of the PM, and residual digestive juices are excreted. The midgut epithelium then returns to its resting condition.

An aggregated array of rough endoplasmic reticulum is prominent in the midgut cells of mature, nonfed *Aedes* mosquitoes.[284] Although the term "whorl" generally is used to designate this structure, the array is not spirally formed. Rather, it is comprised of concentric rings arranged in plates that are packed into spheres. They disaggregate almost immediately after the gut has filled with blood or other protein-containing solution. Initial aggregation or reaggregation follows stimulation by JH.[285] This whorled architecture suggests a resting condition, but the function of the disaggregated organelles has not yet been established.

The manner in which ixodid ticks digest ingested blood differs radically from that of mosquitoes. The gut epithelium proliferates following engorgement, producing phagocytic cells capable of ingesting whole red cells and digesting this material by lysosomal function. Free protease has not been demonstrated within the gut of ticks.

G. Rate of Digestion

In Diptera, digestion of the blood meal is rapid. It varies with vector species, volume ingested, and ambient temperature but usually requires 2 to 4 days under summer conditions. Blood type, age, insemination, and parity all have been observed to influence the rate of digestion; most of these studies did not control for meal size or separate the lag phase from the remaining active part of the digestive process. Generally, 1 to 2 additional days are required from the time digestion is complete until females deposit their eggs. Hence, the gonotrophic cycle is always longer than the digestive cycle. In the case of chipmunk- and squirrel-fed *Ae. triseriatus*, the gonotrophic cycle lasts as much as 2 days longer than with females fed on other hosts,[286] but more eggs develop as well. This suggests that larger meals may have been taken from these hosts, thereby requiring more time for digestion. The blood of guinea pigs, which *Cx. tarsalis* digests more slowly than other bloods, correspondingly results in minimal fertility.[287]

H. Blood Type: Nutritional Sufficiency and Vector Dependency

The type of blood ingested can have two discrete influences on the digestive process. As just seen, digestion time and the quantity of blood ingested may vary depending on the blood source. Meal size will directly affect fecundity. Alternatively, different blood types may vary in their nutritional qualities and thereby affect fecundity. Various investigators have compared mosquito fecundity on different blood sources. Unfortunately, few have taken meal size into account. Studies in which blood volumes were controlled provide data which are difficult to interpret for the most part. Mosquitoes generally produce more eggs after feeding on exotic animals than on their natural hosts. For example, species that normally feed on mammals often develop more eggs on bird blood. The only consistent pattern is that most species produce fewer eggs on equal volumes of human blood than on most other types tested.[288] Low levels in human blood of the essential amino acid isoleucine have been credited with this reduction. Species with the strongest host-specific tendencies might be expected to show the best correlation between blood type and fecundity. Indeed, certain host-specific ectoparasites may die after ingesting the blood of exotic hosts. *Cs. melanura*, a strongly ornithophilic species, was reluctant to ingest rabbit blood when fed through a capillary tube, and enemas of rabbit blood resulted in far fewer eggs than equivalent amounts of chicken blood.[132] This constitutes an ideal example of nutritional dependency by mosquitoes. Human-biting populations of *Ae. aegypti* and certain *Anopheles* appear to not gain reproductive advantage from the blood of their apparently preferred host. *Argas* ticks, on the other hand, take more blood in the course of feeding on their natural hosts than on other animals.[289]

I. Influence of Infection and Other Blood Characteristics

There is little evidence that the presence of any vertebrate pathogen in blood influences the digestion process. Fecundity differences, however, have been associated with feeding on infected (anemic) than on noninfected hosts. Again, meal size often was poorly controlled, and other studies on the same parasites have shown no reduction in fecundity due to infection or anemia. To date, no one has demonstrated that virus-infected blood affects fecundity of a hematophagic arthropod. *Ae. triseriatus,* engorging on immune chipmunks, were shown to have a lower rate of oral transmission of LAC virus following venereal exposure to the virus.[290] It is not clear how viral antibodies ingested in the blood meal find their way, apparently intact, to sites of viral replication external to the midgut.

J. Temperature Influence

As already suggested, ambient temperature, within certain biological limits, influences the time required for both blood digestion and completion of the gonotrophic cycle. Time between blood meals is an important variable in any model of arbovirus transmission. Since survival rates have a logarithmic effect on transmission, so too does feeding frequency — a function of survival. Survival rates, however, may be reduced as temperatures rise above the optimum. The resulting increased frequency of feeding, associated with shorter gonotrophic cycles at higher temperatures, may be offset by greater mortality. In terms of vectorial capacity, this picture is complicated by findings that certain viruses may infect better and replicate faster at temperatures below the optimum for feeding turnover in the vector.

K. Impact of Delayed Oviposition

In the subtropics and tropics, continuously permissive temperatures permit perennial development, but seasonal rainfall patterns often result in prolonged periods unfavorable for larval development. It is not clear how vector populations and virus cycles perpetuate during adverse periods; thus, reproduction becomes reduced where conditions are locally or temporally unfavorable. An alternative possibility is that gravid females survive with retained eggs for long dry periods. Bursts of oviposition and blood feeding would then follow new rains. In Florida, for example, *Cx. nigripalpus* gravid females can survive 100 days under outdoor conditions during the cool, dry season when breeding normally is interrupted.[291] When offered ovipostion sites, they immediately deposit viable eggs and quest for blood. Indeed, the VEE virus vector, *Cx. portesi,* may delay oviposition for at least 42 days.[106] This then could be a viable mechanism for arbovirus maintenance during short or perhaps even prolonged periods when conditions are unfavorable for vector reproduction and transmission in warm climates.

VIII. SUMMARY AND CONCLUSIONS

Arboviral infection cycles are extraordinarily diverse. They encompass a variety of vertebrates and seemingly poorly adapted, nonspecific vectors tied together in a web stretched too thin to be maintained in an uncertain environment. Although outbreaks of zoonotic arboviral infection may seem sporadic, such infections persist. We have provided a conceptual basis for this intricate, innate diversity by describing behavioral and physiological components of the blood-feeding process of various hematophagic arthropods, the process that is central to transmission of arboviruses. However, blood feeding involving vectors of stable infections is more efficient than it first appears. These natural systems are highly integrated.

Two peculiarities render arboviral infection particularly problematic; they tend to be time-limited in the vertebrate host, thereafter resulting in sterilizing, life-long immunity; their extraordinary communicability tends to exhaust the available supply of nonimmune hosts. Excessive communicability may threaten perpetuation as much as would ineffective transmission.

Infectious reservoir hosts may disproportionately infect vectors, thereby contributing their blood to the arbovirus cycle far in excess of their number. Similarly, infectious vectors may visit reservoir hosts more frequently than do their noninfected siblings, and they feed in a manner that increases opportunity for pathogen transfer. This nonrandomness of viral transmission requires that scientific attention be redirected from populations as a whole to the infected cohorts within those populations. An understanding of the manner in which arboviral agents affect vector-host contact will facilitate improved estimates of vectorial capacity.

REFERENCES

1. **Harwood, R. F.,** Criteria for vector effectiveness, in *Vectors of Disease Agents: Interactions with Plants, Animals and Man,* McKelvey, J.J., Eldridge, B. F., and Maramorosch, K., Eds., Praeger, New York, 1981.
2. **Edman, J., Day, J., and Walker, E.,** Vector-host interplay: factors affecting disease transmission, in *Ecology of Mosquitoes: Proceedings of a Workshop,* Lounibos, L. P., Rey, J., and Frank, J. H., Eds., University of Florida Press, Gainesville, 1985, 273.
3. **Waage, J. K.,** The evolution of insect/vertebrate associations, *Biol. J. Linn. Soc.,* 12, 187, 1979.
4. **Waage, J. K.,** Curse of the vampire: the evolution of blood-sucking insects, *Antenna,* 4, 112, 1980.
5. **Kim, K. C., Ed.,** *Coevolution of Parasitic Arthropods and Mammals,* John Wiley & Sons, New York, 1985.
6. **Price, P. W., Ed.,** *Evolutionary Strategies of Parasitic Insects and Mites,* Princeton University Press, Princeton, N. J., 1980.
7. **Nelson, W. A., Keirans, J. E., Bell, J. F., and Clifford, C. M.,** Host-ectoparasite relationships, *J. Med. Entomol.,* 12, 143, 1975.
8. **Balashov, Y. S.,** Interaction between blood-sucking arthropods and their hosts, and its influence on vector potential, *Ann. Rev. Entomol.,* 29, 137, 1984.
9. **Spielman, A. and Skaff, V.,** Inhibition of metamorphosis and of ecdysis in mosquitoes, *J. Insect Physiol.,* 13, 1087, 1967.
10. **Clements, A.,** *The Physiology of Mosquitoes,* Pergamon Press, London, 1963.
11. **Gwadz, R. W. and Spielman, A.,** Corpus allatum control of ovarian development in *Aedes aegypti, J. Insect Physiol.,* 19, 1441, 1973.
12. **Gwadz, R. W., Lounibos, L. P., and Craig, G. B.,** Precocious sexual receptivity induced by a juvenile hormone analogue in females of the yellow fever mosquito, *Aedes aegypti, Gen. Comp. Endocrinol.,* 16, 47, 1971.
13. **Flanagan, T. R. and Hagedorn, H. H.,** Vitellogenin synthesis in the mosquito: the role of juvenile hormone in the development of responsiveness to ecdysone, *Physiol. Entomol.,* 2, 173, 1977.
14. **Meola, R. W. and Petralia, R. S.,** Juvenile hormone induction of biting behavior in *Culex* mosquitoes, *Science,* 209, 1548, 1980.
15. **Feinsod, F. M. and Spielman, A.,** Nutrient-mediated juvenile hormone secretion in mosquitoes, *J. Insect Physiol.,* 26, 113, 1980.
16. **Spielman, A.,** Effect of synthetic juvenile hormone on ovarian diapause of *Culex pipiens* mosquitoes, *J. Med. Entomol.,* 11, 223, 1974.

17. **Case, T. J., Washino, R. K., and Dunn, R. L.,** Diapause termination in *Anopheles freeborni* with juvenile hormone mimics, *Entomol. Exp. Appl.*, 21, 155, 1977.
18. **Klowden, M. J.,** Coping with inflation: abdominal distension and mosquito reproduction, in *Host Regulated Developmental Mechanisms in Vector Arthropods*, Borovsky, D. and Spielman, A., Eds., University of Florida Press, Gainesville, 1986, 146.
19. **Brown, M., Raikel, A., and Lea, A. O.,** Ultrastructure of midgut endocrine cells in the adult mosquito, *Aedes aegypti*, *Tissue Cell*, 17, 709, 1985.
20. **Lea, A. O. and Van Handel, E. A.,** A neurosecretory hormone-releasing factor from ovaries of mosquitoes that fed on blood, *J. Insect. Physiol.*, 28, 503, 1982.
21. **Borovsky, D.,** Release of egg development neurosecretory hormone in *Aedes aegypti* and *Aedes taeniorhynchus* induced by an ovarian factor, *J. Insect Physiol.*, 28, 311, 1982.
22. **Clements, A. N., Potter, S. A., and Scales, M. D. C.,** The cardiacal neurosecretory system and associated organs of an adult mosquito, *Aedes aegypti*, *J. Insect Physiol.*, 31, 821, 1985.
23. **Lea, A. O.,** Regulation of egg maturation in the mosquito by the neurosecretory system. The role of the corpus cardiacum, *Gen. Comp. Endocrinol.*, Suppl. 3, 602, 1972.
24. **Hagedorn, H. H., Shapiro, J. P., and Hanaoka, K.,** Ovarian ecdysone secretion is controlled by a brain hormone in an adult mosquito, *Nature (London)*, 282, 92, 1979.
25. **Hagedorn, H. H., O'Connor, J. D., Fuchs, M. S., Sage, B., Schlaeger, D.A., and Bohm, M.K.,** The ovary as a source of ecdysone in adult mosquitoes, *Proc. Natl. Acad. Sci. U.S.A.*, 72, 3255, 1975.
26. **Klowden, M. J. and Lea, A. O.,** Abdominal distension terminates subsequent host-seeking behavior of *Aedes aegypti* following a blood meal, *J. Insect Physiol.*, 25, 583, 1979.
27. **Klowden, M. J. and Lea, A. O.,** Humoral inhibition of host-seeking in *Aedes aegypti* during oocyte maturation, *J. Insect Physiol.*, 25, 231, 1979.
28. **Davis, E. E. and Bowen, M. F.,** Humoral regulation of receptor sensitivity and host-seeking behavior in mosquitoes: role of the fat body, in *Host Regulated Developmental Mechanisms in Vector Arthropods*, Borovsky, D. and Spielman, A., Eds., University of Florida Press, Gainesville, 1986, 135.
29. **Sonenshine, D. E.,** Pheromones and other semiochemicals of the acari, *Ann. Rev. Entomol.*, 30, 1, 1985.
30. **Diehl, P. A., Aeschlimann, A., and Obenchain, F. D.,** Tick reproduction: oogenesis and oviposition, in *Physiology of Ticks*, Obenchain, F.D. and Galun, R., Eds., Pergamon Press, Oxford, 1983, chap. 9.
31. **Morris, C. D. and Zimmerman, R. H.,** Epizootiology of eastern equine encephalomyelitis in upstate New York, USA. III. Population dynamics and vector potential of adult *Culiseta morsitans* (Diptera: Culicidae), *J. Med. Entomol.*, 18, 313, 1981.
32. **Scholl, P. J., Porter, C. H., and DeFoliart, G. R.,** *Aedes triseriatus:* Persistence of nulliparous females under field conditions, *Mosq. News*, 39, 368, 1979.
33. **Foster, W. A. and Lea, A. O.,** Sexual behavior in male *Aedes triseriatus* (Diptera: Culicidae): a reexamination, *J. Med. Entomol.*, 12, 459, 1975.
34. **McIver, S. B.,** Sensory aspects of mate-finding behavior in male mosquitoes, *J. Med. Entomol.*, 17, 54, 1980.
35. **Jaenson, T. G. T.,** Attraction to mammals of male mosquitoes with special reference to *Aedes diantaeus* in Sweden, *J. Am. Mosq. Control Assoc.*, 1, 195, 1985.
36. **Hartberg, W. K.,** Observations on the mating behaviour of *Aedes aegypti* in nature, *Bull. WHO*, 45, 847, 1971.
37. **DeFoliart, G. R.,** *Aedes triseriatus:* vector biology in relationship to the persistence of La Crosse virus in endemic foci, in *California Serogroup Viruses*, Calisher, C. H. and Thompson W. H., Eds., Alan R. Liss, New York, 1983, 89.
38. **Roberston, A. S., Patrick, C. D., Semtner, P. J., and Hair, J. A.,** The ecology and behavior of the lone star tick (Acarina: Ixodidae). VII. Pre- and post-molt behavior of engorged nymphs and larvae, *J. Med. Entomol.*, 12, 530, 1975.
39. **Barnard, D. R., Morrison, R, D., and Popham, T. W.,** Light and temperature sensitivity of feeding-related and reproductive processes in *Amblyomma americanum* (Acari: Ixodidae) on cattle, *Environ. Entomol.*, 14, 479, 1985.
40. **Spielman, A.,** Bionomics of autogenous mosquitoes, *Ann. Rev. Entomol.*, 16, 231, 1973.
41. **O' Meara, G. F.,** Gonotrophic interactions in mosquitoes: kicking the blood-feeding habit, *Fla. Entomol.*, 68, 122, 1985.
42. **Craig, G. B.,** Mosquitoes, female monogamy induced by male accessory gland substance, *Science*, 156, 1499, 1967.
43. **Spielman, A., Leahy, M. J., and Skaff, V.,** Seminal loss in repeatedly mated female *Aedes aegypti*, *Biol. Bull.*, 132, 404, 1967.
44. **Provost, M. W. and Haeger, J. S.,** Mating and pupal attendance in *Deinocerites cancer* and comparisons with *Opifex fuscus* (Diptera: Culicidae), *Ann. Entomol. Soc. Am.*, 60, 565, 1967.

45. **Lea, A. O. and Evans, D. G.,** Sexual behavior of mosquitoes: age dependence of copulation and insemination of the *Culex pipiens* complex and *Aedes taeniorhynchus* in the laboratory, *Ann. Entomol. Soc. Am.*, 65, 285, 1972.
46. **Spielman, A., Waterman, R. E., and Mellers, S. M.,** Diversity in pattern of coital contact of mosquitoes: a scanning electron microscopic study, *J. Morphol.*, 142, 187, 1974.
47. **Teesdale, C.,** Studies on the bionomics of *Aedes aegypti* L. in its natural habitats in a coastal region of Kenya, *Bull. Entomol. Res.*, 46, 711, 1955.
48. **Nijhout, H. F. and Craig, G. B.,** Reproductive isolation in *Stegomyia* mosquitoes. III. Evidence for a sexual pheromone, *Entomol. Exp. Appl.*, 14, 399, 1971.
49. **Ribeiro, J. M. C. and Spielman, A.,** Satyr effect: a model predicting parapatry and extinction, *Am. Nat.*, 1986.
50. **Vargo, A. M. and Foster, W. A.,** Gonotropic state and parity of nectar-feeding mosquitoes, *Mosq. News*, 44, 6, 1984.
51. **Magnarelli, L. A.,** Nectar sugars and caloric reserves in natural populations of *Aedes canadensis* and *Aedes stimulans* (Diptera: Culicidae), *Environ. Entomol.*, 12, 1482, 1983.
52. **Haeger, J. S.,** Behavior preceding migration in the salt-marsh mosquito *Aedes taeniorhynchus* (Wiedemann), *Mosq. News*, 20, 136, 1960.
53. **Grimstad, P. R. and DeFoliart, G. R.,** Nectar sources of Wisconsin mosquitoes, *J. Med. Entomol.*, 11, 331, 1974.
54. **Grimstad, P. R.,** Mosquito nectar feeding in Wisconsin in relation to twilight and microclimate, *J. Med. Entomol.*, 11, 691, 1975.
55. **Magnarelli, L. A.,** Diurnal nectar feeding of *Aedes cantator* and *A. sollicitans* (Diptera: Culicidae), *Environ. Entomol.*, 8, 949, 1979.
56. **Magnarelli, L. A.,** Nectar feeding by *Aedes sollicitans* and its relation to gonotrophic activity, *Environ. Entomol.*, 6, 237, 1977.
57. **Foster, W. A.,** Effect of blood feeding on sugar-feeding behavior of mosquitoes, in *Host Regulated Developmental Mechanisms in Vector Arthropods*, Borovsky, D. and Spielman, A., Eds., University of Florida Press, Gainesville, 1986, 163.
58. **Nayar, J. K. and Sauerman, D. M.,** The effects of nutrition on survival and fecundity in Florida mosquitoes. I. Utilization of sugar for survival, *J. Med. Entomol.*, 12, 92, 1975.
59. **Nayar, J. K. and Pierce, P. A.,** Utilization of energy reserves during survival after emergence in Florida mosquitoes, *J. Med. Entomol.*, 14, 54, 1977.
60. **Nayar, J. K. and Pierce, P. A.,** The effects of diet on survival, insemination and oviposition of *Culex nigripalpus* Theobald, *Mosq. News*, 40, 210, 1980.
61. **Nayar, J. K., Provost, M. W., and Hansen, C. W.,** Quantitative bionomics of *Culex nigripalpus* (Diptera:Culicidae) populations in Florida. II. Distribution, dispersal and survival patterns, *J. Med. Entomol.*, 17, 40, 1980.
62. **Nayar, J. K.,** Bionomics and physiology of *Culex nigripalpus* (Diptera: Culicidae) of Florida: an important vector of diseases, *Bull. Fla. Agric. Exp. Stn.*, 82, 1982.
63. **Jones, J. C. and Madhukar, B. V.,** Effects of sucrose on blood avidity in mosquitoes, *J. Insect Physiol.*, 22, 357, 1976.
64. **Van Handel, E.,** Metabolism of nutrients in the adult mosquito, *Mosq. News*, 44, 573, 1984.
65. **Bidlingmayer, W. L. and Hem, D. G.,** Sugar feeding by Florida mosquitoes, *Mosq. News*, 33, 535, 1973.
66. **Magnarelli, L. A.,** Bionomics of *Psorophora ferox* (Diptera: Culicidae): seasonal occurrence and acquisition of sugars, *J. Med. Entomol.*, 17, 328, 1980.
67. **Magnarelli, L. A.,** Nectar-feeding by female mosquitoes and its relation to follicular development and parity, *J. Med. Entomol.*, 14, 527, 1978.
68. **Nasci, R. S. and Edman, J. D.,** *Culiseta melanura* (Diptera: Culicidae): population structure and nectar feeding in a freshwater swamp and surrounding areas in southeastern Massachusetts, USA, *J. Med. Entomol.*, 21, 567, 1984.
69. **Mullens, B. A.,** Age-related adult activity and sugar feeding by *Culicoides variipennis* (Diptera: Ceratopogonidae) in Southern California, USA, *J. Med. Entomol.*, 22, 32, 1985.
70. **Howell, F. G.,** Influence of the daily light cycle on the behavior of *Argas cooleyi* (Acarina: Argasidae), *J. Med. Entomol.*, 13, 99, 1976.
71. **Rechav, Y.,** Migration and disperal patterns of three African ticks (Acari: Ixodidae) under field conditions, *J. Med. Entomol.*, 13, 99, 1976.
72. **Lane, R. S., Anderson, J. R., Yaninek, J. S., and Burgdorfer, W.,** Diurnal host seeking of adult Pacific Coast ticks, *Dermacentor occidentalis* (Acari: Ixodidae), in relation to vegetational type, meteorological factors, and rickettsial infection rates in California, USA, *J. Med. Entomol.*, 22, 558, 1985.
73. **Bidlingmayer, W. L.,** A comparison of trapping methods for adult mosquitoes: species response and environmental influence, *J. Med. Entomol.*, 4, 200, 1967.

74. **Bidlingmayer, W. L.,** The influence of environmental factors and physiological stage on flight patterns of mosquitoes taken in the vehicle aspirator and truck, suction, bait and New Jersey light traps, *J. Med. Entomol.*, 11, 119, 1974.
75. **Clark, G. G., Rohrer, W. H., and Robbins, D. N.,** Diurnal biting activity of *Aedes triseriatus* Complex (Diptera:Culicidae) in a focus of La Crosse virus transmission, *J. Med. Entomol.*, 22, 684, 1985.
76. **Nelson, M. J., Self, L. S., Pant, C. P., and Usman, S.,** Diurnal periodicity of attraction to human bait of *Aedes aegypti* (Diptera: Culicidae) in Jakarta, Indonesia, *J. Med. Entomol.*, 14, 504, 1978.
77. **Edman, J. D. and Haeger, J. S.,** Host-feeding patterns of Florida mosquitoes. V. *Wyeomyia*, *J. Med. Entomol.*, 14, 477, 1977.
78. **Linley, J. R., Hoch, A. L., and Pinheiro, F. P.,** Biting midges (Diptera:Ceratopogonidae) and human health, *J. Med. Entomol.*, 20, 347, 1983.
79. **Hoch, A. L., Peterson, N. E., LeDuc, J. W., and Pinheiro, F. P.,** An outbreak of Mayaro virus disease in Belterra, Brazil. III. Entomological and ecological studies, *Am. J. Trop. Med. Hyg.*, 30, 689, 1981.
80. **Bidlingmayer, W. L.,** Mosquito flight paths in relation to environment. I. Illumination levels, orientation and resting areas, *Ann. Entomol. Soc. Am.*, 64, 1121, 1971.
81. **Bidlingmayer, W. L.,** Effect on moonlight on flight activity of mosquitoes, *Ecology*, 45, 87, 1964.
82. **Edman, J. D.,** Host-feeding patterns of Florida mosquitoes. I. *Aedes, Anopheles, Coquillettidia, Mansonia* and *Psorophora*, *J. Med. Entomol.*, 8, 687, 1971.
83. **Curtis, L. C.,** Observations on mosquitoes at Whitehorse, Yukon Territory (Culicidae:Diptera), *Can. Entomol.*, 85, 353, 1953.
84. **Linthicum, K. J., Davies, F. G., and Kairo, A.,** Observations of the biting activity of mosquitoes at a flooded dambo in Kenya, *Mosq. News*, 44, 595, 1984.
85. **Reisen, W. K. and Aslamkhan, M.,** Biting rhythms of some Pakistan mosquitoes (Diptera:Culicidae), *Bull. Entomol. Res.*, 68, 313, 1978.
86. **Edman, J. D., Webber, L. A., and Kale, H. W.,** Host-feeding patterns of Florida mosquitoes. II. *Culiseta*, *J. Med. Entomol.*, 9, 429, 1972.
87. **Edman, J. D.,** Host-feeding patterns of Florida mosquitoes. III. *Culex (Culex)* and *Culex (Neoculex)*, *J. Med. Entomol.*, 11, 95, 1974.
88. **Edman, J. D.,** Host-feeding patterns of Florida mosquitoes. IV. *Deinocerites*, *J. Med. Entomol.*, 11, 105, 1974.
89. **Edman, J. D.,** Host-feeding patterns of Florida mosquitoes. VI. *Culex (Melanoconion)*, *J. Med. Entomol.*, 15, 521, 1979.
90. **Boreham, P. F. L.,** Some applications of blood meal identifications in relation to the epidemiology of vector-borne tropical diseases, *Trans. R. Soc. Trop. Med. Hyg.*, 69, 83, 1975.
91. **Snow, W. F. and Boreham, P. F. L.,** The host-feeding patterns of some culicine mosquitoes (Diptera:Culicidae) in the Gambia, *Bull. Entomol. Res.*, 68, 695, 1978.
92. **Tempelis, C. H.,** Host-feeding patterns of mosquitoes with a review of advances in analysis of blood meals by serology, *J. Med. Entomol.*, 11, 635, 1975.
93. **Walker, A. R. and Boreham, P. F. L.,** Blood feeding of *Culicoides* (Diptera:Ceratopogonidae) in Kenya in relation to the epidemiology of bluetongue and ephemeral fever, *Bull. Entomol. Res.*, 66, 181, 1976.
94. **Hashiguchi, Y., Gomez, L. E., De Coronel, V., Mimori, T., and Kawabata, M.,** Biting activity of two anthropophilic species of sandflies, *Lutzomyia*, in an endemic area of leishmaniasis in Ecuador, *Ann. Trop. Med. Parasitol.*, 79, 533, 1985.
95. **Christensen, H. A., Arias, J. R., De Vasquez, A. M., and De Freitas, R. A.,** Hosts of sandfly vectors of *Leishmania braziliensis guyanensis* in the Central Amazon of Brazil, *Am. J. Trop. Med. Hyg.*, 31, 239, 1982.
96. **Tesh, R. B., Chaniotis, B. N., Aronson, M. D., and Johnson, K. M.,** Natural host preferences of Panamanian phlebotomine sandflies as determined by precipitin test, *Am. J. Trop. Med. Hyg.*, 20, 150, 1971.
97. **Haddow, A. J.,** Entomological studies from a high tower in Mpanga Forest, Uganda. VI. The biting behaviour of mosquitoes and tabanids, *Trans. R. Entomol. Soc. London*, 113, 315, 1961.
98. **Jones, M. D. R. and Gubbins, S. J.,** Changes in the circadian flight of the mosquito *Anopheles gambiae* in relation to insemination, feeding and oviposition, *Physiol. Entomol.*, 3, 213, 1978.
99. **Jones, M. D. R. and Gubbins, S. J.,** Modification of female circadian flight-activity by a male accessory gland pheromone in the mosquito, *Culex pipiens quinquefasciatus*, *Physiol. Entomol.*, 4, 345, 1979.
100. **Bidlingmayer, W. L.,** The measurement of adult mosquito population changes — some considerations, *J. Am. Mosq. Control Assoc.*, 1, 328, 1985.
101. **Day, J. F. and Carlson, D. B.,** The importance of autumn rainfall and sentinel flock location to understanding the epidemiology of St. Louis encephalitis virus in Indian River County, Florida, *J. Am. Mosq. Control Assoc.*, 1, 305, 1985.
102. **Eldridge, B. F.,** The effect of temperature and photoperiod on blood-feeding and ovarian development in mosquitoes of the *Culex pipiens* complex, *Am. J. Trop. Med. Hyg.*, 17, 133, 1968.

103. **Case, T. J., Washino, R. K., and Dunn, R. L.,** Diapause termination in *Anopheles freeborni* with juvenile hormone mimics, *Entomol. Exp. Appl.*, 2, 155, 1977.
104. **Spielman, A. and Wong, J.,** Environmental control of ovarian diapause in *Culex pipiens, Ann. Entomol. Soc. Am.*, 60, 905, 1973.
105. **Bailey, C. L., Eldridge, B. F., Hayes, D. E., Watts, D. M., Tammariello, R. F., and Dalrymple, J. M.,** Isolation of St. Louis encephalitis virus from overwintering *Culex pipiens* mosquitoes, *Science*, 199, 1346, 1978.
106. **Eldridge, B. F.,** Vector maintenance of pathogens in adverse environments: with special reference to mosquito maintenance of arbovirus, in *Vectors of Disease Agents: Interactions with Plants, Animals and Man*, McKelvey, J. J., Eldridge, B. G., and Maramorasch, K., Eds., Praeger, New York, 1981, 143.
107. **Belozerov, V. N.,** Diapause and biological rhythms in ticks, in *Physiology of Ticks*, Obenchain, F. D. and Galun, R., Eds, Pergamon Press, Oxford, 1982, chap. 13.
108. **McIver, S. B.,** Sensilla of mosquitoes (Diptera:Culicidae), *J. Med. Entomol.*, 19, 489, 1982.
109. **Khan, A. A.,** Mosquito attractants and repellents, in *Chemical Control of Insect Behavior*, Shorey, H. H. and McKelvey, J. J., Eds., John Wiley & Sons, New York, 1977, chap. 18.
110. **Gillies, M. T.,** Some aspects of mosquito behaviour in relation to the transmission of parasites, in *Behavioral Aspects of Parasite Transmission*, Canning, E. and Wright, C., Eds., Academic Press, London, 1972, 69.
111. **Hocking, B.,** Blood sucking behavior of terrestrial arthropods, *Ann. Rev. Entomol.*, 16, 1, 1971.
112. **Gillies, M. T. and Wilkes, T. J.,** Responses of host-seeking *Mansonia* and *Anopheles* mosquitoes (Diptera: Culicidae) in West Africa to visual features of a target, *J. Med. Entomol.*, 19, 68, 1982.
113. **Bidlingmayer, W. L. and Hem, D. G.,** The range of visual attraction and the effect of competitive visual attractants upon mosquito (Diptera: Culicidae) flight, *Bull. Entomol. Res.*, 70, 321, 1980.
114. **Waladde, S. M. and Rice, M. J.,** The sensory basis of tick feeding behaviour, in *Physiology of Ticks*, Obenchain, F. D. and Galun, R., Eds., Pergamon Press, Oxford, 1982, chap. 3.
115. **Day, J. F.,** unpublished data, 1986.
116. **Allan, S. A.,** Studies on Vision and Visual Attraction of the Salt Marsh Horse Fly, *Tabanus nigrovittatus* Macquart, Ph.D. dissertation, University of Massachusetts, Amherst, 1984, 183.
117. **Simmons, K. R.,** Reproductive Ecology and Host-Seeking Behavior of the Black Fly, *Simulium venustum* Say (Diptera: Simuliidae), Ph.D. dissertation, University of Massachusetts, Amherst, 1985, 204.
118. **Gillies, M. T.,** The role of carbon dioxide in host-finding by mosquitoes (Diptera: Culicidae): a review, *Bull. Entomol. Res.*, 70, 525, 1980.
119. **Schreck, C., Smith, N., Carlson, D., Price, G., Haile, D., and Godwin, D.,** A material isolated from human hands that attracts female mosquitoes, *J. Chem. Ecol.*, 8, 429, 1981.
120. **Price, G. D., Smith, N., and Carlson, D. A.,** The attraction of female mosquitoes *(Anopheles quadrimaculatus* Say) to stored human emanations in conjuction with adjusted levels of relative humidity, temperature and carbon dioxide, *J. Chem. Ecol.*, 5, 383, 1979.
121. **Bos, H. J. and Laarman, J. J.,** Guinea pig, lysine, cadaverine and estradiol as attractants for the malaria mosquito *Anopheles stephensi, Entomol. Exp. Appl.*, 18, 161, 1975.
122. **Marks, E. N.,** Mosquitoes (Culicidae) in the changing Australian environment, *Queensl. Nat.*, 20, 101, 1972.
123. **Bell, W. J. and Carde, R. T., Eds.,** *Chemical Ecology of Insects*, Chapman & Hall, London, 1984.
124. **Davis, E. E.,** Development of lactic acid-receptor sensitivity and host-seeking behaviour in newly emerged female *Aedes aegypti* mosquitoes, *J. Insect Physiol.*, 30, 211, 1984.
125. **McKeever, S.,** Observations of *Corethrella* feeding on tree frogs *(Hyla), Mosq. News*, 37, 522, 1977.
126. **Webb, J. P.,** Host-locating behavior of nymphal *Ornithodoros concanensis* (Acarina:Argasidae), *J. Med. Entomol.*, 16, 437, 1979.
127. **Chandler, J. A., Boreham, P. F. L., Highton, R. B., and Hill, M. N.,** A study of the host selection patterns of the mosquitoes of the Kisumu area of Kenya, *Trans. R. Soc. Trop. Med. Hyg.*, 69, 415, 1976.
128. **Trpis, M. and Hausermann, W.,** Genetics of house-entering behaviour in East African populations of *Aedes aegypti, Bull. Entomol. Res.*, 68, 521, 1978.
129. **Tabachnick, W., Munstermann, L., and Powell, J.,** Genetic distinctness of sympatric forms of *Aedes aegypti* in East Africa, *Evolution*, 33, 287, 1979.
130. **McClelland, G. A. H.,** Some man-made mosquito problems in Africa, in *Proc. Tall Timbers Conf. Ecol. Animal Control Habitat Manage.*, 5, 122, 1973.
131. **Kloter, K.,** unpublished data, 1985.
132. **Nasci, R. S. and Edman, J. D.,** Blood-feeding patterns of *Culiseta melanura* (Diptera: Culicidae) and associated sylvan mosquitoes in southeastern Massachusetts, eastern equine encephalitis enzootic foci, *J. Med. Entomol.*, 18, 493, 1981.
133. **Morris, C. D., Zimmerman, R. H., and Edman, J. D.,** Epizootiology of eastern equine encephalomyelitis in upstate New York, USA. II. Population dynamics and vector potential of adult *Culiseta melanura* (Diptera:Culicidae) in relation to distance from breeding site, *J. Med. Entomol.*, 17, 453, 1980.

134. **Bidlingmayer, W. L. and Hem, D. G.,** Mosquito flight paths in relation to the effect of the forest edge upon trap catches in the field, *Mosq. News,* 41, 55, 1981.
135. **Bidlingmayer, W. L.,** Mosquito flight paths in relation to the environment. Effect of vertical and horizontal visual barriers, *Ann. Entomol. Soc. Am.,* 68, 51, 1975.
136. **Bidlingmayer, W. L.,** Visual control of mosquito flight paths, *Proc. Fla. Anti-Mosquito Assoc.,* 51, 44, 1980.
137. **McCrae, A. W. R., Boreham, P. F. L., and Ssenkubuge, Y.,** The behavioural ecology of host selection in *Anopheles implexus* (Theobald) (Diptera: Culicidae), *Bull. Entomol. Res.,* 66, 587, 1976.
138. **Bidlingmayer, W. L. and Hem, D. G.,** Mosquito (Diptera:Culicidae) flight behaviour near conspicuous objects, *Bull. Entomol. Res.,* 69, 691, 1979.
139. **Nasci, R. S.,** Differences in host choice between the sibling species of treehole mosquitoes *Aedes triseriatus* and *Aedes hendersoni, Am. Trop. Med. Hyg.,* 31, 411, 1982.
140. **Novak, R. J. and Rohrer, W.,** Vertical distribution of adult mosquitoes (Diptera:Culicidae) in a northern deciduous forest in Indiana, *J. Med. Entomol.,* 18, 116, 1981.
141. **Camin, J. H. and Drenner, R. W.,** Climbing behavior and host-finding of larval rabbbit ticks *(Haemaphysalis leporispalustris), J. Parasitol.,* 64, 905, 1978.
142. **Kennedy, J. S.,** The visual responses of flying mosquitoes, *Proc. Zool. Soc. London (A),* 109, 221, 1939.
143. **Edman, J. D.,** Orientation of some Florida mosquitoes (Diptera:Culicidae) toward small vertebrates and carbon dioxide in the field, *J. Med. Entomol.,* 15, 292, 1979.
144. **Gillies, M. T. and Wilkes, T. J.,** Field experiments with a wind tunnel on the flight speed of some West African mosquitoes (Diptera:Culicidae), *Bull. Entomol. Res.,* 71, 65, 1981.
145. **Bidlingmayer, W. L., Evans, D. G., and Hansen, C. H.,** Preliminary study of the effects of wind velocities and wind shadows upon suction trap catches of mosquitoes (Diptera:Culicidae), *J. Med. Entomol.,* 22, 295, 1985.
146. **Gillett, J. D.,** Out for blood: flight orientation up-wind in the absence of visual clues, *Mosq. News,* 39, 221, 1979.
147. **Kennedy, J. S.,** Zigzagging and casting as a programmed response to wind-borne odour: a review, *Physiol. Entomol.,* 8, 109, 1983.
148. **David, C. T., Kennedy, J. S., Ludlow, A. R., Perry, J. N., and Wall, C.,** A reappraisal of insect flight towards a distant point source of wind-borne odor, *J. Chem. Ecol.,* 8, 1207, 1982.
149. **Murlis, J. and Jones, C. D.,** Fine-scale structure of odour plumes in relation to insect orientation to distant pheromone and other attractant sources, *Physiol. Entomol.,* 6, 71, 1981.
150. **Nishimura, M.,** How mosquitoes fly to man, *Res. Pop. Ecol.,* 24, 58, 1982.
151. **Mogi, M. and Yamamura, N.,** Estimation of the attraction range of a human bait for *Aedes albopictus* (Diptera, Culicidae) adults and its absolute density by a new removal method applicable to populations with immigrants, *Res. Pop. Ecol.,* 23, 328, 1981.
152. **Ahmadi, A. and McClelland, G. A. H.,** Mosquito-mediated attraction of female mosquitoes to a host, *Physiol. Entomol.,* 10, 251, 1985.
153. **Alekseev, A. N., Rasnitsyn, S. P., and Vitilin, L. M.,** Group attack by females of bloodsucking mosquitoes (Diptera, Culicidae, *Aedes*) I. Discovery of the "invitation effect", *Medit. Parasitol. Parazit. Bolezni Mosc.,* 46, 23, 1977 (in Russian).
154. **Emord, D. E. and Morris, C.D.,** A host-baited CDC trap, *Mosq. News,* 46, 23, 1977.
155. **Schlein, Y., Yuval, B., and Warburg, A.,** Aggregation pheromone released from the palps of feeding female *Phlebotomus papatasi* (Psychodidae), *J. Insect Physiol.,* 30, 153, 1984.
156. **Sonenshine, D. E., Homsher, P. J., Beveridge, M., and Dees, W. H.,** Occurrence of ecdysone in specific body organs of the camel tick, *Hyalomma dromedaria,* and the American dog tick, *Dermacentor variabilis* (Acari:Ixodidae), with notes on their synthesis from cholesterol, *J. Med. Entomol.,* 22, 303, 1985.
157. **LaPointe, D.,** Effect of Host Age on Mosquito Attraction, Feeding, and Fecundity, M.S. thesis, University of Massachusetts, Amherst, 1982, 89.
158. **Blackmore, J. S. and Dow, R. P.,** Differential feeding of *Culex tarsalis* on nestling and adult birds, *Mosq. News,* 18, 15, 1958.
159. **Wood, C. S.,** Preferential feeding of *Anopheles gambiae* mosquitoes on human subjects of blood group O: a relationship between the ABO polymorphism and malaria vectors, *Hum. Biol.,* 46, 385, 1974.
160. **Wood, C. S.,** ABO blood groups related to selection of human hosts by yellow fever mosquitoes, *Hum. Biol.,* 48, 337, 1976.
161. **Thornton, C., Dopre, C., Willson, J. and Hubbard, J.,** Effects of human blood group, sweating and other factors on individual host selection by species A of the *Anopheles gambiae* complex (Diptera, Culicidae), *Bull. Entomol. Res.,* 66, 651, 1976.
162. **Mahon, R. and Gibbs, A.,** Arbovirus infected hens attract more mosquitoes, in *Viral Diseases in South-East Asia and the Western Pacific,* Mackenzie, J. S., Ed., Academic Press, Sydney, 1982, 502.

163. **Turell, M. J., Bailey, C. L., and Rossi, C. A.,** Increased mosquito feeding on Rift Valley fever virus-infected lambs, *Am. J. Trop. Med. Hyg.*, 33, 1232, 1984.
164. **Anderson, R. M., Ed.,** *Population Dynamics of Infectious Diseases*, Chapman & Hall, New York, 1982.
165. **Day, J. F. and Edman, J. D.,** The importance of disease induced changes in mammalian body temperature to mosquito blood feeding, *Comp. Biochem. Physiol*, 77A, 447, 1984.
166. **Snow, W. F.,** The effect of a reduction in expired carbon dioxide on the attractiveness of human subjects to mosquitoes, *Bull. Entomol. Res.*, 60, 43, 1970.
167. **Port, G. R. and Boreham, P. F. L.,** The relationship of host size to feeding by mosquitoes of the *Anopheles gambiae* Giles complex (Diptera:Culicidae), *Bull. Entomol. Res.*, 70, 133, 1980.
168. **Piesman, J., Sherlock, I. A., and Christensen, H. A.,** Host availability limits population density of *Panstrongylus megistus*, *Am. J. Trop. Med. Hyg.*, 32, 1445, 1983.
169. **Snow, W. F.,** Effect of size of cattle bait on the range of attraction of tsetse and mosquitoes, *Insect Sci. Appl.*, 4, 343, 1983.
170. **Boreham, P. F. L., Chandler, J. A., and Jolly, J.,** The incidence of mosquitoes feeding on mothers and babies at Kisumu, Kenya, *J. Trop. Med. Hyg.*, 81, 63, 1978.
171. **Edman, J. D. and Webber, L. A.,** Effect of vertebrate size and density on host-selection by caged *Culex nigripalpus*, *Mosq. News*, 35, 508, 1975.
172. **Helle, T. and Aspi, J.,** Does herd formation reduce insect harassment among reindeer? A field experiment with animal traps, *Acta Zool. Fenni.*, 175, 129, 1983.
173. **Day, J. F. and Curtis, G. A.,** Opportunistic blood-feeding on egg-laying sea turtles by salt marsh mosquitoes, *Fla. Entomol.*, 66, 359, 1985.
174. **Cornet, M. and Chateau, R.,** The use of carbon dioxide in studies on the vectors of sylvatic yellow fever. Preliminary note, *Cah. ORSTOM Entomol. Med. Parasit.*, 9, 301, 1971 (in French).
175. **Browning, T. O.,** The aggregation of questing ticks, *Rhipicephalus pulchellus* on grass stems, with observations on *R. appendiculatus*, *Physiol. Entomol.*, 1, 107, 1976.
176. **Den Boer, J. and Den Boer, M. H.,** Aggregation in the questing tick, *Rhipicephalus pulchellus*, *Physiol. Entomol.*, 5, 107, 1980.
177. **Campbell, A.,** Ecology of the American dog tick, *Dermacentor variabilis* in southwestern Nova Scotia, in *Recent Advances in Acarology*, Vol. 2, Rodriguez, J.G., Ed., Academic Press, New York, 1979, 135.
178. **Campbell, A., Ward, R. M., and Garvie, M. B.,** Seasonal activity and frequency distributions of ticks (Acari: Ixodidae) infesting snowshoe hares in Nova Scotia, Canada, *J. Med. Entomol.*, 17, 22, 1980.
179. **Day, J. F. and Edman, J. D.,** Mosquito engorgement on normally defensive hosts depends on host activity patterns, *J. Med. Entomol.*, 21, 732, 1984.
180. **Walker, E. D.,** Field evidence against rodent burrow entering by *Aedes triseriatus* (Diptera: Culicidae), *Great Lakes Entomol.*, 17, 185, 1984.
181. **Reeves, W. C.,** Mosquito vector and vertebrate host interaction: the key to maintenance of certain arboviruses, in *The Ecology and Physiology of Parasites*, Fallis, A. M., Ed., University of Toronto Press, Toronto, 1971, 223.
182. **Chandler, J. A., Parsons, J., Boreham, P. F. L., and Gill, G. S.,** Seasonal variations in the proportions of mosquitoes feeding on mammals and birds at a heronry in western Kenya, *J. Med. Entomol.*, 14, 233, 1977.
183. **Nasci, R. S.,** Behavioral ecology of variation in blood-feeding and its effect on mosquito-borne diseases, in *Ecology of Mosquitoes: Proceedings of a Workshop*, Lounibos, L. P., Rey, J., and Frank, J. H., Eds., University of Florida Press, Gainesville, 1985, 293.
184. **Edman, J. D. and Taylor, D. J.,** *Culex nigripalpus:* seasonal shift in the bird-mammal feeding ratio in a mosquito vector of human encephalitis, *Science*, 161, 67, 1968.
185. **Gouck, H. K.,** Host preference in various strains of *Aedes aegypti* as determined by an olfactometer, *Bull. WHO*, 47, 680, 1972.
186. **Mukwaya, L. G.,** Genetic control of feeding preferences in the mosquitoes *Aedes (Stegomyia) simpsoni* and *aegypti*, *Physiol. Entomol.*, 2, 133, 1977.
187. **Highton, R. G., Hryan, J. H., Boreham, P. F. L., and Chandler, J. A.,** Studies on the sibling species *Anopheles gambiae* Giles and *Anopheles arabiensis* Patton (Diptera:Culicidae) in the Kisumu area, Kenya, *Bull. Entomol. Res.*, 69, 43, 1979.
188. **Spielman, A. and Rossignol, P. A.,** Insect vectors, in *Tropical and Geographical Medicine*, Warren, K. S. and Mahmoud, A. F., Eds., McGraw-Hill, New York, 1984, chap. 20.
189. **Hess, A. D. and Hayes, R. O.,** Relative potentials of domestic animals for zooprophylaxis against mosquito vectors of encephalitis, *Am. J. Trop. Med. Hyg.*, 19, 327, 1970.
190. **Reisen, W. K. and Boreham, P. F. L.,** Host selection patterns of some Pakistan mosquitoes, *Am. J. Trop. Med. Hyg.*, 28, 408, 1979.
191. **Charlwood, J. D., Dagor, H., and Paru, R.,** Blood-feeding and resting behaviour in the *Anopheles punctulatus* Donitz complex (Diptera:Culicidae) from coastal Papua New Guinea, *Bull. Entomol. Res.*, 75, 463, 1985.

192. **Kay, B. H., Boreham, P. F. L., and Williams, G. M.,** Host preferences and feeding patterns of mosquitoes (Diptera: Culicidae) at Kowanyama, Cape York Peninsula, northern Queesland, *Bull. Entomol. Res.*, 69, 441, 1979.
193. **Walker, E. D. and Edman, J. D.,** Feeding-site selection and blood-feeding behavior of *Aedes triseriatus* (Diptera: Culicidae) on rodent (Sciuridae) hosts, *J. Med. Entomol.*, 22, 287, 1985.
194. **Mullens, B. A. and Gerhardt, R. R.,** Feeding behavior of some Tennessee Tabanidae, *Environ. Entomol.*, 8, 1047, 1979.
195. **Kemp, D. H., Stone, B. F., and Binnington, K. C.,** Tick attachment and feeding: role of the mouthparts, feeding apparatus, salivary gland secretions and the host response, in *Physiology of Ticks*, Obenchain, F. D. and Galun, R., Eds., Pergamon Press, Oxford, 1982, chap. 4.
196. **Davies, J. B.,** Attraction of *Culex portesi* Senevet & Abonnenc and *Culex taeniopus* D & K (Diptera: Culicidae) to 20 animal species exposed in a Trinidad forest. I. Baits ranked by numbers of mosquitoes caught and engorged, *Bull. Entomol. Res.*, 68, 707, 1978.
197. **Christensen, H. A. and Herrer, A.,** Panamanian *Lutzomyia* (Diptera: Psychodidae) host attraction profiles, *J. Med. Entomol.*, 17, 522, 1980.
198. **Wright, J. E. and DeFoliart, G. R.,** Associations of Wisconsin mosquitoes and woodland vertebrate hosts, *Ann. Entomol. Soc. Am.*, 63, 777, 1970.
199. **Service, M. W.,** Feeding behaviour and host preferences of British mosquitoes, *Bull. Entomol. Res.*, 60, 653, 1971.
200. **Jones, J. C. and Pilitt, D. R.,** Blood-feeding behavior of adult *Aedes aegypti* mosquitoes, *Biol. Bull.*, 145, 127, 1973.
201. **Schofield, C. J.,** Population dynamics and control of *Triatoma infestans, Ann. Soc. Belge Med. Trop.*, 65(Suppl. 1), 149, 1985.
202. **Hausfater, G. and Sutherland, R.,** Little things that tick off baboons, *Nat. Hist.*, 2, 55, 1984.
203. **Brooke, M. De L.,** The effect of allopreening on tick burdens of molting eudyptid penguins, *Auk*, 102, 893, 1985.
204. **Schmidtmann, E. T., Jones, C. J., and Gollands, B.,** Comparative host-seeking activity of *Culicoides* (Diptera: Ceratopogonidae) attracted to pastured livestock in central New York State, USA, *J. Med. Entomol.*, 17, 221, 1980.
205. **Day, J. F. and Benton, A. H.,** Population dynamics and coevolution of adult siphonapteran parasites of the southern flying squirrel *(Glaucomys volans volans), Am. Midlands Nat.*, 103, 333, 1980.
206. **MacLeod, J., Colbo, M. H., Madbouly, M. H., and Mwanaumo, B.,** Ecological studies of ixodid ticks (Acari: Ixodidae) in Zambia. III. Seasonal activity and attachment sites on cattle, with notes on other hosts, *Bull. Entomol. Res.*, 67, 161, 1977.
207. **Hayashi, F. and Hasegawa, M.,** Selective parasitism of the tick *Ixodes asanumaei* (Acarina:Ixodidae) and its influence on the host lizard *Eumeces okadae* in Miyake-jima, Izu Islands, *Appl. Entomol. Zool.*, 19, 181, 1984.
208. **Hayashi, F. and Hasegawa, M.,** Infestation level, attachment site and distribution pattern of the lizard tick Ixodes asanumaei (Acarina: Ixodidae) in Aoga-shimla, Izu Islands, *Appl. Entomol. Zool.*, 19, 299, 1984.
209. **Edman, J. D. and Kale, H. W.,** Host behavior: its influence on the feeding success of mosquitoes, *Ann. Entomol. Soc. Am.*, 64, 513, 1971.
210. **Kale, H. W., Edman, J. D., and Webber, L. A.,** Effect of behavior and age of individual ciconiinform birds on mosquito feeding success, *Mosq. News*, 32, 343, 1972.
211. **Edman, J. D., Webber, L. A., and Schmid, A.,** Effect of host defenses on the feeding pattern of *Culex nigripalpus* when offered a choice of blood sources, *J. Parasitol.*, 60, 874, 1974.
212. **Webber, L. A. and Edman, J. D.,** Anti-mosquito behaviour of ciconiiform birds, *Anim. Behav.*, 20, 228, 1972.
213. **Walker, E. D. and Edman, J. D.,** Influence of defensive behaviour of eastern chipmunks and gray squirrels (Rodentia: Sciuridae) on feeding success of *Aedes triseriatus* (Diptera: Culicidae), *J. Med. Entomol.*, 23, 1, 1986.
214. **Day, J. F.,** The Influence of Host Health on Mosquito Engorgement Success, Ph.D. dissertation, University of Massachusetts, Amherst, 1982, 160.
215. **Edman, J. D., Day, J. F., and Walker, E. D.,** Field confirmation of laboratory observations on the differential antimosquito behavior of herons, *Condor*, 86, 91, 1984.
216. **Duffy, D. C.,** The ecology of tick parasitism on densely nesting Peruvian seabirds, *Ecology*, 64, 110, 1983.
217. **Waage, J. K. and Nondo, J.,** Host behavior and mosquito feeding success: an experimental study, *Trans. R. Soc. Trop. Med. Hyg.*, 76, 119, 1982.
218. **Edman, J. D., Webber, L. A., and Kale, H. W.,** Effect of mosquito density on the interrelationship of host behavior and mosquito feeding success, *Am. J. Trop. Med. Hyg.*, 21, 487, 1972.

219. **Nelson, R. L., Tempelis, C. H., Reeves, W. C., and Milby, M. M.,** Relation of mosquito density to bird:mammal feeding ratios of *Culex tarsalis* in stable traps, *Am. J. Trop. Med. Hyg.*, 25, 644, 1976.
220. **Fujito, S., Buei, K., Nakajima, S., Ito, S., Yoshida, M., Sonoda, H., and Kakramura, H.,** Effect of the population density of *Culex tritaeniorhynchus* Giles on bloodsucking rates in cowsheds and pigpens in relation to its role in the epidemic of Japanese encephalitis, *Jpn. J. San. Zool.*, 22, 38, 1971.
221. **Rogers, D. J.,** Tsetse density and behaviour as factors in the transmission of trypanosomes, *Trans. R. Soc. Trop. Med. Hyg.*, 73, 131, 1979.
222. **Walker, E. D. and Edman, J. D.,** The influence of host defensive behavior on mosquito (Diptera:Culicidae) biting persistence, *J. Med. Entomol.*, 22, 370, 1985.
223. **Detels, R., Cates, M., Cross, J., Irving, G., and Wattern, R.,** Ecology of Japanese encephalitis virus in Taiwan in 1968, *Am. J. Trop. Med. Hyg.*, 19, 716, 1970.
224. **Stamm, D. D.,** Studies on the ecology of equine encephalomyelitis, *Am. J. Public Health.*, 48, 328, 1958.
225. **Smith, G., Francy, D., Campos, E., Katona, P., and Calisher, C.,** Correlation between human cases and antibody prevalence in house sparrows during a focal outbreak of St. Louis encephalitis in Mississippi, 1979, *Mosq. News*, 43, 322, 1983.
226. **Yuill, T. M.,** The role of mammals in the maintenance and dissemination of La Crosse virus, in *California Serogroup Viruses*, Calisher, C. H. and Thompson, W. H., Eds., Alan R. Liss, New York, 1983, 77.
227. **Wilkinson, P. R. and Garvie, M. B.,** Notes on the role of ticks feeding on lagomorphs and ingestion of ticks by vertebrates in the epidemiology of Rocky Mountain spotted fever, *J. Med. Entomol.*, 12, 480, 1975.
228. **Day, J. F. and Edman, J. D.,** Malaria renders mice susceptible to mosquito feeding when gametocytes are most infective, *J. Parasitol.*, 69, 163, 1983.
229. **Day, J. F., Ebert, K. M., and Edman, J. D.,** Feeding patterns of mosquitoes (Diptera: Culicidae) simultaneously exposed to malarious and healthy mice, including a method for separating blood meals from conspecific hosts, *J. Med. Entomol.*, 20, 120, 1983.
230. **Wenk, P.,** How bloodsucking insects perforate the skin of their hosts, in *Fleas*, Traub, R. and Starcke, H., Eds., A. A. Balkema, Rotterdam, 1980, 319.
231. **Griffiths, R. B. and Gordon, R. M.,** An apparatus which enables the process of feeding by mosquitos to be observed in tissues of a live rodent, together with an account of the ejection of saliva and its significance in malaria, *Ann. Trop. Med. Parasitol.*, 46, 311, 1952.
232. **Ribeiro, J. M. C., Rossignol, P. A., and Spielman. A.,** Role of mosquito saliva in blood vessel location, *J. Exp. Biol.*, 108, 1, 1984.
233. **Ribeiro, J. M. C. and Garcia, E. S.,** The role of the salivary glands in feeding in *Rhodnius prolixus*, *J. Biol.*, 94, 219, 1981.
234. **Ribeiro, J. M. C., Rossignol, P. A., Spielman, A.,** Salivary gland apyrase determines probing time in anopheline mosquitoes, *J. Insect Physiol.*, 31, 689, 1985.
235. **Hosoi, T.,** Adenosine-5′ phosphates as the stimulating agent in blood for inducing gorging of the mosquito, *Nature (London)*, 181, 1664, 1958.
236. **Galun, R., Avi-Dor, Y., and Bar Zeev, M.,** Feeding response in *Aedes aegypti* stimulation by adenosine triphosphate, *Science*, 124, 1674, 1963.
237. **Friend, W. G. and Smith, J. J. B.,** Factors affecting feeding by bloodsucking insects, *Ann. Rev. Entomol.*, 22, 309, 1977.
238. **Galun, R. and Kindler, S.,** Glutathione as an inducer of feeding in ticks, *Science*, 147, 166, 1965.
239. **Galun, R., Koontz, L. C., Gwadz, R. W., and Ribeiro, J. M. C.,** Effect of ATP analogues on the gorging response of *Aedes aegypti*, *Physiol. Entomol.*, 10, 275, 1985.
240. **Ribeiro, J. M. C., Rossignol, P. A., and Spielman, A.,** *Aedes aegypti* model for blood finding strategy and prediction of parasite manipulation, *Exp. Parasitol.*, 60, 118, 1985.
241. **Rossignol, P. A., Ribeiro, J. M. C., and Spielman, A.,** Increased intradermal probing time in sporozoite-infected mosquitoes, *Am. J. Trop. Med. Hyg.*, 33, 17, 1984.
242. **Grimstad, P. R., Ross, G. E., and Craig, G. B.,** *Aedes triseriatus* (Diptera: Culicidae) and La Crosse virus. II. Modification of mosquito feeding behavior by virus infection, *J. Med. Entomol.*, 17, 11, 1980.
243. **Beach, R., Kiilu, G., and Leeuwenburg, J.,** Modification of sand fly biting behavior by *Leishmania* leads to increased parasite transmission, *Am. J. Trop. Med. Hyg.*, 34, 278, 1985.
244. **Rossignol, P. A., Ribeiro, J. M. C., Jungery, M., Turell, M. J., Spielman, A., and Bailey, C. L.,** Enhanced mosquito blood-finding success on parasitemic hosts: evidence for vector-parasite mutualism, *Proc. Natl. Acad. Sci. U.S.A.*, 82, 7725, 1985.
245. **Spielman, A. and Wong, J.,** Dietary factors stimulating oogenesis in *Aedes aegypti*, *Biol. Bull.*, 147, 443, 1974.
246. **Gwadz, R.,** Regulation of blood meal size in the mosquito, *J. Insect Physiol.*, 15, 2039, 1972.
247. **Oliver, J. H.,** Relationship among feeding, gametogenesis, mating and syngamy in ticks, in *Host Regulated Development Mechanisms in Vector Arthropods*, Borovsky, D. and Spielman, A., Eds., University of Florida Press, Gainesville, 1986, 93.

248. **Piesman, J. and Spielman, A.,** *Babesia microti.* Infectivity of parasites from ticks for hamsters with white-footed mice, *Exp. Parasitol.*, 53, 242, 1982.
249. **Courtney, C. C., Christensen, B. M., and Goodman, W. G.,** Effect of *Dirofilaria immitis* on blood meal size and fecundity in *Aedes aegypti* (Diptera: Culicidae), *J. Med. Entomol.*, 22, 398, 1985.
250. **Klowden, M. J. and Lea, A. O.,** "Physiologically old" mosquitoes are not necessarily old physiologically, *Am. J. Trop. Med. Hyg.*, 29, 1460, 1980.
251. **Magnarelli, L.,** Physiological age of mosquitoes (Diptera:Culicidae) and observations on partial blood-feeding, *J. Med. Entomol.*, 13, 447, 1977.
252. **Mitchell, C. J. and Millian, K. Y.,** Continued host seeking by partially engorged *Culex tarsalis* (Diptera: Culicidae) collected in nature, *J. Med. Entomol.*, 18, 249, 1981.
253. **Mitchell, C. J., Bowen, G. S., Monath, T. P., Cropp, C. B., and Kerschner, J.,** St. Louis encephalitis virus transmission following multiple feeding of *Culex pipiens pipiens* (Diptera: Culicidae) during a single gonotrophic cycle, *J. Med. Entomol.*, 16, 254, 1979.
254. **Lenahan, J. K. and Boreham, P. F. L.,** Studies on multiple feeding by *Anopheles gambiae* s.l. in a Sudan savanna area of north Nigeria, *Trans. P. Soc. Trop. Med. Hyg.*, 73, 418, 1979.
255. **Lenahan, J. K. and Boreham, P. F. L.,** Effect of host movement on multiple feeding by *Aedes aegypti* (L.) (Diptera, Culicidae) in a laboratory experiment, *Bull. Entomol. Res.*, 66, 681, 1976.
256. **Edman, J. D.,** unpublished data, 1985.
257. **Edman, J. D. and Bidlingmayer, W. L.,** Flight capacity of blood-engorged mosquitoes, *Mosq. News*, 29, 386, 1969.
258. **Steelman, C. D.,** Effects of external and internal arthropod parasites on domestic livestock production, *Ann. Rev. Entomol.*, 21, 155, 1976.
259. **LaPointe, D.,** unpublished data, 1984.
260. **Gothe, R., Kunze, K., and Hoogstraal, H.,** The mechanisms of pathogenicity in tick paralysis, *J. Med. Entomol.*, 16, 357, 1979.
261. **Dalmat, H. T.,** The black flies (Diptera, Simuliidae) of Guatemala and their roles as vectors, *Smithsonian Misc. Coll.*, 25, 1, 1955.
262. **Walker, A. R., Fletcher, J. D., and Gill, S. G.,** Structural and histochemical changes in the salivary glands of *Rhipicephalus appendiculatus* during feeding, *Int. J. Parasitol.*, 15, 81, 1984.
263. **Sauer, J. R., and Essenberg, R. C.,** Role of cyclic nucleotides and calcium in controlling tick salivary gland function, *Am. Zool.*, 24, 217, 1984.
264. **Ribeiro, J. M. C., Makoul, G. T., Levine, J., Robinson, D. R., and Spielman, A.,** Antihemostatic, antiinflammatory, and immunosuppressive properties of the saliva of a tick, *Ixodes dammini*, *J. Exp. Med.*, 161, 332, 1985.
265. **Wikel, S. K. and Allen, J. R.,** Immunological basis of host resistance to ticks, in *Physiology of Ticks*, Obenchain, F. D. and Galun, R., Eds., Pergamon Press, Oxford, 1982, chap. 5.
266. **Ackerman, S., Clare, F. B., McGillard, T. W., and Sonnenshine, D. E.,** Passage of host serum components, including antibody, across the digestive tract of *Dermacentor variabilis* (Say), *J. Parasitol.*, 67, 737, 1981.
267. **Brown, S.,** Antibody and cell-mediated immune resistance by guinea pigs to adult *Amblyomma americanum* ticks, *Am. J. Trop. Med. Hyg.*, 31, 1285, 1982.
268. **Brown, S. J.,** Antibody and cell-mediated immune resistance by guinea pigs to adult *Amblyomma americanum* ticks, *Am. J. Trop. Med. Hyg.*, 31, 1285, 1982.
269. **Trager, W.,** Acquired immunity to ticks, *J. Parasitol.*, 25, 57, 1939.
270. **Brown, S. J. and Askenase, P. W.,** Rejection of ticks from guinea pigs by anti-hapten antibody-mediated degranulation of basophils at cutaneous basophil hypersensitive sites: role of mediators other than histamine, *J. Immunol.*, 134, 1160, 1985.
271. **Gold, D., Lengy, J., Lass, N., and Tager, A.,** Studies on *Culex pipiens molestus* in Israel. II. Skin response in man to extracts from the mosquito, *Int. Arch. Allergy*, 31, 274, 1967.
272. **Azad, A. F.,** Acquisition and persistence of murine typhus infection in *Xenopsylla cheopis* after feeding on immune rats, in *Host Regulated Development Mechanisms in Vector Arthropods*, Borovsky, D. and Spielman, A., Eds., University of Florida Press, Gainesville, 1986, 158.
273. **Briegel, H. and Lea, A. O.,** Relationship between protein and proteolytic activity in the midgut of a mosquito, *J. Insect Physiol.*, 21, 1597, 1975.
274. **Berner, R., Rudin, W., and Hecker, H.,** Peritrophic membranes and protease activity in the midgut of the malaria mosquito, *Anopheles stephensi* (Insecta: Diptera) under normal and experimental conditions, *J. Ultrastruct, Res.*, 83, 195, 1983.
275. **Hardy, J. L., Houk, E. J., Kramer, L. D., and Reeves, W. C.,** Intrinsic factors affecting vector competence of mosquitoes for arboviruses, *Ann. Rev. Entomol.*, 28, 229, 1983.
276. **Richards, A. G. and Richards, P. A.,** Origin and composition of the peritrophic membrane of the mosquito, *Aedes aegypti*, *J. Insect Physiol.*, 17, 2253, 1971.

277. **Rudzinska, M. A., Lewengrub, S., Spielman, A., and Piesman, J.,** Invasion of *Babesia microti* into epithelial cells of the tick gut, *J. Protozool.*, 30, 339, 1983.
278. **Berner, R., Rudin, W., and Hecker, H.,** Peritrophic membranes and protease activity in the midgut of the malaria mosquito, *Anopheles stephensi* (Liston) (Insecta: Diptera) under normal and experimental conditions, *J. Ultrastruct. Res.*, 83, 195, 1983.
279. **Gooding, R. H.,** Digestive processes of haematophagous insects. I. A literature review, *Quest. Entomol.*, 8, 5, 1971.
280. **Detra, R. L. and Romoser, W. S.,** Permeability of *Aedes aegypti* larval peritrophic membrane to proteolytic enzyme, *Mosq. News*, 39, 582, 1979.
281. **Howard, L. M.,** Studies on the mechanism of infection of the mosquito midgut by *Plasmodium gallinaceum*, *Am. J. Hyg.*, 75, 287, 1962.
282. **Houk, E. H. and Hardy, J. L.,** Midgut cellular responses to bloodmeal digestion in the mosquito, *Culex tarsalis* Coquillett (Diptera: Culicidae), *Int. J. Insect. Morphol. Embryol.*, 11, 109, 1982.
283. **Colluzi, M. A. and Ascoli, F.,** Effect of cibarial armature of mosquitoes (Diptera, Culicidae) on blood meal hemolysis, *J. Insect. Physiol.*, 28, 885, 1982.
284. **Bertram, D. S. and Bird, R. G.,** Studies on mosquito-borne viruses in their vectors, *Trans. R. Soc. Trop, Med. Hyg.*, 55, 404, 1961.
285. **Rossignol, P. A., Spielman, A., and Jacobs, M. S.,** Rough endoplasmic reticulum in mosquitoes (Diptera: Culicidae): aggregation stimulated by juvenile hormone, *J. Med. Entomol.*, 19, 719, 1982.
286. **Mather, T. N. and DeFoliart, G. R.,** Effect of host blood source on the gonotrophic cycle of *Aedes triseriatus*, *Am. J. Trop. Med. Hyg.*, 32, 189, 1983.
287. **Downe, A. E. R. and Archer, J. A.,** The effects of different blood-meal sources on digestion and egg production in *Culex tarsalis* Coq. (Diptera:Culicidae), *J. Med. Entomol.*, 12, 431, 1975.
288. **Nayar, J. K. and Sauerman, D. M.,** The effects of nutrition on survival and fecundity in Florida mosquitoes. IV. Effects of blood source on oocyte development, *J. Med. Entomol.*, 14, 167, 1977.
289. **Galun, R. and Sternberg, S.,** Effects of host spectra on feeding behaviour and reproduction of soft ticks (Acari: Argasidae), *Bull. Entomol. Res.*, 68, 153, 1978.
290. **Thompson, W. H.,** Lower rates or oral transmission of La Crosse virus by *Aedes triseriatus* venereally exposed after engorgement on immune chipmunks, *Am. J. Trop. Med. Hyg.*, 32, 1416, 1983.
291. **Day, J. F.,** unpublished data, 1985.
292. **Spielman, A.,** unpublished data.

Chapter 7

OCCURRENCE, BIOLOGY, AND PHYSIOLOGY OF DIAPAUSE IN OVERWINTERING MOSQUITOES

Carl J. Mitchell

TABLE OF CONTENTS

I. INTRODUCTION

Winter survival poses special problems for mosquitoes and other poikilothermal animals inhabiting areas of high latitude and temperate areas with prolonged cold periods. With rare exceptions, such as species that have adapted to continuous autogenous reproduction in underground shelters or cohabitation with man and domestic animals during the winter, mosquitoes can survive such adverse conditions only if their life cycle includes a period of dormancy. Winter dormancy in mosquitoes can be classified as diapause or quiescence. The primary function of diapause is seasonal adaptation, but it also serves to synchronize life cycles and determine patterns of voltinism.[1]

The word diapause is derived from Greek and means "rest or interruption of work".[2] It was introduced into the scientific literature during the last century to describe a stage in the embryogenesis of a grasshopper; however, subsequent usage has altered its meaning. Saunders'[3] definition is used here, i.e., "a period of arrest of growth and development which enables the species to overwinter (hibernate) or aestivate, or to synchronize its development cycle to that of the seasons". Diapause is different from quiescence, which is "a state of dormancy directly imposed by adverse factors in the environment (e.g., cold torpor, dehydration)".[3] Diapause may be further characterized as obligatory (occurring during each generation as in univoltine species) or facultative (induced or averted by appropriate external stimuli and characteristic of multivoltine species).

Muller[4] and Mansingh[5] proposed more comprehensive classifications of dormancy in insects. The distinctions given above are usually adequate for discussing cold-weather dormancy in mosquitoes; however, the term oligopause, proposed by Muller[4] and redefined by Mansingh,[5] is sometimes used as well. Oligopause embraces instances of dormancy that are intermediate between classical diapause and quiescence.

Among mosquito species that enter diapause, arrest of growth or development has been documented in each life history stage except the pupa. Only the female imago is known to undergo diapause. The life history stage in which diapause occurs is characteristic of the species and usually diapause is confined to a single stage; however, a few mosquito species may undergo diapause in more than one stage.

The potential relationships between overwintering mechanisms of certain arboviruses and their mosquito vectors and the demonstration of transovarial transmission by mosquitoes of several arboviruses[6] provide ample justification for examining the phenomenon of diapause in overwintering mosquitoes in detail. In addition, the intrinsic biological significance of diapause as an adaptation for winter survival is well illustrated by the family Culicidae. The following is an attempt to outline the geographic limits of cold-weather dormancy in mosquitoes; examine overwintering strategies, and the biology and physiology of species that undergo diapause; and summarize mosquito/arbovirus associations from regions of the world where cold-weather diapause by mosquitoes is the rule. Overwintering mechanisms of medically important arboviruses are considered in detail in the chapters on individual viruses.

II. GEOGRAPHIC OCCURRENCE OF COLD-WEATHER DORMANCY IN MOSQUITOES

Mosquitoes, like many groups of organisms, exhibit their greatest degree of speciation in the tropics. This is related to the diversity of ecologic niches which, in turn, is influenced by temperature. Mosquito species and life history stages of the same species may vary widely in their response to temperature extremes.[7] The lower temperature limits for growth and morphogenic processes of insects in general are in the range of 5 to 10°C.[8] Only those species that have evolved seasonal adaptation mechanisms allowing them to survive moderate-to-severe winters have been successful in colonizing the cold-temperate regions of the

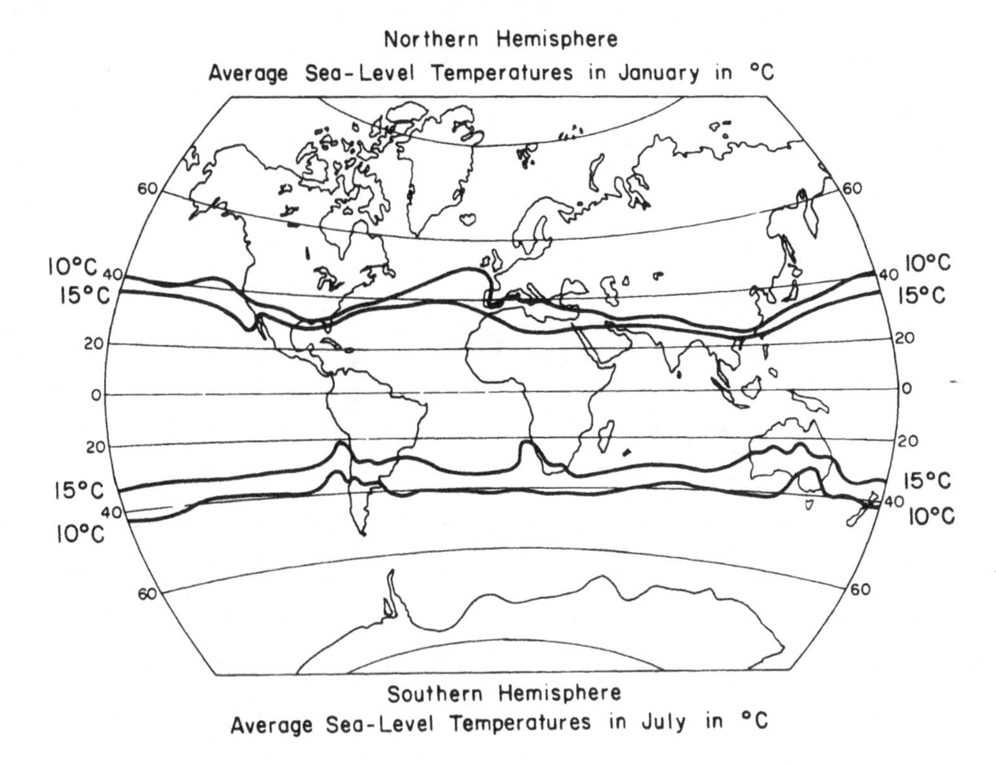

FIGURE 1. Cold-month 10 and 15°C isotherms, adjusted to sea level, in the northern and southern hemispheres.

world. Somewhat arbitrarily, I have selected average cold-month isotherms of 10°C, adjusted to sea level, as the lower limits for continuous mosquito breeding activity and development during the respective winters of the northern and southern hemispheres (Figure 1). There are exceptions;[7] however, most mosquito species occurring north of this isotherm in the northern hemisphere and south of this isotherm in the southern hemisphere will undergo a period of dormancy, either diapause or quiescence, during at least part of the winter season. In addition, average 15°C cold-month isotherms are shown (Figure 1). It is proposed that in warmer areas toward the equator, cold-weather dormancy is an uncommon event. In the areas of intergradation between the 10 and 15°C cold-month isotherms, some species in some areas may breed year-round, whereas others may undergo periods of cold-weather dormancy that are more likely to take the form of quiescence than diapause. There are exceptions, of course. First, the limits of the cold-month isotherms shown in Figure 1 are only rough approximations. Temperatures are greatly influenced by altitude, proximity to the sea, and local weather patterns; however, the scale of the figure precludes the inclusion of most exceptions due to these variables. Second, the family Culicidae is quite heterogeneous, and generalizations concerning the family on a global scale are likely to meet with exceptions. Nevertheless, the cold-month isotherms may provide useful reference points when discussing cold-weather dormancy, especially when considering the arbovirus associations of mosquito species that enter diapause during the winter.

III. EXTRINSIC FACTORS AFFECTING DIAPAUSE IN MOSQUITOES

The environmental factors that induce diapause are not the environmental obstacles themselves but rather environmental cues that reliably predict the onset of an unfavorable season.[3,8-10] It has been shown that in most insects, the principal stimulus for the onset of

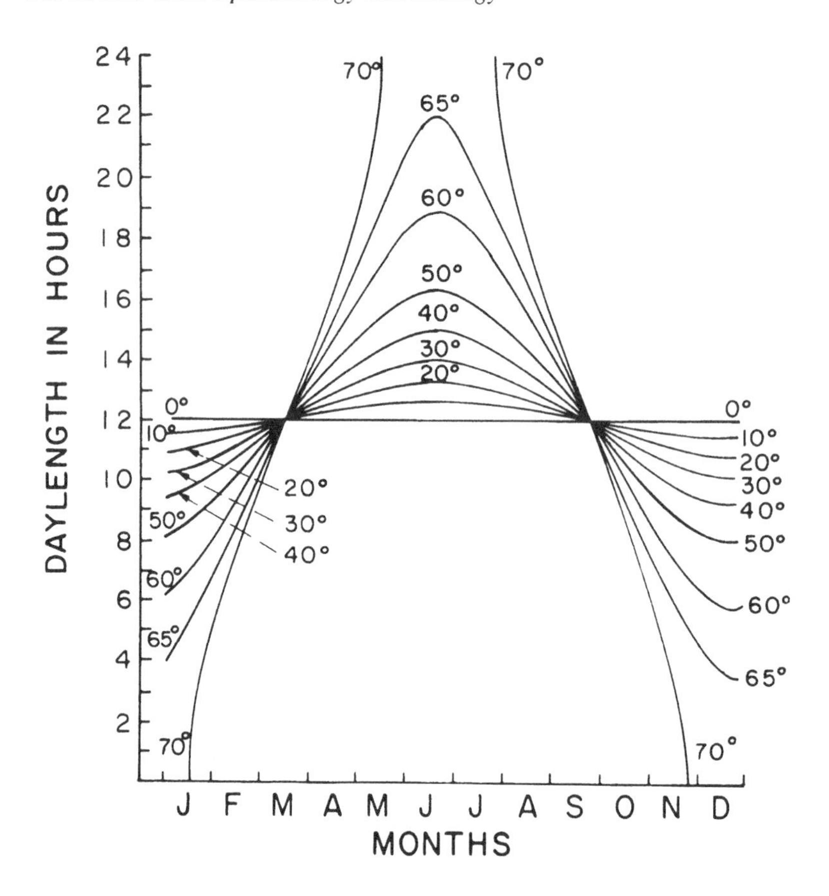

FIGURE 2. Seasonal changes in daylength at different latitudes in the northern hemisphere. (From Danilevskii, A. S., *Photoperiodism and Seasonal Development of Insects*, 1st English ed., Oliver & Boyd, London, 1965. With permission.)

diapause is a change in photoperiod or daylength, although other factors such as temperature, water, and diet may play an important role. Seasonal changes in photoperiod proceed with mathematical accuracy and provide "noise-free" time cues that can be used to govern seasonal activity. It has been shown repeatedly that mosquitoes that undergo diapause during the winter in temperate regions do so in response to short photophases of late summer and autumn. The interaction between temperature and photoperiod on diapause induction in mosquitoes is complex. Generally, however, lower temperatures enhance the effects of short photoperiods, causing greater numbers to enter diapause.

Figure 2 illustrates the effects of seasonal change and latitude on daylength in the northern hemisphere. Seasonal changes in daylength are insignificant at the equator and become more pronounced as latitude increases. Obviously, wide-ranging species must respond differently to daylength at different latitudes. Many insect populations with extensive latitudinal ranges can be divided into geographic races with different critical photoperiods.[9,11] Vinogradova[12] pointed out that the critical photophase inducing diapause in different populations of *Anopheles messeae* and *An. hyrcanus* decreases from north to south. Such clinal variation has since been documented for a number of mosquito species, e.g., *An. freeborni,*[13] *Aedes triseriatus,*[14-17] *Ae. atropalpus,*[18] *Ae. sierrensis,*[19] and *Wyeomyia smithii.*[20] An apparent exception to the general rule concerning clinal variation in critical daylengths at different latitudes was reported by Trimble and Smith.[21] They compared *Toxorhynchites rutilus* colonies from Delaware and Lousiana and found that critical daylengths for inducing larval diapause did not differ for the two colonies at 19°C. However, the threshold daylength for dormancy induction was greater in Delaware larvae, and their response to photoperiod was less affected by high temperature (27°C). The authors suggest that these differences may be alternative adaptations to latitudinal variation in critical daylength and may be related to the timing of seasonal development.

Many insect species have an endogenous physiologic clock that determines whether diapause will occur in response to particular photoperiodic stimuli. Short-day photoperiods interact with the clock and trigger endocrine events that result in diapause. Saunders[3] has developed the concept that diapause induction or aversion in certain insects involves a photoperiodic "counter" mechanism. This counter triggers diapause in response to the accumulation of a fixed number of short-day photoperiod cycles measured by the photoperiodic clock. The fixed number of cycles is referred to as the required day number (RDN) and is temperature compensated. Beach[18] demonstrated the relationships between the RDN, temperature, rates of development of *Ae. atropalpus* during the photosensitive fourth instar and pupal stages, and induction or aversion of diapause. A strain from Ontario (45°N lat.) required 4 days at temperatures as high as 28°C during the photosensitive stages for induction of diapause. A Georgia strain (34°N lat.) required 7 days at 24°C or less. These results indicate that a counter mechanism is operative in diapause induction in a mosquito species. Further, the results suggest that this requirement for different numbers of diel cycles during the photosensitive period may be an important means of modifying the diapause response of populations from different latitudes and geographic areas.

Insects enter diapause when environmental factors are still favorable for growth and development; therefore, once diapause has begun, mechanisms are needed to ensure that diapause will persist until temperatures become sufficiently low to prevent growth or development or until a favorable season arrives.[10] These mechanisms are based largely on photoperiod and temperature. Diapause is largely a dynamic state, decreasing in depth or intensity as the season progresses and as responses to diapause-maintaining factors diminish. Consequently, overwintering diapause may end by midwinter rather than in spring, and the insect may remain quiescent until environmental conditions are favorable for renewed growth and development.

IV. EGG AND LARVAL DIAPAUSE

A. General Characteristics

Diapause in the egg and larval stages takes the form of an arrest in development. In mosquito eggs, this arrest occurs following completion of embryogenesis, i.e., when the embryo is fully formed as a pharate larva. In species that pass the winter in the egg stage, the eggs of univoltine species enter an obligatory diapause, whereas in multivoltine species the diapause is facultative and those eggs laid in late summer or autumn enter diapause. There may be exceptions among species with photosensitive eggs, and eggs laid earlier in the season may enter diapause along with eggs laid in late summer and autumn.[22] A temporary arrest in development, or quiescence, may also occur during the egg stage upon completion of embryonic development. This quiescence is a response to adverse environmental conditions such as drought. The hatching of aedine eggs is a reflex response to low or decreasing levels of dissolved oxygen.[23,24] Prior conditioning, involving a variety of environmental factors, may be required to stimulate hatching.[7] The following schema can be used to distinguish diapause and quiescence in the egg stage. A portion of suitably conditioned quiescent eggs will hatch as soon as the unfavorable conditions have been removed and they are submerged in deoxygenated water. In contrast, eggs undergoing diapause will not hatch until they have been reactivated by exposure to appropriate environmental stimuli such as photoperiod and temperature.[7]

Eggs of all aedine mosquitoes become quiescent if conditions are unfavorable for hatching, including eggs that have broken diapause but remain in unfavorable conditions. The problem of quiescence is further complicated by variable hatching responses among eggs laid at the same time and treated in the same fashion. This results in installment hatching during consecutive flooding and drying cycles[25] and, presumably, is an adaptation to the use of intermittently flooded breeding habitats.

All species of the aedine genera *Aedes*, *Psorophora*, and *Haemagogus* are capable of arresting development in the egg stage.[7] This may be in response to drying as well as a diapause mechanism and can occur in tropical as well as temperate areas. *Psorophora* and *Haemagogus* are restricted to the New World and principally to the tropics and subtropics. *Culiseta morsitans* also lays eggs that are resistant to drying; this Holarctic species overwinters as fourth instar larvae[26] or eggs.[179] *An. walkeri* apparently overwinters in the egg stage in the northern part of its range,[27] but it is not known whether the eggs can withstand drying.

One or more species in each of the following genera are known to pass the winter in the larval stage in temperate regions: *Anopheles, Aedeomyia, Aedes, Armigeres, Opifex, Culex, Culiseta, Coquillettidia, Mansonia, Orthopodomyia, Tripteroides, Wyeomyia,* and *Toxorhynchites*. Excluding those genera that are represented by three or fewer species in temperate regions, the list is reduced to *Anopheles, Aedes, Culex,* and *Culiseta*. Among these, it is only in the genus *Culiseta* that larval overwintering is a significant feature. There is insufficient evidence to distinguish between diapause and quiescence for the majority of species in the various genera among which larval overwintering is known.

B. Induction, Maintenance, and Termination

Shroyer and Craig[22] point out that the term "egg diapause" is not descriptive if applied indiscriminately to all seasonally restricted hatching latencies in mosquitoes. Latencies induced by different stimuli may differ in other respects and may have distinctive physiologic mechanisms. For example, conditions that terminate cold-induced latency may not terminate photoperiod-induced latency. Consequently, Shroyer and Craig[22] propose that the term egg diapause be restricted to photoperiod-induced hatching latencies. Mosquito species in which photoperiodic induction of diapause has been demonstrated or reasonably inferred are listed in Table 1.

Baker's[28] report of egg diapause in a New York strain of *Ae. triseriatus* was the first demonstration of photoperiodism in mosquitoes. Vinogradova[35] was the first to report that photoperiod exposure of the maternal generation determines diapause in eggs of the F_1 generation. She demonstrated that adults of *Ae. togoi* reared from pupae maintained under short photophase conditions laid diapausing eggs. Subsequent cases of photoperiodic control of diapause in *Aedes* eggs of the maternal type have been reported, e.g., *Ae. atropalpus* and *Ae. epactius,*[37] *Ae. vexans,*[33] and *Ae. caspius;*[41] however, the egg stage is photosensitive in *Ae. triseriatus,*[14,47] *Ae. hendersoni,*[22] *Ae. campestris,*[39] and *Ae. canadensis.*[40] It is the pharate larvae, rather than the developing embryo, that is photosensitive in *Ae. triseriatus,*[22] *Ae. hendersoni,*[22] *Ae. campestris,*[39] and *Ae. taeniorhynchus.*[45] This has ecologic significance since eggs laid early in the season may enter diapause as well as those laid during late summer and autumn. Information is lacking concerning the nature of photosensitive receptors in mosquito eggs. Kalpage and Brust[38] demonstrated that only the anterior end of *Ae. atropalpus* eggs is photosensitive. They speculated that light might enter the egg through the micropyle.

In the northern part of its range, *Ae. triseriatus* overwinters primarily in the egg stage; however, Love and Welchel[46] found that in the autumn, fourth instar larvae from their colony in Georgia exposed to the natural photophase did not pupate. Based on relative photoperiodic responses and censuses of overwintering populations from 30 to 40°N lat., Holzapfel and Bradshaw[15] concluded that larval diapause in this species is mainly a back-up for embryonic diapause and that its major adaptive significance is the modulation of later winter and spring development. *Ae. togoi* overwinters in the egg stage in southern Siberia,[35] but may overwinter as both eggs and fourth instar larvae in Nagasaki, Japan.[36]

In *Wy. smithii,* diapause usually occurs in the third larval instar; however, exposure of diapausing third instar larvae to incomplete diapause-terminating stimuli may result in a high incidence of fourth instar diapause.[71] The initiation of fourth instar diapause in the spring serves to further retard development of a proportion of the overwintering population until favorable conditions are assured.

In many insects, thresholds of photoperiod sensitivity can be raised or lowered by temperature, and in some, the effect of daylength is partially or completely masked except at medium temperatures.[2] Mosquitoes are no exception; the diapause-inducing effects of short photophases are averted in *Ae. atropalpus,*[37] *Ae. dorsalis,*[42,43] and *Ae. vexans*[34] by exposing the photosensitive stages to constant high temperatures (30 to 32°C). Whether these observations have any relevance to natural situations is questionable. This lack of expression of diapause in the zone of sublethal temperatures may represent a breakdown of normal developmental processes.[8] However, tests conducted with *Ae. taeniorhynchus* demonstrated that short-day treatments failed to induce diapause in eggs maintained at 27°C.[45]

Reference was made earlier to the study of Beach[18] on *Ae. atropalpus* from Georgia, where temperatures above 24°C accelerated development of the photosensitive aquatic stages to the point that the requirement for seven or more short-day photoperiod cycles was not met, thus averting diapause. This strategy permits late-season broods to take full advantage of existing conditions by avoiding short-day-induced diapause if temperatures are high and other requirements for growth are available. Larval diapause in *Ae. triseriatus* may also result from photoperiodic treatment of the larvae themselves under certain conditions.[49] This is dependent upon slowing development of the larval instars by a low-quantity diet or low temperature, or both, until the short-day photoperiod has time to exert an affect. In nature, larvae might well be confronted by low-temperature or low-diet conditions in late summer or fall. Therefore, those individuals that were not programmed for diapause during the egg stage might be afforded a second chance for winter survival.

Maintenance and termination of diapause in mosquitoes that overwinter as eggs and larvae are controlled largely by photoperiod and temperature (Table 2). The interactions are varied and complex. Long-day treatment of eggs previously stored at cold temperatures results in maximal termination of egg diapause in *Ae. sollicitans*[44] and *Ae. triseriatus.*[17] Similarly, larval diapause in *Tx. rutilus* is most rapidly terminated by exposure of chilled larvae to long days.[72] Long photophases and exposure to either high or low temperatures may terminate diapause in the laboratory; however, in nature, photoperiod and/or temperature may only regulate the rate of decrease in diapause depth or intensity.[10] Few insect species have been shown to require a specific stimulus to end diapause. *Wy. smithii* may be an exception in that the critical photoperiod for maintaining diapause does not change during autumn and winter, but spring daylengths that exceed the critical photoperiod terminate diapause.[68]

C. Cold Tolerance and Behavior of Diapausing Larvae

In temperate zone insects, cold hardiness may involve cold acclimation, supercooling by manufacture of glycerol or other polyhydric alcohols, or the ability to withstand bodily freezing.[73,74] Little information is available for assigning mosquito species to any of these categories. Diapausing larvae of *Wy. smithii* may survive when the water in their pitcher-plant habitat freezes solid, but they are intolerant of prolonged periods at subzero temperatures.[68,69,75] At −5°C in the laboratory, 60% died within 8 weeks, and larval mortality in the field averaged 45% after 4 months with average ground temperatures of approximately −3.7°C.[69]

Diapausing *Wy. smithii* larvae appear as active as nondiapausing larvae and also appear to be feeding constantly; however, the amount of food ingested is negligible in comparison

Table 1
MOSQUITO SPECIES IN WHICH PHOTOPERIODIC INDUCTION OF DIAPAUSE HAS BEEN DEMONSTRATED OR REASONABLY INFERRED[a]

	Species	Sensitive stage	Diapause stage	Ref.
An. (Ano.)	*atroparvus*	Larva — adult	Adult	12
	barberi		Larva	28
	freeborni	Larva — adult	Adult	13, 29, 30
	hyrcanus		Adult	12
	messeae	Larva — adult	Adult	12
	plumbeus		Larva	31
	punctipennis	Larva — adult	Adult	32
An. (Cel.)	*superpictus*		Adult	12
Ae. (Adm.)	*vexans*	Adult and egg	Egg	33, 34
Ae. (Fin.)	*togoi*	Egg — adult	Egg and 4th instar larva	35, 36
Ae. (Och.)	*atropalpus*	4th instar Larva and pupa	Egg	18, 37, 38
	campestris	Egg	Egg	39
	canadensis	Egg	Egg	40
	caspius	Adult	Egg	41
	dorsalis	Adult and egg	Egg	42, 43
	epactius		Egg	37
	sierrensis	Larva	4th instar larva	19
	sollicitans	Egg — adult	Egg	44
	taeniorhynchus	Egg	Egg	45
Ae. (Pro.)	*hendersoni*	Egg	Egg	22
	triseriatus	Egg — larva	Egg and larva	14—17, 22, 28, 46—49
Ae. (Stg.)	*albopictus*		Egg	50
Cx. (Cux.)	*peus*	Pupa — adult	Adult	51
	pipiens	Egg — adult	Adult	12, 52—56
	restuans	Pupa — adult	Adult	57, 58
	tarsalis	Pupa — adult	Adult	59, 60
	tritaeniorhynchus	Larva — adult	Adult	61, 62
Cs.(Cli.)	*melanura*		Larva	63, 64

Cs. (Cus.)	*inornata*	Larva	Adult	65, 66
Or. signifera			Larva	28
Ps.(Jan.)	*ferox*	Maternal generation + E_{qq}	Egg	40
Tx. (Lyn.)	*rutilus*	Larva	4th instar larva	21, 67
Wy. (Wyo.)	*smithii*	Early instar larvae	3rd instar larva	20, 68—71
		3rd instar	4th instar larva	

[a] This list is modified and extended from those of Saunders,[3] Danilevskii,[8] and Beck.[9]

Table 2
EXTRINSIC FACTORS INVOLVED IN DIAPAUSE TERMINATION IN OVERWINTERING MOSQUITO EGGS AND LARVAE

Species		Temperature		Photoperiod	Ref.
		Chilling	Warming		
Ae.(Och.)	*atropalpus*		+	+	38
	campestris		+	+	39
	dorsalis	+	+	+	42, 43
	sollicitans	+	+	+	44
	taeniorhynchus	+	+	+	45
Ae. (Pro.)	*triseriatus*	+		+	15, 17, 28, 47
Cs. (Cli.)	*melanura*[a]		+	+	63
Tx. (Lyn.)	*rutilus*	+		+	67, 72
Wy. (Wyo.)	*smithii*			+	68—71

[a] High protein diet also provided.

to nondiapausing larvae.[69] Clay and Venard[76] observed that diapausing *Ae. triseriatus* larvae move freely and feed but are less active and consume less food than nondiapausing larvae.

Diapausing larvae of *An. plumbeus, An. barberi,* and *Orthopodomyia signifera* display negative phototropic reactions.[28,77] *Armigeres* larvae have been collected during the winter from the mud in the botton of breeding sites in Japan.[78] Hayes[79] found that diapausing *Cs. melanura* larvae, restrained in pipes sunk into their breeding habitat, moved deeper into the muck as winter progressed, some to depths of 6 in.

D. Hormonal Control of Diapause in Eggs and Larvae

Studies on diapause in the mosquito egg and larva have concentrated on identifying environmental cues and threshold levels necessary to elicit appropriate responses. Few attempts have been made to determine how these environmental cues are translated into individual responses. In embryonic diapause in other insects, environmental stimuli received by the nervous system are integrated and transformed into specific chemical signals by the endocrine system for regulation of induction and termination of diapause.[80] Until recently, the lack of molting hormone, ecdysone, was though to explain the endocrine basis of diapause in those insects that diapause as larvae. However, it has been reported that juvenile hormone (JH) positively regulates larval diapause in some insects,[81] and a diapause hormone has been described from the silkworm, *Bombyx mori.*[80]

Clay and Venard[76] terminated larval diapause in *Ae. triseriatus* by treatment with ecdysterone and inokosterone. Their objective was to explore mosquito control possibilities; however, the results suggest that molting hormone deficiency may be one characteristic of larval diapause in this species.

V. DIAPAUSE IN OVERWINTERING ADULT MOSQUITOES

A. General Characteristics

Hibernation as adult, inseminated, nulliparous females is the usual manner of overwintering for most *Anopheles, Culex,* and *Culiseta* in cool-temperate regions. It is rare or nonexistent among temperate species of Aedini and Mansoniini and has been reported for a single species of Uranotaeniini collected in southern France.[82,83] Adult diapause usually involves an assemblage of characteristics encompasing both physiologic and behavioral components. For those species that enter diapause and do not simply become quiescent, preparation for hibernation by inseminated, adult female mosquitoes includes: (1) reduced biting drive, (2) ovarian diapause, and (3) hypertrophy of fat body. Exceptions among certain *Anopheles* that undergo gonotrophic dissociation[84-86] are discussed later. Most species of mosquito that hibernate as adults are multivoltine. In Alaska, however, *Cs. impatiens* and *Cs. alaskaensis* enter diapause during the season of emergence, overwinter as inseminated females, then take their first blood meal and oviposit when they are 1 year or more in age.[87,88]

B. Photoperiodic Induction and Sensitive Stages

Mosquitoes that enter diapause as adults do so in response to short photophases experienced by developmental stages during late summer and autumn (Table 1). However, at least one species, *Cs. inornata,* exhibits facultative reproductive diapause in reponse to short daylengths in the northern part of its range where it hibernates[65] and in response to long daylengths in the sourthern part of its range where it aestivates during the hot-dry season.[66] The photosensitive stages may vary with species. Also, the effects of photoperiod on certain developmental stages may be cumulative. Exposure of *An. messae* to a short day from the beginning of the fourth instar induces diapause in 57% of the resultant females, whereas exposure from the beginning of the third instar results in diapause in all females.[12] It is well established that ovarian diapause can be induced in several *Culex* species, including *Cx. pipiens,* by exposing the pupal stage to short photophases.[51,55-57] However, Sanburg and Larsen[54] present convincing evidence against there being a single critical photosensitive stage in *Cx. pipiens*. Their results indicate that all stages respond to daylength and that the response in any stage is dependent upon the photophases to which previous stages have been exposed. Results of various combinations of life history stages exposed to 10- or 15-hr photophases indicate that the experience of every developmental stage has a significant effect on ovarian follicle size in adult females. Also, Danilevskii[8] pointed out that temperature fluctuations (high during photophase and low during scotophase), as found in nature, tend to enhance the reaction to short photophase. Therefore, the practice of producing diapausing *Culex* mosquitoes for experimental studies by rearing the larvae at a constant high temperature (27°C) and a photoperiod of 16L:8D and inducing diapause by exponsing the pupae to short photophase[51,56,57,89] may not result in mosquitoes representative of natural populations that are exposed to decreasing daylengths and fluctuating, cooler temperatures from egg to adult during late summer and autumn.

C. Ovarian Diapause

The ovarian follicles of mosquitoes are of the polytrophic type with the nurse cells contained in a single egg chamber with the oocyte. Each follicle contains one oocyte and seven nurse cells. In follicles that have just separated from the germarium, the oocyte is undifferentiated from the nurse cells, and all eight cells appear alike. Each follicle is surrounded by a follicular epithelium, formed when the follicle separates from the germarium. The developing follicle passes through a number of stages, described by Christophers[90] as stages I through V. Mer[91] distinguished two further stages, N and I-II. Development of ovarian follicles to Christophers' stage II is indicative of the resting stage and represents the completion of previtellogenic development in anautogenous females.

Christophers' classification scheme was based on anopheline mosquitoes, but it is applicable and has been used for other genera as well. Two modifications based on this system and frequently used for determining ovarian diapause in *Culex* mosquitoes require mention. Kawai[62] described in detail the development of ovarian follicles in *Cx. tritaeniorhynchus* and recognized 13 stages. He proposed that the completion of previtellogenic development, or resting stage, occurred when the primary follicles are in stage N (containing eight undifferentiated cells surrounded by a completed epithelium) or Ia (one oocyte and seven nurse cells are differentiated, but the nucleus of the oocyte is not yet apparent). Other investigators have considered the upper limit for a diapausing follicle to be stage N.[51,92] Spielman and Wong[55] measured the most advanced follicles and considered a follicle:germarium ratio of 1.5:1, or less, to be indicative of diapause.

Recently, a great deal of interest has focused on the elucidation of hormonal mechanisms necessary for regulating vitellogenesis in mosquitoes.[93-97] Although little work has been done with hibernating mosquitoes, there are obvious corollaries between vitellogenesis studies and hormonal influences on ovarian diapause. Studies by Clements,[98] Gillett,[99] and Larsen and Bodenstein[100] suggested that a blood meal results in the secretion of a hormone by the neurosecretory cells of the brain which activates the corpora allata (CA) to secrete JH. Lea[101-103] used microsurgical techniques to manipulate the CA and the medial neurosecretory cells (MNC) to demonstrate that JH is released prior to blood feeding. Also, a new hormone produced by the MNC was found and named egg development neurosecretory hormone (EDNH). This hormone is released after the blood meal and stimulates vitellogenesis. Lea[103] and Meola and Lea[104] showed that EDNH is produced by the MNC and stored in the corpus cardiacum (CC) until a blood meal triggers its release into the hemolymph.

Spielman et al.[105] showed that feeding or injecting 20-0H-ecdysone into females of several mosquito species stimulated yolk deposition in the absence of a blood meal. This led to the proposal that EDNH stimulated the previtellogenic ovary to secrete ecdysone.[106,107] Ecdysone is then converted to 20-0H-ecdysone, which presumably induces the fat body to synthesize vitellogenin.[106] The validity of the presumed physiologic role or ovarian ecdysone in mosquitoes relies heavily on the in vitro experiments of Hagedorn and co-workers.[106,108] Extremely high doses of 20-0H-ecdysone were required, and the amount of vitellogenin synthesized was only a fraction of that obtained in blood-fed females.[109] Subsequently, it was shown that physiologic doses of 20-0H-ecdysone can initiate and support vitellogenesis in isolated abdomens of blood-fed *Ae. aegypti* if the abdomens are pretreated with JH.[110] However, a vitellogenic role for JH after blood feeding is contradicted by earlier studies. When *Ae. aegypti* females were allatectomized 3 to 5 days after adult eclosion (a time period sufficient for their ovaries to attain their resting stage), subsequent blood feeding still resulted in mature eggs,[101] and these were viable.[180] Therefore, the question of whether JH is required after blood feeding for vitellogenesis in *Ae. aegypti* is unresolved.

JH deficiency is a universal characteristic of diapause in adult insects.[111] Thus, the CA play a central role in regulating adult diapause. CA activity is, in turn, normally regulated by another control center, presumably the brain. CA are often smaller in diapausing adult insects than in adults that are reproductively active.[111] The reverse situation was reported for hibernating *An. maculipennis.*[112] It was reported that the CA increase in size during diapause and differ in staining properties from those of gonoactive females. Following the first blood meal in the spring, the CA assumed the size and staining properties of those of summer females.

Juvenile hormones and their analogs have been tested extensively for their effects on diapause in adult insects.[111] Typically, very high doses are required to elicit a response. Spielman[113] terminated ovarian diapause in *Cx. pipiens* by treating specimens with a synthetic JH. Case et al.[114] showed that methoprene, a JH mimic, stimulated both biting behavior and diapause termination in *An. freeborni.* Mitchell[115] stimulated biting behavior and terminated ovarian diapause in *Cx. tarsalis* by topical application of methoprene. Also, Meola and Petralia[116] demonstrated that biting behavior in nondiapausing *Cx. pipiens* and *Cx. quinquefasciatus* can be blocked by removal of the CA shortly after adult emergence and can be restored by reimplantation of CA or injection of a synthetic JH. This may be analogous to a diapause situation.

Without exception, adult diapause in insects has been attributed to JH deficiency. However, the discovery of specific diapause proteins in *Leptinotarsa decemlineata* beetles and the suggestion that induction of these proteins may require both the absence of JH and the presence of ecdysteroids indicates the possibility of other hormonal influences as well.[111]

D. Gonotrophic Dissociation

The term "gonotrophic dissociation" was coined by Swellengrebel[84] to describe a condition existing in *An. atroparvus* in the Netherlands during the winter months. The females shelter in cattle sheds warmed by the animals and take occasional blood meals during the winter without developing their ovaries. Some females may develop their fat bodies but not to the same degree as species such as *An. messeae* that do not undergo gonotrophic dissociation. The latter is said to exhibit gonotropic concordance, a situation in which females neither blood feed nor develop eggs.[84] Guelmino[85] suggested an antigonadotrophic role for the fat body in hibernating females but did not present any data to support this hypothesis.

Additional studies on anophelines have confirmed and extended the results of Swellengrebel.[84] In Russia, gonotrophic dissociation occurs in *An. atroparvus* and *An. superpictus* but not in *An. messeae* and *An. hyrcanus*.[12,117] In the western U.S., extensive studies have been conducted to determine the physiologic condition of overwintering *An. freeborni*.[13,29,30,118] Diapause in this species is characterized principally by ovarian diapause and fat body hypertrophy; however, a few females will take blood and exhibit gonotrophic dissociation.

The question of whether *Culex* mosquitoes do take a blood meal and then survive the winter in cool-temperate areas has received a great deal of attention. If this occurs, it could be an important mechanism for overwinter maintenance of a number of arboviruses. Evidence has been presented supporting the occurrence of this phenomenon in *Cx. pipiens*,[56,89,119,120] *Cx. tritaeniorhynchus*,[62,121] and *Cx. restuans*.[122] However, the weight of evidence suggests that blood feeding, on the one hand, and ovarian diapause and fat body hypertrophy, on the other, are almost always mutually exclusive events and that gonotrophic dissociation rarely occurs in *Culex* mosquitoes in nature.[53,54,59,115,123-128] Eldridge[92] has suggested that the term "gonotrophic dissociation" should not be applied to *Culex* mosquitoes, "because even a prehibernating blood meal is probably rare by diapausing females, let alone repeated blood meals during hibernation".

E. Host-Seeking and Blood-Feeding Behavior

The question of whether *Culex* and other nonanopheline mosquitoes exhibit gonotrophic dissociation in nature is intimately associated with their host-seeking behavior. It is well established that diapausing *Culex* mosquitoes can be induced to take a blood meal if the host-seeking step is by-passed and they are placed in contact or proximity to a host[115,124,129] and that some mosquitoes will not develop their ovaries if held at a short photophase and an average temperature of 18°C or below.[89,115,124,129] The crucial question is, do diapausing *Culex* mosquitoes display host-seeking behavior in nature?

Host seeking and blood feeding in mosquitoes are behavioral responses to different stimuli and are controlled by different sensory receptors.[130] The blood-feeding proclivities of mosquitoes under laboratory conditions may be a poor guide to what happens in nature, because the early stages of the host-finding process are omitted.[115,131] Convincing evidence that diapausing *Culex* mosquitoes display host-seeking behavior in nature has not been forthcoming. Similar conclusions were reached by Hudson[65] in his studies of diapausing *Cs. inornata* in Alberta, Canada. He indicated that the demostration of gonotrophic dissociation in this species in the laboratory cannot be considered to have any epidemiologic significance until it is shown that diapausing females take blood meals in nature. Hudson[132] also demonstrated that even when diapausing *Cs. inornata* fed, the blood meals were small and ejected prematurely and that trypsin activity in the midgut was lacking. Similar observations on premature blood meal expulsion have been made with diapausing *Cx. tarsalis* from Colorado.[181] The question then arises — if diapausing females do not host seek and, when induced to feed by by-passing the host-seeking step are not able to digest the blood meal, may not gonotrophic dissociation in such females simply be laboratory artifact? Trypsin induction in *Ae. aegypti* is promoted by ecdysone,[133] presumably produced in the ovary. Both the neurosecretory system and the ovaries are required for maximal tryptic activity in normal *Ae. aegypti* females.[134] One could speculate that the chain of hormonal events associated with vitellogenesis is interrupted in diapausing mosquitoes and that trypsin induction does not proceed normally.

The practice of producing diapausing mosquitoes for experimental studies by rearing the larvae at a high, constant temperature and long photophase and then transferring the pupae to diapause-inducing conditions was referred to earlier. Perhaps differences resulting from such artificial conditions might be reflected in the depth or intensity of diapause and may explain some of the conflicting reports concerning host-seeking and blood-feeding behavior in diapausing *Culex* produced in the laboratory,[89] as compared with specimens collected in nature.[115,124] Differences in the degree of fat body development and in survival under simulated winter conditions also have been observed between field collected and colony produced diapausing *Cx. tarsalis*.[60,135] These differences may have been due, in part, to prolonged colonization and strain differences related to geographic origin; however, colony-rearing methods may also have been contributing factors.

F. Fat Body Hypertrophy and Metabolic Activity

Diapausing insects usually are biochemically distinct from their nondiapausing conterparts. Fat, carbohydrate, and protein reserves often are higher, and specific proteins may be unique to diapausing individuals.[111] Fat body hypertrophy is a striking characteristic of *Culex* and *Culiseta* and several *Anopheles* that overwinter as adult females.[12,29,65,84,92,136,137] Van Handel[138] discussed the metabolism of dietary carbohydrates (nectar) and proteins (blood) and their relationship to the behavior of the adult mosquito. Female mosquitoes accumulate large amounts of fat from sugar meals, and these reserves are used for survival at rest but not for vigorous flight activity. Whether dietary sugar will be stored as glycogen that will be available for flight, or as fat, is regulated by a hormone from the MNC of the brain in nondiapausing *Ae. sollicitans* and *Ae. taeniorhynchus*.[138,141] Feeding a single dose of sugar resulted in a small amount of glycogen and a large amount of fat being deposited within 24 hr. When the MNC were surgically removed before the meal, the opposite occurred. Reimplantation of the MNC reduced the glycogen storage level to that of intact controls but did not restore the ability to make and store more fat. The CA have no effect on glycogen and fat storage, and neither the CA nor the MNC affected utilization of glycogen or fat. Although extrapolating from these results and making inferences about diapausing females in other genera is risky, the involvement of the MNC in glycogen and fat storage levels during migratory flights and preparation for hibernation seems a likely possibility.

In nature, fat body hypertrophy of mosquitoes preparing for hibernation results primarily from feeding on plant sugars.[7] Glucose and fructose are the predominant compounds identified in crop liquids of prehibernating and hibernating *Cx. tarsalis*.[142,143] Harwood and Takata[144] demonstrated that temperature and photoperiod stimulate lipid deposition in *Cx. tarsalis* prior to hibernation. Buxton[136] pointed out over 50 years ago that the percentage of fat in diapausing *Cx. pipiens* dropped from 61% of dry body weight in October to 18% in March. He also found that the amount of water present gradually decreased as well. Tekle[145] demonstrated a greater than 60% loss of fat reserves in *Cx. pipiens* during 6 weeks of induced hibernation. Many anophelines synthesize lipids as a source of overwintering energy. In *An. freeborni*, 75% of the total lipids are composed of triglycerides.[146-148] In the Sacramento

Valley of California, *An. freeborni* adults reached a maximum lipid content in October, after which there was a steady decline during the overwintering period. The same pattern occurred in *Cx. tarsalis* adults, except that field populations reached their maximum lipid content in November.[148]

The JH titer indirectly influences metabolism by dictating either reproduction (high energy consuming processes) or diapause (low energy consuming processes).[111] The fat body, which is the major site of metabolic activity in insects, is thus greatly affected by the developmental program. The fat body also is pivotal in determining which proteins are to be synthesized. For example, in many insects, a high JH titer results in the fat body synthesizing an abundance of vitellogenin protein which is incorporated into the developing oocyte.[111] In some adult insects, specific diapause proteins are synthesized in response to short photophases. Information is not available regarding protein synthesis in diapausing mosquitoes.

G. Hibernacula and Winter Survival

Insect migration and dormancy have similar physiological, ecological, and behavioral components that are under hormonal control.[149] Both provide for escape from unfavorable conditions — one in space, the other in time. Also, both are associated with JH deficiency. Reproductive diapause is characteristic of insects preparing for migration or hibernation, and some species may undertake migratory flights prior to entering hibernation. *An. sacharovi* has been reported to make such flights.[150] In California's Central Valley, the increase in the number of *Cx.tarsalis* females in foothill areas during late summer and autumn coincides with a decrease in numbers in the valley floor. This is strong circumstantial evidence for the occurrence of migratory flights by prehibernating *Cx. tarsalis* females.[151,152] *Anopheles freeborni* may undertake similar flights from breeding sites to hibernacula.[153]

A wide variety of naturally occurring and man-made shelters may serve as hibernacula for overwintering adult mosquitoes. However, relatively few of the species that are believed to hibernate as adult females have been collected from overwintering sites. Descriptions of the microenvironment of mosquito hibernacula are further biased by the accessibility of such hibernacula to man as well as to mosquitoes. Temperature is obviously a critical element in determining suitability of overwintering sites. Whether supercooling occurs in hibernating mosquitoes is speculative and may be unnecessary in species that overwinter as adults. Hopla[88] failed to identify glycerol compounds from overwintering *Cs. alaskaensis* collected in Alaska at 65°N lat. Instead, he found that adult females survive by sheltering in the bases of dense stands of grass that become well insulated by snow cover. Sheltered sites that accumulate snow characteristically remain at 0°C.[74] Interestingly, Mouchet et al.[83] reported the recovery of a variety of hibernating mosquito species from reed piles and dense vegetation of reed swamps in southern France (44°N lat.) where hibernacula temperatures probably are similar to those found in *Cs. alaskaensis* hibernacula in Alaska. Harwood and Takata[144] suggested that higher levels of unsaturated fatty acids found in diapausing *Cx. tarsalis* might be associated with cold tolerance.

Conditions of high relative humidity are necessary for most mosquito species that overwinter as adults. Hibernating *Cx. pipiens* and *Cx. tarsalis* collected in the field and tested in the laboratory survive best under conditions of high humidity and with temperatures at or slightly below freezing.[135,154] In Kern County, California, where the period of diapause or quiescence is not of long duration, more than 90% of female *Cx. tarsalis* collected from shelters during late autumn survived without blood or carbohydrate feeding for 35 days under high humidity conditions at 9°C and a photoperiod of 10L:14D.[126] In the field, most females collected during winter were empty, inseminated, and nulliparous, indicating a cessation of gonotrophic activity.[59,125,127] Reisen et al.[127] indicated that hibernating states of *Cx. tarsalis* in Kern County, determined by ovarian morphometry, were represented by reproductive diapause, quiescence, and perhaps, oligopause.

Water conservation and water balance mechanisms of diapausing adult mosquitoes have not been described. The relative humidity of *Cx. tarsalis* hibernacula is generally high,[155-157] and otherwise similar sites with low humidities are not utilized. Likewise, mortality during hibernation has been attributed to decreasing or low humidities.[155,157] Feeding on free water would seem risky in view of the potential for encountered freezing temperatures. Undoubtedly, metabolic water produced by lipid catabolism plays a role in the water balance of diapausing adults, but this remains to be investigated in mosquitoes.

Light intensity may be important in the distribution of adults in hibernacula. In artificial marl-caves in the Netherlands, *Cx. pipiens* tended to occupy the twilight zone in a region 10 to 55 m from the cave entrance.[158]

Few studies have been done to monitor survival rates of diapausing mosquitoes through an entire winter. Mitchell[157] recovered 8.7% of a marked population of *Cx. tarsalis* from a Colorado mine 3 to 5 months after marking a diapausing population with a fluorescent dust and releasing them back into the mine in late autumn. Bailey et al.[120] studied the survival of *Cx. pipiens* in Maryland. They concluded that a significant number of diapausing females that were given a prehibernation blood meal by enforced feeding do not develop eggs and can survive the winter at rates comparable to diapausing nonbloodfed females.

Readiness to take a blood meal in mosquitoes collected outdoors has been used to show the end of diapause in *Cx. tarsalis,*[59] *Cx. pipiens,*[55] *An. freeborni,*[29] and *An. punctipennis.*[32] Little information is available on the factors responsible for maintenance and termination of diapause in adult mosquitoes, although the mechanisms are probably based largely on photoperiod and temperature as they are in other insects. Tauber et al.[10] have stressed the dynamic state of diapause and the fact that depth or intensity decreases as the season progresses and that many insects terminate diapause by midwinter rather than spring. Mosquitoes probably follow a similar pattern. Since JH deficiency is a characteristic of adult diapause in all insects studied and the presence of JH is associated with gonoactivity, it follows that diapause termination must, in some way, be associated with the reactivation of the CA. The mechanisms by which this occurs are obscure but probably rely on the detection of environmental stimuli by the brain. As mentioned earlier, JHs and their analogs have been used to terminate ovarian diapause in *Cx. pipiens,*[113] *An. freeborni,*[114] and *Cx. tarsalis.*[115] However, it has been suggested that the stimulation of reproduction achieved with JH or its analogs can best be perceived as an interruption of diapause rather than termination since, in some insects, adults eventually revert to diapause if they remain in diapause-inducing conditions.[111] This has not been tested with mosquitoes.

VI. MOSQUITO OVERWINTERING AND ARBOVIRUS MAINTENANCE

Among the 504 registered arthropod-borne and selected vertebrate viruses are 228 that have been isolated from mosquitoes and classified as known, possible, or probable arboviruses.[159] These arboviruses, like their mosquito vectors, show their greatest diversity in the tropics, and only 51 (22%) have been isolated from mosquitoes within the temperate regions delineated by the 10°C cold-month isotherms shown in Figure 1. These mosquito/arbovirus associations, their geographic distribution, and information concerning overwintering strategies of the mosquitoes are summarized in Table 3. Standard references usually are given for information on overwintering strategies of the mosquitoes; original literature citations can be found in the standard references. References to the mosquito/arbovirus associations and their geographic distribution can be found in Karabatsos.[159]

The majority of mosquito/arbovirus associations from areas where mosquitoes enter diapause to escape cold temperatures are found in the Holarctic region since rather limited land masses lie below the 10°C cold-month isotherm of the southern hemisphere (Figure 1). In South America, western equine encephalitis (WEE) virus activity has been reported in Rio

Table 3
DISTRIBUTION OF MOSQUITO/ARBOVIRUS ASSOCIATIONS THAT OCCUR WITHIN THE LIMITS OF 10°C COLD-MONTH ISOTHERMS AND OVERWINTERING STRATEGIES OF THE MOSQUITOES

				Mosquito overwintering	
	Species	Distribution[a]	Associated arboviruses[b]	Strategies[c]	Ref.
An. (Ano.)	*crucians*	1	CV, EEE, LAC, SLE, TVT	L	26
	franciscanus	1	MD	A	160
	freeborni	1	CV, MD, WEE	A	26
	hyrcanus	3	TAH	A	161
	maculipennis	2, 3	CVO/BAT	A	162
	punctipennis	1	CF, JC, LAC, SSH	A	26
	quadrimaculatus	1	CV, JC, SLE	A	26
	walkeri	1	CV	E	26
An. (Cel.)	*annulipes*	4	TIL, TRU	L	163
Ae. (Aed.)	*cinereus*	1	CV, SSH	E	26
Ae. (Adm.)	*vexans*	1, 2, 3	CE, CV, EEE, JC, KEY, LAC, MD, SLE, SSH, TVT, WEE/TAH/GET, SAG, AKA	E	26
Ae. (Och.)	*abserratus*	1	JC	E	26
	atlanticus	1	EEE, KEY, LAC, TVT	E	26
	aurifer	1	KEY, SSH	E	26
	campestris	1	CV, WEE	E	26
	camptorhynchus	4	TER	?	see 163
	canadensis	1	CV, EEE, JC, KEY, LAC, SSH	E	26
	cantans	2	WN	E	162
	cantator	1	JC	E	26
	caspius	2, 3	TAH/BKN	E	162
	cataphylla	1	SSH	E	26
	communis	1, 2, 3,	CV, JC, LAC, SSH, TVT/INK/SIN	E, L	26
	dianteus	2	TAH	E	26
	dorsalis	1	CE, CV, LAC, MD, TVT, WEE	E	26
	excrucians	1	NOR, SSH	E	26
	fitchii	1	SSH	E	26
	flavescens	1	CV, WEE	E	26
	hexodontus	1	NOR, SSH	E	26
	implicatus	1	SSH	E	26
	intrudens	1	SSH	E	26
	melanimon	1	CE, JC, SLE, WEE	E	164
	mitchellae	1	EEE	E[d]	
	nigripes	1	SSH	E	26
	nigromaculis	1	CE, CV, EEE, MD, WEE	E	26
	punctor	1, 2	JC, NOR, SSH/INK	E	26
	sollicitans	1	CV, JC, KEY, LAC	E	26
	spencerii	1	CV, WEE	E	26
	sticticus	1, 2	EEE/TAH	E	26
	stimulans	1	CV, JC, SSH	E	26
	taeniorhynchus	1	CV, EEE, NOR	E	26, 45
	thelcter	1	JC	E[d]	
	trivittatus	1	BOC, CV, JC, LAC, SSH, TVT	E	26
	vigilax	4	BF, EH, GG, RR, TER, YAC	E[d]	

Table 3 (continued)
DISTRIBUTION OF MOSQUITO/ARBOVIRUS ASSOCIATIONS THAT OCCUR WITHIN THE LIMITS OF 10°C COLD-MONTH ISOTHERMS AND OVERWINTERING STRATEGIES OF THE MOSQUITOES

				Mosquito overwintering	
	Species	Distri-bution[a]	Associated arboviruses[b]	Strategies[c]	Ref.
Ae. (Pro.)	*triseriatus*	1	CV, JC, KEY, LAC, SSH, TVT	E, L	14, 46
Ae. (Psk.)	*bancroftianus*	4	BF	L	163
Ps. (Gra.)	*columbiae*	1	CV, JC, SA, WEE	E	26 (as *confinnis*)
	discolor	1	JC	E	26
	signipennis	1	CE, LOK, MD, SA, TVT, WEE	E	26
Ps. (Pso.)	*howardii*	1	LAC	E	26
Cx. (Bar.)	*modestus*	2, 3	WN, LED/KYZ	A	83
Cx. (Cux.)	*annulirostris*	4	BF, EH, KOK, KOO, KUN, MVE, PR, RR, SIN, TER	A[d]	
	pervigilans	5	WHA	A[d]	
	peus	1	SLE, TUR, WEE	A	51
	pipiens	1, 2, 3, 4	CV, LAC, SLE, TUR, TVT, WEE/TAH/JBE, AINO/BF	A	26
	pseudovishnui	3	AINO	A[d]	
	restuans	1	EEE, HP, SLE, WEE	A	122
	salinarius	1	EEE, FLA, SLE	A	122
	tarsalis	1	CE, CV, FLA, GLO, HP, LLS, LOK, MD, SLE, TUR, UMA, WEE	A	26
	tritaeniorhynchus	3	AINO, AKA, JBE, SAG	A	62, 78
Cx. (Lop.)	*cylindricus*	4	BF	?	
Cx. (Mai.)	*hortensis*	3	BKN	?	
Cx. (Mel.)	*erraticus*	1	SLE, WEE	A	165
Cx. (Ncx.)	*territants*	1	EEE, FLA, SLE	A	26
Cs. (Cli)	*melanura*	1	EEE, FLA, HJ, HP	L	26
	tonnoiri	5	WHA	A	166
Cs. (Cuc.)	*morsitans*	1, 2	EEE/SIN[178]	E, L	26, 179
Cs. (Cus.)	*alaskaensis*	1	NOR	A	26
	annulata	2	TAH	A, L	161
	impatiens	1	SSH	A	26
	inornata	1	CE, CV, JC, JS, MD, NOR, SSH, TUR, SLE, TVT, WEE	A	26
Cq. (Coq.)	*linealis*	4	BF, RR, TRU	L[d]	
	perturbans	1	CV, EE, JC, TVT	L	26
	richiardii	2	CVO	L	161
Ma. (Mnd.)	*uniformis*	4	RR	L[d]	

[a] 1 = North America, 2 = Europe, 3 = Asia, 4 = Australia, and 5 = New Zealand.
[b] Slash marks separate continental distribution; underlined viruses isolated from mosquito pools of mixed species.
[c] E = egg, L = larva, A = adult, ? = unknown. In warmer areas outside the 10°C cold-month isotherms, overwintering strategies for some species may differ from those listed in the table.
[d] Probable.

Negro Province, Argentina, at 41°S lat.;[167] however, all isolations from mosquitoes have come from areas north of the 10°C isotherm.[168] Therefore, information is lacking concerning mosquito/arbovirus associations in areas of South America where vector mosquitoes pass the winter in a state of diapause. In Africa, studies by Jupp[169,170] suggest that ovarian diapause does not occur in *Culex* spp. in the temperate highveld region of South Africa, but that these mosquitoes overwinter by quiescence. *Culiseta longiareolata* apparently also overwinters in South Africa by quiescence.[171] The mosquito/arbovirus associations from Australia and New Zealand that have been reported south of the 10°C cold-month isotherm are listed in Table 3. Many of the associations listed for Australia are from the Murray River basin and the coast of New South Wales, both of which fall marginally, if at all, within the boundaries of the 10°C July isotherm. Nevertheless, the continuous distribution of vector mosquitoes into adjacent cooler regions poses the possibility of diapausing mosquitoes being involved in the overwinter maintenance of some of these viruses.

The 51 arboviruses included in Table 3 have been isolated from 74 mosquito species within the limits of the 10°C cold-month isotherms. The mosquitoes involved are assigned to 7 genera and 20 subgenera. In general, the species of *Anopheles, Culex,* and *Culiseta* overwinter principally as inseminated adult females (22 species), *Aedes* and *Psorophora* as eggs (39 species), and *Coquillettidia* and *Mansonia* as larvae (5 species). There are exceptions, as well as a few species that overwinter in more than one life history stage or whose overwintering strategy is unknown.

Although the possible involvement of mosquito species in arbovirus overwintering mechanisms is not limited to the demonstrated associations shown in Table 3, this is a logical data base from which to begin the search. To date, 15 of the 51 arboviruses that have been isolated from mosquitoes in temperate areas also have been shown to be transmitted transovarially in nature or in the laboratory. Nine arboviruses, California encephalitis (CE), eastern equine encephalitis (EEE), Japanese encephalitis (JE), Jamestown Canyon (JC), Keystone (KEY), La Crosse (LAC), snowshoe hare (SSH), Tahnya (TAH), and trivittatus (TVT) have been recovered from naturally infected mosquito larvae in temperate areas; six others, Kokobera (KOK), Kunjin (KUN), Murray Valley encephalitis (MVE), Ross River virus (RRV), St. Louis encephalitis (SLE), and San Angelo virus (SA) have been shown to be transovarially transmitted in experimental studies.[172] It is perhaps significant that with the exceptions of EEE and JE, all of the viruses recovered from naturally infected mosquito larvae are California serogroup viruses that are transmitted mainly by *Aedes* mosquitoes that overwinter in the egg stage. *Aedes* species, and most other mosquitoes as well, that inhabit areas of high latitude are univoltine and may take only a single blood meal during each generation. Therefore, in the far north, they only serve as vectors for arboviruses that are transmitted transovarially. *Aedes* species also have been implicated in the transovarial transmission (TOT) of certain arboviruses in tropical areas.[172,173] The demonstrated importance of TOT of arboviruses by *Aedes* mosquitoes and the long-term survival of diapausing and quiescent *Aedes* eggs make species in this genus ideal candidates for overwinter maintenance of other arboviruses with which they have been associated in temperate areas (Table 3).

Although isolations of WEE,[155,174] JE,[175] SLE,[176] and TAH[177] viruses have been made from overwintering adult female *Culex* mosquitoes, the significance of these isolations is unknown. Such infections must be extremely rare events, or current assay systems are unable to detect cryptic infections. The failure to experimentally demonstrate the TOT of alphaviruses by *Culex* mosquitoes make it unlikely that TOT by *Culex* is an important maintenance mechanism for this virus group. The rarity of blood feeding among prehibernating *Culex,* coupled with the low probability that such mosquitoes would find viremic hosts to feed upon in late summer or autumn, makes overwinter maintenance of arboviruses by this mechanism also unlikely.

VII. FUTURE RESEARCH

To list in detail the lacunae in our knowledge concerning overwintering diapause in mosquitoes and the many areas for future research would be stating the obvious. Scarcely any aspect has been comprehensively investigated. Understandably, initial studies have concentrated on identifying environmental cues and threshold levels necessary to elicit appropriate responses. These studies have yielded a wealth of valuable information, and much remains to be done in these areas. However, information concerning transduction of environmental stimuli by the neuroendocrine system of embryos, larvae, pupae, and adults is nonexistent. The hormonal basis for egg and larval diapause in mosquitoes has not been investigated, and most information on this subject in adults is sketchy and based on inferences from studies done with nondiapausing mosquitoes. A comprehensive study on the endocrinology and metabolism of a diapausing mosquito has yet to be undertaken. It is a formidable task to be sure; however, advances during the last decade using radioimmunoassays (RIA), high perfomance liquid chromatography, (HPLC), mass spectrometry (MS), rocket immunoelectrophoresis, enzyme-linked immunosorbent assays (ELISA) using monoclonal antibodies, and tissue culture incubations have permitted rapid gains to be made in understanding the endocrinology of nondiapausing mosquitoes. There is still a place for the delicate microsurgical manipulations for identifying cause-and-effect relationships in diapausing mosquitoes.

Comparative studies on the physiology of different species of diapausing mosquito are essential. For example, it would be revealing to compare tryptic activity and blood meal digestion plus CA activity and JH titers in gonotrophic dissociation and concordance cohorts of a species such as *An. freeborni*. Presumably, concordance females could be induced to take a blood meal by by-passing the host-seeking step. Similar comparisons between anophelines that exhibit gonotrophic dissociation and culicines that do not would be instructive.

Field studies should not be neglected. The crucial question of whether certain mosquitoes display host-seeking behavior and take a blood meal prior to entering hibernation can only be answered definitively by novel approaches in the field. There are practical reasons for studying the biology and physiology of diapause in overwintering mosquitoes. Modern insect control methods that depend on a systems analysis approach to pest management must be based on a sound knowledge of insect seasonality. Also, several third-generation pesticides take advantage of physiologic attributes, often hormonal, of insects. Interrupting diapause in prehibernating mosquitoes would be an effective control mechanism if delivery systems could be developed that would assure exposure of a significant portion of the mosquito population to hormone mimics that are currently available.

ACKNOWLEDGMENTS

I am grateful to the following persons: Dr. Christine Dahl (Uppsala), Dr. Peter G. Jupp (Johannesburg), Dr. Brian H. Kay (Brisbane), and Dr. Jean Mouchet (Bondy) for providing information about overwintering strategies of mosquitoes in their respective geographic areas; Dr. William K. Reisen (Bakersfield) and Dr. Bruce F. Eldridge (Corvallis) for supplying copies of manuscripts in press; and Dr. Arden O. Lea (Athens) and Dr. Thomas P. Monath (Ft. Collins) for critically reading this manuscript and making helpful comments.

REFERENCES

1. **Dingle, H., Ed.,** *Evolution of Insect Migration and Diapause*, Springer-Verlag, New York, 1978, 51.
2. **Lees, A. D.,** *The Physiology of Diapause in Arthropods*, Cambridge University Press, London, 1955, chap. 1 and 2.
3. **Saunders, D. S.,** *Insect Clocks*, 2nd ed., Pergamon Press, New York, 1982, xiv, xv, 139, 271, 356, 357.
4. **Muller, H. J.,** Formen der Dormanz bei Insecten, *Nova Acta Leopold.*, 35, 7, 1970.
5. **Mansingh, D.,** Physiological classification of dormancies in insects, *Can. Entomol.*, 103, 983, 1971.
6. **Tesh, R. B.,** Transovarial transmission of arboviruses in their invertebrate vectors, in *Current Topics in Vector Research*, Vol. 2, Harris, K. F., Ed., Praeger, New York, 1984, chap. 3.
7. **Clements, A. N.,** *The Physiology of Mosquitoes*, Macmillan, New York, 1953, chap. 6 and 11.
8. **Danilevskii, A. S.,** *Photoperiodism and Seasonal Development of Insects*, 1st English ed., Oliver & Boyd, London, 1965, 2, 32, 33, 109, 116, 251.
9. **Beck, S. D.,** *Insect Photoperiodism*, 2nd ed., Academic Press, New York, 1980, chap. 7 and 8.
10. **Tauber, M. J., Tauber, C. A., and Masaki, S.,** *Seasonal Adaptations of Insects*, Oxford University Press, Oxford, 1985.
11. **Danilevskii, A. S., Goryshin, N. I., and Tyshchenko, V. P.,** Biological rhythyms in terrestrial arthropods, *Ann. Rev. Entomol.*, 15, 201, 1970.
12. **Vinogradova, E. B.,** An experimental investigation of the ecological factors inducing imaginal diapause in bloodsucking mosquitoes (Diptera, Culicidae), *Entomol. Rev.*, 39, 210, 1960.
13. **Depner, K. R. and Harwood, R. F.,** Photoperiodic responses of two latitudinally diverse groups of *Anopheles freeborni* (Diptera:Culicidae), *Ann. Entomol. Soc. Am.*, 59, 7, 1966.
14. **Kappus, K. D. and Venard, C. E.,** The effects of photoperiod and temperature on the induction of diapause in *Aedes triseriatus* (Say), *J. Insect Physiol.*, 13, 1007, 1967.
15. **Holzapfel, C. M. and Bradshaw, W. E.,** Geography of larval dormancy in the tree-hole mosquito, *Aedes triseriatus* (Say), *Can. J. Zool.*, 59, 1014, 1981.
16. **Sims, S. R.,** Larval dispause in the eastern tree-hole mosquito, *Aedes triseriatus:* latitudinal variation in induction and intensity, *Ann. Entomol. Soc. Am.*, 75, 195, 1982.
17. **Shroyer, D. A. and Craig, G. B., Jr.,** Egg diapause in *Aedes triseriatus* (Diptera:Culicidae): geographic variation in photoperiodic response and factors influencing diapause termination, *J. Med. Entomol.*, 20, 601, 1983.
18. **Beach, R. F.,** The required day number and timely induction of diapause in geographic strains of the mosquito *Aedes atropalpus*, *J. Insect Physiol.*, 24, 449, 1978.
19. **Jordan, R. G. and Bradshaw, W. E.,** Geographic variation in the photoperiodic response of the western tree-hole mosquito, *Aedes sierrensis*, *Ann. Entomol. Soc. Am.*, 71, 487, 1978.
20. **Bradshaw, W. E.,** Geography of photoperiodic response in a diapausing mosquito, *Nature (London)*, 262, 384, 1976.
21. **Trimble, R. M. and Smith, S. M.,** Geographic variation in the effects of temperature and photoperiod on dormancy induction, developmental time and predation in the tree-hole mosquito, *Toxorhynchites rutilis septendrionalis* (Diptera:Culicidae), *Can. J. Zool.*, 57, 1612, 1979.
22. **Shroyer, D. A. and Craig, G. B., Jr.,** Egg hatchability and diapause in *Aedes triseriatus* (Diptera: Culicidae): temperature- and photoperiod-induced latencies, *Ann. Entomol. Soc. Am.*, 73, 39, 1980.
23. **Gjullin, C. M., Hegarty, C. P., and Bollen, W. B.,** The necessity of a low oxygen concentration for the hatching of *Aedes* mosquito eggs, *J. Cell. Comp. Physiol.*, 17, 193, 1941.
24. **Judson, C. L.,** The physiology of hatching of aedine mosquito eggs: hatching stimulus, *Ann. Entomol. Soc. Am.*, 53, 688, 1960.
25. **Breeland, S. G. and Pickard, E.,** Field observations on 28 broods of floodwater mosquitoes resulting from controlled floodings of a natural habitat in the Tennessee Valley, *Mosq. News*, 27, 343, 1967.
26. **Carpenter, S. J. and La Casse, W. J.,** *Mosquitoes of North America*, University of California Press, Berkeley, 1955.
27. **Hurlbut, H. S.,** Further notes on the overwintering of the eggs of *Anopheles walkeri* Theobald with a description of the eggs, *J. Parasitol.*, 24, 521, 1938.
28. **Baker, F. C.,** The effect of photoperiodism on resting, treehole, mosquito larvae, *Can. Entomol.*, 67, 149, 1935.
29. **Washino, R. K.,** Physiological condition of overwintering female *Anopheles freeborni* in California, *Ann. Entomol. Soc. Am.*, 63, 212, 1970.
30. **Washino, R. K., Gieke, P. A., and Schaefer, C. H.,** Physiological changes in the overwintering females of *Anopheles freeborni* in California, *J. Med. Entomol.*, 8, 279, 1971.
31. **Vinogradova, E. B.,** The role of photoperiodism in the seasonal development of *Anopheles plumbeus* Steph., *Dokl. Akad. Nauk. S.S.S.R.*, 142, 481, 1962.

32. **Washino, R. K. and Bailey, S. F.,** Overwintering of *Anopheles punctipennis* (Diptera:Culicidae) in California, *J. Med. Entomol.*, 7, 95, 1970.
33. **Wilson, F. R. and Horsfall, W. R.,** Eggs of floodwater mosquitoes. XII. Installment hatching of *Aedes vexans* (Diptera:Culicidae), *Ann. Entomol. Soc. Am.*, 63, 1644, 1970.
34. **McHaffey, D. G.,** Photoperiod and temperature influences on diapause in eggs of the floodwater mosquito *Aedes vexans* (Meigen) (Diptera:Culicidae), *J. Med. Entomol.*, 9, 564, 1972.
35. **Vinogradova, E. B.,** An experimental study of the factors regulating induction of imaginal diapause in the mosquito *Aedes togoi* Theob., *Entomol. Obozr.*, 44, 527, 1965.
36. **Mogi, M.,** Studies on *Aedes togoi* (Diptera:Culicidae). I. Alternative diapause in the Nagasaki strain, *J. Med. Entomol.*, 18, 477, 1981.
37. **Anderson, J. F.,** Influence of photoperiod and temperature on the induction of diapause in *Aedes atropalpus* (Diptera:Culicidae), *Entomol. Exp. Appl.*, 11, 321, 1968.
38. **Kalpage, K. S. P. and Brust, R. A.,** Studies on diapause and female fecundity in *Aedes atropalpus*, *Environ. Entomol.*, 3, 139, 1974.
39. **Tauthong, P. and Brust, R. A.,** The effect of photoperiod on diapause induction, and temperature on diapause termination in embryos of *Aedes campestris* Dyar and Knab (Diptera:Culicidae), *Can. J. Zool.*, 55, 129, 1977.
40. **Pinger, R. R. and Eldridge, B. F.,** The effect of photoperiod on diapause induction in *Aedes canadensis* and *Psorophora ferox* (Diptera, Culicidae), *Ann. Entomol. Soc. Am.*, 70, 437, 1977.
41. **Vinogradova, E. B.,** Photoperiodic and temperature induction of egg diapause in *Aedes caspius caspius* Pall. (Diptera, Culicidae), *Parasitologia*, 9, 385, 1975.
42. **McHaffey, D. G. and Harwood, R. F.,** Photoperiod and temperature influences on diapause in eggs of the floodwater mosquito, *Aedes dorsalis* Meigen (Diptera, Culicidae), *J. Med. Entomol.*, 7, 631, 1970.
43. **Mulligan, F. S., III,** Direct induction of embryonic diapause in colonized *Aedes dorsalis*, *Ann. Entomol. Soc. Am.*, 73, 589, 1980.
44. **Anderson, J. F.,** Induction and termination of embryonic diapause in the salt marsh mosquito, *Aedes sollicitans* (Diptera:Culicidae), *Conn. Agric. Exp. Stn. Bull.*, 711, 1, 1970.
45. **Parker, B. M.,** Effects of photoperiod on the induction of embryonic diapause in *Aedes taeniorhynchus* (Diptera:Culicidae), *J. Med. Entomol.*, 22, 392, 1985.
46. **Love, G. J. and Whelchel, J. G.,** Photoperiodism and the development of *Aedes triseriatus* (Diptera, Culicidae), *Ecology*, 35, 340, 1955.
47. **Vinogradova, E. B.,** The effect of photoperiodism upon the larval development and the appearance of diapausal eggs in *Aedes triseriatus* Say (Diptera, Culicidae), *Parasitologia*, 1, 19, 1967.
48. **Wright, J. E. and Venard, C. E.,** Diapause induction in larvae of *Aedes triseriatus*, *Ann. Entomol. Soc. Am.*, 64, 11, 1971.
49. **Clay, M. E. and Venard, C. E.,** Larval diapause in the mosquito *Aedes triseriatus:* effects of diet and temperature on photoperiodic induction, *J. Insect Physiol.*, 18, 1441, 1972.
50. **Wang, R. L.,** Observations on the influence of photoperiod on egg diapause in *Aedes albopictus*, *Acta Entomol. Sin.*, 15, 75, 1966.
51. **Skultab, S. and Eldridge, B. F.,** Ovarian diapause in *Culex peus* (Diptera:Culicidae), *J. Med. Entomol.*, 22, 454, 1985.
52. **Tate, P. and Vincent, M.,** The biology of autogenous and anautogenous races of *Culex pipiens* L. (Diptera: Culicidae), *Parasitology*, 28, 115, 1936.
53. **Oda, T. and Wada, Y.,** Developmental stages of *Culex pipiens pallens* sensitive to photoperiodic conditions, *Trop. Med.*, 14, 198, 1972.
54. **Sanburg, L. L. and Larsen, J. R.,** Effect of photoperiod and temperature on ovarian development in *Culex pipiens pipiens*, *J. Insect Physiol.*, 19, 1173, 1973.
55. **Spielman, A. and Wong, J.,** Environmental control of ovarian diapause in *Culex pipiens*, *Ann. Entomol. Soc. Am.*, 66, 905, 1973.
56. **Eldridge, B. F.,** The effect of temperature and photoperiod on blood-feeding and ovarian development in mosquitoes of the *Culex pipiens* complex, *Am. J. Trop. Med. Hyg.*, 17, 133, 1968.
57. **Eldridge, B. F., Johnson, M. D., and Bailey, C. L.,** Comparative studies of two North American mosquito species, *Culex restuans* and *Culex salinarius:* response to temperature and photoperiod in the laboratory, *Mosq. News*, 36, 506, 1976.
58. **Madder, D. J., Surgeoner, G. A., and Helson, B. V.,** Induction of diapause in *Culex pipiens* and *C. restuans* (Diptera:Culicidae) in southern Ontario, *Can. Entomol.*, 115, 877, 1981.
59. **Bellamy, R. E. and Reeves, W. C.,** The winter biology of *Culex tarsalis* (Diptera:Culicidae) in Kern County, California, *Ann. Entomol. Soc. Am.*, 56, 314, 1963.
60. **Harwood, R. F. and Halfhill, E.,** The effect of photoperiod on fat body and ovarian development of *Culex tarsalis* (Diptera:Culicidae), *Ann. Entomol. Soc. Am.*, 57, 596, 1964.
61. **Eldridge, B. F.,** The influence of daily photoperiod on blood-feeding activity of *Culex tritaeniorhynchus* Giles, *Am. J. Hyg.*, 77, 49, 1963.

62. **Kawai, S.,** Studies on the follicular development and feeding activity of the females of *Culex tritaeniorhynchus* with special reference to those in autumn, *Trop. Med.*, 11, 145, 1969.
63. **Hayes, R. O. and Maxfield, H. K.,** Interruption of diapause and rearing larvae of *Culiseta melanura* (Coq.), *Mosq. News*, 27, 458, 1967.
64. **Maloney, J. M. and Wallis, R. C.,** Response of colonized *Culiseta melanura* to photoperiod and temperature, *Mosq. News*, 36, 190, 1976.
65. **Hudson, J. E.,** Induction of diapause in female mosquitoes, *Culiseta inornata*, by a decrease in daylength, *J. Insect Phsyiol.*, 23, 1377, 1977.
66. **Barnard, D. R. and Mulla, M. S.,** Effects of photoperiod and temperature on blood feeding, oogenesis and fat body development in the mosquito, *Culiseta inornata*, *J. Insect Physiol.*, 23, 1261, 1977.
67. **Bradshaw, W. E. and Holzapfel, C. M.,** Biology of tree-hole mosquitoes: photoperiodic control of development in northern *Toxorhynchites rutilus* (Coq.), *Can. J. Zool.*, 53, 889, 1975.
68. **Smith, S. M. and Brust, R. A.,** Photoperiodic control of the maintenance and termination of larval diapause in *Wyeomyia smithii* (Coq.) (Diptera:Culicidae) with notes on oogenesis in the adult female, *Can. J. Zool.*, 49, 1065, 1971.
69. **Evans, K. W. and Brust, R. A.,** Induction and termination of diapause in *Wyeomyia smithii* (Diptera:Culicidae), and larval survival studies at low and subzero temperatures, *Can. Entomol.*, 104, 1937, 1972.
70. **Bradshaw, W. E. and Lounibos, L. P.,** Photoperiodic control of development in the pitcher-plant mosquito, *Wyeomyia smithii*, *Can. J. Zool.*, 50, 713, 1972.
71. **Lounibos, L. P. and Bradshaw, W. E.,** A second diapause in *Wyeomyia smithii:* seasonal incidence and maintenance by photoperiod, *Can. J. Zool.*, 53, 215, 1975.
72. **Bradshaw, W. E. and Holzapfel, C. M.,** Interaction between photoperiod, temperature, and chilling in dormant larvae of the tree-hole mosquito, *Toxorhynchites rutilis* Coq., *Biol. Bull. Mar. Biol. Lab. Woods Hole*, 152, 147, 1977.
73. **Baust, J. G. and Rojas, R. R.,** Review — insect cold hardiness: facts and fancy, *J. Insect Physiol.*, 31, 755, 1985.
74. **Danks, H. V.,** Modes of seasonal adaptation in the insects. I. Winter survival, *Can. Entomol.*, 110, 1167, 1978.
75. **Owen, W. B.,** The mosquitoes of Minnesota, with special reference to their biologies, *Tech. Bull. Minn. Agric. Exp. Stn.*, No. 126, 1937.
76. **Clay, M. E. and Venard, C. E.,** Diapause in *Aedes triseriatus* (Diptera:Culicidae) larvae terminated by molting hormones, *Ann. Entomol. Soc. Am.*, 64, 968, 1971.
77. **Roubaud, E. and Colas-Belcour, J.,** Observations sur la biologie de l'*Anopheles plumbeus*. I. Le comportement larvaire, *Soc. Pathol. Exot. Bull.*, 25, 763, 1932.
78. **Sasa, M.,** Zoophilism, hibernation and appearance of mosquitoes of Japan, *Jpn. Med. J.*, 2, 99, 1949.
79. **Hayes, R. O.,** Studies on eastern encephalitis in Massachusetts during 1960, *Proc. N. J. Mosq. Exterminators Assoc.*, 48, 59, 1961.
80. **Yamashita, O. and Hasegawa, K.,** Embryonic diapause, in *Comprehensive Insect Physiology, Biochemistry, and Pharmacology*, Vol. 1, Kerkut, G. A. and Gilbert, L. I., Eds., Pergamon Press, New York, 1985, chap. 11.
81. **Chippendale, G. M.,** Hormonal regulation of larval diapause, *Ann. Rev. Entomol.*, 22, 121, 1977.
82. **Mouchet, J. and Rageau, J.,** Observations sur les moustiques de la Camargue et du Bas-Rhone. I. L'hibernation d' *Uranotaenia unguiculata* Edwards, 1913 (Diptere, Culicidae), *Bull. Soc. Pathol. Exot.*, 58, 246, 1965.
83. **Mouchet, J., Rageau, J., and Chippaux, A.,** Hibernation de *Culex modestus* Ficalbi (Diptera, Culicidae) en Camargue, *Cah. ORSTOM Ser. Entomol Med. Parasitol.*, 7, 35, 1969.
84. **Swellengrebel, N. H.,** La dissociation des fonctions sexuelles et nutritives: (dissociation gono-trophique) d'*Anopheles maculipennis* comme cause du paludisme dans les Pays-Bas et ses rapports avec "l'infection domiciliaire", *Ann. Inst. Pasteur*, 43, 1370, 1929.
85. **Guelmino, D. J.,** The physiology of *Anopheles maculipennis* during hibernation. An attempt to interpret the phenomenon of gonotrophic dissociation, *Ann. Trop. Med. Parasitol.*, 45, 161, 1951.
86. **Washino, R. K.,** The physiological ecology of gonotrophic dissociation and related phenomena in mosquitoes, *J. Med. Entomol.*, 13, 381, 1977.
87. **Frohne, W. C.,** Natural history of *Culiseta impatiens* (wlk.), (Diptera, Culicidae), in Alaska, *Trans. Am. Microsc. Soc.*, 72, 103, 1953.
88. **Hopla, C. E.,** The natural history of the genus *Culiseta* in Alaska, *Proc. N. J. Mosq. Exterminators Assoc.*, 57, 56, 1970.
89. **Eldridge, B. F. and Bailey, C. L.,** Experimental hibernation studies in *Culex pipiens* (Diptera:Culicidae): reactivation of ovarian development and blood feeding in prehibernating females, *J. Med. Entomol.*, 15, 462, 1979.
90. **Christophers, S. R.,** The development of the egg follicle in anophelines, *Paludism*, 2, 73, 1911.

91. **Mer, G. G.,** Experimental study on the development of the ovary in *Anopheles elutus* Edw. (Diptera:Culicidae), *Bull. Entomol. Res.*, 27, 351, 1936.
92. **Eldridge, B. F.,** Diapause and related phenomena in *Culex* mosquitoes: their relation to arbovirus disease ecology, in *Current Topics in Vector Research,* Vol. 3, Harris, K. F., Ed., Praeger, New York, 1987, chap. 1.
93. **Lea, A. O.,** The control of reproduction by a blood meal: the mosquito as a model for vector endocrinology, *Acta Trop.*, 232, 112, 1975.
94. **Hagedorn, H. H.,** The control of vitellogensis in the mosquito, *Aedes aegypti, Am. Zool.*, 14, 11207, 1974.
95. **Hagedorn, H. H.,** The role of ecdysteroids in reproduction, in *Comprehensive Insect Physiology, Biochemistry, and Pharmacology,* Vol. 8, Kerkut, G. A. and Gilbert, L. I., Eds., Pergamon Press, New York, 1985, chap. 7.
96. **Borovsky, D.,** Control mechanisms for vitellogenin synthesis in mosquitoes, *Bioessays,* 1, 264, 1984.
97. **Fuchs, M. S. and Kang, S. H.,** Ecdysone and mosquito vitellogensis: a critical appraisal, *Insect Biochem.*, 11, 627, 1981.
98. **Clements, A. N.,** Hormonal control of ovary development in mosquitoes, *J. Exp. Biol.*, 33, 211, 1956.
99. **Gillett, J. D.,** Initiation and promotion of ovarian development in the mosquito, *Aedes (Stegomyia) aegypti (Linneaeus), Ann. Trop. Med. Parasitol.*, 50, 375, 1956.
100. **Larsen, J. R. and Bodenstein, D.,** The humoral control of egg maturation in the mosquito, *J. Exp. Zool.*, 140, 343, 1959.
101. **Lea, A. O.,** Some relationships between environment, corpora allata, and egg maturation in *Aedine* mosquitoes, *J. Insect Physiol.*, 9, 793, 1963.
102. **Lea, A. O.,** Egg maturation in mosquitoes not regulated by the corpora allata, *J. Insect Physiol.*, 15, 537, 1969.
103. **Lea, A. O.,** Regulation of egg maturation in the mosquito by the neurosecretory system: the role of the corpus cardiacum, *Gen. Comp. Endocrinol.*, 3, 602, 1972.
104. **Meola, R. W. and Lea, A. O.,** Humoral inhibition of egg development in mosquitoes, *J. Med. Entomol.*, 9, 99, 1972.
105. **Spielman, A., Gwadz, R. W., and Anderson, W. A.,** Ecdysone-initiated ovarian development in mosquitoes, *J. Insect Physiol.*, 17, 1807, 1971.
106. **Hagedorn, H. H., O'Connor, J. D., Fuchs, M. S., Sage, B., Schlaeger, D. A., and Bohm, M. K.,** The ovary as a source of 20-OH-ecdysone in an adult mosquito, *Proc. Natl. Acad. Sci. U.S.A.*, 72, 3255, 1975.
107. **Hagedorn, H. H., Shapiro, J. P., and Hanaoka, K.,** Ovarian ecdysone secretion is controlled by a brain hormone in an adult mosquito, *Nature (London),* 282, 92, 1979.
108. **Fallon, A. M., Hagedorn, H. H., Wyatt, G. R., and Laufer, H.,** Activation of vitellogenin synthesis in the mosquito, *Aedes aegypti,* by ecdysone, *J. Insect Physiol.*, 20, 1815, 1974.
109. **Borovsky, D. and Van Handel, E.,** Does ovarian ecdysone stimulate mosquitoes to synthesize vitellogenin?, *J. Insect Physiol.*, 25, 861, 1979.
110. **Borovsky, D. Thomas, B. F., Carlson, D. A., Whisenton, L. R., and Fuchs, M. S.,** Juvenile hormone and 20-hydroxy-ecdysone as primary and secondary stimuli of vitellogenesis in *Aedes aegypti, Arch. Inst. Biochem. Physiol.*, 2, 75, 1985.
111. **Denlinger, D. L.,** Hormonal control of diapause, in *Comprehensive Insect Physiology, Biochemistry, and Pharmacology,* Vol. 8, Kerkut, G. A. and Gilbert, L. I., Eds., Pergamon Press, New York, 1985, chap. 11.
112. **Detinova, T. S.,** On the influence of glands of internal secretion upon the ripening of the gonads and the imaginal diapause in *Anopheles maculipennis, Zool. Zh.*, 34, 291, 1945.
113. **Spielman, A.,** Effect of synthetic juvenile hormone on ovarian diapause of *Culex pipiens* mosquitoes, *J. Med. Entomol.*, 11, 223, 1974.
114. **Case, T. J., Washino, R. K., and Dunn, R. L.,** Diapause termination in *Anopheles freeborni* with juvenile hormone mimics, *Entomol. Exp. Appl.*, 21, 155, 1977.
115. **Mitchell, C. J.,** Diapause termination, gonoactivity, and differentiation of host-seeking behavior from blood-feeding behavior in hibernating *Culex tarsalis* (Diptera:Culicidae), *J. Med. Entomol.*, 5, 386, 1981.
116. **Meola, R. W. and Petralia, R. S.,** Juvenile hormone induction of biting behavior in *Culex* mosquitoes, *Science,* 209, 1548, 1980.
117. **Danilevskii, A. S. and Glinyananya, E. I.,** The dependence of gonotrophic cycle and imaginal diapause of blood-sucking mosquitoes on variation in daylength, *Uch. Zap. Leningr. Gos. Univ. Ser. Biol. Nauk,* 46, 34, 1958.
118. **McKenna, R. J. and Washino, R. K.,** Parity of fall-winter populations of *Anopheles freeborni* in the Sacramento Valley, California. A preliminary report, *Proc. Pap. Ann. Conf. Calif. Mosq. Control Assoc.*, 38, 94, 1970.

119. **Eldridge, F. B.,** Environmental control of ovarian development in mosquitoes of the *Culex pipiens* complex, *Science,* 151, 826, 1966.
120. **Bailey, C. L., Faran, M. E., Gargan, T. P., and Hayes, D. E.,** Winter survival of blood-fed and nonblood-fed *Culex pipiens* L., *Am. J. Trop. Med. Hyg.,* 31, 1054, 1982.
121. **Takahashi, M.,** Appearance of the gonotrophic dissociation in *Culex tritaeniorhynchus* under semi-experimental conditions, *Jpn. J. Sanit. Zool.,* 21, 18, 1970.
122. **Eldridge, B. F., Bailey, C. L., and Johnson, M. D.,** A preliminary study of the seasonal geographic distribution and overwintering of *Culex restuans* Theobald and *Culex salinarius* Coquillet (Diptera:Culicidae), *J. Med. Entomol.,* 9, 233, 1972.
123. **Oda, T. and Kuhlow, F.,** Seasonal changes in gonoactivity of *Culex pipiens pipiens* in northern Germany and its response to daylength and temperature, *Tropenmed. Parasitol.,* 25, 175, 1974.
124. **Mitchell, C. J.,** Differentiation of host-seeking behavior from blood-feeding behavior in overwintering *Culex pipiens* (Diptera:Culicidae) and observations on gonotropic dissociation, *J. Med. Entomol.,* 20, 157, 1983.
125. **Nelson, R. L.,** Parity in winter populations of *Culex tarsalis* Coquillett in Kern County, California, *Am. J. Hyg.,* 80, 242, 1964.
126. **Reisen, W. K., Meyer, R. P., and Milby, M. M.,** Overwintering studies on *Culex tarsalis* in Kern County, California: survival and the experimental induction and termination of diapause, *Ann. Entomol. Soc. Am.,* 79, 664, 1986.
127. **Reisen, W. K., Meyer, R. P., and Milby, M. M.,** Overwintering studies on *Culex tarsalis* in Kern County, California: temporal changes in abundance and reproductive status with comparative observations on *Cx. quinquefasciatus, Ann. Entomol. Soc. Am.,* 79, 677, 1986.
128. **Wilton, D. P. and Smith, G. C.,** Ovarian diapause in three geographic strains of *Culex pipiens* (Diptera:Culicidae), *J. Med. Entomol.* 22, 524, 1985.
129. **Arntfield, P. W., Gallaway, W. J., and Brust, R. A.,** Blood feeding in overwintering *Culex tarsalis* (Diptera:Culicidae) from Manitoba, *Can. Entomol.,* 114, 85, 1982.
130. **McIver, S. B.,** Sensilla of mosquitoes (Diptera:Culicidae), *J. Med. Entomol.,* 19, 489, 1982.
131. **Hocking, B.,** Blood-sucking behavior of terrestrial arthropods, *Ann. Rev. Entomol.,* 16, 1, 1971.
132. **Hudson, J. E.,** Follicle development, blood feeding, digestion, and egg maturation in diapausing mosquitoes, *Culiseta inornata, Entomol. Exp. Appl.,* 25, 136, 1979.
133. **Briegel, H. and Lea, A. O.,** Ecdysone, the ovarian hormone and intestinal proteases in mosquitoes, *Experientia,* 33, 813, 1975.
134. **Briegel, H. and Lea, A. O.,** Influence of the endocrine system on tryptic activity in female *Aedes aegypti, J. Insect Physiol.,* 25, 227, 1979.
135. **Anderson, A. W. and Harwood, R. F.,** Cold tolerance in adult female *Culex tarsalis* (Coquillett), *Mosq. News,* 26, 1, 1966.
136. **Buxton, P. A.,** Changes in the composition of adult *Culex pipiens* during hibernation, *Parasitology,* 27, 263, 1935.
137. **Service, M. W.,** Some environmental effects on blood-fed hibernating *Culiseta annulata* (Diptera:Culicidae), *Entomol. Exp. Appl.,* 11, 286, 1968.
138. **Van Handel, E.,** Metabolism of nutrients in the adult mosquito, *Mosq. News,* 44, 573, 1984.
139. **Van Handel, E. and Lea, A. O.,** Medial neurosecretory cells as regulators of glycogen and triglyceride synthesis, *Science,* 149, 298, 1965.
140. **Van Handel, E. and Lea, A. O.,** Control of glycogen and fat metabolism in the mosquito, *Gen. Comp. Endocrinol.,* 14, 381, 1970.
141. **Lea, A. O. and Van Handel, E.,** Suppression of glycogen synthesis in the mosquito by a hormone from the medial neurosecretory cells, *J. Insect Physiol.,* 16, 319, 1970.
142. **Schaefer, C. H. and Miura, T.,** Sources of energy utilized by natural populations of the mosquito, *Culex tarsalis,* for overwintering, *J. Insect Physiol.,* 18, 797, 1972.
143. **Reisen, W. K., Meyer, R. P., and Milby, M. M.,** Patterns of fructose feeding by *Culex tarsalis* (Diptera: Culicidae), *J. Med. Entomol.,* 23, 366, 1986.
144. **Harwood, R. F. and Takata, N.,** Effect of photoperiod and temperature on fatty acid composition of the mosquito, *Culex tarsalis, J. Insect Physiol.,* 11, 711, 1965.
145. **Tekle, A.,** The physiology of hibernation and its role in the geographical distribution of populations of the *Culex pipiens* complex, *Am. J. Trop. Med. Hyg.,* 9, 321, 1960.
146. **Schaefer, C. H. and Washino, R. K.,** Changes in the composition of lipids and fatty acids in adult *Culex tarsalis* and *Anopheles freeborni* during the overwintering period, *J. Insect Physiol.,* 15, 395, 1969.
147. **Schaefer, C. H. and Washino, R. K.,** Synthesis of energy for overwintering in natural populations of the mosquito, *Culex tarsalis, Comp. Biochem. Physiol.* 35, 503, 1970.
148. **Schaefer, C. H., Miura, T., and Washino, R. K.,** Studies on the overwintering biology of natural populations of *Anopheles freeborni* and *Culex tarsalis* in California, *Mosq. News,* 31, 153, 1971.
149. **Dingle, H.,** Migration strategies of insects, *Science,* 175, 1327, 1972.

150. **Gillett, J. D.,** *The Mosquito: Its Life, Activities, and Impact on Human Affairs,* Doubleday, Garden City, N.Y., 1972, chap. 5.
151. **Bailey, S. F., Eliason, D. A., and Hoffman, B. L.,** Flight and dispersal of the mosquito, *Culex tarsalis* Coquillett, in the Sacramento Valley of California, *Hilgardia,* 37, 73, 1965.
152. **Kliewer, J. W., Miura, T., and Chapman, H. C.,** Seasonal occurrence and physiology of *Culex tarsalis* in foothills of Fresno County, California, *Ann. Entomol. Soc. Am.,* 62, 13, 1969.
153. **Bailey, S. F. and Baerg, D. C.,** The flight habits of *Anopheles freeborni* Aitken, *Proc. Pap. Ann. Conf. Calif. Mosq. Control Assoc.,* 35, 55, 1967.
154. **Mail, A. G. and McHugh, R. A.,** Relation of temperature and humidity to winter survival of *Culex pipiens* and *Culex tarsalis, Mosq. News,* 21, 252, 1961.
155. **Blackmore, J. S. and Winn, J. F.,** A winter isolation of western equine encephalitis virus from hibernating *Culex tarsalis* Coq., *Soc. Exp. Biol. Med. Proc.,* 91, 146, 1956.
156. **Chapman, H. C.,** Abandoned mines as overwintering sites for mosquitoes, especially *Culex tarsalis* Coq., in Nevada, *Mosq. News,* 21, 234, 1961.
157. **Mitchell, C. J.,** Winter survival of *Culex tarsalis* (Diptera:Culicidae) hibernating in mine tunnels in Boulder County, Colorado, USA, *J. Med. Entomol.,* 16, 482, 1979.
158. **Kuchlein, J. H. and Ringelberg, J.,** Further investigations on the distribution of hibernating *Culex pipiens pipiens* L. (Diptera, Culicidae) in artificial marlcaves in South Limburg (Netherlands), *Entomol. Exp. Appl.,* 7, 25, 1964.
159. **Karabatsos, N.,** *International Catalog of Arboviruses,* 3rd ed., Centers for Disease Control, U.S. Public Health Service, Ft. Collins, Colo., 1985, 71.
160. **Bohart, R. M. and Washino, R. K.,** *Mosquitoes of California,* 3rd ed., University of California Press, Berkeley, 1978, 20.
161. **Dahl, C. and White, G. B.,** Culicidae, in *Limnofauna Europaea,* 2nd ed., Illes, J., Ed., Gustav Fischer Verlag, Stuttgart, 1978, 392, 394.
162. **Horsfall, W. R.,** *Mosquitoes, Their Bionomics and Relation to Disease,* Ronald Press, New York, 1955, 103, 423, 426.
163. **Dobrotworsky, N. V.,** *The Mosquitoes of Victoria,* Melbourne University Press, Melbourne, 1965, 48, 117, 163.
164. **Harmston, F. C. and Lawson, F. A.,** *Mosquitoes of Colorado,* U.S. Department of Health, Education and Welfare, Washington, D.C., 1967, 26.
165. **Knight, K. L. and Wonio, M.,** *Mosquitoes of Iowa (Diptera:Culicidae),* Spec. Rep. No. 61, Iowa State Deptarment of Zool. and Entomology, Iowa City, 1969, 34.
166. **Dumbleton, L. J.,** Developmental stages and biology of *Culiseta tonnoiri* (Edwards) and a note on *Culex pervigilans* Bergroth (Diptera:Culicidae), *N.Z. J. Sci.,* 8, 137, 1965.
167. **Barrero Oro, J. G., de Garre, M. E. G., Arechabala, J. M., and Lord, R. D.,** Conversiones serologicas para el virus de encefalitis equina del oeste en dos enfermos del brote de Viedma, *Terceras Journadas Argentinas de Microbiologia, Tucuman, Libro de Resumenes de Communicaciones Libres,* 1973, 11.
168. **Mitchell, C. J., Monath, T. P., Sabattini, M. S., Cropp, C. B., Daffner, J. F., Calisher, C. H., Jakob, W. L., and Christensen, H. A.,** Arbovirus investigations in Argentina, 1977—1980. II. Arthropod collections and virus isolations from Argentine mosquitoes, *Am. J. Trop. Med. Hyg.,* 34, 945, 1985.
169. **Jupp, P. G.,** Preliminary studies on the overwintering stages of *Culex* mosquitoes (Diptera:Culicidae) in the highveld region of South Africa, *J. Entomol. Soc. S. Afr.,* 32, 91, 1969.
170. **Jupp, P.G.,** Further studies on the overwintering stages of *Culex* mosquitoes (Diptera:Culicidae) in the highveld region of South Africa, *J. Entomol. Soc. S. Afr.,* 32, 91, 1969.
171. **Van Pletzen, R. and Van der Linde, T. C. de K.,** Studies on the biology of *Culiseta longiareolata* (Macquart) (Diptera:Culicidae), *Bull. Entomol. Res.,* 71, 71, 1981.
172. **Tesh, R. B.,** Transovarial transmission of arboviruses in their invertebrate vectors, in *Current Topics in Vector Research,* Vol. 2, Harris, K. F., Ed., Praeger, New York, 1984, chap. 3.
173. **Linthicum, K. J., Davies, F. G., Kairo, A., and Bailey, C. L.,** Rift Valley fever virus (family Bunyaviridae, genus *Phlebovirus*). Isolations fom Diptera collected during an inter-epizootic period in Kenya, *J. Hyg. (Cambridge),* 95, 197, 1985.
174. **Reeves, W. C., Bellamy, R. E., and Scrivani, R. P.,** Relationships of mosquito vectors to winter survival of encephalitis vectors. I. Under natural conditions, *Am. J. Hyg.,* 67, 78, 1958.
175. **Rosen, L., Shroyer, D. A., and Lien, J. H.,** Transovarial transmission of Japanese encephalitis virus by *Culex tritaeniorhynchus* mosquitoes, *Am. J. Trop. Med. Hyg.,* 29, 711, 1980.
176. **Bailey, C. L., Eldridge, B. F., Hayes, D. E., Watts, D. M., Tammariello, R. F., and Dalrymple, J. M.,** Isolation of St. Louis encephalitis virus from overwintering *Culex pipiens* mosquitoes, *Science,* 199, 1346, 1978.
177. **Chippaux, A., Rageau, J., and Mouchet, J.,** Hibernation de l'arbovirus Tahyna chez *Culex modestus* Fic. en France, *C. R. Acad. Sci. Paris,* 270, 1648, 1970.

178. **Jaenson, T. G. T., Niklasson, B., and Henriksson, B.,** Seasonal activity of mosquitoes in an Ockelbo disease endemic area in central Sweden, *J. Am. Mosq. Control. Assoc.*, 2, 18, 1986.
179. **Morris, C. D., Zimmerman, R. H., and Magnarelli, L. A.,** The bionomics of *Culiseta melanura* and *Culiseta morsitans dyari* in Central New York State (Diptera:Culicidae), *Ann. Entomol. Soc. Am.*, 69, 101, 1976.
180. **Lea, A. O.,** personal communication
181. **Mitchell, C. J. and Briegel, H.,** unpublished data.

Chapter 8

DIAPAUSE IN TICK VECTORS OF DISEASE

Daniel E. Sonenshine

TABLE OF CONTENTS

I. INTRODUCTION

Diapause is a state of interrupted activity, developmental processes, or reproductive activity which may occur at some period during the life cycle. This phenomenon is widespread among arthropods and many other invertebrates. During this apparently dormant period, the animal remains in a more or less arrested state, with greatly reduced metabolism and absence of locomotor activity. Diapause is but one of the many ecophysiological responses by which the development, growth, and reproduction of an arthropod is synchronized with the seasonal changes in its environment, as well as the availability of food and other essential resources. Once the animal enters the diapause state, it remains in this condition for an extended period, often weeks or months, until new environmental stimuli induce its termination. In this respect, diapause is distinctly different from daily or other short-term periodic rhythms.

Ticks and other *Acari* use diapause in much the same way as do insects — to reduce expenditure of energy during environmentally unfavorable periods, i.e., when food is scarce or when other biotic parameters affecting their survival are unfavorable. This physiological mechanism is common in both the Ixodidae, or hard ticks, and the Argasidae (soft ticks). The evolution of this phenomenon may have been stimulated by the drastic climatic changes that occurred during the transition from the Mesozoic to the early Cenozoic era of the earth's history. Hoogstraal and Chung-Kim[1] suggest that the ancestral Ixodidea evolved as obligate parasites of Paleozoic or early Mesozoic reptiles. During this period, characterized by warm, humid climatic conditions and an abundance of large glabrous animals, hosts were readily available throughout the year. Presumably, these ancestral ticks lacked well-defined seasonal adaptations and this simple pattern is still retained in some tropical and subtropical species, e.g., the one-host tick, *Boophilus calcaratus,* (a vector of Crimean-Congo hemorrhagic fever, CCHF) parasitizing cattle in semitropical areas of the southern U.S.S.R.[2,3] Following the radical changes in the early Cenozoic era and the replacement of the reptiles by birds and mammals as the dominant vertebrate life forms, new adaptations for successful host finding and developmental strategies evolved. Undoubtedly, diapause was one of these adaptations without which the Ixodidea would have remained confined to a few isolated tropical biotopes, if they had survived at all.

Several different types of diapause occur in ticks. Perhaps the most common is host-seeking diapause (behavioral diapause), a phenomenon in which individuals no longer seek and attack vertebrate hosts. Ticks in this stage will refuse opportunities for feeding even when placed in direct contact with hosts. Less common is morphogenetic diapause, a physiological state in which immatures and females feed but do not proceed with normal developmental processes or oviposition. To clarify the different physiological responses involved in these responses, the term morphogenetic diapause will be restricted to cessation of developmental processes, e.g., molting, maturation of reproductive organs, etc., and a new term, ovipositional diapause, will be used to describe those responses in which replete, mated females delay egg laying.

The induction and termination of the various types of diapause are believed to be initiated by exogenous stimuli expressed over an extended period, e.g., gradually increasing daylength or long-term absence of hosts. Many species are influenced by changing photoperiod, or the combined effects of directional changes in photoperiod, total incident solar radiation, temperature, and possibly other abiotic factors. In other species, particularly those restricted to migratory hosts such as birds or bats, diapause is correlated with the time of departure and return of their hosts to the nesting site.

This review examines the varied types of diapause in the two major families of ticks, and the role that diapause plays in their survival strategies. In addition, the biotic and abiotic factors responsible for inducing and terminating diapause, and the physiological mechanisms that control this process are examined.

II. BEHAVIORAL OR HOST-SEEKING DIAPAUSE

The most common type of diapause in ticks is host-seeking diapause. Ticks entering this physiological state do not exhibit questing, or host-seeking behavior. Many diapause immediately after emergence. In those species that climb vegetation during questing, i.e., the host-seeking phase, unsuccessful individuals descend the stems to the leaf litter and duff layer of the forest floor, or the vegetative mat of the ecotone and meadow habitats. As environmental conditions become increasingly less favorable, they penetrate into the soil, where insulation against temperature and humidity extremes is greatest. Here they remain until more favorable environmental conditions (discussed later) return. Ticks in this form of diapause are inactive. However, if aroused, they are capable of locomotion. Nevertheless, they will not attack hosts that are present nearby, and most individuals will refuse to feed when placed on a warm-blooded host.

To understand the principles guiding this type of diapause, we examine several examples from the Ixodidae in depth, specifically (1) the American dog tick, *Dermacentor variabilis* (Say), (2) species of the genus *Ixodes,* especially the sheep tick, *Ix. ricinus* L., and (3) the winter tick, *D. albipictus*. Comparisons with other argasid and other ixodid tick species, as well as with insects, is noted where appropriate.

A. Ixodidae: Diapause in Response to Declining Photoperiod and Temperature

1. Metastriata: D. variabilis (American Dog Tick)

The American dog tick provides an example of the type of diapause response that is widespread among most genera of hard ticks (Ixodidae), i.e., behavioral diapause induced by gradually diminishing daylength and declining temperatures. The only virus isolated from *D. variabilis* is sawgrass *(Rhabdovirus,* Rhabdoviridae).[3] This species is distributed throughout most of the eastern U.S. and parts of southern Canada, extending from Florida and the Gulf Coast states, northward as far as latitude 44°N in southwestern Nova Scotia. Although disjunct populations occur in several areas of the north central U.S., southern Canada, and even the westernmost states, the distribution of continuous breeding populations is generally consistent with the eastern deciduous forest biome.[4] In the northern part of the tick's range, this distribution is believed to be limited by mid-winter temperatures, especially the 30°F January isotherm. This is thought to be due to the critical sensitivity of the diapausing larval and adult stages to periods of sustained air temperatures below 0°C (although the effects on the soil temperatures in the tick's microclimate may be considerably higher). A 1-year life cycle is characteristic of tick populations found in the southern part of this range, whereas a 2-year life cycle is common in the northernmost latitudes.[5]

D. variabilis is a seasonal species. Larvae emerge from their overwintering diapause to feed on small mammals, molt into nymphs, and feed again on small mammal hosts, almost always within the same spring or summer season. Molting of the fed nymphs follows, again almost always within the same spring or summer season. However, adult host-seeking behavior is determined by the direction of changing environmental conditions, particularly photoperiod, leading to either vigorous questing behavior or diapause. The same may be true for larvae, although most individuals in this life stage hatch so late in the summer or early autumn that few of them choose to commence host-seeking activity.

Diapause is the biological response that enables these ticks to synchronize their feeding, development, and reproduction with the periods of optimal climatic conditions, thereby avoiding lethal environmental extremes. The first evidence of this response was reported by Smith and Cole,[6] who found that larval and nymphal (but not adult) *D. variabilis* exposed to increasing photoperiods exhibited a greater willingness to feed than ticks exposed to declining photoperiods. They suggested photoperiodic induction of diapause, since air temperatures in the field, when larval activity ceased, were much warmer than in the spring, when larval activity began again. Moreover, they noted that by experimentally manipulating the photoperiod prior to feeding of the immatures on mice, ticks exposed to long photoperiods were much more aggressive, i.e., more willing to feed than ticks exposed to short photoperiods. Subsequently, Atwood and Sonenshine[7] suggested that total solar energy received at the ground surface provided the essential environmental control that terminated diapause in the spring, while declining photoperiod influenced its inception in the later summer or autumn. Correlating the onset of larval tick activity with observed solar radiation near Richmond, Va., they noted that in the spring, larval host-seeking activity commenced when the incident solar radiation received in the microhabitat exceeded 200 g cal/cm^{-2}/day continuously over a period of many days or weeks (Figure 1). This effect was, of course, subject to local habitat conditions, e.g., the degree of light penetration through the forest canopy, absence of snow cover, and perhaps other unknown factors. The specific part of the radiant energy spectrum which is most effective in exciting host seeking is unknown. Environmental data suggest that in Holarctic regions, the shorter wavelengths, i.e., the ultraviolet (UV) and blue green, would constitute a greater proportion of the radiant energy spectrum in the spring, when these wavelengths would be less subject to refraction than the red and near infrared (IR) wavelengths. Experiments by Sonenshine[8] on larval feeding in response to exposure to specific radiant energy spectra support the hypothesis that these shorter wavelengths are most effective in stimulating feeding. Attachment by *D. variabilis* larvae exposed to UV (long wave) was almost double that of larvae exposed to a broad energy spectrum, and almost three times that of ticks exposed to IR radiation. No photoperiod-associated response was found with nymphal ticks. The threshold initiation of larval host-seeking activity was found to occur at least 2 weeks earlier in Virginia than in Massachusetts and 4 to 6 weeks earlier than in Nova Scotia, where similar studies on the ecology of *D. variabilis* have been done.[5-10] In the absence of snow cover, some larvae may emerge to feed in the southern parts of the species range in January or February. However, sustained larval host-seeking activity does not commence until the solar energy threshold has been exceeded continuously for many days, usually not until March. Once it is initiated, host-seeking activity continues unabated over a 3 to 4 month period, with no apparent diminution in response to day-to-day variations in incident solar energy, until all of the available larvae feed or die (Figure 2). Host-seeking activity of adult *D. variabilis* is also controlled by diapause. In Virginia, adult ticks emerge from their overwintering diapause to seek hosts in April or early May, and increasing numbers of adults can be captured on flags until the peak in late June or early July (Figure 3). Atwood and Sonenshine[7] demonstrated that sustained adult tick host-seeking activity commenced when the incident solar energy received at ground level exceeded 469 g cal/cm^{-2}/day, and daylength exceeded 13.2 hr (Figure 4). Sustained host-seeking activity continued until daylength declined to 13.3 hr, at which time the incident solar energy was 313 g cal/cm^{-2}/day. These authors suggested that short wavelength radiation and photoperiod represent the dominant factors responsible for inducing and/or terminating diapause in *D. variabilis* adults. Thus, adult activity was largely restricted to the period when the photoperiod exceeded approximately 13 hr/day and the average incident solar radiation intensity exceeded 300 g cal/cm^{-2}/day. No relationship was found betweeen tick activity and air temperature changes.

Additional evidence of the effects of photoperiod and solar radiation on host-seeking behavior in this species was described by Sonenshine,[11] who noted that radioisotope tagged larvae reared under laboratory conditions continued questing and feeding behavior when released into the field in August. The larvae had been exposed to long daylength photoperiods in an incubator prior to their release, using daylight fluorescent lamps that produced approximately 75 fc light intensity. Thus, these larvae had been conditioned to subthreshold energy intensities, and were most probably in diapause when they were brought to the field. Release in the field exposed the laboratory-reared ticks to intense solar radiation, approxi-

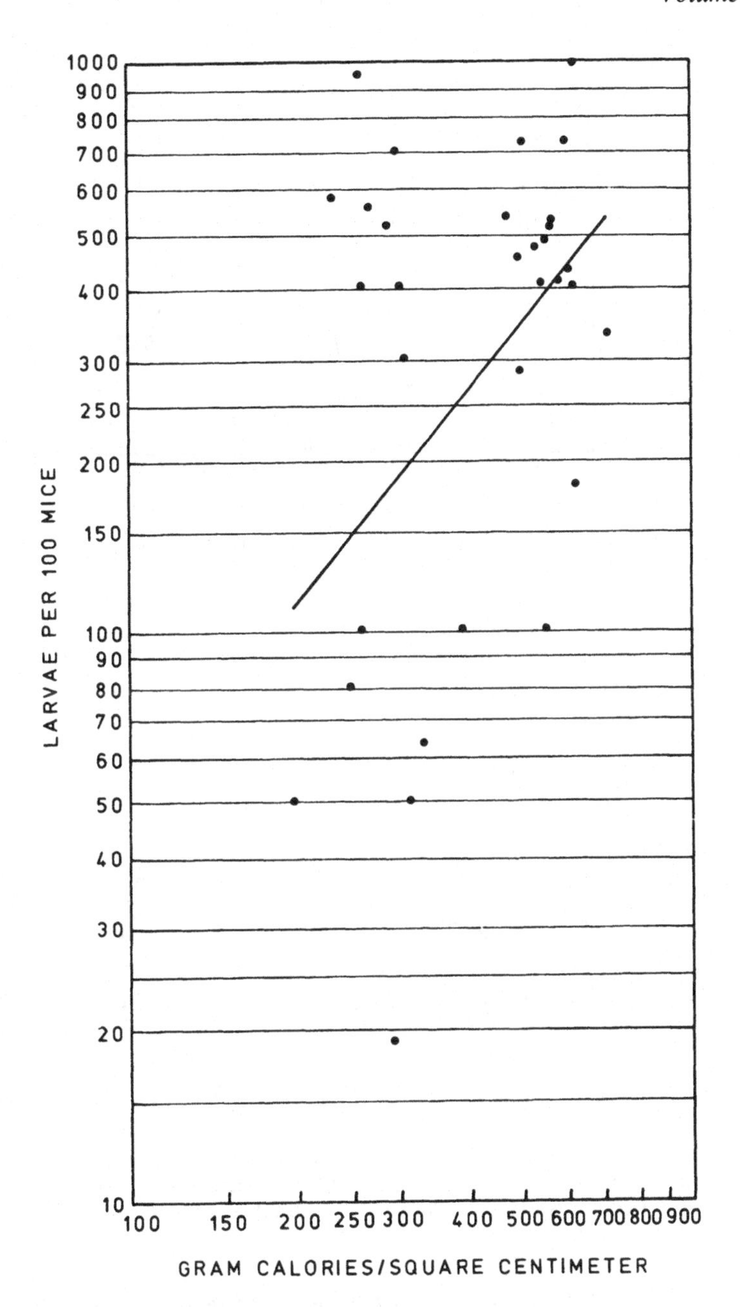

FIGURE 1. Relationship between *D. variabilis* larval host seeking activity and increasing solar radiation intensity. (From Atwood, E. L. and Sonenshine, D. E., *Ann. Entomol. Soc. Am.*, 60, 354, 1967. With permission.)

mately 300 g cal/cm^{-2}/day, or far in excess of their previous experience. When released, they were allowed to escape into the surface vegetation, which placed them well above the vegetative mat of the meadow or leaf litter of the forest floor, i.e., exposed to intense radiation. These larvae responded with aggressive host-seeking activity, which continued for approximately 6 weeks. During this particular late summer-autumn period, the laboratory-reared larvae constituted 89.6% of all larvae found on hosts. Clearly, the laboratory-reared larvae responded to the dramatic changes in their environment by terminating diapause. In contrast, few wild-caught larvae were found on the same hosts as the laboratory-reared ticks,

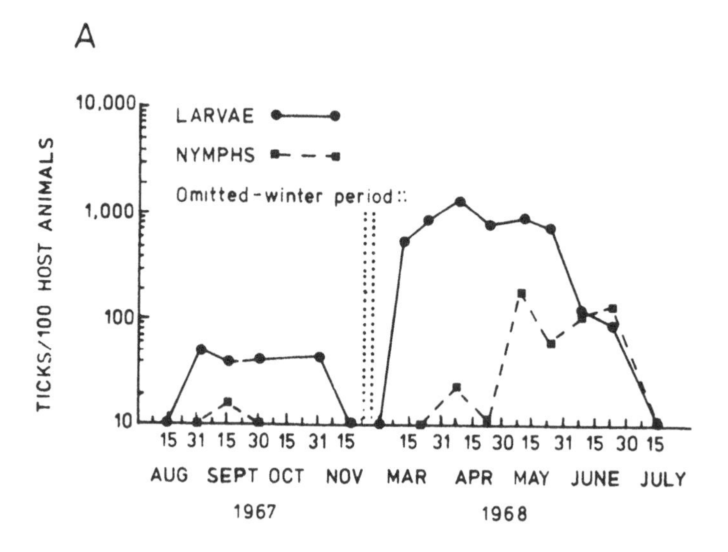

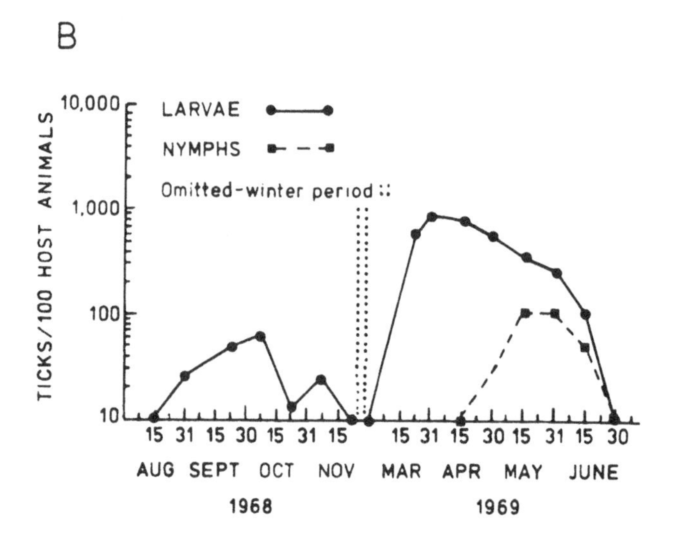

FIGURE 2. Seasonal activity of *D. variabilis* larvae and nymphs at a study site near Richmond, Va., comparing the fall and spring activity periods in two successive years. (From Sonenshine, D. E., *Ann. Entomol. Soc. Am.*, 65, 1164, 1972. With permission.)

despite the fact that female repletion and egg production would have introduced a massive new cohort of freshly hatched, vigorous young individuals. In fact, the number of wild-caught larvae on hosts during the summer and autumn months was less than 10% of the number found on such hosts during the spring months.[12] When one examines these statistics, summarized by Sonenshine,[11] for egg production, hatching, and feeding activity (Table 1), it is apparent that something has intervened to prevent this huge new population of young, physiologically competent larvae from seeking hosts. This summer-autumn population (pre-diapause population) must be larger than the spring (postdiapause) population, which consists of individuals that have survived many months of starvation, cold, and dessication. Moreover, no new larvae can be added to the population during the late autumn, winter, or early spring months. Nevertheless, the abundance of larvae on hosts in the spring is about tenfold greater than on hosts in autumn. Thus, the only explanation for these events is diapause.

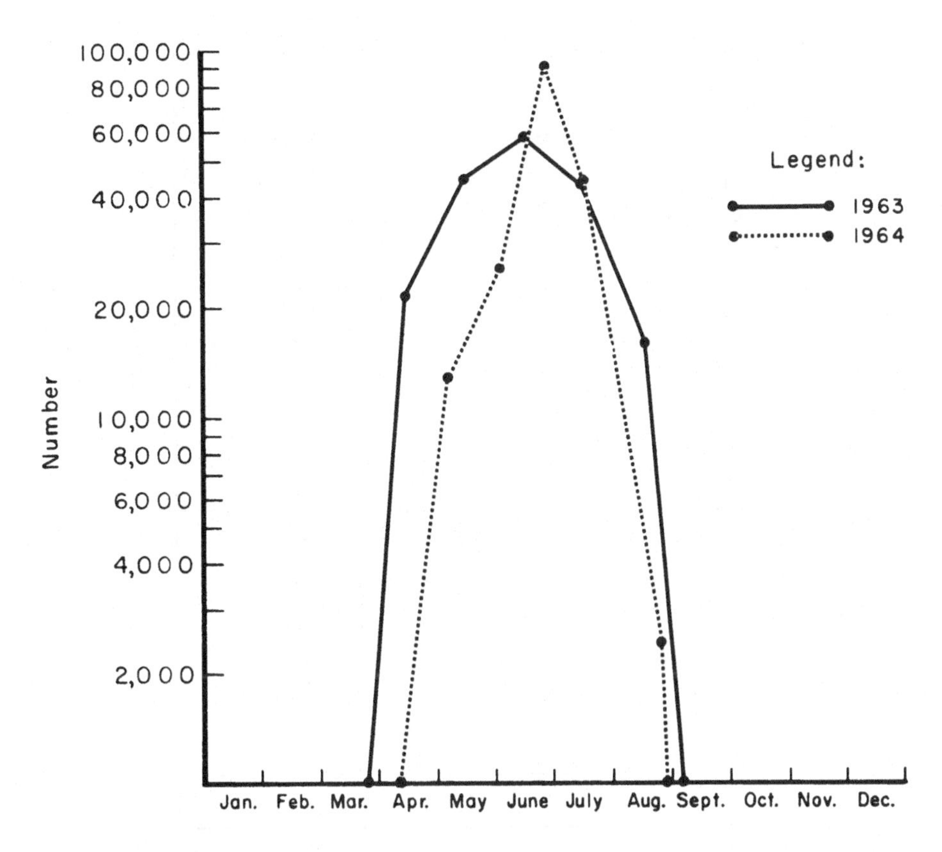

FIGURE 3. Seasonal activity of *D. variabilis* adults at a study site near Richmond, Va., in two successive years. (From Atwood, E. L. and Sonenshine, D. E., *Ann. Entomol. Soc. Am.*, 60, 354, 1967. With permission.)

The effects of diapause and the specific thresholds that are used to initiate or terminate this response are to produce a 1-year cycle, from larva to adult, in the southern part of the species range, vs. a 2-year life cycle in the extreme northern part of the species range. In Virginia, larvae emerge to commence host seeking in mid-March, and the seasonal peak is reached in mid-April in most years[11] (Figure 2). Nymphal ticks are rarely seen at this time. However, nymphs appear in increasing numbers during April, and the seasonal peak occurs in May. Thus, in the relatively mild climate and long daylength that characterizes springtime conditions in this region, larval and nymphal development can proceed rapidly. Consequently, nymphal molting yields large numbers of young adults in May or June. Increasing daylength on approaching the summer solstice and maximum incident solar energy excites host-seeking behavior by these vigorous young adults, and these join the questing postdiapause individuals that survived from the preceding year in a steady surge toward the seasonal activity peak in late June or July. Thus, adult host-seeking activity is unimodal. Host-seeking by the overwintering adults, representing less than 10% of the adult population, is quickly surpassed by the influx of young adults, almost all of which engage in questing behavior. Later in the summer, after the seasonal peak, unsuccessful adults enter diapause. That adults are able to switch from questing to diapause and back to questing is supported by the capture of individuals marked in the preceding year. Survival of these ticks is limited by the dessication stress and depleted energy reserves resulting from unsuccessful questing, and most are unable to survive despite the relatively mild winter conditions typical of this region.

In Nova Scotia, at the northern edge of the species range, *D. variabilis* has a 2-year cycle.[5] In this cool, moist climate moderated by the Gulf Stream, larvae emerge from their

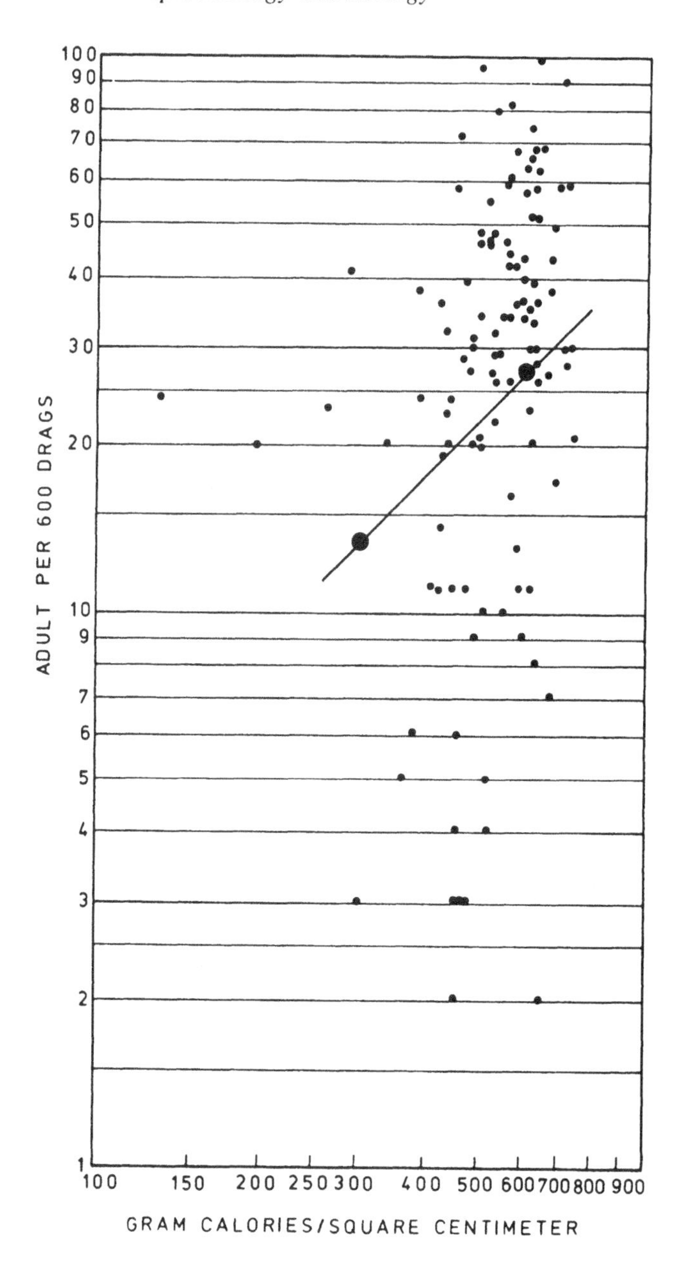

FIGURE 4. Relationship between *D. variabilis* adult host-seeking activity and increasing solar radiation intensity. (From Atwood, E. L. and Sonenshine, D. E., *Ann. Entomol. Soc. Am.*, 60, 354, 1967. With permission.)

winter diapause in May (or rarely, in April). With the increasingly moderating temperatures, molting occurs rapidly and the nymphal feeding peak follows soon afterwards in June or July. Adults, however, do not emerge until late in the summer, when declining photoperiods and reduced incident solar energy induces diapause. Thus, the delayed onset of larval tick activity results in a generation of adults that will enter diapause without experiencing the stress of host-seeking activity. This adult population functions as a cohort, i.e., a similarly conditioned group of individuals exhibiting uniform behavioral responses, and emerges to commence host seeking in the following May. Although precise figures on overwinter

Table 1
SIMPLIFIED LIFE TABLE FOR THE AMERICAN DOG TICK, *DERMACENTOR VARIABILIS*, IN MIXED FOREST-OLD FIELD HABITAT

Life stage	Max abundance unfed tick/acre ± 2 S.E.	Total yield fed ticks/acre	Feeding success (%)	Dominant survival mortality/ factors
Eggs (summer)	161,839 ± 47,064	—	—	Desication
Prediapause larvae (summer-fall)	129,497 ± 37,734	—	—	Abiotic factors
Postdiapause larvae (Spring)	39,687 ± 11,657	797	1.76	Host numbers
Nymphs (spring)	670 ± 429	59(?)	8.80(?)	Host numbers/abiotic factors
Adults (spring-summer)	596 ± 132	74	13.00	Host numbers/abiotic factors
Females (spring-summer)	285 ± 66	26	9.10	Host numbers/abiotic factors

From Sonenshine, D.E., *Ann. Entomol. Soc Am.*, 65, 1164, 1972. With permission.

survival are lacking, it is probably very high since population densities of over 20,000/ha were reported. These postdiapause adults feed and produce eggs, virtually all of which hatch late in the summer, producing the next cohort of diapausing larvae for the following spring.

In Massachusetts and eastern Long Island, adults may occasionally exhibit a bimodal activity pattern. McEnroe[9] suggested that this is due to the existence of two distinct cohorts. Cohort 1, adults emerging from their overwinter diapause, commence activity in late April or May. This is the dominant period of adult tick activity and, in some years, it is the only activity period. Cohort 2 represents young adults that emerged from the nymphal molt in early summer, early enough to commence host-seeking activity. The remainder of this second cohort, emerging later in the summer, enter diapause. If weather conditions allow early larval feeding, sufficient numbers of cohort 2 adults are produced early in the summer so as to generate a significant second peak of adult tick activity. McEnroe[9,10] proposed that soil temperature rather than photoperiod was the dominant variable affecting the termination of diapause. According to this author, questing activity is driven by the tick's water sorption capabilities, which are reduced or nonfunctional at low temperatures. According to McEnroe, 17°C constitutes the threshold for sustained water sorption activation, and ticks do not emerge from the saturated environment of the soil and vegetative mat until they are able to sustain their water balance. No similar explanation of diapause termination has been reported in any other arthropod group, and it appears doubtful that this is the dominant factor in ticks.

2. Prostriata: Ixodes ricinus and Other Species of the Genus Ixodes

Ix. ricinus is the principal vector of louping ill in the U.K. and Ireland and of tick-borne encephalitis (TBE; also called central European encephalitis, CEE) throughout most of the European continent.[3,13] This tick and closely related species also transmit Tettnang, Tribec, Uukuniemi, and Eyach viruses.[14] In the U.S.S.R., *Ix. ricinus* and the Siberian Taiga tick, *Ix. persulcatus* are the principal vectors of Russian spring summer encephalitis (RSSE), as well as vectors of Kemerovo virus, Uukuniemi, and possibly Omsk hemorrhagic fever[3,13,14] All three viral agents are representatives of the genus *Flavivirus*. *Ix. ricinus* is also implicated in the transmission of CCHF *(Nairovirus,* Bunyaviridae).[3] In the U.S. and Canada, *Ix. cookei* has been implicated as one of several vector of Powassan virus *(Flavivirus,* Flaviviridae).[13]

The genus *Ixodes* provides enormous diversity in habitats, patterns of host-seeking behavior, gonadal maturation, and mating behavior. In general, species of this genus utilize the two distinct host-finding strategies common in ticks, (1) the nest or burrow inhabiting, or nidiculous type, and (2) the open landscape, ambush, or non-nidiculous type. Species of the subgenus *Pholeoixodes* illustrate the former, while species of the subgenus *Ixodes* generally follow the latter. Thus, *Ixodes (Pholeoixodes) rugicollis,* a nest- or burrow-inhabiting parasite of foxes, mustellids, and other small carnivores in Europe, does not show evidence of seasonal periodicity or ovipositional diapause.[15] Auburt[15] suggests that this species has a nonsynchronous life cycle, since there is no advantage to a seasonal pattern: "in the protective environment of the nest, climatic conditions are not a synchronizing factor (as they are for non-*Pholeoixodes* ticks)." This generalization, however, cannot be applied to all species of the subgenus; e.g., *Ix. (Pholeoixodes) trianguliceps,* a burrow-inhabiting parasite of small mammals, has a well-developed seasonal cycle.[16] No viruses have been isolated from either of these burrow-inhabiting parasites.

The sheep tick, *Ix. ricinus,* ranges from Ireland, the U.K., and southern Scandinavia southward to Spain, Portugal, Algeria, and Morocco and eastward across central and eastern Europe and the European U.S.S.R. to northern Iran.[17] According to Hoogstraal,[17] *Ix. ricinus* has a 4-year life cycle in the northern areas of its range, but this is reduced to 3 or even 2 years in the warmer climates. The distribution of this tick is strongly influenced by the need for a microclimate with near saturated air, usually provided by a dense mat of thick, rotting leaf mold and understory vegetation in the forest or the dense mat of roots, rotting stems and detritus below the luxuriant growth of ferns, rushes, and grasses common in the unimproved pastures.[18]

The classical studies of Milne,[19,20] Lees and Milne,[21] and others in the U.K. demonstrated a bimodal seasonal pattern for this species.[22] In England and Wales, adults engage in host-seeking behavior in the spring, with peak abundance usually in May. Following a midsummer decline, a second phase of host-seeking activity occurs in the late summer and early autumn, with the seasonal peak usually in September. The spring activity peak is almost always much greater than the late summer-autumn activity peak. Further north, in Scotland, seasonal activity is delayed, and a unimodal activity pattern occurs with the peak in July. The nymphal peaks generally follow the same pattern as that observed with the adults, and the activity peaks for the two life stages may fall within a few days of one another. The larval activity peak, however, is considerably delayed, typically appearing in late May or early June; here again, a second peak occurs in early fall. Host data for sheep generally reflect this activity pattern, although collections from rabbits, birds, and other wildlife show some deviations. Thus, adults, nymphs, and possibly even larvae diapause and overwinter with varying degress of success. The spring adult activity generates numerous eggs which may hatch during the summer, or alternatively, may undergo morphogenetic diapause with hatching delayed until the following spring or summer. Although opinions differ as to whether larvae also exhibit behavioral diapause, delayed development of the eggs appears to be the dominant factor determining the appearance of larval activity. In the milder southern climates, eggs produced by autumn-fed adults diapause (morphogenetic diapause) over the winter[23,24] and their hatching leads to the dominant summer activity peak, while adults fed earlier lay eggs that may hatch without diapause[23] or hatch the following spring.[24] In the harsher northern or continental climates, only a single peak may occur. Fed larvae molt in middle or late summer, and nymphs follow the same activity pattern as the larvae, overwinter in either the unfed state or perhaps even as engorged nymphs. Surviving nymphs reappear in the spring, feed and molt to young adults which diapause and overwinter before continuing activity in the following spring. Thus, most *Ix. ricinus* require 2 years to complete the life cycle from larva to unfed adult, and many require a third year for adult feeding, mating, and reproduction.

Diapause is believed to provide the physiological response that determines whether individuals seek host or delay their host-seeking activity. Evidently, all three life stages can and do diapause. The response reflects a short daylength response pattern, i.e., declining photoperiod and diminshing incident solar radiation reduces activity and induces the diapause state. The reverse, i.e., increasing daylength, terminates diapause. Other explanations have been offered, but as Arthur[22] has noted, the evidence presented to support them appears inadequate or inconsistent. Moreover, the response patterns illustrated by populations of this species reflect the host-seeking diapause characteristic of temperate zone ticks.

In Austria, Pretzmann et al.[25-27] also described a bimodal seasonal activity pattern for *Ix. ricinus* in a focus of TBE, where the ticks were found to feed on a wide spectrum of mice, voles, squirrels, and other small mammal hosts (Figure 5). This study, carried out over a 3-year period, demonstrated that the two activity peaks for larvae occurred in June and September, while those for nymphs occurred in May and September. In both cases, the spring activity peak was the largest. Immature tick activity was greatly reduced during midsummer, and all activity terminated in November. The adult activity peaks paralleled those for the nymphs. The same bimodal activity patterns were reflected in the attack rates for *Ix. ricinus* on mice and voles examined in the study site. Thus, the seasonal activity patterns observed in this locality resembled that described in England and Wales, with the same bimodal pattern for all three life stages, despite the disparity in geographic location, altitude, and climatic conditions. It is also noteworthy that these similarities occurred despite the great difference in habitat types between the two localities. The foci studied in England and

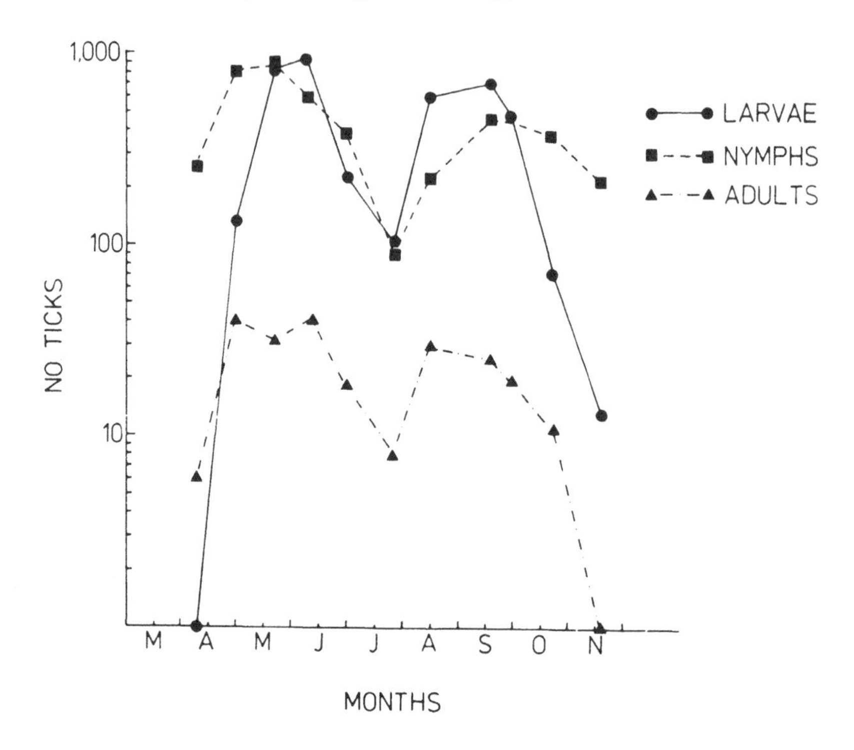

FIGURE 5. Seasonal activity of the different life stages of the European sheep tick, *Ix. ricinus*, as determined by dragging at a focus of tickborne encephalitis in lower Austria. (From Loew, J. et al., *Zentralbl. Bakteriol. Parasitenkd. Infektionskr. Hyg.*, 194, 133, 1964. With permission.)

Wales comprised unimproved moorland grazing areas, all near sea level, whereas the focus used in Austria was at an altitude of 600 to 800 m and was dominated by spruce-pine forest communities. Pretzmann et al.[25] concluded that the biphasic periodicity they described indicated that all three life stages diapause. Furthermore, computing overwinter success from their drag captures, they concluded that nymphs had the greatest overwintering success, 20%, while larvae had the least success, only 5% (no figures were given for the adults).

In Ireland, Gray[24] observed distinct differences in larval seasonal activity which were believed to be related to the period of activity of the preceding parental generation, expressed via morphogenetic diapause of their eggs. A relatively small spring activity peak was derived from spring-fed adults in the preceding year. In contrast, the much larger mid-summer (July) seasonal peak, according to Gray,[24] is derived from adults fed during the preceding autumn. Larvae fed in July molt to nymphs, whereas those fed in August undergo developmental diapause. Thus, developmental rather than behavioral diapause may be the important diapause factor in the development of this species in Ireland. Gray[67] suggests that the higher temperatures and solar radiation in the open British and Irish habitats results in earlier and more synchronized activity than in the harsher climates and forested habitats of this species in other localities.

Experimental evidence in support of the hypothesis that *Ix. ricinus* enters diapause in response to gradually declining daylength was described by Belozerov.[28-30] Belozerov[28] demonstrated that exposure of nymphs to long daylength periods (20 L:4 D) resulted in greatly increased attachment success: 60 and 65% in 4- and 7-month-old nymphs, respectively, vs. only 10 and 27% success in similarly aged nymphs exposed to short daylength photoperiods (12 L:12 D). This effect is temperature dependent, since increases in temperature led to aggressive feeding behavior in both treatment groups. However, transferring nymphs from short- to long-day photoperiods, both at 18°C, led to gradually increasing aggressiveness, with greatly increased attachment success; thus, the attack rate increased from only 10% before the transfer to 96% over a 45-day period after the transfer. Combining increased photoperiod with higher temperature accelerated the switch to aggressive behavior. Increasing the temperature without also increasing the photoperiod was much less effective. However, transfer from long- to short-day photoperiods had no effect. Belozerov concluded that aggressive host-seeking and feeding behavior, once initiated, was irreversible. To further elucidate the effects of photoperiod, Belozerov[29] exposed young nymphs, 2 to $2^1/_2$ months postmolting to long or short-day photoperiods, all at 18°C. Those exposed to long-day regimes (18 to 24 hr light) showed attack rates of 32 to 38%. In contrast, those exposed to short-day regimes (0 to 16 hr light) had much lower rates, only 8 to 21%. Increasing the temperature to 25°C led to increased activity in both groups, but the same relationship remained: 46 to 72% of the long-day nymphs fed, while only 17 to 32% of the short-day nymphs fed. Belozerov concluded that increasing photoperiod and increasing temperature act synergistically to excite aggressive host-seeking behavior in *Ix. ricinus* nymphs, with increasing photoperiod constituting the predominant stimulus.

In summary, aggressive attack behavior is three to ten times greater in long- rather than in short-day treated nymphs. The reduced activity in the short-day treated nymphs constitutes behavioral diapause. Feeding behavior is irreversible. It is important to note that these responses are most apparent in young nymphs. In older nymphs, especially those more than 6 months old, aggressiveness increases regardless of previous conditioning.

According to Belozerov,[28] *Ix. ricinus* larvae do not show behavioral diapause. Rather, they respond with morphogenetic diapause. Differences in the critical threshold were observed for populations from different geographic regions. Thus, in larvae collected in Leningrad, the reaction threshold was 17 to 18 hr light, whereas for larvae from Moldavia, it was 15 to 16 hr light. These findings, however, appear to be in conflict with data from other regions. For example, data from Nosek[31] suggests that in Slovakia and lower Austria, the dominant seasonal peak for the larvae is in May or June, and the second seasonal peak occurs in August or September.

A closely related species, *Ix. persulcatus,* also exhibits behavioral diapause. This tick has a more restricted distribution and occurs primarily in the northern boreal forests in eastern Europe, European U.S.S.R., and Sibeira. Although its life cycle is similar to that of *Ix. ricinus,* its seasonal activity cycle is usually unimodal. Activity commences in April,when melting snow exposes much of the vegetation, reaches a peak in May, and terminates in July,[32] although vigorous activity may continue throughout July and August.[33] Rarely, a second seasonal peak is seen in September.

Another example of seasonal periodicity in the subgenus *Ixodes* occurs in *Ix. dammini.* This species is the major vector of Lyme disease in the northeastern U.S., a vector of the malaria-like *Babesia microti,* and a serious pest. This tick exhibits clearly defined seasonal perodicity, with a uni- or bimodal pattern, depending upon the locality. In Connecticut, adults are found in greatest numbers on deer in the fall, primarily in October and November; a smaller peak also occurs in the spring, during April and May.[34,35] Larvae and nymphs also show seasonal periodicity. In Connecticut, a biphasic seasonal cycle was found when white footed mice, *Peromyscus leucopus,* were examined. The seasonal peaks for larvae feeding on these mice occurred in June (smaller) and again in August through October (larger). Nymphal activity generally resembled that of the larvae, but the seasonal peak in June was larger than the peak in September.[36] In Massachusetts, the seasonal cycle for this species is unimodal, with the larval seasonal peak July and August, with the nymphal peak in May and June (no adults were collected in this study).[37] These data clearly reflect behavior (i.e., host-seeking) diapause for the larvae, but whether it is due to declining autumn and winter temperatures, snow cover, declining photoperiod, or some other environmental (or host) stimulus is unknown. The early spring peak for nymphs also suggests some form of diapause, but whether it represents behavioral diapause (delayed host-seeking by nymphs that emerged from larvae fed in the summer) or morphogenetic diapause (delayed molting of engorged larvae) is unknown.

B. Ixodidae: Diapause in Response to Increasing Photoperiod and Temperature

1. D. albipictus

No viruses are known to have been isolated from this species. This species provides another example of behavioral diapause. However, this example is remarkable because the ticks initiate their behavior in response to declining rather than increasing photoperiods. They feed and reproduce during the autumn and winter months, hence the name, "winter tick". This species is a serious pest of deer, elk, moose, and other wildlife as well as livestock throughout much of North America. The life history and seasonal responses of this species were described in detail by Howell.[38] In California, larvae hatch in late winter and early spring. However, the emerging larvae, exposed to increasing daylength, diapause rather than feed. They remain in diapause throughout the spring and early summer. Gradually, as daylength decreases in late summer, larvae emerge from diapause, climb the vegetation, and commence questing activity. The earliest that larvae were found on hosts was August 27th. Questing behavior generally begins in September. To determine whether larvae were actually in diapause prior to September or merely sheltering from the hot, dry climate, Howell collected larvae from the field in March and attempted to feed them on a bull, guinea pigs, and various other animals. All failed to feed prior to September 18. After mid-September, larvae placed on cattle and horses fed readily.

Once attached to its host, *D. albipictus* feeds, molts, and reattaches in each life stage, including the adults (one-host cycle). Nymphs appear on the host in October and adults in October and November under California conditions. In Virginia, Sonenshine etal.[39] reported 556 males, 402 females, 378 nymphs, and only 6 larvae on deer *Odocoileus virginianus* examined during the hunting season, from November 19 to December 31, over a 3-year period. This distribution resembles that seen in California. Whether a second period of tick activity occurred in the spring is unknown, since no further collecting was done after the end of the hunting season in December.

Following feeding, fecund *D. albipictus* females drop to lay their eggs in the vegetation. By following the infestations on a group of horses allowed to roam freely in hillside pastures, Howell demonstrated the occurrence of two seasonal peaks of activity, the smaller in September to October, the second and much larger peak in December to February. No ticks were found after the end of March. Howell recognized that the first seasonal peak resulted from the "breaking of the inhibiting factor", i.e., diapause. The second seasonal peak follows after oviposition and hatching in the relatively mild winter climate of the San Francisco Bay area of California. The emerging larvae, exposed to short daylength in November to January, now commence host seeking immediately. Since this population is much larger than that which survived the preceding spring and summer, this winter peak is much larger than that observed in the autumn. Larvae that fail to find hosts by April enter diapause, to reemerge the following September. Elsewhere in the U. S. and Canada, where the winters are more severe, oviposition and hatching are delayed and there is no biphasic seasonal activity for these ticks. However, the same general pattern holds with all host-seeking and feeding activity confined to cool autumn and winter months.

An unusual case of nymphal diapause was observed in *D. albipictus* infesting moose in Alberta, Canada.[40] Larvae were found on these animals from September to November. The ticks engorged and molted into nymphs without delay. The nymphs, however, remained unengorged, despite the ready availability of the host, until January or February. Nymphs began to feed in January, and peak engorgement occurred in March or April. This is a clear example of nymphal diapause which is especially remarkable because of the fact that the diapausing individuals are on the host, hidden in the animal's pelage. A similar phenomenon can also occur in *D. marginatum* and *D. silvarum*.[41] Glines and Samuel[40] suggest that this nymphal diapause enables the parasites to synchronize their feeding with the time when the host's nutrition will improve. They note that subarctic grazing animals "are on their lowest plane of nutrition in late winter and spring when food is less available". They also call attention to the often severe alopecia induced by the feeding of these ticks, and suggest that delayed feeding might minimize this effect at the time when it would be most deleterious.

C. Argasidae

At least 49 tick-borne viral agents are known from 34 species of argasid ticks, including several that are probably only casual associations.[17]

All known argasid ticks are nidicoles, living in or near the nest, burrow, cave, or man-made shelter used by their hosts.They typically shelter in "crevices in wood or stone, or near the soil surface, in or near nesting or resting sites to which suitable hosts periodically return", e.g., "arboreal bird rookeries, marine bird colonies on the ground, mud-nest or san-burrow colonies of swallows and martens, vulture nests, bat caves, and other caves, dens and lairs as well as tree shaded soil in semi-deserts and deer beds in forest".[1] Those argasids that parasitize nonmigratory hosts depend on the frequent return of these animals, even if the visits to the shelter are erratic. In such cases, there is no advantage to behavioral or ovipositional diapause and none has evolved. An example of a typical argasid life cycle without diapause is that illustrated by the Egyptian *Ornithodoros (Pavlovskyella) erraticus*. Viruses isolated from this species include Qalyub *(Nairovirus*, Bunyaviridae) and Artashat (serologically ungrouped, Bunyaviridae).[17] This common parasite of the Nile delta infests animal burrows, cemeteries, stables, sandy mounds, and even palm trees . Its chief hosts are various rats, hedgehogs and other wild rodents, ground nesting birds, lizards, and occasionally man. The ticks are active throughout the year, without any apparent behavioral diapause.[42] Moderately reduced abundance during the winter probably reflects diminished activity due to temperature rather than actual diapause.

III. AESTIVATION

In climates where there is a distinct rainy season, one might expect the evolution of a diapause response that would enable ticks to avoid the long, hot, dry season; however, no conclusive evidence of such a response has been reported. Equatorial East Africa provides a suitable environment where such a phenomenon might be detected, since well-defined wet and dry seasons occur, but there is little change in daylength or air temperatures in areas of uniform altitude. Clifford et al.[43] monitored the seasonal activity of a variety of ixodid ticks on hares, *Lepus capensis* and *L. crawshayi* in grassland and mixed scrub grassy savannah habitats of Kenya. Although there were variations in the abundance of the different species, there was no clear correlation with the rainy or dry season. A reduction in tick abundance was observed during the height of the rainy season in April, but this may have been attributed to the avoidance of long, wet grass by the hares or a repositioning of their questing sites by the ticks. Moreover, in contrast to the long quiescent period typical of most diapause responses, the reduction in tick abundance on the hares was very brief, approximately 1 month. These findings do not support the hypothesis of aestivation diapause in these ticks.

IV. MORPHOGENETIC DIAPAUSE

This response is perhaps the least common of the various types of diapause in ticks. Belozerov[44] identified 17 species of *Ixodes, Haemaphysalis,* and *Hyalomma* in which fed larvae, fed nymphs, or individuals of both immature stages fail to metamorphose immediately after feeding. All are species of the Palearctic region. According to Belozerov,[23,44] morphogenetic diapause also affects the embryonating eggs. As noted previously, Gray [67] considered that egg diapause may be important in explaining the seasonal dynamics of *Ix. ricinus* in Ireland.

An apparent example of this phenomenon is that given by Hoogstraal and Chung-Kim[1] for *Ornithodoros (Alveonasus) lahorensis*. This species is widespread in the U.S.S.R., southeastern Europe, and across central and western Asia where it is a serious pest of livestock. It is also a vector of Crimean-Congo hemorrhagic fever (CCHF; *Nairovirus*, Bunyaviridae).[17] The ticks shelter in barns, stables, and other man-made animal shelters. Larvae attach to hosts during the autumn and winter, when most of the animals are in their stables. Attached ticks feed and remain on the host, molting and feeding again to the first, second, and third stage nymphs. Subsequently, third instar nymphs drop to the ground, but remain hidden in the shelters until spring, when they molt to adults. Adults feed in the spring and oviposit during the spring and summer months. No evidence of behavioral or ovipositional diapause has been reported. Another example of this type of diapause is seen in the life cycle of *Ix. triangulipceps*, discussed below.[16]

V. OVIPOSITIONAL DIAPAUSE

This type of diapause is found in several ixodid ticks. It is the primary form of diapause in argasid ticks.[17]

A. *Ixodidae*

Although behavioral diapause (host-seeking diapause) is the most common form of seasonal adaptation in these ticks, ovipositional diapause also occurs in some species. An example is the Palearctic tick, *D. marginatus*. *D. marginatus* inhabits forest shrub, marsh, semidesert, and alpine communities from southwestern Siberia and Kazakastan in the U.S.S.R., northern Afghanistan and Iran westward across northern European Russia, Central Europe, and France. It is an important vector of CCHF (Razdan, Bhanja), Dhori (Bunyaviridae), Russian Spring-summer encephalitis (RSSE; *Flavivirus*, Flaviridae), Omsk hemorrhagic fever (OHF), and West Nile (WN) virus (Togaviridae).[3] *D. reticulatus* also transmits RSSE, TBE, and OHF.

D. margiantus and the closely related *D. reticulatus (D. pictus* of Soviet authors) uses ovipositional diapause as a means of synchronizing population growth with favorable climatic conditions. The onset of ovipositional diapause begins in the late summer-autumn season. In contrast, females that engorge in the late winter or spring season oviposit without delay. Belozerov[44] conducted experiments with laboratory-reared populations held at a uniform-temperature and relative humidity but exposed to either (1) short daylength photoperiods, 9 L:15 D, or (2) long daylength photoperiods, 18 L:6 D, before and after feeding. Females held as immatures and young unfed adults under long-day photoperiods had very long delays in oviposition (Table 2), from 106 to 361 days. In contrast, females from immatures held at 9 L: 15 D ovoposited much earlier, from 12 to 14 days (most oviposited within 12 to 30 days); older females oviposited more quickly than younger females. Little relation was found between engorged female bodyweight and the length of the preoviposition period, i. e., diapause was not affected by the size of the size of the blood meal. Blood meal volume was also affected; females held under long daylength conditions imbibed significantly less than those held under short daylength conditions. Changing the length of the photoperiod after feeding had no effect. Thus, the switch to ovipositional diapause was entrained prior to feeding, perhaps even in the immatures. The threshold for ovipositional diapause was 14 hr photophase.[45] Once entrained, this type of diapuse is irreversible.

Table 2
EFFECT OF PHOTOPERIODIC CONDITIONS ON THE PERIOD OF PRE-OVIPOSITION DEVELOPMENT IN *D. MARGINATUS* FEMALES KEPT AT 18°C[66]

Photo-period	Female age (months)	Begin feeding	Length of attachment (days)	Wt of fed females (mg)	Av. length pre-ovipos. per. (days)	Commenced oviposition ± 30 days		Commenced oviposition + 30 days	
						No. females	Period	No. females	Period
9 hr light	1½	9/26	11.7 (8—18)	256.5 (103—417)	82.1	3	14 (12—17)	17	94.2 (81—114)
	2	9/26	15.1 (12—18)	261.0 (128—417)	66.3	4	18.5 (12—20)	14	80.0 (49—108)
	2½	1/15	11.0 (10—12)	191.5 (111—338)	34.0	9	20.7 (16—26)	6	54.1 (34—65)
18 hr	1½	9/26	16.1 (12—20)	132.0 (67—259)	222.0	—	—	8	222.0 (171—276)
	1½	10/03	9.6 (9—11)	162.6 (92—350)	211.4	—	—	15	211.4 (106—361)
	2	10/03	14.0 (10—18)	185.0 (91—329)	215.0	—	—	7	215.0 (169—276)
	2½	10/03	12.0 (10—16)	210.0 (1—380)	202.4	—	30	15	202.4 (147—280)

Experiments also demonstrated that ovipositional diapause was not temperature dependent, i.e., changing the temperature independent of the photoperiod conditioning did not determine whether diapause was induced. However, increasing the temperature increased the duration of the diapause response, i. e., higher temperatures combined with long photoperiods increased the period of oviposition delay. Thus, temperature is a secondary moderator of the response, but it acts synergistically with photoperiod.

In its natural habitat in northern Europe and the northern U.S.S.R., *D. marginatus* exhibits both behavioral and ovipositional diapause. Adults are active in the late winter and early spring, diapause during late spring and aearly summer, and become active again in the late summer or autumn season[45] Long daylength photoperiods signalling the forthcoming winter, inhibit host-seeking behavior just as they inhibit oviposition. In contrast, short daylength photoperiods in late winter and spring encourage activity, portending the onset of mild temperatures, and females feeding during the second seasonal activity cycle delay oviposition until the following spring. Larvae hatch from eggs in spring and development proceeds throughout the spring and summer months. Adults emerging during the summer diapause until autumn or even the following spring. Regardless of when they feed, all egg production is delayed until spring. Thus, both behavioral and ovipositional diapause are expressed in this species. The same diapause responses were observed in tick populations from two different regions, Volgograd and Daghestand.[45] Recently, Gilot and Pautou[46] described a similar life cycle pattern for *D. marginatus* in the French Alps,with the onset of adult activity as early as March and the absence of adults during June, July, and most of August.

These adaptations enable this species to synchronize its population growth with periods of optimal climatic conditions. In the subarctic range of this species, seasonal changes are extreme, and a minimum 2-year life cycle is advantageous to its success. Behavioral diapause enables the ticks, especially the adult females, to synchronize the timing of host-seeking and feeding behavior with the optimal periods for population expansion. Ovipositional diapause insures that egg production and hatching will also be synchronized so that all new progeny commence development during the mild spring period.

A similar pattern of development, also with ovipositional diapause, characterizes the life cycle of the closely related *D. reticulatus*.[45]

B. *Argasidae*

Many argasid ticks, especially those associated with migratory hosts, have a well-defined ovipositional diapause. I observed this in my own studies on the biology of the bat tick, *Ornithodors kelleyi*,[47] a species of no known medical or economic importance (no viruses are known from this tick). In the eastern U.S., insectivorous bats, e.g., the little brown bat, *Myotis lucifugus*, hibernate in cool, damp caves during the autumn and winter, from October to March or April. In the spring, the bats emerge and disperse, roosting in basements, attics, church steeples, barns, and abandoned buildings, as well as in natural shelters. Ticks hidden in cracks and crevices in these habitats emerge to feed on the returning animals, which sleep in these secluded sites during the daylight hours. I observed that laboratory-reared females that engorged on post-hibernating bats in the spring oviposited within 10 to 12 days. In contrast, colony source ticks allowed to feed on bats taken in late autumn (November or December) from where they had been hibernating did not oviposit until 111 to 135 days after the blood meal, despite the fact that they were held under identical temperature, humidity, and dark conditions as the ticks fed in the spring. This example of ovipositional diapause is especially remarkable because of the absence of photoperiod as a potential stimulus. Thus, some component of host blood may provide the stimulus for inhibiting reproductive processes in these engorged females, and this action continues until most of the blood meal is digested. This ability of *O. kelleyi* to synchronize its reproductive biology with the habits of its specific bat hosts may be an important factor contributing to its success in nature.

A classic study of ovipositional diapause was done by Khalil and her colleagues[48,49] on *Argas (P.) arboreus,* a parasite of herons and other roosting birds. The ticks live in the heron rookeries and consequently find hosts only during the months when these migratory birds are in residence. Ovipositional diapause provides the means for synchronizing parasite population growth with the time of maximum host availability. In Egypt, female *A. arboreus* delay oviposition until spring. When specimens collected in July were allowed to feed on domestic pigeons, most >95% oviposited within 40 days. However, specimens collected in October delayed oviposition for many months, 120 to 220 days after feeding; 95 to 100% of the females were in diapause. Indeed, the intensity of the response was so pronounced that some of the females emerging from diapause required a second blood meal before they would oviposit. Another example of this phenomenon occurs in *A. cooleyi* infesting migratory gulls, which appear to have a photoperiod induced diapause that enables the ticks to synchronize oviposition with the nesting period of these migratory hosts.[50] These ticks transmit Mono Lake, Sunday Canyon, and Sixgun City viruses.[17]

VI. OTHER SEASONAL ADAPTATIONS INVOLVING DIAPAUSE

An especially interesting example of seasonal adaptations occurs in the Palearctic ixodid tick, *Ix. trianguliceps,* a parasite of mice, voles, shrews, and other burrow-inhabiting small mammals,[16] in which both morphogenetic and ovipositional diapause controls tick population growth. Seasonal changes in temperature rather than photoperiod appear to provide the environmental signals that induce or terminate diapause. In England, each life stage has well-defined seasonal peaks. Larval activity is bimodal, with peaks in June or July (minor) and November (major). Nymphal activity peaks in July, while adult activity shows three seasonal feeding periods, with maximum abundance in the fall. This activity is regulated by both ovipositional and morphogenetic diapause. Autumn-engorged females delay oviposition until the following spring. Similarly, molting of autumn-engorged larvae and nymphs is also postponed until spring or summer. Thus, tick population growth is synchronized with the period of maximum host population expansion, which occurs during the autumn and winter months.

Ix. trianguliceps is believed to survive solely in the burrows and underground nests of its small mammal hosts[16] where seasonal photoperiodic changes cannot be perceived. Consequently, Randolph[16] (cautiously) concluded that temperature variations exert the major regulatory effect on the growth and development of this species. Although one must be very cautious about identifying diapause (as opposed to low temperature torpor) as the physiological process responsible for these effects, Randolph's data provide a compelling case for the role of both morphogenetic and ovipositional diapause, regulated by temperature, as the mechanism for synchronizing tick population dynamics.

VII. PHYSIOLOGICAL BASIS OF DIAPAUSE

In insects, diapause is characterized not only by inactivity, but by low oxygen consumption, absence of cell divisions, low RNA synthesis, low rates of protein synthesis, and fat body growth.[51,52] At least some of these phenomena are observed in ticks. In *D. marginatus* females in ovipositional diapause, oxygen consumption is reduced to as little as one tenth the amount consumed by nondiapausing females, 20 to 80 $mm^3/g/hr$ vs. 220 to 360 $mm^3/g/hr$.[41] During diapause, hemolymph protein content, both in quantity and diversity, increases whereas nondiapausing females show reduced protein content as vitellogenic oocytes withdraw proteins needed for egg development. Belozerov[41] suggests that these differences are due to the absence of pinocytic activity by the ovary.

In insects, diapause ensues as a result of the animal's capacity to receive and transduce appropriate signals into an appropriate neural program that can be stored and subsequently retrieved at a later date.[52] This implies an ability to accumulate and store these sensory inputs. Photoperiod serves as the environmental cue for induction and termination of diapause in an immense variety of insects.[52] Temperature is probably also important, although usually in a secondary role, with low temperatures acting synergistically to increase diapause incidence and alter its threshold. According to Denlinger,[52] a hierarchy of stimuli and internal regulatory events must occur before diapause will ensue, including maternal experience, daylength photoperiod, and temperatures experienced by the same and preceding life stages, and internal hormonal changes. Failure to meet these criteria, or meet them in the appropriate sequence "erases the program and channels the insect towards non-diapause".

The precise manner in which environmental information is perceived and transduced is still poorly understood. In insects, the pigment molecule responsible for transduction of the light signal is probably a carotenoid, and photoreceptors are located in the brain. Blackening the visual centers or thermocautery of these centers fails to prevent diapause in most species. The brain is also the storage center for this information. In some species, specific neurosecretory cells have been implicated as the storage centers. Neurosecretory substance (NSS) increases in the NSS cells during diapause, and axons from the brain to the corpus cardiacum (CC) become packed with NSS granules at the end of diapause.

Insect endocrinologists implicate ecdysone and juvenile hormone (JH) as the major mediators controlling the induction of termination of diapause.[52] Larval or pupal diapause is initiated when ecdysteroid titers fall to very low levels or disappear. This condition is induced by high concentrations of JH, which inhibits the production of prothoracicotropic hormone (PTTH), thereby inhibiting ecdysone production and/or secretion by the prothoracic glands.[51] In effect, this results in shutdown of the prothoracic glands.[51] In contrast, diapause can be terminated in larvae or pupae by injection of ecdysteroids or prothoracic gland extracts. Allatectomy, removing the suppressor effect of JH, also terminates diapause. Thus, high JH titers are implicated in diapause induction in the immature stages, and low JH titers (or JH absence) allow its termination. During diapause of larvae or pupae, NSS (possibly including PTTH) accumulates in the NSS cells of the brain, but the NSS is not released. These substances accumulate in the axons leading to the active glands. Diapause terminates when JH declines greatly or disappears, allowing the NSS (including PTTH) to be released and stimulate production of ecdysone. In adult insects, reproductive diapause is associated with the absence or low titers of JH, or gonadotropic hormone. JH is the primary gonadotropic homone. It activates the fat body, the major site of metabolic activity in many adult insects, thereby prompting high energy consumption, and vitellogenin synthesis. JH also activates the ovary, so that the developing oocytes incroporate vitellogenin from the hemolymph. Removal of the stimulatory effects of JH effectively terminates these activities, and induces diapause.

In some insects, e.g., the silkworm moth *Bombyx mori,* diapause hormone, secreted by a single pair of neurosecretory cells in the subesophageal ganglion, acts on the ovarioles which produce diapause embryos at the time of germ band formation (morphogenetic diapause). The diapause hormone accelerates trehalose utilization, which results in accumulation of glycogen in the eggs progammed for diapause, and uptake of 3-hydroxykynurenine, which is converted to omochrome, producing the dark pigmentation characteristic of diapausing eggs.[51]

Much less is known about the physiological basis of diapause regulation in ticks. In the case of nondiapausing individuals, photoreceptors in the synganglion (but, apparently, not in the eyes) respond to specific radiant energy "inputs", which are counted and this information is stored in some manner.[41] Presumably, the accumulation of nerve impulses from these specific receptor neurons eventually reaches a critical threshold, exciting growth of various neurosecretory centers. NSS increases and passes to specific but presently undetermined endocrine organs that regulate molting (i.e., metamorphosis), or increase metabolic activity. These activities are probably regulated by ecdysone, and the affected target center is probably the endocrine organ that produces this steroid. Belozerov[41] suggests the existence of an activation hormone, which he terms PAH presumably analogous to the PTTH of insects that excites the ecdysone-producing glands.

Individuals exposed to environmental conditions opposite to those initiating the conditions described above accumulate few sensory "inputs". Under these conditions, endocrine production is delayed or terminated, and the animal enters diapause. Belozerov[41] proposed the existence of another hormone, the inhibiting hormone (PIH) which is presumed to block the connection between the receptor system and the PAH. Thus, according to this hypothetical model, both PAH and PIH are neuroendocrine hormones, and their specific titers are determined by the type of stimulus to which the tick is programmed genetically to respond. Under appropriate conditions, PAH predominates and questing behavior, development, and other biological functions proceed. PIH predominates and diapause ensues when, for example, *D. marginatus* is exposed to long daylength photoperiods.

Convincing evidence of the role of neurosecretory substance in diapause in ticks was described by Ioffe,[53] who reported marked increases in the abundance and size of NSS cells in the synganglion of *D. reticulatus*. The quantity of NSS doubled from lowest levels in May to peak concentrations in August. This increase parallels the gradual increase in daylength over this period, with a lag of about 1 month. It is especially interesting that the increases involved all of the NSS centers, not just certain specialized centers that occur in insects.

Much less is known about the hormonal regulation of development and reproduction in ticks. Ecdysteroids, especially ecdysone and 20-hydroxyecdysone, have been found in a wide variety of tick species, [54-57] and their presence may be universal in the *Acari*. However, no gland homologous or even analogous to the prothoracic gland has been found. Several recent studies implicate the synganglion or structures adjacent to it.[58,59] Numerous studies implicate the existence of a juvenoid,[60] but no specific JH-like hormone has been positively identified. In vitro culture of glands or other organs, well advanced in insects where it has served as an important tool for study of hormonal effects, is virtually unknown in studies with ticks. Until this needed knowledge is available, direct comparisons with the regulation of diapause in insects is difficult or impossible. In general, however, diapause in ticks resembles the patterns seen in insects. The same environmental stimuli predominate, and a similar sequence of neuroendocrine events, at least in the production of neurosecretion, is also suggested by the limited available evidence.

VIII. DIAPAUSE AND ECOLOGY OF TICK-BORNE VIRUSES

Ticks transmit at least 126 arboviruses.[17] This remarkable capability is probably the result of certain biological attributes that distinguish ticks from other hematophagous arthropods. In ticks, most of the internal organs do not undergo extensive histolysis during the development of the immature stages. Consequently, transstadial transmission of viral infections is common. Transovarial transmission of viruses is also widespread, although certain notable exceptions have been recorded, e.g., QRF virus in *A. arboreus*.[13] Another contributing factor that enhances tick vector competence is the enormous blood meal consumed by most tick species, often as much as 2 or 3 mℓ in some ixodid species. This voracity increases the likelihood of infection even when host viremias are low. It is noteworthy that numerous viruses were isolated from Panamanian ticks, but none were isolated from nearly 12,000 specimens of other parasitic Panamanian acarines.[61]

Diapause affects the seasonal occurrence of tick-borne arboviral infections, fostering epizootic outbreaks coincident with the onset of tick host-seeking behavior. In nature, tick-borne viruses persist principally in their tick vectors with the warm blooded bird and mammal hosts acting chiefly as amplifiers; dissemination occurs when uninfected ticks feed on viremic hosts (some arboviruses also persist in their vertebrate hosts, e.g., POWE virus in the gray squirrel, or Omsk hemorrhagic fever virus in shrews, voles, and muskrats).[13,17] This concept of the vertebrate host population as the amplifier component for viral mutiplication in a complex ecosystem was reviewed by Macleod[18] in his discussion of louping ill in *Ix. ricinus* and vertebrate hosts in the U.K. Diapause, especiallly host-seeking diapause, ensures that the massive population of postdiapause ticks behave as a single cohort, emerging more or less simultaneously to attack susceptible hosts.

An example of diapause-regulated epizootic spread of tick-borne viral infections is found in the ecology of QRF virus infecting cattle egrets, *Bubulcus i. ibis,* near Cairo, summarized by Hoogstraal.[17] QRF infection expands rapidly in the spring, following the return of the birds from their winter migration and the emergence of *A. arboreus* nymphs and adults from their winter diapause. The incidence of infection falls during summer, but rises again in early autumn. This epizootic spread of infection is related to the breeding cycle of the egrets, since the chicks and young birds are the most susceptible population groups. Thus, diapause fosters rapid amplification of the disease in the spring, the period when increasing tick host-seeking behavior and the expansion of the susceptible host populations coincide. Infections in man and in other vertebrates may be incidental to this *A. arboreus*-egret biocoenose, or it may be indicative of a much broader range of infection capabilities for this poorly studied viral agent.[17] Similar bird host expansion and tick attack cycles may lead to mass die offs of bird hosts, such as that described by Hoogstraal and Feare[62] for sooty terns *(Sternata fuscata)* in the Seychelles or by Clifford[63] for these birds at Cape Province, South Africa. These epizootics may, of course, be influenced by the exceptional densities of nesting marine birds and the concomitant expansion of the associated tick populations. In the Dry Tortugas, off the coast of Florida, tick density *(O. denmarki)* near the roosting sites of brown noddy terns *(Anous stolidus stolidus)* and sooty terns *(Sternata fuscata fuscata)* reached 20,000/ft^2, whereas none were found in the same sampling areas only 3 years earlier.[64] The importance of these marine epizootic diseases, including several agents of known potential infectivity for man, in coastal areas near human population centers constitutes significant public health risks that have been largely overlooked.

IX. SUMMARY

A variety of regulatory schemes occur in insects and other arthropods, and ticks are no exception. These range in complexity from no apparent regulation to seasonal cycles that are exquisitely synchronized with the optimum period of host availability and climatic conditions favorable to development and reproduction. Photoperiod is probably the most important environmental factor controlling the induction and termination of diapause, with temperature playing a significant but secondary role. Behavioral diapause is perhaps the most widespread response, but numerous examples of ovipositional diapause and even morphogenetic diapause abound. The regulatory center is believed to be in the brain, where numerous neurosecretory centers in diapausing ticks (in constrast to only a very few in insects) produce increased quantities of unknown neuroendocrine compounds, which in turn excite or inhibit the production of ecdysteroids and juvenoids. In insects, high JH and low ecdysteroid titers result in diapause in larvae and pupae, while withdrawal of JH results in diapause in adults. Presumably, the same sequence of hormonal controls occurs in ticks, although this remains to be determined.

The similarities in diapause responses and regulatory controls between ticks and insects are impressive; however, certain noteworthy differences appear to exist, and these should be investigated. The possible existence of an ovipositional diapause-inducing hormone in

A. arboreus infesting migratory birds[65] appears to be unique, quite different from the diapause hormone of Lepidoptera which only affects the development of the embryos. The identity and mode of action of this unusual tick hormone should be established. Diapause induction or termination in the life cycle of ticks infesting migratory bats suggests another type of regulatory event, independent of photoperiod or temperature, and perhaps regulated by host blood components instead of abiotic parameters. Clearly, ticks, because of their distinct chemical language and highly specialized adaptations, offer new opportunities for understanding this remarkable phenomenon.

REFERENCES

1. **Hoogstraal, H. and Chung-Kim, K.,** Tick and mammal coevolution, with emphasis on *Haemaphysalis*, in *Coevolution of Parasitic Arthropods and Mammals*, Chung-Kim, K., Ed., John Wiley & Sons, New York, 1985, 505.
2. **Balaschov, Y. S.,** Bloodsucking ticks, (Ixodoidea)-vectors of diseases of man and animals *Misc. Publ. Entomol. Soc. Am.*, 2, 159, 1972 (in Russian, NAUKA Publishers, Leningrad, 1967).
3. **Hoogstraal, H.,** Ticks, in *Parasites, Pests and Predators*, Gaafar, S. M., Howard, W. E., and Marsh, R. E., Eds., Elsevier, Amsterdam, 1985, 347.
4. **Sonenshine, D. E.,** Zoogeography of the American dog tick, *Dermacentor variabilis*, in *Recent Advances in Acarology*, Vol. 2, Rodriguez, J. G., Ed., Academic Press, New York, 1979, 123.
5. **Garvie, M. B., McKiel, J. A., Sonenshine, D. E., and Campbell, A.,** Seasonal dynamics of Amercian dog tick, *Dermacentor variabilis* (Say) in southwestern Nova Scotia, *Can. J. Zool.*, 56, 28, 1978.
6. **Smith, C. N. and Cole, M. N.,** The effect of length of day on the activity and hibernation of the American dog tick, *Dermacentor variabilis* (Say) (Acarina:Ixodidae), *Ann. Entomol. Soc. Am.*, 34, 426, 1941.
7. **Atwood, E. L. and Sonenshine, D. E.,** Activity of the American dog tick, *Dermacentor variabilis* (Say) in relation to solar energy changes, *Ann. Entomol. Soc. Am.*, 60, 354, 1967.
8. **Sonenshine, D. E.,** Radiant energy and tick activity, in *Weather and Parasitic Animal Disease*, Tech Note No. 159, Gibson, T. E., Ed., World Meteorological Organization, Geneva, 1978, 117.
9. **McEnroe, W. D.,** The regulation of adult American dog tick, *Dermacentor variabilis* (Say), seasonal activity and breeding potential (Ixodidae:Acarina), *Acarologia*, 16, 651, 1974.
10. **McEnroe, W. D.,** *Dermacentor variabilis* (Say) in eastern Massachusetts, in *Recent Advances in Acarology*, Vol. 2, Rodriguez, J. G., Ed., Academic Press, New York, 1979, 145.
11. **Sonenshine, D. E.,** Ecology of the American dog tick, *Dermacentor variabilis*, in a study area in Virginia. I. Studies on population dynamics using radioecological methods, *Ann. Entomol. Soc. Am.*, 65, 1164, 1972.
12. **Sonenshine, D. E., Atwood, E. L., and Lamb, J. T., Jr.,** The ecology of ticks transmitting Rocky Mountain spotted fever in a study in Virginia, *Ann. Entomol. Soc. Am.*, 59, 1234, 1966.
13. **Hoogstraal, H.,** Viruses and ticks, in *Viruses and Invertebrates*, Gibbs, A. J., Ed., Elsevier/North-Holland, Amsterdam, 1973, 349.
14. **Ackerman, R., Rehse-Kupper, B., Cassals, J., Rehse, E., and Danielova, V.,** Isolation of Eyach virus from ixodid ticks, in *Arctic and Tropical Arboviruses*, Kurstak, E., Ed., Academic Press, New York, 1979.
15. **Aubert, M. F. A.,** Breeding of the ixodid tick, *Ixodes (Pholeoixodes) rugicollis* (Acari:Ixodidae) under laboratory conditions, *J. Med. Entomol.*, 18, 324, 1981.
16. **Randolph, S.,** Seasonal dynamics of a host-parasite system: *Ixodes trianguliceps* (Acari:Ixodidae) and its small mammal hosts, *J. Anim. Ecol.*, 44, 425, 1975.
17. **Hoogstraal, H.,** Argasid and Nuttallielid ticks as parasites and vectors, *Adv. Parasitol.*, 24, 135, 1985.
18. **Macleod, J.,** Ticks and disease in domestic stock in Great Britain, in *Aspects of Disease Transmission by Ticks*, Vol. 1, Arthur, D. R., Ed., Zoological Society, London, 1962, 29.
19. **Milne, A.,** The ecology of the sheep tick, *Ixodes ricinus* L. Microhabitat economy of the adult tick, *Parasitology*, 40, 14, 1950.
20. **Milne, A.,** The ecology of the sheep tick, *Ixodes ricinus* L. Spatial distribution, *Parasitology*, 40, 35, 1950.
21. **Lees, A. D. and Milne, A.,** The seasonal and diurnal activity of individual sheep ticks *(Ixodes ricinus* L.), *Parasitology*, 41, 189, 1951.
22. **Arthur, D. R.,** *Ticks and Disease*, Pergamon Press, Oxford, 1960, 445.

23. **Belozerov, V. N.,** Diapause in eggs of the tick *Ixodes ricinus L.* and its dependence on the photoperiodic conditions of maintenance of unfed females, *Vestn. Leningr. Univ. Biol.*, 9, 33, 1973.
24. **Gray, J. S.,** Studies on the larval activity of the tick, *Ixodes ricinus* L. in Co. Wicklow, Ireland, *Exp. Appl. Acarol.*, 1, 307, 1960.
25. **Pretzmann, G., Loew, J., and Radda, A.,** Investigations in a natural focus of tick-borne encephalitis (TBE) in lower Austria — 3rd communication: attempt at a demonstration in toto of the cycle of TBE in the natural endemic focus *Zentralbl. Bakteriol. Parasitenkd. Infekionskr. Hyg.*, 190, 299, 1963.
26. **Pretzmann, G., Radda, A., and Loew, J.,** Studies pertaining to the ecology of *Ixodes ricinus* L. in an endemic area of early summer meningoencephalitis (ESME) in the district of Neunkirchen (Lower Austria), *Z. Morph. Okol. Tierre*, 54, 397, 1964.
27. **Loew, J., Radda, A., Pretzmann, G., and Studynka, G.,** Investigations in a natural focus of tickborne encephalitis (TBE) in lower Austria, etc., *Zentralbl. Bakteriol. Parasitenkd. Infekionskr. Hyg.* 194, 133, 1964.
28. **Belozerov, V. N.,** Photoperiodic regulation of development and behavior of *Ixodes ricinus* L. larvae and nymphs from different populations and its changes owing to age of the ticks, *Tezisy Dokl. Pervoe Akarol. Sov. (Moscow)*, 26, 1966.
29. **Belozerov, V. N.,** Nymphal diapause in the tick *Ixodes ricinus* L. (Ixodidae). III. Photoperiodic reaction in unfed nymphs, *Parzitologiya*, 4, 139, 1970.
30. **Belozerov, V. N.,** Nymphal diapause in the tick *Ixodes ricinus* L. (Ixodidae). IV. Effects of changes in photoperiodic maintenance regimes on aggressiveness of unfed nymphs, *Parzitologiya*, 5, 3, 1971.
31. **Nosek, J.,** Effects of microclimate on *Ixodes ricinus*, in *Weather and Parasitic Animal Disease*, Tech. Note No. 159, Gibson, T. E., Ed., World Meteorological Organization, Geneva, 1978, 105.
32. **Vansulin, S. A., Smyslova, T. O., and Solina, L. T.,** Distribution and biological properties of *Ixodes persulcatus* ticks (Ixodidae) in the health resort zone of Leningrad, *Parazitologiya*, 15, 498, 1981.
33. **Boyko, V. A. and Ivliev, V. G.,** Biocenotic associations between birds and ixodid ticks in natural foci of tick borne encephalitis in a forest-steppe zone along middle reaches of the Volga, in *Transcontinental Associations of Migratory Birds and Their Role in Transmission of Arboviruses*, Cherepanov, A. I., Ed., Novosibirsk, Moscow, 1972, 326.
34. **Carey, A. B., Krinsky, W. I., and Main, A. J.,** *Ixodes dammini* (Acari:Ixodidae) and associated ixodid ticks in south-central Connecticut, USA, *J. Med. Entomol.*, 17, 87, 1980.
35. **Main, A. J., Spruance, H. E., Kloter, K. O., and Brown, S. E.,** *Ixodes dammini* (Acari:Ixodidae) on white-tailed deer *(Odocoileus virginianus)* in Connecticut, *J. Med. Entomol.*, 18, 487, 1981.
36. **Main, A. J., Carey, A. B., Carey, M. G., and Goodwin, R. H.,** Immature *Ixodes dammini* (Acari:Ixodidae) on small animals in Connecticut, USA, *J. Med. Entomol.*, 19, 655, 1982.
37. **Wilson, M. L. and Spielman, A.,** Seasonal activity of immature *Ixodes dammini* (Acari:Ixodidae), *J. Med. Entomol.*, 22, 408, 1985.
38. **Howell, D. E.,** The ecology of *Dermacentor albipictus* (Packard), in *Proc. 6th Pacific Sci. Congr.*, 4, 439, 1939.
39. **Sonenshine, D. E., Lamb, J. T., Jr., and Anastos, G.,** Distribution, Hosts and Seasonal Activity of Virginia Ticks, Virginia Academy of Science, Vol. 16, 1965, 26.
40. **Glines, M. V. and Samuel, W. M.,** The development of the winter tick, *Dermacentor albipictus* and its effect on the hair coat of moose, *Alces* alces of central Alberta, Canada, in *Acarology VI*, Vol. 2, Griffiths, D. A. and Bowman, C. E., Eds., Ellis Horwood, Chichester, U.K., 1984, 1208.
41. **Belozerov, V. N.,** Diapause and biological rhythms in ticks, in *Physiology of Ticks*, Obenchain, F. D. and Galun, R., Eds., Pergamon Press, Oxford, 1982, 469.
42. **Khalil, G. M., Helmy, N., Hoogstraal, H., and El-Said, A.,** Seasonal dynamics of *Ornithodoros (Pavlovskyella) erraticus* (Acari:Ixodoidea:Argasidae) and the spirochete *Borrelia crocidurae* in Egypt, *J. Med. Entomol.*, 21, 536, 1984.
43. **Clifford, C. M., Flux, J. E. C., and Hoogstraal, H.,** Seasonal and regional abundance of ticks (Ixodidae) on hares (Leporidae) in Kenya, *J. Med. Entomol.*, 13, 40, 1976.
44. **Belozerov, V. N.,** Daylength as a factor determining oviposition delay in female *Dermacentor marginatus* Sulz., *Med. Parazitol. (Moskva)*, 32, 521, 1963.
45. **Belozerov, V. N. and Kvitko, G.,** Main feature of photoperiodic reaction in *Dermacentor marginatus* Sulz. ticks (Ixodoidea), *Zool. Zh.*, 44, 363, 1965.
46. **Gilot, B. and Pautou, G.,** Repartition et ecologie de *Dermacentor marginatus* (Sulzer, 1776) (Ixodoidea) dans les Alpes Francaises et leur avant-pays, *Acarologia*, 24, 261, 1983.
47. **Sonenshine, D. E. and Anastos, G.,** Observations on the life history of the bat tick, *Ornithodoros kelleyi* (Acarina:Argasidae), *J. Parasitol.*, 46, 449, 1960.
48. **Khalil, G. M.,** The subgenus *Persicargas* (Ixodoidea:Argasidae:*Argas*). XIX. Preliminary studies on diapause in *A. (P.) arboreus* Kaiser, Hoogstraal and Kohls, *J. Med. Entomol.*, 11, 363, 1974.
49. **Khalil, G. M.,** The subgenus *Persicargas* (Ixodoidea:Argasidae:Argas). XXVI. *Argas (P.) arboreus:* effect of photoperiod on diapause induction and termination, *Exp. Parasitol.*, 40, 232, 1976.

50. **Schwann, T. G. and Winkler, D. W.,** Ticks parasitizing humans and California gulls at Mono Lake, California, U.S.A., in *Acarology VI*, Vol. 2, Griffiths, D. A. and Bowman, C. E., Eds., Ellis Horwood, Chichester, U. K., 1984, 1193.
51. **Krishna Kumaran, A.,** Introduction: evolution of regulatory controls in insect life cycles, in *Endocrinology of Insects*, Vol. 1 Downer, R. G. H. and Laufer, H., Eds., Pergamon Press, Oxford 1983, 353.
52. **Denlinger, D.,** Hormonal control of diapause, in *Comprehensive Insect Physiology, Biochemistry, and Pharmacology*, Vol. 8, Kerdut, G. A. and Gilbert, L. I., Eds., Pergamon Press, New York, 1985.
53. **Ioffe, I. D.,** Seasonal changes in neurosecretion contents of neurosecretory cells *Dermacentor pictus* Herm. ticks (Ixodoidea, Acarina), *Med. Parzitol. (Moskva)*, 34, 57, 1965.
54. **Delbecque, J. P., Diehl, P. A., and O'Connor, J. D.,** Presence of ecdysone and ecdysterone in the tick *Amblyomma hebraeum* Koch, *Experientia*, 34, 1379, 1978.
55. **Diehl, P. A. and Aeschlimann, A.,** Tick reproduction: oogenesis and oviposition, in *Physiology of Ticks*, Obenchain, F. D. and Galun, R., Eds., Pergamon Press, Oxford, 1982, 277.
56. **Dees, W. H., Sonenshine, D. E., and Breidling, E.,** Ecdysteroids in the American dog tick, *Dermacentor variabilis* (Acari:Ixodidae) during different periods of tick development, *J. Med. Entomol.*, 21, 514, 1984.
57. **Dees, W. H., Sonenshine, D. E., and Breidling, E.,** Ecdysteroids in the camel tick, *Hyalomma dromedarii* (Acari:Ixodidae), and comparison with sex pheromone activity, *J. Med. Entomol.*, 22, 22, 1985.
58. **Binnigton, K.,** Ultrastructural evidence for the endocrine nature of the lateral organs of the cattle tick, *Boophilus microplus*, *Tissue Cell*, 13, 475, 1981.
59. **Sonenshine, D. E., Homsher, P. J., Beveridge, M., and Dees, W. H.,** Occurrence of ecdysteroids in specific body organs of the camel tick, *Hyalomma dromedarii* and the American dog tick, *Dermacentor variabilis*, with notes on their synthesis from cholesterol, *J. Med. Entomol.*, 22, 303, 1985.
60. **Solomon, K. R., Mango, C. K. A., and Obenchain, F. D.,** Endocrine mechanisms in ticks. Effects of insect hormones and their mimics on development and reproduction, in *Physiology of Ticks*, Obenchain, F. D. and Galun, R., Pergamon Press, Oxford, 1982, 399.
61. **Yunker, C. E., Brennan, J. M., Hughes, L. E., Phillip, C. B., Clifford, C. M., Peralta, P. H., and Vogel, J.,** Isolation of viral and rickettsial agents from Panamanian *Acarina*, *J. Med. Entomol.*, 12, 250, 1975.
62. **Hoogstraal, H. and Feare, C. J.,** Ticks and tickborne viruses, in *Biogeography and Ecology of the Seychelles Island*, Stoddard, D. R., Ed., W. Junk, The Hague, 1984, 267.
63. **Clifford, C. M.,** Tickborne viruses of sea birds, in *Arctic and Tropical Arboviruses*, Kurstak, E., Ed., Academic Press, New York, 1979.
64. **Yunker, C. E.,** Tickborne viruses associated with seabirds in North America and related islands, *Med. Biol.*, 53, 312, 1975.
65. **Khalil, G. M. and Shankaky, N. M.,** Hormonal control of diapause in the tick, *Argas arboreus*, *J. Insect Physiol.*, 22, 1659, 1976.
66. **Belozerov, V. N.,** Day length as a factor in determining the delay of egg laying in *Dermacentor marginatus* (Sulz.) females, *Med. Parazitol.*, 32, 521, 1963.
67. **Gray, J. S.,** personal communication.

Chapter 9

WEATHER, VECTOR BIOLOGY, AND ARBOVIRAL RECRUDESCENCE

Paul Reiter

TABLE OF CONTENTS

I. BASIC PRINCIPLES

A. Recrudescence

Many arboviral diseases are characterized by long periods of invisibility, when little or no evidence of their existence can be detected. At erratic intervals, sometimes separated by several decades, there is a sudden recrudescence, often developing into an explosive epidemic. Retrospective studies of such epidemics frequently suggest, by association, weather-related factors which could have been responsible for triggering this recrudescence, and these associations offer fertile ground for speculative explanations. Nevertheless, the timing of recrudescence remains enigmatic and notoriously unpredictable. In practical terms this is a serious problem; all too frequently, health or veterinary authorities are unaware of the existence of an epidemic until many weeks after its commencement, and may be unable to implement countermeasures until after the majority of infections have already been contracted.

Recrudescence is by no means unique to arboviral infections. In the aftermath of widespread transmission of all viral diseases, the immune system acts within populations as a common defense factor, the herd immunity, which restrains the recurrence of high rates of transmission until a sufficiently large number of nonimmune individuals are recruited, either by natural increase or by immigration from other populations. Recurrent epidemics of measles and influenza viruses are classic examples of this mechanism. However, in the case of the arboviruses, the obligatory involvement of one or more species of arthropod introduces a wide range of additional factors which greatly increase the complexity of transmission. In this chapter we are concerned with weather-related aspects of the natural history of the vector which may influence recrudescence.

B. Climate, Weather, and Arthropods

Climate is the long-term summation of the atmospheric elements — radiation, temperature, precipitation, humidity, and wind — and their variations. The global interactions of the components of climate are highly complex and resolve into a number of long-term meteorological cycles. Best known among these are the southern high-pressure oscillation (the "el niño" phenomenon) and the ice ages. Description of climate is therefore time-dependent, and climatic indices vary from decade to decade and century to century.

Short-term variation within climate is termed weather. The changeability of weather varies widely at different latitudes. It is greatest in the mid-latitudes, where a continuous succession of high and low pressure centers results in a constantly shifting weather pattern, and least in the tropics, where day-to-day and month-to-month changes are so small as to be almost synonymous with climate.

Climate is a major component in the environment of all arthropods, and indeed all living organisms. All species live within defined climatic limits, although the actual limits of their distribution are only partly determined by climate. Weather also exerts a profound effect on arthropods. Within their climatic limits, all the atmospheric elements constantly effect every aspect of behavior, development, and dispersal, while at the boundaries of their climatic limits, relatively minor deviations from the ambient norm can be catastrophic.

It is important to realize the immediacy of these effects. It is possible to measure temperature to a degree by counting the frequency with which crickets chirp: the organism operates in direct linkage to its atmospheric environment, and has only limited physiological mechanisms for responding to short-term fluctuations. Indeed, for most arthropods, even in the humid tropics, daily cycles of temperature and humidity range far beyond the span of their optimal environment, and mean daily conditions are far less important than specific conditions at a select portion of the day. Strict circadian cycles, generally regulated by time cues based on the light regime, dominate their behavior to ensure that crucial activities coincide with the regular occurrence of optimal conditions. If these conditions are not met, behavioral strategies, usually involving evasion and inactivity, are the dominant response; the organism defers normal behavior until the next occasion that its activity cycle coincides with acceptable conditions.

C. Climatic Seasonality and Arthropod Biology

Few arthropods exhibit continuous, year-round activity, and there is a wide variety of seasonal arrangements of active and dormant phases in their life cycles. In temperate zones, winter presents a conspicuous obstacle, and there is a wealth of studies describing dormancy in this season. Far less well known, but widespread and common even at high latitudes, are dormancy phases in other seasons. For example, many insects restrict their active phases to a short fraction of the summer and spend the remainder in a phase of inactivity analogous to overwintering.[1] In the tropics, activity may be restricted to a portion of the wet or the dry season, or both.[2] In addition to these preprogrammed periods of inactivity, transient events, such as periods of low temperature or heavy rainfall may interrupt normal feeding and reproductive behavior during an active season, or may interrupt or terminate inactivity during a dormant season.[3]

In most species, programmed dormancy is expressed in one stage of the life cycle. Thus, *Aedes* mosquitoes survive winter or drought seasons in the egg stage whereas *Culex* and *Anopheles* species survive as adults. Exceptions, such as ticks, generally require 2 or more years to complete a life cycle (see Chapters 7 and 8).

The alternative to seasonal dormancy is to continue reproduction throughout the year, with opportunistic bursts of population increase whenever conditions are suitable. In these circumstances, climatic seasonality dictates abundance by the availability of food or breeding sites; however, caution is required when assuming that all members of a species are continually active all of the time. Polymodal strategies, whereby part of a population continues activity while the remainder is dormant are not uncommon. An example is the "bet-hedging" exhibited in the staggered hatching of *Aedes* eggs. A less well-known example is dormancy in *Cx. quinquefasciatus;* in the tropics this species breeds throughout the year, yet there is evidence[4] that during the dry season a large fraction of the adult population becomes inactive. A problem with this type of dormancy is that when all stages of the life cycle occur simultaneously, its presence may be difficult to demonstrate.

D. Vectorial Capacity, Weather, and Activity vs. Inactivity

The entomological features of transmission can be expressed in terms of the vectorial capacity, the daily rate of potentially infective contacts between vector and host:

$$C = \frac{ma^2p^n}{(-\log_e p)} \tag{1}$$

where m is the number of vectors per host, a is the number of blood meals taken by a vector per host per day, p is the daily survival probability of the vector, and n is the number of days between infection of the vector and the time it becomes capable of infecting a new host.[5,6]

The concept of vectorial capacity was developed for models of vector-borne parasitic disease, but can also be applied to arboviral disease. There is only one crucial difference: the ability of the immune system of the host to react rapidly to purge its system of viruses, and to maintain effective barriers against their reintroduction. In contrast to the chronic nature of parasitic infections, which may persist in a host for several decades, the duration of arboviral viremia is very short, usually a matter of days. In addition, the potential for repeated reinfections by a single species of parasite is often high, whereas infections with a particular arbovirus are generally a once-in-a-lifetime event. For this reason, whenever the blood-feeding activity of an arbovirus vector is interrupted for longer than the sum of the host's incubation period and the duration of viremia, the virus effectively ceases to exist in the host, and its survival is entirely dependent on the fortunes of the vector.

The simplest form of vector-borne transmission involves the direct transfer of pathogen from one host to another, without replication in the vector. This mechanical transmission is dependent on a vector which takes a number of blood meals on more than one host in rapid succession, and a pathogen which can survive the interval between feeds on the vector, generally on its mouthparts. In this case, p can be regarded as close to unity and n as a fraction of a day. The vectorial capacity is therefore high, but is dependent on the ability of small amounts of pathogen to infect new hosts. Although mechanical transmission can be demonstrated in the laboratory, it is much more difficult to prove in the field. Weather conditions which favor high vector populations and active feeding behavior will presumably increase vectorial capacity.

In more complex forms of transmission, multiplication of the pathogen in the vector ensures that the infective dose is far larger than the quantity which was originally ingested. This generally involves multiplication and movement of pathogen through the gut wall, via the hemolymph, to other organs. Infection of the salivary glands is particularly important because the injection of infected salivary fluid during feeding is the most common mode of transfer of pathogen to a new host. It is the time (n) required for this obligatory incubation that makes longevity the key factor in transmission of all vector-borne disease, for no matter how favorable all other parameters may be, transmission cannot occur unless the vector survives to feed after this incubation period, and small changes in the daily survival probability p will radically alter the frequency of new infections.

After the incubation period, longevity determines how many infective feeds a vector can make. In arboviral disease, because viremia is of short, fixed duration, this is of greater significance than in vector-borne parasitic disease, for the lower the transmission rate, the more the survival of the virus depends on the longevity of the vector.

Some of the effects of climatic seasonality on m and a are readily identified. Maximum vector/host contact will occur in seasons when vectors are active and their populations are high, and epidemics occur at these times. Conversely, low vector populations or inactive vectors imply low contact and an obligatory seasonal maintenance phase.

It is logical to extend such associations in our search for weather-induced recrudescence following a prolonged maintenance phase, but caution is required. For example, during a period of above normal temperatures, given adequate breeding sites and nutrition, the rate of development of mosquito larvae and pupae will be above normal. More will survive to adulthood, and ovarian development and the extrinsic incubation period will be shorter; factors m, a, and n will all tend to a higher vectorial capacity (C). However, the key factor (p) will be adversely affected, because an increase in the frequency of hazardous events such as host-seeking, feeding, and oviposition will increase the mortality rate. Estimates of p for *Cx. pipiens* s.l. in Memphis ranged from 0.65 (average life expectancy 2.3 days) at the height of summer to above 0.97 (average life expectancy 32.8 days) in the winter.[7] This extreme case illustrates how the longevity of a vector, inversely related to the frequency of risk, can be reduced by a shorter gonotrophic cycle. Higher temperatures can thus increase vector-host contact but reduce vectorial capacity. In this context we must remember that circadian control, which restricts adult activity to specific times of day, dictates that the gonotrophic period can only change by discrete units of time, generally 24 hr. Thus, a small change of temperature at the threshold of the circadian "gate" can have a major effect on p, and hence vectorial capacity.

The summertime estimate of p mentioned above was based on a standard method, the measurement of the percentage of parous insects in field-collected samples by ovarian dissection. The winter estimate was based on weekly counts of marked mosquitoes resting in a tunnel. Ovarian dissection would have been irrelevant in the latter study, because insects were not feeding, and remained nulliparous throughout the study. This illustrates an important point: a vector population which experiences a delay in its activity, by short-term quiescence due to adverse weather, a shortage of oviposition sites or suitable hosts, or a longer, programmed period of dormancy, may present a low physiological age structure but in reality have a high survival rate. Once again, climate or weather factors which prolong the gonotrophic cycle by delaying activity can enhance vectorial capacity and constitute a potential trigger for recrudescence.

Another factor related to quiescence is synchrony. Recruitment to an inactive phase may be sudden in response to an environmental change, or gradual, as individuals attain a particular point in their life cycle. By contrast, the stimuli which terminate inactive phases act on the population as a whole, thereby inducing synchrony in the activated cohort. If incubation of a virus was completed before or during inactivity, then this synchrony will result in an abrupt onset of transmission by all infected individuals.

The arguments set out in this section imply that the inactivity of vectors may be of equal or greater significance than their activity. This paradox may throw light on some of the enigmas of arboviral recrudescence, but the very nature of the covert behavior involved means that direct evidence from the field is difficult to obtain. The problem is compounded by the unpredictability of recrudescence itself: comprehensive long-term field studies are rarely in progress when renewed transmission occurs, and the lag-time inherent in clinical surveillance systems and their interpretation mean that weeks or even months may pass before the event is recognized. The result is that vector biologists almost invariably initiate intensive studies when epidemics are already history, and we know far more about the age structure, vectorial capacity, behavior, and dynamics of vector populations after epidemics than in the crucial periods when they begin.

The analysis of weather patterns themselves presents additional problems. Attempts to correlate weather and biological events require the choice of a suitable time scale, resulting in descriptors like "warmer than normal" months or "wetter than normal" seasons. While it is clearly useful to attempt to resolve hourly or daily data into such units, they frequently give no indication of important detail such as the daily range of temperatures involved, the number of times that rainfall occurred, or the number of short-term anomalies which interrupted the overall pattern observed. Conversely, long-term, year-to-year events escape attention if analysis is focused on shorter units or on calendar years: the springtime population of a vector may be greatly influenced by breeding success in the previous autumn, and a disastrous year may have consequences which extend for many years afterward. In the next section, all these points will be illustrated by examining recrudescence in a single disease, St. Louis encephalitis (SLE).

II. AN EXAMPLE: WEATHER AND URBAN ST. LOUIS ENCEPHALITIS

A. The St. Louis Epidemic of 1933

The largest outbreak of SLE ever recorded in a single city was also the first to be recognized, that of St. Louis, Mo., in 1933. The earliest case was noted on July 7 and the last in late October, with the peak on August 27. The epidemic received great publicity around the world, and was treated as a major emergency.

The disease was new and transmission theories abounded. It was noted that because the summer was the driest on record, there was very little flow in the open drainage ditches used for sewage disposal; these developed a particularly offensive stench and were breeding unusually large numbers of mosquitoes. Because there were no obvious indications of contagion between cases or links to water, food, or milk supply, and because mosquitoes were abundant, many people felt that mosquito transmission was the answer. The investigators themselves did not think this likely, but decided on a thorough investigation of the possibility, if only to "obtain...information...in the negative sense". Exhaustive attempts were made to demonstrate transmission by *Ae. aegypti, Anopheles quadrimaculatus,* and *Cx. pipiens* s.l., the latter chosen because it was the most common species, but no transmission was observed. It was concluded that the results spoke against mosquito transmission. The diffuse distribution of cases was cited as supportive evidence because this was considered analogous to that of poliomyelitis and quite unlike that of previous epidemics of yellow fever.

It is fascinating to read the official report of the epidemic[8] in the light of another document, commissioned by the Surgeon General in September 1933 but not published until 25 years later.[9] Its author, L.L. Lumsden, conducted a classic epidemiological study from which he produced abundant evidence that SLE was indeed mosquito-borne and that the vector was *Cx. pipiens* s.l. breeding in sewage-polluted water. He hypothesized that the virus entered the mosquito from the sewage, and that the exceptionally dry weather had precipitated the epidemic by boosting the mosquito population.

B. Patterns of Weather and Outbreaks of Urban SLE

Many authors have commented on weather factors in their descriptions of subsequent SLE outbreaks. In a review of these, Monath[10] compared epidemic to nonepidemic years in terms of deviations from normal monthly temperatures and precipitation. A pattern emerged for epidemic years in most localities: above-normal temperatures in January, February, and May through August, and below-normal temperatures in April. January and February were abnormally wet, but July was abnormally dry.

The epidemic in St. Louis was not included in Monath's analysis, but the weather of 1933 fits his pattern precisely. Indeed, the extremes of that year were so remarkable that they deserve further mention; for this purpose I draw liberally from contemporary weather descriptions, published weekly by the U.S. Department of Agriculture.[11] The winter of 1932—1933 was the second warmest on record, with temperatures in December and January approximating conditions appropriate for April. Abnormally high temperatures persisted until late March, but farming activities were severely delayed because of incessant rainfall. April was cool, May was slightly warmer than average, but the rain continued at twice the normal rate so that by the beginning of June conditions were critical — continued wetness had made field operations impossible. At this point a dramatic change occurred; rainfall for the next 3 months was the lowest ever recorded, June was the hottest on record, and monthly average temperatures for July and August were also well above normal.

Lumsden and his contemporaries drew attention to the coincidence of SLE and an abnormal lack of rainfall, but made no mention of conditions in the preceding 5 months, which were also very unusual. My purpose in giving so detailed a description has not been to highlight their omission — without knowledge of the natural history of the mosquito and the complex zoonosis which is involved, this information could not have seemed relevant — but to underline some of the pitfalls of summarizing and drawing conclusions from meteorological data in studies of arboviral recrudescence. However, I first review some of the speculations which have been offered to explain the association between SLE outbreaks and weather patterns.

C. Speculations on Entomological Reasons for Associations between Weather Patterns and Urban SLE Outbreaks

Serologic studies on birds indicate that in nonepidemic years, enzootic transmission of urban SLE occurs in spring and autumn, but in epidemic years there is a sudden recrudescence in summer, generally in late June or July. Human cases become apparent some weeks later; this is the basis of SLE surveillance by sentinel flocks.[12]

During urban epidemics *Cx. pipiens* s.l. appears to be the most abundant potential vector, and more isolations of SLE virus have been made from this than any other species. Field and laboratory evidence indicate that two other mosquitoes, *Cx. restuans* Theobald and *Cx. salinarius* Coquillett are also involved in transmission, but these are not commonly collected after recrudescence. In Florida, all three species are present, but another vector, *Cx. nigripalpus* Theobald, is more abundant during epidemics.[13]

Cx. pipiens s.l. overwinter as adults. It has been suggested that a warm, wet winter could be conducive to high survival during quiescence, resulting in a rapid buildup of populations in the spring. In addition, if SLE overwinters in the mosquito, high survival rates of the vector would favor the virus as well. However, no difference in the mortality rate of overwintering *Cx. pipiens* s.l. was observed in Memphis in the winters of 1981—1982 and 1982—1983,[7] although the first was colder than normal and the latter was the fourth warmest on record. Indeed, since depletion of fat reserves, predation, and fungal disease are the principal causes of winter death in this species,[14] it would seem that mortality should be more likely to increase rather than decrease in abnormally warm, wet conditions. Moreover, despite many attempts to prove the contrary, there is very little evidence that SLE virus overwinter in *Cx. pipiens* s.l.

A cool April has been explained as delaying the nesting activity of birds, thereby inducing synchrony between high mosquito populations and the availability of nonimmune nestlings. However, low temperatures will also delay the multiplication of vectors, and in many areas synchrony between mosquito production and nesting activity occurs in a normal April. A normal or warm April followed by a warm May would therefore seem more advantageous to continued virus amplification.

Warm weather in May might favor transmission by accelerating the life cycle of the mosquito and the multiplication of the virus, but here we are faced with an anachronism: in much of the area where urban SLE occurs there is little or no *Cx. pipiens* s.l. activity in May. The first major population peak occurs some 2 months later, from early to mid-July in Memphis[15] and mid- to late July in the northern states and Canada.[16] For this reason it is not easy to implicate *Cx. pipiens* s.l. in the pre-epidemic recrudescence of SLE, nor to account for weather-associated factors which might precipitate this.

By contrast, *Cx. restuans* is abundant in springtime but rarely evident after recrudescence. In the adult stage *Cx. pipiens* s.l. and *Cx. restuans* are so alike that many authors consider field-caught specimens to be indistinguishable. For these reasons there is little precise information on the population dynamics of either species. However, in the larval stages they are readily separated, and recent studies have therefore used quantitative egg-raft collections to monitor adult activity.[17] They share a preference for breeding in foul-smelling water and in many ways can be regarded as ecological homologs, except for the temporal separation of their activity. It is this last detail that may be of crucial importance to the recrudescence of SLE: when the adult population of *Cx. restuans* is at its annual peak, there is a sudden cessation of activity and the species enters a state of dormancy. This disappearance of adult *Cx. restuans* is dramatic, and few egg-rafts appear at ovitraps during the summer. Resumption of activity in the autumn can be equally dramatic, and may coincide with the return of cooler weather.[15] If temperatures permit, the species continues breeding into the winter months, although there is some evidence of a winter diapause mechanism in its northern range.[18]

An exception to this pattern was noted in Memphis in 1983: a major peak of *Cx. restuans* activity occurred suddenly in July, 8 weeks after the normal May peak, followed by a smaller but well-defined peak in mid-August. The remarkable feature of these peaks was that they coincided with frontal systems which brought unseasonably cold weather during an abnormally hot, dry summer.[15] Studies in Vero Beach, Fla. have indicated a similar pattern: both *Cx. restuans* and *Cx. salinarius* suddenly cease to oviposit in late April or early May, but resume oviposition at any time that temperatures drop below 20°C.[26]

The coincidence of *Cx. restuans* activity and SLE transmission in spring and autumn suggests that this species may be a major vector and that in nonepidemic years quiescent adults function as a bridge over the summer period. Warm weather in May would boost populations before quiescence, and reactivation by periods of unseasonably cool summer weather would cause the sudden, synchronous release of virus at a time when it would be relayed to the burgeoning *Cx. pipiens* population. A high survival rate in the inactive *Cx. restuans* would ensure a high transmission rate, and an oscillating summer weather pattern would be more conducive to recrudescence than one of consistent high temperature.

As already mentioned, June 1933 was the hottest on record, but closer scrutiny reveals a remarkable anomaly. After 2 weeks of abnormally hot weather, temperatures plunged. A minimum reading of 10°C was recorded in St. Louis, contrasting with minima as high as 24°C the previous week. Frost damage was reported over considerable areas including the lake region and the northern Ohio River Valley. In the midwest, oat crops were abandoned and mowed for hay, and in the southern and eastern states the growth of sugar, cotton, and tobacco was retarded. Abnormally hot weather returned in the last week of June, with minimum temperatures as high as 29°C, but was interrupted by another unusually cold week around the 4th of July, with minima of 18 to 20°C. For the remainder of the summer there were at least three more cycles of cold weather, with minima as low as 13°C.[19] Thus, the requirement for an oscillating pattern of summer temperatures predicted by the *Cx. restuans* hypothesis is fulfilled by the St. Louis data. Moreover, the large geographic region affected by these transient weather phenomena accounts for the concurrence of widely dispersed SLE outbreaks which is frequently observed in epidemic years. Similar spells of unseasonably cold weather in warmer than normal summers can also be identified in years when SLE transmission was high. Conversely, in hot summers when such cool spells were absent transmission has never been high.

The St. Louis data clearly illustrates the limitations of summarizing weather data in seasonal or monthly terms, and this applies equally to other times of the year. Arthropods can exploit warm periods in an otherwise "cold" spring. High total rainfall may suggest a wet summer but be too intermittent to provide suitable groundwater breeding sites. A brief killing frost can obviate the effects of a warm autumn.

Finally, the St. Louis example also encompasses long-term factors which may have influenced vector populations and SLE recrudescence. The winter of 1932—1933 was the 12th abnormally warm winter in the preceding 13. The hot, dry summer was one of a series in a period which came to be known as the "dust bowl years", and the drought of 1930 was even more severe than that of 1933, as were those of 1934 and 1936.

III. OTHER EXAMPLES OF ASSOCIATIONS BETWEEN WEATHER AND ARBOVIRAL RECRUDESCENCE

SLE has served to illustrate the complexity of weather/vector/arbovirus interactions. Throughout this book, examples of weather-associated recrudescence in many other arboviral diseases can be found. It is unnecessary to review these in detail, but brief mention of some is justified.

A. Rainfall and Temperature

References to rainfall and temperature are common in the literature. Japanese encephalitis (JE) is perhaps the most studied, and appears to have similar weather associations to SLE. Other mosquito-borne viruses include eastern equine encephalomyelitis (EEE), western equine encephalitis (WEE), Murray Valley encephalitis (MVE), Rift Valley fever (RVF), and dengue. Tick-borne fevers include Crimean-Congo hemorrhagic fever (CCHF) and Omsk hemorrhagic fever (OHF). In these and many other cases, a direct influence of rainfall and temperature on vector abundance and vector survival has often been suggested.

B. Humidity

Oviposition by *Cx. nigripalpus* is inhibited when relative humidity at sunset is below 90%. Provost[20] suggested that a prolonged delay of oviposition by drought could trigger an SLE epidemic by increasing the probability of completion of extrinsic incubation. In principle, this was the forerunner of the *Cx. restuans* hypothesis of urban SLE epidemics described in the previous section.

A seasonal shift in host selection by *Cx. nigripalpus* from birds in winter and spring to mammals in summer and autumn has also been attributed to humidity: high humidities apparently enable mosquitoes to leave their woodland habitat (where birds are abundant) to feed on mammals in open fields.[21] Clearly, this shift may be important as a coupling mechanism between avian and mammalian cycles of SLE transmission. A similar pattern of host preference has been observed in *Cx. tarsalis* Coquillet, the principal vector of rural SLE in the western U.S.[22]

In many regions of the hot, dry African savannah there is no surface water for up to 8 months of the year. During part of this dry season, mean monthly temperatures exceed 40°C and mean monthly relative humidities drop below 10%. A study in one such area in the Sudan[23] revealed that adult *An. gambiae* s.l. can survive in these conditions for 9 months by sheltering in human dwelling places, dry wells, and animal burrows and taking an occasional blood meal. This adaptation to extremely dry conditions results in year-round malaria transmission, and is terminated by the advent of moister conditions. I mention it here as an extreme example of the quiescence/transmission phenomenon in relation to humidity.

C. Wind

Wind plays an important role in insect dispersal and migration, and there is a large literature on the possible role of infected, wind-borne vectors in arboviral recrudescence.[25,26] The difficulty of explaining the "disappearance" of arboviruses for long periods of time makes wind-borne reintroduction from endemic areas an attractive possibility. In addition, there is firm evidence for the spread of plant viruses by plant-sucking insects, and the appearance of adult *Simulium damnosum* sl. infected with *Onchocerca volvulus* in the Sahel at the start of the West African rainy season appears to substantiate migration of a human pathogen on the Intertropical Convergence Front.

Wind patterns offer fertile ground for speculation. There is evidence that on a number of occasions, African horse sickness and bluetongue virus may have spread to nonendemic areas in wind-borne *Culicoides* midges. Study of surface synoptic charts indicates a coincidence between the tracks of tropical storms and outbreaks of Ibaraki and bovine ephemeral fever in Japan, the inference being that infected insects are launched into these weather systems as they pass over Asian countries where the diseases are endemic. Seasonal winds associated with the annual movements of the Intertropical Convergence Zone suggest a regular vehicle for the reintroduction of JE to Japan and Korea from the tropics, and there are several reports of *Cx. tritaeniorhynchus* (the major vector of JE in that region) being collected at sea. A similar pattern has also been suggested for all the encephalitides in North America, including SLE.

A major problem with these ideas is that wind is not an isolated weather phenomenon; the movement of air masses is a major cause of changes of temperature, precipitation, humidity, and other parameters, and, as we have seen throughout this chapter, a case can be built for the involvement of any or all of these in arboviral recrudescence. Contemporary techniques for identifying geographical strains of viruses will resolve many of these speculations. To date, such evidence does not support the suggestion that SLE virus migrates from the tropics to North America.

D. Host Behavior and Other Factors

Although not strictly of vector origin, a few other factors deserve mention. During unusually hot weather, people may prefer to live and sleep out of doors, thus increasing exposure to mosquitoes and other vectors. On the other hand, when air conditioning is available, people may spend more time inside buildings, protected from vectors by closed doors and windows. In regions without piped water supply, increased domestic storage of water during drought may promote breeding by peridomestic vectors such as *Ae. aegypti*. In many countries, activities like berry picking, mushroom gathering, recreational camping, or hunting, may be greatly influenced by weather, and attract large numbers of people to areas where ticks and other vectors may be common. Finally, nectar flow in plants, which is an important food source for all mosquitoes, is strongly influenced by weather conditions.

IV. CONCLUSION

There is clearly a wealth of evidence that weather plays a role in arboviral recrudescence, and this can be explained in terms of vector biology. Behind most attempts to rationalize this role is the hope that weather forecasting and weather analysis might eliminate the element of surprise in arboviral epidemics. This is an attractive proposition, but the sobering truth is that arboviral epidemiology is a complex, multifactorial process, and that coincidental events involving some or all variables are the true precipitating factors for recrudescence.

A better understanding of the weather/arthropod portion of the equation is nonetheless worth pursuing, not only because of its potential contribution to epidemic forecasting, but also to improve our insight into the advisability of a vector control response. For example, if epidemics are triggered by brief bursts of activity in vectors which are otherwise hidden in inaccessible places, then a major investment in aerial adulticiding measures will not be justified unless operations can be linked to accurate short-term weather forecasts or effected in a season when the vector is more active.

What is clearly required is a comprehensive, in-depth monitoring of the recrudescence process, a blow-by-blow account which cannot begin when transmission has already accelerated, but must be in place and running during the interepidemic phase, however long this may last. All the factors contributing to vectorial capacity must be accounted for. New methods of age determination are required which indicate chronological rather than physiological age. Because infected insects are rare, even during peak transmission periods, emphasis needs to be placed on adequate sample size, even if this means very large collections. At the same time, the quantitative aspects of sampling must be strictly maintained, so that the dynamics of populations can be accurately portrayed. Improved sampling methods may be indicated, and in many cases need to be devised to target the infected cohorts of the vector population rather than aiming at the maximum number of specimens. Meteorological observations must be directly relevant to study areas, and include adequate appraisal of the microhabitats involved.

The effort required for a comprehensive project of this type may seem too great for the average research team to contemplate, but participation of several interested groups could spread the load of laboratory work involved. In addition, the initial phase could be viewed as archival; material would be routinely collected and stored under suitable conditions until an epidemic had come and gone. At that point, intensive processing would begin, working back from the time of known transmission. Whatever strategy is developed, an aggressive approach is clearly required to convert the wealth of random speculation on the weather/recrudescence relationship into useful information.

REFERENCES

1. **Masaki, S.,** Summer diapause, *Ann. Rev. Entomol.,* 25, 1, 1980.
2. **Denlinger, D. L.,** Dormancy in tropical insects, *Ann. Rev. Entomol.,* 31, 239, 1986.
3. **Beck, S. D.,** Insect thermoperiodism, *Ann. Rev. Entomol.,* 28, 91, 1983.
4. **Gillett, J. D.,** *Mosquitos,* Weidenfeld & Nicolson, London, 1971, 77.
5. **Macdonald, G.,** The analysis of equilibrium in malaria, *Trop. Dis. Bull.,* 49, 813, 1952.
6. **Garrett-Jones, C.,** The human blood index of malaria vectors in relation to epidemiological assessment, *Bull. WHO,* 30, 241, 1964.
7. **Reiter, P.,** Survival rates of overwintering *Culex pipiens* s.l. in Memphis, Tennessee, U.S.A., manuscript in preparation.
8. **Cumming, H. S. et al.,** Report on the St. Louis outbreak of encephalitis, Public Health Bull. No. 214, U.S. Treasury Department, Public Health Service, Washington, D.C., 1935.
9. **Lumsden, L. L.,** St. Louis encephalitis in 1933. Observations on epidemiological features, *Publ. Health Rep.,* 73, 340, 1958.
10. **Monath, T. P.,** Epidemiology, in *St. Louis Encephalitis,* Monath, T. P., Ed., American Public Health Association, Washington D.C., 1980, 239.
11. U.S. Department of Commerce and U.S. Department of Agriculture, Weekly Weather and Crop Bulletin Ser., 1933.
12. **Bowen, G. S. and Francy, D. B.,** Surveillance, in *St. Louis Encephalitis* Monath, T. P., Ed., American Public Health Association, Washington, D.C., 1980, 473.
13. **Mitchell, C. J., Francy, D. B., and Monath, T. P.,** Arthropod vectors, in *St. Louis Encephalitis,* Monath, T. P., Ed., American Public Health Association, Washington, D.C., 1980, 313.
14. **Sulaiman, S. and Service, M. W.,** Studies on hibernating populations of the mosquito *Culex pipiens* L. in southern and northern England, *J. Nat. Hist.,* 17, 849, 1983.
15. **Reiter, P. and Francy, D. B.,** Weather-related variations in seasonal activity patterns of urban *Culex* mosquitoes and their implications in the epidemiology of St. Louis encephalitis, manuscript in preparation.
16. **Madder, D. J., Surgeoner, G. A., and Helson, B. V.,** Number of generations, egg production, and developmental time of *Culex pipiens* and *Culex restuans* (Diptera:Culicidae) in southern Ontario, *J. Med. Entomol.,* 20, 275, 1983.
17. **Reiter, P.,** A standard procedure for the quantitative surveillance of certain *Culex* mosquitoes by egg raft collection, *J. Am. Mosq. Control Assoc.,* 2, 219, 1986.
18. **Eldridge, B. F., Bailey, C. L., and Johnson, M. D.,** A preliminary study of the seasonal geographic distribution and overwintering of *Culex restuans* Theobald and *Culex Salinarius* (Diptera:Culicidae), *J. Med. Entomol.,* 9, 233, 1972.
19. U.S. Weather Bureau, Climatological Data: Missouri Section, Vol. 37, 1933.
20. **Provost, M. W.,** The Natural History of *Culex nigripalpus,* Fla. State Bd. Health Monogr. No. 12, 46, 1969.
21. **Edman, J. D.,** Host-feeding patterns of Florida mosquitoes. III. *Culex (Culex)* and *Culex (Neoculex), J. Med. Entomol.,* 11, 95, 1974.
22. **Tempelis, C. H., Reeves, W. C., Bellamy, R. E., and Lofy, M. F.,** A three-year study of the feeding habits of *Culex tarsalis* in Kern County, California, *Am. J. Trop. Med. Hyg.,* 14, 170, 1965.
23. **Omer, S. M. and Cloudsley-Thompson, J. L.,** Survival of female *Anopheles gambiae* Giles through a 9-month dry season in Sudan, *Bull. WHO,* 42, 319, 1970.
24. **Sellers, R. F.,** Weather, host and vector — their interplay in the spread of insect-borne animal virus diseases, *J. Hyg. (Cambridge),* 85, 65, 1980.
25. **Pedgley, D. E.,** Wind-borne spread of insect transmitted diseases of animals and man, *Philos. Trans. R. Soc. Lond. Ser. B.,* 302, 463, 1983.
26. **Haeger, J. S.,** personal communication.

Chapter 10

VERTEBRATE HOST ECOLOGY

Thomas W. Scott

TABLE OF CONTENTS

I. INTRODUCTION

Vertebrates are one of three components that are necessary for transmission of an arthropod-borne virus (arbovirus); the three essential parts are the vertebrate host, the virus, and the arthropod vector. The principal role of a vertebrate host is to become viremic; i.e., to circulate virus in its peripheral blood. If the viremia is sufficiently high, the virus will infect blood-feeding vectors. The vertebrate-vector-vertebrate exchange of virus is perpetuated by arthropods that imbibe viremic blood, become infected, and then biologically transmit virus by biting susceptible vertebrates during refeeding. Because a single viremic vertebrate may serve as a source of infection for multiple vectors, vertebrates play an important role in the amplification of virus transmission activity.

This chapter reviews the interactions of vertebrates with viruses, arthropods, and environmental factors, and the ways in which these events affect arbovirus transmission. The discussion includes research conducted from the early 1900s through 1986 and focuses on nonhuman vertebrates. Human epidemiology is beyond the scope of this chapter. Moreover, only rarely, such as with dengue (DEN) or yellow fever (YF) viruses, are humans important contributors to virus transmission.[1] Most arboviruses that infect humans are maintained in zoonotic cycles in which humans are incidentally infected but not required for maintenance of transmission.

Literature concerning the role of vertebrates in arbovirus cycles is voluminous. Citations have therefore been selected that fortify key issues. Many of these articles or books also summarize pertinent literature and should be examined for those references. For additional reviews on vertebrate hosts of arbovirus diseases, see References 2 to 19.

II. HISTORICAL REVIEW

Arbovirology as a specific area of public health research began with the YF studies of Carlos Finlay and Major Walter Reed at the beginning of the 20th century.[20] During 1901, Reed and his associates showed that subcutaneous inoculation of serum from patients with YF into susceptible volunteers resulted in disease.[20] Thus, they demonstrated that the agent was present in the blood of the infected patient.

In 1927, A. F. Mahaffy made another important discovery when he inoculated blood from an infected African man into a rhesus monkey *(Macaca rhesus)*[20] and the monkey died 4 days later. This experiment established that animals other than humans could be infected with the YF agent and opened the way for a myriad of field and laboratory investigations. During the same year, Stokes, Bauer, and Hudson showed that *Aedes aegypti* mosquitoes could transmit YF virus among monkeys.[20]

Most researchers in the early 1900s believed that YF virus was confined to urban areas by a transmission cycle involving humans and *Ae. aegypti.* That attitude, however, soon changed with the discovery of the jungle or sylvan portion of the YF virus transmission cycle. After investigating an epidemic in Colombia during 1907, Roberto Franco and his co-workers concluded that patients had contracted YF virus infections in the forest.[20] This observation was largely ignored until 1932 when Fred L. Soper asserted that YF virus transmission was not exclusively the result of a human-*Ae. aegypti*-human cycle. Instead, YF was maintained in tropical forests by other mosquito species and nonhuman vertebrate hosts, humans being accidental recipients of the virus.[20] Subsequent studies in South America and Africa showed that nonhuman primates were the principal vertebrate hosts in the sylvan cycle of YF virus.[20]

Early research on YF virus revealed a transmission pattern that would appear repeatedly as additional arbovirus cycles were delineated. That pattern involved zoonotic transmission among nonhuman vertebrates by blood-sucking arthropods, with the occasional infection of humans who entered areas of active virus transmission.

At the same time that researchers were studying YF virus-infected primates in South America and Africa, other arbovirus diseases and their vertebrate hosts were beginning to be investigated elsewhere in the world. For example, during 1926, the role of humans as a source of DEN virus for the vector *Ae. aegypti* was described.[3,21] Meyer and his associates isolated western equine encephalomyelitis (WEE) virus from horses in the San Joaquin Valley of California during 1930.[6] Three years later, eastern equine encephalomyelitis (EEE) virus was first recovered from sick and dying horses on the east coast of the U.S.[1] In 1938, Ten Broeck[22] suggested that birds might be vertebrate hosts of EEE virus. In 1940, vertebrates were first associated with the transmission of St. Louis encephalitis (SLE) virus during an epidemiological investigation in Yakima Valley, Washington.[17] Studies by Reeves and Hammon[6] during the early 1940s revealed that songbirds were the vertebrate hosts of WEE and SLE viruses in California.

As the number of recognized arboviruses increased to the current figure, which exceeds 500,[23] the contributions of other vertebrate hosts to specific transmission cycles were described. In spite of the observation that blood-sucking arthropods feed on all four classes of terrestrial vertebrates,[24,25] birds and mammals have been found to be the principal vertebrate sources of virus for vectors. Reptiles and amphibians have been infrequently or inconclusively associated with arbovirus transmission.[7,12,26-28] However, for reasons that are discussed later, their role in virus overwintering should be studied more thoroughly.

Most arbovirus studies that addressed vertebrate host ecology focused on the analysis of blood for virus or antibody, the distribution and abundance of potential hosts, vector-host interactions, or experimental infections. After 85 years of arbovirus research, there is still much to learn concerning how vertebrates affect arbovirus transmission and, conversely, how vectors and viruses affect vertebrates.

III. FACTORS AFFECTING THE ROLE OF VERTEBRATE HOSTS IN ARBOVIRUS TRANSMISSION CYCLES

Various factors influence the extent to which vertebrates participate in arbovirus transmission cycles. In general, an important vertebrate host is part of the following scenario. The host is present in large numbers and is readily accessible to vectors in time and space. The host is attractive to arthropod vectors and allows vectors to feed upon it. The host is susceptible to virus infection, experiences low mortality from infection, and becomes viremic with a titer of sufficient magnitude and duration to infect susceptible blood-feeding vectors. The life history of the host proceeds in such a way that immunologically susceptible individuals enter the population at times of active transmission. Host herd immunity remains low. Host mobility may also be important; for example, sedentary hosts may be easy for vectors to locate, and migratory or mobile hosts could expand the distribution of the virus.

All vertebrates do not contribute equally to virus transmission, and the role of a given species, group of species, or individuals within a species in a transmission cycle varies for different viruses and geographic locations. In fact, local or even microenvironmental factors can strongly influence a vertebrate's role in virus transmission. The best single indicator of active participation in a transmission cycle is frequent virus isolation from free-ranging vertebrates. Many viruses are seldom recovered from wild vertebrates, however, because viremias are usually brief. Consequently, antibody surveys are often conducted to obtain evidence of infection. Ideally, results from these investigations are compared to laboratory studies on viremia and antibody responses of experimentally inoculated vertebrates.

Experimental transmission studies with the suspected host and vector species should be attempted to validate assumptions about the host's role in transmission. It is also important to demonstrate that arthropod vectors regularly obtain blood meals from the suspected host. Antibody studies alone should be interpreted with caution for two reasons. First, vertebrates

Table 1
SUMMARY OF TERMS USED TO DESCRIBE PARTICIPATION OF VERTEBRATES IN ARBOVIRUS TRANSMISSION CYCLES

Term[a]	Definition
Reservoir host Enzootic host Maintenance host Amplifying host Primary host	Species required for virus transmission. These species constitute the vertebrate portion of cycles. They are susceptible to virus infection, have a viremia of sufficient magnitude and duration to regularly infect susceptible blood-feeding vectors, are attractive to vectors and allow them to feed upon them, have a life history that generates immunologically susceptible individuals when virus is being actively transmitted, and are present at high population densities during times of virus transmission.
Incidental host Secondary host	Species not consistently involved in enzootic transmission for one or more of the following reasons: they do not produce a viremia sufficient to regularly infect vectors, they are not regularly fed upon by enzootic vectors, their population densities are too low to maintain continuous transmission, or peaks in the population do not occur when virus is being actively transmitted.
Disseminating host	Species that move virus from an area of active transmission to another location. They usually possess characteristics of a reservoir host. This transfer activity could supplement current transmission, exchange virus between enzootic foci, or introduce virus into areas where transmission had not previously occurred.
Dead-end host	Species that have no involvement in virus transmission for one or more of the following reasons: they do not have a viremia of sufficient magnitude to infect vectors, they are not fed upon by vectors, or they are disassociated in time and space from virus transmission.

[a] Synonyms are listed under each term.

can produce antibodies without becoming viremic or after having a viremia that was too low to infect vectors. These vertebrates would not contribute to virus transmission.[1,29] Second, antibody titers can decline to undetectable levels.[30]

After conducting these types of analyses, the relative contributions of different species or individuals within a species can be defined. Terms that are used to describe a vertebrate's role in arbovirus cycles are summarized in Table 1. Table 1 shows that the degree of vertebrate participation typically varies from participation that is critical to transmission maintenance, to participation only under certain specific conditions, to no participation at all.[17,31]

A. Attractiveness of Vertebrate Hosts to Arthropods

The first phase of the vector-host interaction that can lead to virus transmission is the attraction of vertebrates to host-seeking arthropods. An important vertebrate host must be readily available to vectors for acquisiton or transmission of a virus infection. The attributes of vertebrates that attract arthropods have been most thoroughly studied for mosquitoes, although attractants for other insects and ticks have also been examined (see Chapter 7).[25,32]

Investigations of the attraction of vectors to hosts have used several methodologies: olfactometers, baited traps, behavioral observations of responses to attractants, removal of arthropods directly from hosts, or arthropod blood-meal analyses.[32-38] Traps that capture arthropods but prevent them from contacting hosts and olfactometers provide the best measures of attractiveness. These techniques examine arthropods' responses prior to their attempts to blood feed — a time when arthropods might modify their behavior or when vertebrates might try to behaviorally or physiologically repel them.

Because the cues emanated by vertebrates and subsequently detected by arthropods are diverse,[25,32] generalizations concerning host attraction are difficult to make. Such generalizations are useful, however, because they summarize important components of the vector-host interaction.[25] For example, host-seeking behavior of arthropods comprises a sequence of events, each of which is initiated by a particular stimulus or set of stimuli affiliated with the vertebrate host.[25,32] For visually active arthropods, such host cues as color, intensity, contour, movement, and size may be important for host location.[25,32,36,39] Sound has also been shown to be an important cue, especially for ticks.[37,40] Emanations from vertebrates that have been studied as attractants include CO_2, humidity, temperature, amino acids, sex hormones, carbamino compounds, lactic acid, and aggregation pheromones released by feeding arthopods.[25,36,37,39,41-43]

None of these factors alone is as attractive as the combined factors presented by the entire host animal. This phenomenon may explain why it has been so difficult to determine empirically exactly what attracts arthropods to vertebrates.[25] In addition, arthropods probably use multiple cues to locate vertebrate hosts. Consequently, we have a better understanding of the hosts that vectors feed on than why vectors are attracted to vertebrates in the first place.

B. Permissiveness of Vertebrate Hosts to Arthropod Blood Feeding

After an arthropod has located a vertebrate host, it must settle on that host and attempt to obtain a blood meal. Vertebrates, however, are not always receptive to ectoparasites' feeding attempts and may repel them either behaviorally or immunologically.

Most investigations of vertebrate defensive behavior against blood-sucking arthropods have been conducted with mosquitoes and birds or mosquitoes and small mammals (see Chapter 7). Initial studies revealed that hosts can modify mosquito feeding patterns by reducing mosquitoes' ability to imbibe blood.[44] Subsequent research showed the following:

1. Tolerant hosts were fed on more often than were nontolerant hosts.[45]
2. Host defensive behavior reduced the biting persistence of mosquitoes[46] as well as the amount of blood that mosquitoes imbibed.[47] Laboratory[44,48] and field[49] studies categorized and quantified antimosquito behaviors of several bird species, including the head shake, foot stamp, and wing flip.
3. Adult birds were generally, but not always, more defensive toward mosquitoes than were nestlings.[50,51]
4. Hosts could learn to increase their efficiency at behaviorally repelling mosquitoes.[52]
5. When hosts were behaviorally active, mosquito feeding success was lowest.[53]
6. Increasing the number of host-seeking mosquitoes decreased mosquito feeding success because hosts became more defensive.[54]

There have been at least four studies designed to determine the effect of host health on mosquito blood-feeding success. After conducting field and laboratory experiments, Mahon and Gibbs[55] concluded that Sindbis (SIN) virus-infected chickens attracted more mosquitoes than did noninfected control birds. Turell and co-workers[56] demonstrated that after simultaneous exposure to infected and noninfected lambs, there were more mosquitoes containing Rift Valley fever (RVF) virus than there were mosquitoes in which virus was undetectable. They did not, however, determine the mechanism underlying their observation. Although Day and Edman[57] did not study an arbovirus, they did show that mice infected with *Plasmodium* spp. became lethargic and did not behaviorally repel mosquitoes as often as did control mice. Mosquitoes fed most often on infected mice, and infected mice were the least behaviorally defensive, when gametocytes in their blood were most infective to mosquitoes. Results from these three studies support the idea that mosquitoes feed most often on infected hosts, and therefore the number of infected mosquitoes in the environment is larger than if all hosts were fed upon equally. This increase could correspondingly increase the probability of successful virus transmission and the likelihood of epidemic disease.

Recently, Edman and Scott[51] tested this idea with WEE and SLE viruses in house sparrows (*Passer domesticus*). Their results did not support the hypothesis. Neither virus reduced the defensiveness of adult sparrows toward mosquitoes. Infected adults inoculated with $10^{3.5}$ BHK $TCID_{50}$ of virus showed no signs of illness. Although nestling sparrows were more susceptible to clinical disease than were adults, especially with WEE virus, they did not demonstrate antimosquito behavior. Thus, virus infection did not increase mosquito feeding on adults or nestlings due to a modification in host antimosquito behavior.

Edman and Scott[51] did, however, observe increased defensiveness by nestling sparrows at approximately 12 to 16 days of age. Future studies on this topic should examine the association between the ontogeny of host defensive behavior and increased susceptibility to clinical disease of young vertebrates as well as the impact of infection on host attraction. Virus infection at the appropriate age might increase the period of time young hosts allow mosquitoes to feed on them. Similarly, although host antimosquito behavior was not modified in the above study, attraction might be, and attraction was not examined.

Collectively, studies on hosts' defensive reactions to arthropod annoyance vary with regard to five host-associated criteria.[50,51] Host species is the most important factor, followed by body size, host age, individual variation within a species, and health status. Although the biting density of arthropods is important, these terms describe the most important host-intrinsic factors.

In addition to behaviorally repelling potential arthropod vectors, vertebrates exhibit other kinds of resistance. Early studies showed that cattle could have either innate or acquired resistance to feeding by ixodid ticks.[58] Innate resistance can be passed to offspring and appears to have a genetic basis. The proposed mechanisms for this kind of resistance include physical features of the cattle, cell-mediated immunity, and diet.

Acquired resistance to tick feeding has also been observed in cattle,[59] but has been most thoroughly documented for guinea pigs and rabbits.[58] After previous exposure to ixodid ticks, some vertebrates mount an immune-mediated response that reduces the following: number of attached ticks, tick feeding success, tick moulting success, number of tick eggs laid, and female tick fertility.[58] In guinea pigs, immunity is a response to tick salivary, and possibly gut, antigens; IgGl antibodies and immune cells recruit mixed leukocyte populations to cutaneous sites of tick attachment and feeding.[58] Interaction between basophils, the cell type that is most often dominant, and eosinophils produces the observed rejection. Details of the role of basophils in this reaction are not well known, but a recent study suggets that basophil mediators other than histamine are involved in resistance.

The impact of vertebrate immunity on rapidly feeding arthropods has not been explained as thoroughly as it has been for ixodid ticks, which remain attached and feeding on their hosts for several days. Some responses of vertebrates to fast-feeding vectors are similar to those dicussed above for ixodid ticks. For example, mosquito death rate increased and fecundity decreased after mosquitoes fed on hosts that had been previously immunized and hypersensitized, respectively.[58] Fertility and survival were reduced for sand flies *(Lutzomyia longipalpis)* after they fed on guinea pigs that their cohorts had fed on earlier.[58] However, multiple exposure of a host to sand flies and argasid ticks did not increase the host's ability to reduce arthropod feeding success or reduce the amount of blood the arthropod imbibed.[58,60] Apparently, immune-mediated resistance in vertebrates is less effective against rapidly feeding vectors like mosquitoes, sand flies, and argasid ticks because leukocyte recruitment to bite sites requires more time than blood feeding by these fast feeders.[60,61]

A laboratory study by Feinsod and co-workers[62] presents a provocative hypothesis concerning the possible effect of acquired host immunity to blood-feeding arthropods on susceptibility to arbovirus infection. They speculated that some vertebrates that have been naturally fed upon by arthropod vectors may have some protection from virus infection. Those vertebrates could have antibodies to antigens located in the envelope of the virus. Indeed, certain arboviruses incorporate proteins from the host's plasma membranes into their envelope. Moreover, the above researchers showed that SIN virus that had been propagated twice in *Ae. aegypti* mosquitoes was neutralized more efficiently by sera from guinea pigs vaccinated with mosquito extracts or fed upon by mosquitoes than was virus that had been passed in a mammalian cell culture (VERO cells).

Because most studies on the permissiveness of vertebrates to arthropod blood feeding have been conducted in the laboratory, the full effect of innate and acquired resistance on vector- or virus-host interactions in nature is not known. Field studies should be conducted to validate laboratory observations, to determine the general applicability of these processes, and to assess the epidemiological impact of resistance on transmission of arboviruses.

C. Distribution and Abundance of Vertebrate Hosts

In order for a vertebrate host to regularly attract arthropods and allow blood feeding, the host must be present in appropriate places and times as well as in sufficient numbers. If vertebrates are separated from vectors in time and space or are rare, vectors will have difficulty locating them, and the probability is low that they will make a significant contribution to virus amplification. Vertebrates that are important to the amplification of an arbovirus usually are abundant and are present when and where vectors are present.

Host distribution and abundance, therefore, are important factors that can affect the spatial and temporal arrangement of arbovirus transmission. This concept is demonstrated by several virus transmission cycles with dimensions that are defined by colonies of breeding birds. In the western U.S., Fort Morgan (FM) virus is transmitted by the nest-dwelling swallow bug *(Oeciacus vicarius)* among cliff swallows *(Hirundo pyrrhonota)* and house sparrows that nest in large, compact colonies.[63] Virus transmission occurs at nesting sites during the approximately 4-month period when birds are present and reproductively active. Transmission is positively correlated with the abundance of nestling birds.

The nesting activities of other birds similarly define the time and place for arbovirus transmission. In the North Atlantic, Kemerovo (Great Island and Bauline) and Sakhalin (Avalon) group viruses are transmitted by the tick *Ixodes uriae* within colonies of nesting sea birds.[64] A similar phenomenon has been reported for Mono Lake virus and gulls in California,[65] and for Japanese encephalitis (JE) virus and ciconiiform birds in Japan.[66,67] In all three cases, transmission is limited or accelerated by the presence of compact populations of nesting avian hosts.

Within these colonies of avian hosts, vertebrate infections are not evenly distributed. Scott and associates[63] observed that two or more FM virus-infected nestling house sparrows were frequently clustered within certain nests. The same clustering effect was reported for nestling house sparrows infected with WEE virus in Hale County, Texas[68] and humans infected with DEN virus in Puerto Rico.[69] These findings suggest that at any given time, arbovirus transmission occurs in foci composed of clusters of infected hosts. The size of a focus and the distribution of infected hosts within them are dynamic; both will change over time as virus amplification increases or decreases.

Unfortunately, the temporal and spatial dimensions of transmission for most arboviruses are more difficult to define than for those associated with colonial birds. For viruses with widely dispersed hosts, locating and examining large numbers of animals is difficult. The specific effects, therefore, of host distribution and abundance on transmission of these viruses is not as well understood. In general, however, the appearance of susceptible animals within the population and the size and density of the host population are critical factors in determining the amount of virus transmission.

If a vertebrate host is essential for maintenance of virus transmission and all other factors remain the same, the amount of transmission should be positively associated with bursts and crashes in that host species' population density. The above-cited studies with colonial birds are dramatic examples of this idea. When dense and large populations of nesting birds are present, transmission occurs. When birds leave nesting sites, transmission ceases. The transmission activity of viruses that have small mammals as hosts may similarly fluctuate in accordance with shifts in host population density. Endemic benign nephropathy (EBN) is apparently not transmitted by an arthropod, but it is a zoonotic virus disease of humans that is maintained in small mammals, such as *Clethrionomys* spp. In Finland and Sweden, the number of cases of EBN is positively associated with increases and decreases in the abundance of these rodent hosts.[70-72]

Examples of a positive correlation between fluctuations in rodent populations and the frequency of arbovirus transmission have been difficult to document, even though this association may strongly influence some arbovirus cycles. One noteworthy example concerns rodent populations in the Bush Bush Forest of Trinidad, which dropped during 1964 to almost undetectable levels, while at the same time mosquito vector populations remained high.[73] From all sources examined, only one kind of arbovirus was isolated, whereas seven different viruses had been recovered in previous years. Investigators concluded that the principal vertebrate host, forest-dwelling rodents, was rare and that an insufficient number of alternative hosts was available to maintain detectable transmission of the other viruses.

In addition to becoming viremic and amplifying viruses, vertebrates can also contribute to disease transmission by serving as abundant, noninfected blood sources for the reproduction of arthopod vector populations. As noted by Simpson,[13] cattle are important sources of blood meals (but not virus) for adult ticks that transmit Kyasanur Forest (KF) disease and Russian spring-summer encephalitis (RSSE) viruses. Movement of cattle into areas where these diseases are transmitted could increase tick populations and correspondingly increase the risk of disease transmission to susceptible species.

D. Susceptibility of Vertebrate Hosts to Virus Infection

In order to make a significant contribution to transmission cycles, a vertebrate must develop a viremia that will infect enough vectors to perpetuate vector-host transmission. Hosts that do not contribute to transmission are often called "dead-end hosts" because they do not become viremic or do not have a viremia of sufficient magnitude to infect arthropod vectors.[13] Several intrinsic factors influence the host's viremic response, including species-specific variation, genetics, age, health, and immune status.[74,75] Extrinsic factors include the type of virus inoculated, its virulence characteristics, and the dose of the inoculum.[75]

Variation in susceptibility to infection within a species can be strongly influenced by host genetics. Early studies by Sawyer and Lloyd[76] and Webster[77] showed that lines of mice could be bred to be resistant to YF, louping ill, or SLE virus. Recent studies with C3H mice have shown that resistance is inherited as an autosomal dominant allele.[78] Virus in resistant mice replicates to lower titers than in susceptible mice. Infection spreads more slowly and is usually self-limiting. Although the mechanism is not thoroughly understood, it is known that a functioning immune system is required for resistance, and that the abundance of defective interfering virus particles is positively associated with reduced susceptibility.[75,78] One explanation is that the resistance-gene product modifies virus replication by interfering with the synthesis of virus RNA.[79] In this way, resistant mice may be able to hinder virus replication so that they reduce extraneural virus replication, slow down neuroinvasion, and allow their immune systems enough time to successfully suppress the disease.[75,78]

Although there are few field studies that clearly demonstrate a genetic basis for vertebrate susceptibility or resistance, the following three studies suggest that the concept may be an important factor in the outcome of naturally acquired infections. In a study with RVF virus and sheep, Fagbami and his associates[80] concluded that West African dwarf sheep might be naturally resistant to infection as compared to other breeds. Another example concerns the 1981 epidemic of DEN in Cuba where the incidence of DEN hemorrhagic fever was higher in whites than in blacks.[81] Finally, Darnell and co-workers[82] showed that wild mice collected in California and Maryland possessed the flavivirus resistance gene.

In susceptible vertebrates, viremias are usually brief, lasting 2 to 5 days, and apparently this is sufficient time to regularly infect vectors.[7,12] Some viruses in certain hosts, however, circulate for extended periods of time;[7] for example:

1. SLE viremias in sloths *(Choloepus hoffmanni* and *Bradypus variegatus)* examined in Panama lasted 7 to 27 days.[83]
2. Turtles and snakes inoculated with EEE virus had viremias as long as 28 to 53 days.[26,84]
3. Texas tortoises *(Gopherus berlandieri)* inoculated with WEE virus were viremic for as long as 105 days. Ambient temperature affected the magnitude and duration of their viremias.[85]
4. A porcupine *(Erethizon dorsatum)*[86] and a deer mouse *(Peromyscus maniculatus)* were viremic with Colorado tick fever (CTF) virus for 20 to 50 days.[7]

Hibernation appears to accentuate the phenomenon of unusually long viremia by suppressing or altering the immune system.[7] Experimentally infected bats were viremic with SLE and JE viruses for 15 to 30 days.[87] If bats were hibernating while they were experimentally infected, this time interval was increased to 42 days for EEE virus,[88] 90 days for VEE (Venezuelan equine encephalitis) virus,[89] and 107 days for JE virus.[90] Similarly, hibernating snakes inoculated with WEE virus remained infected throughout one winter.[7]

Most evaluations of vertebrate viremia and antibody responses have been conducted in the laboratory, and very few laboratory observations have been validated in the field. Of course, researchers cannot experimentally infect free-ranging vertebrates for reasons of biological safety, but the virus titer of blood from naturally infected vertebrates can be determined. Field validation is necessary to determine the epidemiological relevance of laboratory studies on pathobiology, viremia response, or arthropod infection and transmission. For example, McLean[91] showed that laboratory studies with wild animals could be misleading. Stressed cotton rats *(Sigmodon hispidus)* became infected with Keystone virus more easily and had a viremia of greater duration and magnitude than did unstressed rats. Stress could occur from the unnatural conditions of captivity.

For many reservoir hosts, virus infections are mild or clinically inapparent.[13] If infections were lethal, they could reduce the size of the host population, limit the potential for host reproduction, and reduce the introduction of susceptible animals into the transmission cycle. Under these circumstances, if the virus is restricted to a narrow range of host species, it could reduce the probability of its own transmission. However, if the virus has a broad host range (i.e., it can switch species of vertebrate hosts), it can severely affect certain host species without becoming extinct itself.

A vertebrate responds to infection by producing antibodies that protect it from future infection with the same or even with related viruses.[74,92] A previously infected vertebrate is then permanently removed from the pool of potential viremic hosts. An unfortunate exception to this generalization is the apparent immune enhancement that occurs in humans who acquire DEN hemorrhagic fever/DEN shock syndrome (DHF/DSS).[75,93,94] Halstead[93] has advanced the idea that DHF/DSS almost exclusively occurs in a small portion of the persons who are sequentially infected with different strains of DEN virus. Although our understanding of the factors that lead to DHF/DSS is incomplete, laboratory studies with cell cultures, monkeys, and human blood suggest that the central concept is IgG antibody-dependent enhancement

Table 2
SUMMARY OF PASSIVE ANTIBODY TRANSFER FROM FEMALE BIRDS TO THEIR OFFSPRING

Virus to which antibodies were directed	Bird species in which antibody transfer was observed	Max. duration of detectable passive antibody	Antibody assay	Ref.
WEE	Pigeon *(Columba livea)*	ND[a]	Neut[b]	103
	Dove *(Streptopelia* spp.)	8 weeks	Neut	104
	Chicken *(Gallus gallus)*	4 weeks	Neut	104
EEE	Pigeon	33 days	Neut	105
	White ibis *(Eudocimus alba)*	ND	Neut	105
SLE	Pigeon	ND	Neut	103
	Dove	4 weeks	Neut	104
	Chicken	2 weeks	Neut	104
	Chicken	4 weeks	HI[c]	106
JE	Chicken	4 weeks	Neut	107
	Japanese cormorant *(Phalacrocorax capillatus)*	ND	Neut	107
	Black-crowned night heron *(Nycticorax nycticorax)*	ND	Neut	107
	Little egret *(Egretta garzetta)*	ND	Neut	107
	Plumed egret *(Egretta intermedia)*	ND	Neut	107

[a] Duration of detectable antibody not determined.
[b] Mouse neutralization test.
[c] Hemagglutination inhibition test.

of DEN virus replication in monocytes. Immune enhancement occurs when anti-DEN IgG-DEN virus complexes attach to monocytes that bear Fc receptors.[75,78,93,94] Infected monocytes are then targets for removal by the immune system, a process that may be T-cell mediated. Chemical mediators of shock are released, and unknown factors increase vascular permeability, which results in shock and hemorrhaging.[75,78,95] Recent epidemiologic studies on DHF/DSS in Thailand support the antibody-dependent enhancement hypothesis.[96]

Additional studies may show that immune enhancement is important for other arbovirus diseases. The phenomenon has been reported repeatedly for several different arboviruses in cell cultures,[79] and Cardosa and colleagues[97] have described another type of enhancement that is dependent on the CR3 complement receptor.

In general, however, immunity protects vertebrates from subsequent infection and, therefore, if herd immunity is high, vector-host transmission can be interrupted because infected arthropods seldom feed on susceptible or infected hosts. When VEE virus entered Texas during 1971, vaccination of horses, the principal vertebrate host in that area, may have been an important factor in the abortion of that epidemic.[98,99]

In addition to being protected by acquired immunity, some vertebrates are protected from infection by passively obtained maternal antibodies, for example through the egg of birds,[100,101] or colostrum of mammals.[102] Table 2 summarizes five studies that demonstrated the transfer of passive antibodies by female birds to their progeny.

In mammals, Issel[102] showed that white-tailed deer fawns *(Odocoileus virginianus)* could obtain neutralizing antibodies against Jamestown Canyon (JC) virus from their mothers' colostrum. These antibodies protected fawns from experimental infection with JC virus and persisted for 8 to 23 weeks. Similarly, studies with bluetongue (BT) virus have shown that domestic lambs can be passively protected against virus infection for up to 6 months.[108] After passive antibody has waned, these animals are susceptible to infection and can become viremic; an infection that occurs while the young host is protected by passive immunity does not confer an active immune response.

Although observed repeatedly, protection through passive immunity is not universal. In a study of naturally infected house sparrows, Holden and colleagues[101] concluded that passive immunity did not interfere with the transmission of WEE virus among nestling sparrows in Texas. Moreover, a recent study by Ludwig and co-workers[100] showed that 5- to 7-day-old and 14- to 16-day-old nestling house sparrows that had obtained passive antibody from their mothers exhibited viremic enhancement; i.e., their viremias were of greater magnitude and duration than those of controls. The observed viremic enhancement in sparrows may be similar to what has been reported for DHF/DSS in humans. In fact, some researchers speculate that human infants with passive antibodies to one DEN virus strain may have an increased susceptibility to DHF/DSS.[75,93]

The impact of vertebrate immunity on virus transmission is not limited to host-virus interactions. It can also modify vector-virus interactions. Thompson[109] showed that female *Ae. triseriatus* that had imbibed blood from La Crosse (LAC) virus-immune eastern chipmunks *(Tamias striatus)* were less likely to become venereally infected with LAC virus than were females that had imbided nonimmune blood. Similarly, Patrican and co-workers[110] observed that *Ae. triseriatus* did not transmit LAC virus after imbibing an artificial blood meal of deer blood containing neutralizing antibodies to JC virus that had been mixed with LAC virus.

E. Age of Vertebrate Hosts

Within a host population, some individuals may contribute more to arbovirus transmission than others. For several virus transmission cycles, young hosts have been shown to be a more frequent source of virus for vectors than were older conspecifics.[63,67,68,111]

Several explanations are possible for this increased contribution of young animals. First, older hosts have had a greater opportunity to contract an infection, respond immunologically, and be removed from the pool of susceptible animals. Second, young hosts are often more susceptible to virus infections, experience more severe disease, and develop viremias of higher intensity and longer duration than do older conspecifics.[74,75] Finally, very young vertebrates are often more likely to allow an arthropod to feed on them because they are less behaviorally developed and less defensive toward arthropods than are adults.[50,51,112,113]

F. Social Organization of Vertebrate Hosts

There is little information concerning the effect of vertebrate social organization on the maintenance and transmission of arbovirus infections. Consequently, much of the discussion in this section is speculation based on theory.

The potential for infectious microbiological agents like arboviruses to affect vertebrate evolution is obvious. By causing illness, viruses can affect the ability of individuals to survive and reproduce.[78,79] More dramatically, they can cause extinction.[114,115] In fact, Williams[116] considers the transfer of diseases to be the most important disadvantage of social behavior.

One important factor in determining the social organization of birds and mammals that serve as hosts of arboviruses is the distribution of a limiting resource, such as food or nesting sites.[117] Horn[118] showed that if food is evenly distributed, birds will spread themselves out and defend territories. If food is unevenly distributed (occurring in patches), birds will roost and nest in colonies and forage in groups. Thus, through the indirect pressure of a limiting resource, host social organization can affect the distribution and availability of vertebrates for virus amplification.

The colonial behavior of vertebrate hosts described above carries with it theoretical costs and benefits.[117] The transmission of FM virus serves as an example. Cliff swallows are colonial, ephemeral, and reproductively synchronized.[63,119] These birds reduce their exposure to ectoparasites and virus by producing large numbers of susceptible nestlings in a brief time period at unpredictable locations.[119] Swallow bugs have difficulty locating these hosts and reproducing rapidly enough to infest and transmit virus to a large number of swallow nestlings. However, a nestling's potential for infection is great if swallows or house sparrows initiate nesting colonies at sites already infested with virus-infected bugs or if house sparrows continue to nest at colony sites after swallows have departed and bug populations are large.[119] Under the appropriate conditions, reproductive synchronization can be beneficial by reducing nestling exposure to bugs and virus. Conversely, high host densities can incur a cost by increasing the probability of infestation and infection.[63,120]

Although host social organization can affect virus transmission, most ecologists doubt that infectious diseases are the sole or even principal selective pressure that has fashioned territorial behavior.[117,121]

The effect of vertebrate dominance status on arbovirus transmission cycles is a topic that has not yet been examined. Does social status affect the behavior, physical condition, or location of hosts in such a way that they are more or less likely to become infected and contribute to virus transmission? If this phenomenon occurs for arboviruses, then some individuals may be more important in virus transmission cycles than others because of their social rank. Christian[122] has suggested that for vertebrates there is an interaction among low social rank, stress, and subsequent increased susceptibility to disease.

G. Movement

Movement by viremic vertebrates can change the geographical distribution of arboviruses. If a host becomes infected in one area and then moves to another where it then infects blood-feeding arthropods, a new focus of transmission may develop. New enzootic foci may be initiated where virus has not previously occurred or where a virus has become extinct, or introduced virus could supplement existing transmission activity.

Because of their ability to rapidly fly long distances, birds and bats have been examined most often as vertebrate disseminators of arboviruses. There are dramatic examples of birds' potential for transporting arboviruses while they migrate and are viremic (see Section VI). The frequency of this kind of event and whether or not virus transmission is established by moving or migrating hosts is not known. Even if movement of arboviruses via flying hosts is rare, over time it could have a significant impact on the distribution and evolution of viruses.

The dissemination of virus by infected birds after nesting activities have been completed probably has a more important effect on virus distribution than does annual reintroduction of virus to initiate transmission.[17] Scherer and co-workers[67] concluded that JE virus was amplified in rookeries of ciconiiform birds. When conditions were right, viremic birds disseminated virus to other locations in Japan, which were populated by humans, and initiated virus transmission at these new locations. Subsequently, epidemic disease was reported in these areas. The suspected southern movement of EEE and Highlands J (HJ) viruses in eastern North America has been reported on several occasions.[123,124] In fact, during one autumn, Lord and Calisher[124] recovered 16 isolates of EEE virus and 2 of HJ virus from birds that were presumed to be migrating south in the eastern U.S.

It should also be noted that arboviruses could be introduced to new areas by the activities of humans. Humans infected with DEN virus and who travel on jet aircraft are the principal means whereby this virus is spread among continents. Similarly, the international exchange of livestock could contribute to the spread of disease. Regulations on the importation of livestock are in part designed to prevent the introduction of viruses like BT, African swine fever (ASF), and RVF.

H. Climate and Habitat

Vertebrate participation in transmission cycles often varies seasonally. Changes in the abundance of vertebrate hosts and arthropod vectors in areas of virus transmission are influenced by temperature in temperate areas and rainfall in the wet tropics. It is clear, however, that temperate seasonal changes more dramatically affect the abundance of vertebrates than do changes in tropical weather.[115]

Hosts, such as songbirds involved in the maintenance of EEE, WEE, and SLE viruses, annually move in and out of temperate foci. Cold temperatures force the migration of large numbers of temperate breeding birds to the tropics.[115] Thus, birds may leave an area of virus transmission at one latitude and enter a different transmission cycle at another latitude.

Climatic changes can in addition affect the behavior of mammals. Groundhogs *(Marmota monax),* hosts of Powassan (POW) virus,[125] and ground squirrels, hosts of CTF virus,[126] alternate between being behaviorally active during warm seasons and being dormant in their burrows during winter hibernation.

Habitat also affects the distribution of vertebrate hosts and can be used to predict areas of arbovirus transmission.[127,128] In a study of the structure of an arbovirus ecosystem, Carey and associates[128] used multivariate analyses to describe a Colorado habitat where CTF virus was transmitted. They concluded that transmission was maintained by the interaction of golden-mantled ground squirrels *(Spermophilus lateralis)* and least chipmunks *(Tamias minimus)* with wood ticks *(Dermacentor andersoni).* The Richardson's ground squirrel *(S. richardsonii),* which occupied areas of deep soil, confined the distribution of golden-mantled ground squirrels to rocky slopes with shallow soil. Least chipmunks ocurred in the same habitat as golden-mantled ground squirrels, but the numbers of least chipmunks were limited by interspecific agonistic behavior from golden-mantled ground squirrels. The authors also discussed habitat parameters for describing the presence of vertebrate hosts, ticks, and virus.

This kind of study is important because it is predictive — the information generated can be used to locate foci of CTF transmission for experimental studies or for disease control. For example, by avoiding areas of expected CTF occurrence, planners of new recreational site locations could reduce campers' exposure to disease.

IV. VERTICAL AND ORAL TRANSMISSION BY VERTEBRATE HOSTS

In addition to vector-host transmission, arboviruses can be transmitted by infected female vertebrates to their fetuses. Transplacental infection occurs when female hosts become infected during pregnancy; most infected fetuses are aborted or born dead.[129] JE has been transmitted in this way to humans,[130] swine,[131] mice,[132] and bats;[133] SLE virus to mice;[134] Wesselsbron disease virus to cattle;[135] VEE virus to humans,[136] mice,[137] and horses;[138] WEE virus to humans;[139] Ross River virus (RRV) to mice;[140]BT virus to sheep[129,141] and cattle;[142] CTF virus to mice;[143] Akabane virus to cattle, goats, and sheep;[129] and Nairobi sheep disease (NSD) virus to sheep and goats.[129] Although no virus was isolated from kits, litter size was reduced and stillbirths were increased when pregnant rabbits were inoculated with LAC virus.[19]

There is no convincing field evidence that vertical transmission of virus in vertebrates contributes to the maintenance of arbovirus transmission. In a hyperendemic area of WEE and SLE virus transmission in Texas, no virus was isolated from house sparrow and blackbird eggs that were artificially hatched in the laboratory.[101] In the laboratory, however, Gibbs and co-workers,[144] observed that lambs from ewes inoculated with BT virus were born viremic and that virus could be recovered from them for 2 months. BT virus has also been recovered from bovine semen, suggesting that the virus might be transmitted vertically through the venereal route.[145] A similar phenomenon has been reported for JE virus in swine.[75]

A provocative extension of the above experimental studies is that transplacental or transovarian infection could be a mechanism for maintaining virus during times when virus is not actively transmitted.[19,138,142] Infections could become latent and not recognized as foreign by the immune system in young animals, with viremia and infection of vectors occurring at a later date (see Section VI).

Hosts can also become infected by ingesting virus-infected vertebrates or arthropods. This process has been demonstrated in the laboratory with LAC virus.[19] Red foxes *(Vulpes fulva)* that ate LAC virus-infected eastern chipmunks or newborn laboratory mice later became viremic. Similarly, eastern chipmunks that ate LAC virus-infected mosquitoes became viremic. With EEE virus, however, Main[88] was unable to infect *Myotis* bats by force feeding them EEE virus suspensions or mealworms *(Tenebrio molitor)* inoculated with EEE virus.

The contribution of oral vertebrate infection to the maintenance and amplication of arboviruses is not known. Oral transmission of EEE virus by pecking among penned gamebirds has been recognized for more than 30 years,[146] but its role in enzootic EEE virus transmission or the transmission of other arboviruses is not as clear as the gamebirds example. The process might be important for cycles like that of enzootic VEE virus because this virus is transmitted by mosquitoes to small rodents,[147,148] and rodents are characteristically defensive toward host-seeking mosquitoes.[45,51] Part of the rodent's defensive behavior might include capturing and eating infected mosquitoes, which could result in a rodent infection.

Oral infection might also be important for several tick-transmitted flaviviruses, such as POW virus, that have been isolated from the milk of infected lactating goats.[149] Young goats or humans may become infected by drinking milk containing virus.

V. REASSORTMENT OF VIRUS GENOMES

Beaty and co-workers[150] have shown that bunyaviruses can reassort their three-segmented genomes and produce new strains of virus in mosquito vectors. Thus, these investigators demonstrated a mechanism of virus evolution in arthropods.

No studies have demonstrated in vivo reassortment in vertebrates. However, bunyaviruses have been shown to reassort in mammalian cell cultures, and there is some evidence from field collected specimens that reassortment occurs in nature.[150,151]

The role that vertebrates could play in the maintenance of new virus strains originating from reassortment has been investigated by Seymour and associates.[152] Chipmunks and snowshoe hares *(Lepus americanus)* were inoculated with LAC virus, snowshoe hare virus (SSH), or one of several reassortant viruses. All but one reassortant virus caused viremia, and the researchers concluded that the middle-sized (M) segment of the virus genome was most important for infection and production of neutralizing antibody. This study suggests that by becoming viremic and infecting mosquitoes, vertebrates could participate in the transmission of new reassortant viruses.

Comparative studies with reassortant and parent viruses have examined the ability of viruses to infect and cause disease in vertebrates. Using viruses in the California serogroup, Tignor and associates[153] studied virus virulence and tissue tropism in the laboratory mouse. They showed that the M segment coded for virulence and that virus caused encephalitis after it entered peripheral nerve endings and moved transynaptically to the brain.

VI. OVERWINTERING

In locations where virus transmission is not continuous throughout the year, vertebrates could possibly participate in the maintenance or overwintering of arboviruses. Mechanisms for viruses to survive during these periods of inactivity have been reviewed by Reeves.[7] The most plausible mechanisms that involve vertebrates include the annual reintroduction of virus and the chronic latent infection of vertebrates with relapsing viremias.

The annual reintroduction of viruses from tropical areas of continuous transmission to temperate foci of intermittent activity has been an area of active research for the past three decades. Birds were regarded as the most likely vertebrate carriers because of their seasonal migration patterns in and out of tropical areas and because they have been implicated as hosts of several arbovirus diseases.[7,17,124,154,155]

Although this hypothesis has not been rejected, neither has it been strongly supported. Several researchers have recovered virus from migrating birds, but none has shown that this process occurs annually. Watson and associates isolated Bahig, Ingwavuma, Q 3255, West Nile (WN), and several unidentified viruses from the blood of birds migrating across the Mediterranean Sea.[154] Takahashi and co-workers isolated JE virus from an arctic warbler *(Phyllosciopus borealis)* migrating into Japan.[154] Calisher and associates[156] isolated Mayaro virus from birds migrating to Louisiana.

Calisher and co-workers[154] conducted the most supportive study of virus reintroduction with EEE virus in birds migrating to North America from their South American wintering grounds. Viruses isolated from a blackpoll warbler *(Dendroica striatas)* and wood thrush *(Hylocichla mustelina)* sampled in Louisiana during May were identified as the South American strain of EEE virus. Casals[157] has shown that EEE virus strains in North and South America are antigenically distinct. These researchers therefore concluded that the birds had been infected in a South American focus of EEE virus transmission and then flew while viremic to Louisiana.

Calisher and co-workers[154] point out, however, that because North and South American strains of EEE virus are distinguishable, it is difficult to conceive of a mechanism whereby a South American virus is introduced annually, and then consistently switches to the same North American serotype. Dickerman and associates[155] concur in their analysis of migrating birds as disseminators of the epizootic strains of VEE virus: because of the geographic segregation of VEE virus strains, avian transport is probably not a regular event.

Furthermore, songbirds migrate from southern wintering grounds and arrive in North American breeding grounds during April and May.[111] However, isolations of EEE virus in the middle to northern U.S. are not recovered until June or July.[111,158] Most avian hosts do not migrate rapidly enough to leave their wintering grounds infected and arrive, still viremic, at their breeding sites.[7,17]

Based on current data, it is doubtful that birds regularly introduce virus to northern foci. The southern movement of viruses like EEE virus in birds that have recently completed breeding activities is probably a more common event than northern introduction of virus prior to the breeding season (see Section III.G). In fact, southern movement of virus in birds may contribute to EEE virus outbreaks that occur on Caribbean islands in temporal association with bird migration. Birds migrating from South to North America might on occasion be responsible for initiating early spring outbreaks in southern states (e.g., Florida or Louisiana). Alternative hypotheses for virus introduction include the following: (1) migrating birds have chronic infections that relapse under the stress of migration[7] or (2) virus is maintained in semitropical locations such as Florida by continuous transmission and then foci of transmission are slowly spread northward by movement of infected, migrating birds.[154]

Chronic infections, unusually long viremias, and atypical immunological responses have been reported for several vertebrates. Chamberlain and co-workers[159] recovered SLE virus from the gizzard of a cowbird 38 days after inoculation; JE virus was recovered from the organs of a pigeon after 39 days;[160] RSSE was recovered from several birds and rodents for 16 to 25 days;[7] and SLE virus was recovered from the brains of experimentally inoculated house sparrows 23 days after infection of nestlings and 81 days after infection of adults.[100] Monath[75] reviewed several other flaviviruses that produce persistent infections in vertebrates, including tick-borne encephalitis (TBE) virus that was recovered from monkeys 2 years after inoculation. He suggested that for a virus to avoid elimination by the immune system, it may be important for latent infections to occur in lymphoreticular tissues. Although these unusually long host-virus associations are interesting, none have been clearly demonstrated as a mechanism for the overwintering of an arbovirus.

Studies with BT virus in cattle and WEE virus in garter snakes *(Thamnophis* spp.) suggest that overwintering of those arboviruses involves chronically infected vertebrates. Luedke and co-workers[142] concluded that a Hereford bull exposed to BT virus *in utero* became chronically infected. Bites from the gnat (*Culicoides variipennis*) that transmits BT virus stimulated a viremia in the bull, which in turn infected gnats imbibing its blood. Although recurring viremia stimulated by vector feeding is an intriguing hypothesis, the concept remains unconfirmed by researchers in other laboratories.

In garter snakes *(Thamnophis* spp.) infected with WEE virus, Reeves,[7] Thomas et al.,[27] and Gebhardt et al.[161] reported that viremias were protracted in duration. However, these researchers were unable to isolate WEE virus from snakes sampled in the field.[27] In another study, Sudia and associates[162] isolated two WEE virus strains from 349 Texas tortoises sampled in south Texas. During the same collection period, sera from 2406 birds and 1690 small mammals did not contain detectable WEE virus. In Saskatchewan, Burton and co-workers[28] similarly recovered WEE virus from poikilotherms; blood from 2 of 202 garter snakes *(Thamnophis* spp.) and 6 of 155 leopard frogs *(Rana pipiens)* contained detectable virus during initial bleedings. In addition, 27% of the snake sera and 21% of the frog sera neutralized virus.

Although these studies document arbovirus infections in reptiles and amphibians,[28,161,162] these classes of vertebrates have been studied so infrequently that based on current data, it is difficult to clearly show that they are a regular component of arbovirus overwintering.[7] An important problem is that enzootic mosquito vectors apparently seldom feed on reptiles and amphibians, while mosquitoes that regularly feed on these species do not appear to be an integral part of virus transmission cycles. Nevertheless, data cited above are intriguing. Thus, the role of reptiles and amphibians in arbovirus cycles as an important mechanism for virus overwintering deserves further evaluation.

VII. EVOLUTION OF THE ROLE OF VERTEBRATE HOSTS IN ARBOVIRUS TRANSMISSION CYCLES

One of the most fascinating areas of arbovirology is the evolution of these complex host-parasite systems. Many researchers speculate that most arboviruses, depending on their taxonomic grouping, began as insect parasites and then secondarily infected vertebrates. Researchers frequently conclude that virulence of the virus in the vertebrate host is a measure of the degree to which the virus and vertebrate have coevolved. Resistance to disease could provide individual hosts with an intraspecific advantage.[75,78,114] Virions that increase their exposure to vectors by minimizing host mortality and minimize the negative effects of the vertebrate immune response could increase their probability of transmission; this is especially true of viruses with a narrow host range.[114]

Janzen[163] has defined coevolution as reciprocal changes by individuals in two populations; i.e., changes in individuals in one population are in response to individuals in another population and vice versa. Janzen points out that at any time the observed response of a host to a parasite may be strongly influenced by previous interactions. For arboviruses, the host or virus that most strongly influenced current traits may no longer be present in the habitat or may no longer participate in the transmission cycle. The coevolution of virus-host interactions may be just as likely to progress to independence as to develop a finely tuned benign relationship. We know very little about the past relationships of arboviruses with their vertebrate hosts, and without that information, it is not necessarily correct to assume that the two coevolved.

Regardless of the way in which virus-host relationships evolved, the potential for arboviruses to act as a strong selective force on vertebrate evolution has been documented repeatedly. Examples of vertebrate host mortality associated with arbovirus infection include the following: mantled howler monkeys *(Alouatta villosa)* infected with YF virus on Barro Colorado Island,[117] songbirds infected with EEE virus in Michigan,[164] horses infected with VEE virus in South and Central America,[98] nestling house sparrows infected with FM virus in Colorado,[63] and forest-dwelling monkeys infected with KF virus in India.[165] Because resistance to arbovirus infections of mice has been shown to be inherited as an autosomal dominant allele and arbovirus infection of susceptible vertebrates can result in permanent disorders or death, vertebrates that possess a resistance gene may have a distinct selective advantage over susceptible conspecifics.[78,79,166]

Monath[75] provides an example of how this might have occurred with avian hosts and louping ill virus. Fenner and Ratcliffe[167,168] demonstrated that over a period of years, European rabbits *(Oryctolagus cuniculus)* in Australia became increasingly resistant to myxoma virus. Although this is not an arbovirus, the rabbit-myxoma virus interaction clearly demonstrates the potential impact of viruses on vertebrate evolution.

Hosts can also affect virus evolution. The ability of a virus to replicate within a vertebrate host may affect the relative success of new virus strains that arise from mutation or genome reassortment. Host movement and distribution could affect viruses by introducing them into new areas where they would be maintained among different vectors and hosts, or by geographically isolating viruses so that over time, they develop into new strains.

VIII. ASSOCIATIONS OF VERTEBRATE HOSTS WITH HUMANS AND DOMESTIC ANIMALS: EFFECT ON EPIDEMIC DISEASE

As noted earlier, most arboviruses are maintained in zoonotic cycles. Occasionally, viruses leave these systems, causing epidemic disease in humans or domestic animals. The proximity of humans or domestic animals to enzootic vertebrate hosts is often positively associated with the frequency of disease.

There are numerous examples of this association, including the following:

- In the early work with YF virus, it was shown that wood cutters who entered the tropical forests of South America were at high risk of infection from virus circulating in the jungle transmission cycle.[2]
- YF virus-infected monkeys in Africa that wander into plantations can infect mosquitoes. Infected mosquitoes can then transmit the disease to human plantation workers.[2]
- WN virus infects rock doves *(Columbia livia)* and house sparrows that live in close association with humans.[8]
- JE virus frequently infects swine,[169] which in some locations live in the same quarters as their human owners.
- SLE virus has been repeatedly isolated in North America from house sparrows and rock doves[16,17,170,171] that live in close association with humans.
- Ground squirrels and chipmunks, which serve as hosts for CTF virus, prefer habitats frequented by campers in the North American Rocky Mountains.[126,128]
- A suspected mode of RVF virus transmission in Africa is the aerosolation of virus during the slaughter or necropsy of infected domestic animals.[172]
- SF virus can be introduced to domestic swine in Africa if they live in close association with infected wart hogs *(Phacochoerus aethiopicus)* or bush pigs *(Potamochoerus porcus)*.[173]
- Songbirds that live in close association with horses or exotic gamebirds have been associated with epizootics of EEE virus.[164]

- In Russia, humans contracted TBE infections by drinking infected goats' milk.[174] Serologically related viruses, such as POW virus, may also be transmitted to humans in this way.[149]

Constant efforts to control this kind of host association might reduce the risk of disease by eliminating large numbers of potential reservoir hosts or by forcing transmission foci to areas that are not as close to humans or domestic animals. However, reducing host populations during epidemics or at times of active transmission may be counterproductive. Without enzootic hosts to feed on, infected arthropods might increasingly feed on humans or domestic animals, and the result could be an increase in disease.[16,17]

IX. FUTURE INVESTIGATIONS

Although much has been learned about the ecology of vertebrates in arbovirus transmission cycles during the past century, more detailed studies that examine specific, contemporary hypotheses are needed. As our understanding of vertebrates' participation has increased, so has our realization that their role in these cycles is complex and that our current knowledge is fragmented.

An improved understanding of fundamental processes in vertebrate arbovirus ecology is needed to develop effective, multifaceted disease control programs. Most current arbovirus control strategies focus on vector control, vaccine development, or diagnosis and therapy. A better understanding of nonhuman vertebrate hosts should be added to this arsenal for long-term interruption of disease transmission.

Expanding our basic understanding of vertebrates in arbovirus transmission cycles requires a modern synthesis of knowledge in ecology, entomology, virology, and epidemiology, as well as in sociobiology, physiology, genetics, immunology, and pathology. Too many studies address only the more straightforward aspects of vertebrate participation in virus transmission, such as virus isolation and seroconversion rates of wild populations, or viremia and antibody responses of experimentally infected animals. These studies provide important baseline information, but more complex issues such as the mechanisms, direction, and rate of host-virus and host-vector evolution need to be addressed.

Advancing the study of vertebrate host ecology in arbovirology is an exciting challenge. Researchers can apply new molecular techniques, such as recombinant DNA methodology, monoclonal antibodies, and enzyme immunoassays, to questions concerning host-virus and host-vector interactions. To determine the epidemiological relevance of recent developments in molecualr virology, collaborative studies will be needed between experts on the molecular and ecological aspects of arbovirus transmission. Interdisciplinary studies like these are the most efficient way to make significant improvements in our understanding and, ultimately, control of these important diseases.

ACKNOWLEDGMENTS

Judith M. Grumstrup-Scott edited and typed this manuscript throughout its development. The text was improved by the reviews of G. G. Clark, A. J. Main, R. G. McLean, M. J. Raupp, T. G. Schwann, and D. M. Watts.

REFERENCES

1. **Chamberlain, R. W.,** Arbovirology — then and now, *Am. J. Trop. Med. Hyg.*, 31, 430, 1982.
2. **Strode, G. K.,** *Yellow Fever*, McGraw-Hill, New York, 1951, 639.
3. **Sabin, A. B.,** Research on dengue during World War II, *Am. J. Trop. Med. Hyg.*, 30, 1952.
4. **Kissling, R. E.,** Host relationship of the arthropod-borne encephalitides, *Ann. N.Y. Acad. Sci.*, 70, 320, 1958.
5. **Downs, W. G.,** Birds in relation to arthropod-borne viruses in Trinidad, *Proc. Int. Ornithol. Congr.*, 13, 581, 1963.
6. **Reeves, W. C. and Hammon, W. McD.,** *Epidemiology of the Arthropod-Borne Viral Encephalitides in Kern County, California, 1943—1952*, Vol. 4, University of California Press, Berkeley, 1962, 257.
7. **Reeves, W. C.,** Overwintering of arboviruses, *Prog. Med. Virol.*, 17, 193, 1974.
8. **Work, T. H.,** Virology in the biology of birds, *Proc. Int. Ornithol. Congr.*, 13, 570, 1963.
9. **Stamm, D. D.,** Susceptibility of bird populations to eastern, western, and St. Louis encephalitis viruses, *Proc. Int. Ornithol. Congr.*, 13, 591, 1963.
10. **Stamm, D. D.,** Relationships of birds and arboviruses, *Auk*, 83, 84, 1966.
11. **McClure, H. E.,** Birds and the epidemiology of Japanese encephalitis, *Proc. Int. Ornithol. Congr.*, 13, 604, 1963.
12. **Chamberlain, R. W.,** Arboviruses, the arthropod-borne animal viruses, *Curr. Top. Microbiol. Immunol.*, 42, 38, 1968.
13. **Simpson, D. I. H.,** Arboviruses and free-living wild animals, *Symp. Zool. Soc. London*, 24, 13, 1969.
14. **Theiler, M. and Downs, W. G.,** *The Arthropod-Borne Viruses of Vertebrates*, Yale University Press, New Haven, 1973, 443.
15. **Hoogstraal, H.,** The epidemiology of tick-borne Crimean-Congo hemorrhagic fever in Asia, Europe, and Africa, *J. Med. Entomol.*, 15, 307, 1979.
16. **McLean, R. G. and Scott, T. W.,** Avian hosts of St. Louis encephalitis virus, in *Proc. 8th Bird Control Seminar*, Jackson, W. B., Ed., Bowling Green State University, Bowling Green, Ohio, 1979, 143.
17. **McLean, R. G. and Bowen, G. S.,** Vertebrate hosts, in *St. Louis Encephalitis*, Monath, T. P., Ed., American Public Health Association, Washington, D.C., 1980, 381.
18. **Karstad, L. H.,** Arboviruses, in *Infectious Diseases of Wild Mammals*, Davis, J. W., Karstad, L. H., and Trainer, D. O., Eds., Iowa State University Press, Ames, 1981, 54.
19. **Yuill, T. M.,** The role of mammals in the maintenance and dissemination of La Crosse virus, in *California Serogroup Viruses*, Calisher, C. H. and Thompson, W. H., Eds., Alan R. Liss, New York, 1983, 77.
20. **Warren, A. J.,** Landmarks in the conquest of yellow fever, in *Yellow Fever*, Strode, G. K., Bugher, J. C., Kerr, J. A., Smith, H. H., Smithburn, K. C., Taylor, R. M., Theiler, M., Warren, A. J., and Whitman, L., Eds., McGraw-Hill, New York, 1951, 5.
21. **Siler, J. F., Hall, M. W., and Hitchens, F. H. K.,** Dengue, its history, epidemiology, mechanism of transmission, etiology, clinical manifestations, immunity and prevention, *Philippine J. Sci.*, 44, 1, 1926.
22. **Ten Broeck, C.,** Birds as possible carriers of the virus of equine encephalomyelitis, *Arch. Pathol.*, 25, 759, 1938.
23. **Karabatsos, N., Ed.,** *International Catalogue of Arboviruses Including Certain Other Viruses of Vertebrates*, 3rd ed., American Society of Tropical Medicine and Hygiene, San Antonio, Tex., 1985, 1147.
24. **Downes, J. A.,** The ecology of blood-sucking diptera: an evolutionary perspective, in *Ecology and Physiology of Parasites*, Fallis, A.M., Ed., University of Toronto Press, Toronto, 1971, 232.
25. **Hocking, B.,** Blood-sucking behavior of terrestrial arthropods, *Ann. Rev. Entomol.*, 16, 1, 1971.
26. **Smith, A. L. and Anderson, C. R.,** Susceptibility of two turtle species to eastern equine encephalitis virus, *J. Wildl. Dis.*, 16, 615, 1980.
27. **Thomas, L. A., Patzer, E. R., Cory, J. C., and Coe, J. E.,** Antibody development in garter snakes (*Thamnophis* spp.) experimentally infected with western equine encephalitis virus, *Am. J. Trop. Med. Hyg.*, 29, 112, 1980.
28. **Burton, A. N., McLintock, J., and Rempel, J. G.,** Western equine encephalitis virus in Saskatchewan garter snakes and leopard frogs, *Science*, 154, 1029, 1966.
29. **Scott, T. W., McLean, R. G., Francy, D. B., Mitchell, C. J., and Card, C. S.,** Experimental infections of birds with Turlock virus, *J. Wildl. Dis.*, 19, 82, 1983.
30. **Hayes, C. G. and Wallis, R. C.,** Ecology of western equine encephalomyelitis in the eastern United States, *Adv. Virus Res.*, 21, 37, 1977.
31. **Smith, C. E. G.,** The spread and maintenance of infections in vertebrates and arthropods, *J. Invert. Pathol.*, 18, 1, 1971.
32. **Friend, W. G. and Smith, J. J. B.,** Factors affecting feeding by bloodsucking insects, *Ann. Rev. Entomol.*, 22, 309, 1977.
33. **Washino, R. K. and Tempelis, C. H.,** Mosquito host bloodmeal indentification: methodology and data analysis, *Ann. Rev. Entomol.*, 28, 179, 1983.

34. **Tempelis, C. H.**, Host-feeding patterns of mosquitoes, with a review of advances in analysis of blood meals by serology, *J. Med. Entomol.*, 6, 635, 1975.
35. **Usinger, R. L.**, *Monograph of Cimicidae (Hemiptera-Heteroptera)*, Entomological Society of America, College Park, Md., 1966, 537.
36. **Brown, A. W. A.**, The attraction of mosquitoes to hosts, *JAMA*, 196, 249, 1966.
37. **Webb, J. P.**, Host-locating behavior of nymphal *Ornithodoros concanensis* (Acarina:Argasidae), *J. Med. Entomol.*, 5, 437, 1979.
38. **Hoogstraal, H. and Aeschlimann, A.**, Tick-host specificity, *Bull. Soc. Entomol. Suisse*, 55, 5, 1982.
39. **McIver, S. B.**, Sensilla of mosquitoes (Diptera:Culicidae), *J. Med. Entomol.*, 19, 489, 1982.
40. **Webb, J. P., George, J. E., and Cook, B.**, Sound as a host-detection cue for the soft tick *Ornithodoros concanensis*, *Nature (London)*, 265, 443, 1977.
41. **Howell, F. G.**, The roles of host-related stimuli in the behavior of *Argas cooleyi* (Acarina:Argasidae), *J. Med. Entomol.*, 11, 715, 1975.
42. **Gillies, M. T.**, The role of carbon dioxide in host-finding by mosquitoes (Diptera:Culicidae): a review, *Bull. Entomol. Res.*, 70, 525, 1980.
43. **Schlein, Y., Yuval, B., Warburg, A.**, Aggregation pheromone released from the palps of feeding female *Phlebotomus papatasi* (Psychodidae), *J. Insect Physiol.*, 30, 153, 1984.
44. **Edman, J. D. and Kale, H. W.**, Host behavior: its influence on the feeding success of mosquitoes, *Ann. Entomol. Soc. Am.*, 64, 513, 1971.
45. **Edman, J. D., Webber, L. A., and Schmid, A. A.**, Effect of host defenses on the feeding pattern of *Culex nigripalpus* when offered a choice of blood sources, *J. Parasitol.*, 60, 874, 1974.
46. **Walker, E. D. and Edman, J. D.**, Influence of defensive behavior of eastern chipmunks and gray squirrels (Rodentia:Sciuridae) on feeding success of *Aedes triseriatus* (Diptera:Culicidae), *J. Med. Entomol.*, 23, 1, 1986.
47. **Klowden, M. J. and Lea, A. O.**, Effect of defensive host behavior on the blood meal size and feeding success of natural populations of mosquitoes (Diptera:Culicidae), *J. Med. Entomol.*, 15, 514, 1979.
48. **Webber, L. A. and Edman, J. D.**, Anti-mosquito behavior of ciconiiform birds, *Anim. Behav.*, 20, 228, 1972.
49. **Edman, J. D., Day, J. F., and Walker, E. D.**, Field confirmation of laboratory observations on the differential antimosquito behavior of herons, *Condor*, 86, 91, 1984.
50. **Kale, H. W., Edman, J. D., and Webber, L. A.**, Effect of behavior and age of individual ciconiiform birds on mosquito feeding success, *Mosq. News*, 32, 343, 1972.
51. **Edman, J. D. and Scott, T. W.**, Host defensive behavior and the feeding success of mosquitoes, *Insect Science and Its Applications*, in press, 1987.
52. **Waage, J. K. and Nondo, J.**, Host behavior and mosquito feeding success: an experimental study, *Trans. R. Soc. Trop. Med. Hyg.*, 76, 119, 1982.
53. **Day, J. F. and Edman, J. D.**, Mosquito engorgement on normally defensive hosts depends on host activity patterns, *J. Med. Entomol.*, 21, 732, 1984.
54. **Edman, J. D., Webber, L. A., and Kale H. W., II**, Effect of mosquito density on the interrelationship of host behavior and mosquito feeding success, *Am. J. Trop. Med. Hyg.*, 21, 487, 1972.
55. **Mahon, R. and Gibbs, A.**, Arbovirus infected hens attract more mosquitoes, in *Viral Diseases in Southeast Asia and Western Pacific*, MacKenzie, J. S., Ed., Academic Press, Sydney, 1982, 502.
56. **Turell, M. J., Bailey, C. L., and Rossi, C. A**, Increased mosquito feeding on Rift Valley fever virus-infected lambs, *Am. J. Trop. Med. Hyg.*, 33, 1232, 1984.
57. **Day, J. F. and Edman, J. D.**, Malaria renders mice susceptible to mosquito feeding when gametocytes are most infective, *J. Parasitol.*, 69, 163, 1983.
58. **Brown, S. J.**, Immunology of acquired resistance to ticks, *Parasitol. Today*, 6, 166, 1985.
59. **Brown, S. J., Barber, R. W., and Askenase, P. W.**, Bovine resistance to *Amblyomma americanum* ticks: an acquired immune response characterized by cutaneous basophil infiltrates, *Vet. Parasitol.*, 16, 147, 1984.
60. **Johnston, C. M. and Brown, S. J.**, Cutaneous and systemic cellular responses induced by the feeding of the argasid tick *Ornithodoros parkeri*, *Int. J. Parasitol.*, 6, 621, 1985.
61. **Johnston, C. M. and Brown, S. J.**, *Xenopsylla cheopis:* cellular expression of hypersensitivity in guinea pigs, *Exp. Parasitol.*, 59, 81, 1985.
62. **Feinsod, F. M., Spielman, A., and Waner, J. L.**, Neutralization of Sindbis virus by antisera to antigens of vector mosquitoes, *Am. J. Trop. Med. Hyg.*, 24, 533, 1975.
63. **Scott, T. W., Bowen, G. S., and Monath, T. P.**, A field study on the effects of Fort Morgan virus, an arbovirus transmitted by swallow bugs, on the reproductive success of cliff swallows and symbiotic house sparrows in Morgan County, Colorado 1976, *Am. J. Trop. Med. Hyg.*, 33, 981, 1984.
64. **Main, A. J., Downs, W. G., Shope, R. E., and Wallis, R. C.**, Avian arboviruses of the Witless Bay Seabird Sanctuary, Newfoundland, Canada, *J. Wildl. Dis.*, 12, 182, 1976.

65. **Johnson, H. N. and Casals, J.,** Arboviruses associated with marine bird colonies of the Pacific Ocean region, in *Transcontinental Connections of Migratory Birds and Their Role in the Dissemination of Arboviruses,* Cherepanov, A. I., Ed., NAUKA, Novosibirsk, 1972.
66. **Buescher, E. L., Scherer, W. F., McClure, H. E., Moyer, J. T., Rosenberg, M. Z., Yoshii, M., and Okada, Y.,** Ecological studies of Japanese encephalitis virus in Japan, IV. Avian infection, *Am. J. Trop. Med. Hyg.,* 8, 678, 1959.
67. **Scherer, W. F., Buescher, E. L., and McClure, H. E.,** Ecologic studies of Japanese encephalitis, V. Avian factors, *Am. J. Trop. Med. Hyg.,* 8, 689, 1959.
68. **Holden, P., Hayes, R. O., Mitchell, C. J., Francy, D. B., Lazuick, J. S., and Hughes, T. B.,** House sparrows, *Passer domesticus* (L.), as hosts of arboviruses in Hale County, Texas. I. Field studies, 1965—1969, *Am. J. Trop. Med. Hyg.,* 22, 244, 1973.
69. **Waterman, S. H., Novak, R. J., Sather, G. E., Bailey, R. E., Rios, I., and Gubler, D. J.,** Dengue transmission in two Puerto Rico communities in 1982, *Am. J. Trop. Med. Hyg.,* 34, 625, 1985.
70. **Lahdevirta, J.,** Nephropathia epidemica in Finland, *Ann. Clin. Res.,* 3, 1, 1971.
71. **Nystrom, K.,** Incidence and Prevalence of Endemic Benign (Epidemic) Nephropathy in AC County, Sweden, in Relation to Population Density and Prevalence of Small Rodents, UMEA University Medical Dissertations No. 30, University of Umea, Sweden, 1977.
72. **Nystrom, K.,** Epidemiology of HFRS (endemic benign nephropathy — EBN) in Sweden, *Scand. J. Infect. Dis. Suppl.,* 35, 92, 1982.
73. **Worth, C. B., Downs, W. G., Aitken, T. H. G., and Tikasingh, E. S.,** Arbovirus studies in Bush Bush Forest, Trinidad, W.I., September 1959—December 1964. IV. Vertebrate populations, *Am. J. Trop. Med. Hyg.,* 17, 269, 1968.
74. **Nathanson, N.,** Pathogenesis, in *St. Louis Encephalitis,* Monath, T. P., Ed., American Public Health Association, Washington, D.C., 1980, 201.
75. **Monath, T. P.,** Pathobiology of the flaviviruses, in *The Togaviridae and Flaviviridae,* Schlesinger, S. and Schlesinger, M. J., Eds., Plenum Press, New York, 1986.
76. **Sawyer, W. A. and Lloyd, W.,** The use of mice in tests of immunity against yellow fever, *J. Exp. Med.,* 54, 533, 1931.
77. **Webster, L. T.,** Inherited and acquired factors in resistance to infection. I. Development of resistant and susceptible lines of mice through selective breeding, *J. Exp. Med.,* 57, 793, 1933.
78. **Brinton, M. A.,** Replication of flaviviruses, in *The Togaviridae and Flaviviridae,* Schlesinger, S. and Schlesinger, M. J., Eds., Plenum Press, New York, 1986, 375.
79. **Brinton, M. A., Blank, K. J., and Nathanson, N.,** Host genes that influence susceptibility to viral diseases, in *Concepts in Viral Pathogenesis,* Notkins, A. L. and Oldstone, M. B. A., Eds., Springer-Verlag, New York, 1984.
80. **Fagbami, A. H., Tomori, O., and Fabiyi, A.,** Experimental Rift Valley fever in West African dwarf sheep, *Res. Vet. Sci.,* 18, 334, 1975.
81. **Guzman, M. G., Kouri, G., Bravo, J., Soler, M., and Vaspuez, S.,** Dengue hemorrhagic fever in Cuba 1981. II. Clinical investigations, *Trans. R. Soc. Trop. Med. Hyg.,* 78, 239, 1984.
82. **Darnell, M. B. and Koprowski, H.,** Genetically determined resistance to infection with group B arboviruses. I. Distribution of the resistance gene among various mouse populations and characteristics of gene expression *in vivo, J. Infect. Dis.,* 129, 240, 1974.
83. **Seymour, C., Kramer, L. D., and Peralta, P. H.,** Experimental St. Louis encephalitis virus infection of sloths and cormorants, *Am. J. Trop. Med. Hyg.,* 32, 844, 1983.
84. **Hayes, R. O., Daniels, J. B., Maxfield, H. K., and Wheeler, R. E.,** Field and laboratory studies on eastern encephalitis in warm- and cold-blooded vertebrates, *Am. J. Trop. Med. Hyg.,* 4, 595, 1964.
85. **Bowen, G. S.,** Prolonged western equine encephalitis viremia in the Texas tortoise *(Gopherus berlandieri), Am. J. Trop. Med. Hyg.,* 26, 171, 1977.
86. **Burgdorfer, W. and Eklund, C. M.,** Studies on the ecology of Colorado tick fever virus in western Montana, *Am. J. Hyg.,* 69, 127, 1959.
87. **Sulkin, S. E., Allen, R., and Sims, R.,** Studies of arthropod-borne virus infections in *Chiroptera.* I. Susceptibility of insectivorous species to experimental infection with Japanese B and St. Louis encephalitis viruses, *Am. J. Trop. Med. Hyg.,* 12, 800, 1963.
88. **Main, A. J.,** Eastern equine encephalomyelitis virus in experimentally infected bats, *J. Wildl. Dis.,* 15, 467, 1979.
89. **Corristan, E. C., LaMotte, L. C., and Smith, D. G.,** Susceptibility of bats to certain encephalitis viruses, *Fed. Proc.,* 15(Abstracts), 1956.
90. **LaMotte, L. C.,** Japanese B encephalitis in bats during simulated hibernation, *Am. J. Hyg.,* 67, 101, 1958.
91. **McLean, R. G.,** Potentiation of Keystone virus infection in cotton rats by glucocorticoid-induced stress, *J. Wildl. Dis.,* 18, 141, 1982.
92. **Vorndam, A. V.,** Immunization, in *St. Louis Encephalitis,* Monath, T. P., Ed., American Public Health Association, Washington, D.C., 1980.

93. **Halstead, S. B.,** Immunopathological parameters of togavirus disease syndromes, in *The Togaviruses: Biology, Structure, Replication*, Schlesinger, R. W., Ed., Academic Press, New York, 1980, 107.
94. **Halstead, S. B.,** Dengue haemorrhagic fever — a public health problem and a field for research, *Bull. WHO*, 58, 1, 1980.
95. **Pang, T.,** Immunoepidemiology helps to unravel the mysteries of dengue haemorrhagic fever, *Immunol. Today*, 4, 334, 1983.
96. **Sangkawibha, N., Rojanasuphot, S., Ahandrik, S., Viriyapongse, S., Jatanasen, S., Salitul, V., Phanthumachinda, B., and Halstead, S. B.,** Risk factors in dengue shock syndrome: a prospective epidemiologic study in Rayong, Thailand. I. The 1980 outbreak, *Am. J. Epidemiol.*, 120, 653, 1984.
97. **Cardosa, M. J., Porterfield, J. S., and Gordon, S.,** Complement receptor mediates enhanced flavivirus replication in macrophages, *J. Exp. Med.*, 158, 258, 1983.
98. **Sudia, W. D. and Newhouse, V. F.,** Epidemic Venezuelan equine encephalitis in North America: a summary of virus-vector-host relationships, *Am. J. Trop. Med. Hyg.*, 101, 1, 1975.
99. **Eddy, G. A., Martin, D. H., Reeves, W. C., and Johnson, K. M.,** Field studies of an attenuated Venezuelan equine encephalomyelitis vaccine (strain TC 83), *Infect. Immun.*, 5, 160, 1972.
100. **Ludwig, G. V., Cook, R. S., McLean, R. G., Francy, D. B., and Beaty, B. J.,** Viremic enhancement due to transovarially acquired antibodies to St. Louis encephalitis virus in birds, *J. Wildl. Dis.*, in press, 1986.
101. **Holden, P., Francy, D. B., Mitchell, C. J., Hayes, R. O., Lazuick, J. S., and Hughes, T. B.,** House sparrows, *Passer domesticus* (L.), as hosts of arboviruses in Hale County, Texas, II. Laboratory studies with western equine encephalitis virus, *Am. J. Trop. Med. Hyg.*, 22, 254, 1973.
102. **Issel, C. J.,** Maternal antibody to Jamestown Canyon virus in white-tailed deer, *Am. J. Trop. Med. Hyg.*, 23, 242, 1974.
103. **Sooter, C. A., Schaeffer, M., Gorrie, R., and Cockburn, T. A.,** Transovarian passage of antibodies following naturally acquired encephalitis infection in birds, *J. Infect. Dis.*, 95, 165, 1954.
104. **Reeves, W. C., Sturgeon, J. M., French, E. M., and Brookman, B.,** Transovarian transmission of neutralizing subtances to western equine and St. Louis encephalitis viruses by avian hosts, *J. Infect. Dis.*, 95, 168, 1954.
105. **Kissling, R. E., Eidson, M. E., and Stamm, D. M.,** Transfer of maternal neutralizing antibodies against eastern equine encephalomyelitis virus in birds, *J. Infect. Dis.*, 95, 179, 1954.
106. **Bond, J. O., Lewis, F. Y., Jennings, W. L., and MacLeod, I. E.,** Transovarian transmission of hemagglutination-inhibition antibody to St. Louis encephalitis virus in chickens, *Am. J. Trop. Med. Hyg.*, 14, 1085, 1965.
107. **Buescher, E. L., Scherer, W. F., Rosenberg, M. Z., Kutner, L. J., and McClure, H. E.,** Immunologic studies of Japanese encephalitis virus in Japan. IV. Maternal antibody in birds, *J. Immunol.*, 83, 614, 1959.
108. **Howell, P. G. and Verwoerd, D. W.,** Blue-tongue virus, *Virol. Monogr.*, 9, 35, 1971.
109. **Thompson, W. H.,** Lower rates of oral transmission of La Crosse virus by *Aedes triseriatus* venereally exposed after engorgement on immune chipmunks, *Am. J. Trop. Med. Hyg.*, 32, 1416, 1983.
110. **Patrican, L. A., DeFoliart, G. R., and Yuill, T. M.,** Oral infection and transmission of La Crosse virus by an enzootic strain of *Aedes triseratius* feeding on chipmunks with a range of viremia levels, *Am. J. Trop. Med. Hyg.* 34, 992, 1985.
111. **Dalrymple, J. M., Young, O. P., Eldridge, B. F., and Russell, P. K.,** Ecology of arboviruses in a Maryland freshwater swamp. III. Vertebrate hosts, *Am. J. Epidemiol.*, 96, 129, 1972.
112. **Blackmore, J. S. and Dow, R. P.,** Differential feeding of *Culex tarsalis* on nestling and adult birds, *Mosq. News*, 18, 15, 1958.
113. **Kale, H. W., Edman, J. D., and Webber, L. A.,** Effect of behavior and age of individual ciconiiform birds on mosquito feeding success, *Mosq. News*, 32, 232, 1972.
114. **Mayr, E.,** *Animal Species and Evolution*, Harvard University Press, Cambridge, Mass., 1963, 662.
115. **MacArthur, R. H.,** *Geographical Ecology: Patterns in the Distribution of Species*, Harper & Row, New York, 1972, 251.
116. **Williams, G. C.,** *Adaptation and Natural Selection: A Critique of Some Current Evolutionary Thought*, Princeton University Press, Princeton, N.J., 1966, 273.
117. **Wilson, E. O.,** *Sociobiology: The New Synthesis*, Harvard University Press, Cambridge, Mass., 1975, 575.
118. **Horn, H. S.,** The adaptive significance of colonial nesting in the Brewer's blackbird *(Euphagus cyanocephalus)*, *Ecology*, 49, 682, 1968.
119. **Loye, J. E. and Hopla, C. E.,** Ectoparasites and microorganisms associated with the cliff swallow in west-central Oklahoma. II. Life history Patterns, *Bull. Soc. Vector Ecol.*, 8, 79, 1983.
120. **Chapman, B. R.,** The Effects of Nest Ectoparasites on Cliff Swallow Populations, Ph.D. thesis, Texas Tech University, Lubbock, 1973.
121. **Ricklefs, R. E.,** *Ecology*, Chiron Press, Newton, Mass., 1973, 780.

122. **Christian, J. J.,** The potential role of the adrenal cortex as affected by social rank and population density in experimental epidemics, *Am. J. Epidemiol.*, 87, 255, 1968.
123. **Stamm, D. D. and Newman, R. J.,** Evidence of southward transport of arboviruses from the U.S. by migratory birds, *An. Microbiol.*, 11, 123, 1963.
124. **Lord, R. D. and Calisher, C. H.,** Further evidence of southward transport of arboviruses by migratory birds, *Am. J. Epidemiol.*, 92, 73, 1970.
125. **Hoogstraal, H.,** Changing patterns of tickborne diseases in modern society, *Ann. Rev. Entomol.*, 26, 75, 1981.
126. **Bowen, G. S., McLean, R. G., Shriner, R. B., Francy, D. B., Pokorny, K. S., Trimble, J. M., Bolin, R. A., Barnes, A. M., Calisher, C. H., and Muth, D. J.,** The ecology of Colorado tick fever in Rocky Mountain National Park in 1974. II. Infections in small mammals, *Am. J. Trop. Med. Hyg.*, 30. 490, 1981.
127. **Carey, A. B.,** Discriminant analysis: a method of identifying foci of vector-borne diseases, *Am. J. Trop. Med. Hyg.*, 28, 750, 1979.
128. **Carey, A. B., McLean, R. G., and Maupin, G. O.,** The structure of a Colorado tick fever ecosystem. *Ecol. Monogr.*, 50, 131, 1980.
129. **Parsonson, J. M., Della-Porta, A. J. and Snowdon, W. A.,** Developmental disorders of the fetus in some arthropod-borne virus infections, *Am. J. Trop. Med. Hyg.*, 30, 660, 1981.
130. **Chaturvedi, U. C., Mathur, A., Chandra, A., Das, S. K., Tandon, H. O., and Singh, U. K.,** Transplacental infection with Japanese encephalitis virus, *J. Infect. Dis.*, 141, 712, 1980.
131. **Burns, K. F.,** Congenital Japanese B encephalitis infection of swine, *Proc. Soc. Exp. Biol. Med.*, 75, 621, 1950.
132. **Mathur, A., Arora, K. L., and Chaturvedi, U. C.,** Transplacental Japanese encephalitis virus (JEV) infection in mice during pregnancies, *J. Gen. Virol.*, 59, 213, 1982.
133. **Sulkin, S. E., Sims, R., and Allen, R.,** Studies of arthropod-borne virus infections in *Chiroptera*. II. Experiments with Japanese B and St. Louis encephalitis virus in the gravid bat. Evidence of transplacental transmission, *Am. J. Trop. Med. Hyg.*, 13, 475, 1964.
134. **Andersen, A. A. and Hanson, R. P.,** Experimental transplacental transmission of St. Louis encephalitis virus in mice, *Infect. Immun.*, 2, 320, 1970.
135. **Coetzer, J. A. W., Theodoridis, A., Herr, S., and Kritzinger, L.,** Wesselsbron disease: a cause of congenital porencephaly and cerebellar hypoplasia in calves, *Ondersterpoort J. Vet. Res.*, 46, 165, 1979.
136. **Wenger, F.,** Necrosis cerebral masiva del feto en casos de encefalitis equina Venezolana, *Invest. Clin. (Maracaibo)*, 21, 13, 1967.
137. **Spertzel, R. O., Crabbs, C. L., and Vaughn, R. E.,** Transplacental transmission of Venezuelan equine encephalomyelitis virus in mice, *Infect. Immun.*, 6, 339, 1972.
138. **Justines, G., Sucre, H., and Alvarez, O.,** Transplacental transmission of Venezuelan equine encephalitis virus in horses, *Am. J. Trop. Med. Hyg.*, 29, 653, 1980.
139. **Fuccillo, D. A. and Sever, J. L.,** Viral teratology, *Bacteriol. Res.*, 37, 19, 1973.
140. **Aaskov, J. G., Davies, C., Tucker, M., and Dalglish, D. A.,** Arboviruses and transplacental transmission, in *Arbovirus Research in Australia: Proceedings 2nd Symposium July 1979*, St. George, T. D. and French, E. L., Eds., CSIRO Division of Animal Health and Queensland Institute of Medical Research, Brisbane, Australia, 1979, 139.
141. **Young, S. and Cordy, D. R.,** An ovine fetal encephalopathy caused by bluetongue vaccine virus, *J. Neuropathol. Exp. Neurol.*, 23, 635, 1964.
142. **Luedke, A. J., Jones, R. H., and Walton, T. E.,** Overwintering mechanism for bluetongue virus: biological recovery of latent virus from a bovine by bites of *Culicoides variipennis*, *Am. J. Trop. Med. Hyg.*, 26, 313, 1977.
143. **Harris, R. E., Morahan, P., and Coleman, P.,** Teratogenic effects of Colorado tick fever virus in mice, *J. Infect. Dis.*, 131, 397, 1975.
144. **Gibbs, E. P. J., Lawman, M. J. P., and Herniman, K. A. J.,** Preliminary observations on transplacental infection of bluetongue virus in sheep — a possible overwintering mechanism, *Res. Vet. Sci.*, 27, 118, 1979.
145. **Breckon, R. D., Luedke, A. J., and Walton, T. E.,** Bluetongue virus in bovine semen: virus isolation, *Am. J. Vet. Res.*, 41, 439, 1980.
146. **Holden, P.,** Transmission of eastern equine encephalomyelitis in ring-necked pheasants, *Proc. Soc. Exp. Biol. Med.*, 88, 607, 1955.
147. **Scherer, W. F., Dickerman, R. W., Ordonez, J. V., Seymour, C., III, Kramer, L. D., Jahrling, P. B., and Powers, C. D.,** Ecologic studies of Venezuelan encephalitis virus and isolations of Nepuyo and Patois viruses during 1968—1973 at a marsh habitat near the epicenter of the 1969 outbreak in Guatemala, *Am. J. Trop. Med. Hyg.*, 25, 151, 1976.
148. **Monath, T. P.,** Arthropod-borne encephalitides in the Americas, *Bull. WHO*, 57, 513, 1979.
149. **Woodall, J. P. and Roz, A.,** Experimental milk-borne transmission of Powassan virus in the goat, *Am. J. Trop. Med. Hyg.*, 26, 190, 1977.

150. **Beaty, B. J., Holterman, M., Tabachnick, W., Shope, R. E., Rozhon, E. J., and Bishop, D. H. L.,** Molecular basis of bunyavirus transmission by mosquitoes: role of the middle-sized RNA segment, *Science*, 211, 1433, 1981.
151. **Bishop, D. H. L., Beaty, B. J., and Shope, R. E.,** Recombination and gene coding assignments of bunyaviruses and arenaviruses, *Ann. N.Y. Acad. Sci.*, 354, 84, 1980.
152. **Seymour, C., Amundson, T. E. A., Yuill, T. M., and Bishop, D. H.,** Experimental infection of chipmunks and snowshoe hares with La Crosse and showshoe hare viruses and four of their reassortants, *Am. J. Trop. Med. Hyg.*, 32, 1147, 1983.
153. **Tignor, G. H., Burrage, T. G., Smith, A. L., Shope, R. E., and Bishop, D. H. L.,** California serogroup gene structure-function relationships: virulence and tissue tropism, in *California Serogroup Viruses*, Calisher, C. H. and Thompson, W. H., Eds., Alan R. Liss, New York, 1980, 129.
154. **Calisher, C. H., Maness, K. S. C., Lord, R. D., and Coleman, P. H.,** Identification of two South American strains of eastern equine encephalomyelitis virus from migrant birds captured on the Mississippi Delta, *Am. J. Epidemiol.*, 94, 172, 1971.
155. **Dickerman, R. W., Martin, M. S., and Dipaola, E. A.,** Studies of Venezuelan encephalitis in migrating birds in relation to possible transport of virus from South to Central America, *Am. J. Trop. Med. Hyg.*, 29, 269, 1980.
156. **Calisher, C. H., Gutierrez, E., Maness, K. S. C., and Lord, R. D.,** Isolation of Mayaro virus from a migrating bird captured in Louisiana in 1967, *PAHO Bull.*, 8, 243, 1974.
157. **Casals, J.,** Antigenic variants of eastern equine encephalitis virus, *J. Exp. Med.*, 119, 547, 1964.
158. **Williams, J. E., Young, O. P., Watts, D. M., and Reed, T. J.,** Wild birds as eastern (EEE) and western (WEE) equine encephalitis sentinels, *J. Wildl. Dis.*, 7, 188, 1971.
159. **Chamberlain, R. W., Kissling, R. E., Stamm, D. D., and Sudia, W. D.,** Virus of St. Louis encephalitis in three species of wild birds, *Am. J. Hyg.*, 65, 110, 1957.
160. **Chunikhin, S. P. and Takahashi, M.,** An attempt to establish the chronic infection of pigeons with Japanese encephalitis virus, *Jpn. J. Sanit. Zool.*, 22, 155, 1971.
161. **Gebhardt, L. P., Stanton, G. J., Hill, D. W., and Collett, G. C.,** Natural overwintering hosts of the virus of western equine encephalitis, *N. Engl. J. Med.*, 271, 172, 1964.
162. **Sudia, W. D., McLean, R. G., Newhouse, V. F., Johnston, J. G., Miller, D. L., Trevino, H., Bowen, G. S., and Sather, G.,** Epidemic Venezuelan equine encephalitis in North America in 1971: vertebrate field studies, *J. Epidemiol.*, 101, 36, 1975.
163. **Janzen, D. H.,** When is it coevolution?, *Evolution*, 34, 611, 1980.
164. **McLean, R. G., Frier, G., Parham, G. L., Francy, D. B., Monath, T. P., Campos, E. G., Therrien, A., Kerschner, J., and Calisher, C. H.,** Investigations of the vertebrate hosts of eastern equine encephalitis during an epizootic in Michigan, 1980, *Am. J. Trop. Med. Hyg.*, 34, 1190, 1985.
165. **Work, T. H. and Trapido, H.,** Summary of preliminary report of investigations of the Virus Research Centre on an epidemic disease affecting forest villagers and wild monkeys of Shimoga District, Mysore, *Indian J. Med. Sci.*, 11, 340, 1957.
166. **Brinton, M. A. and Nathanson, N.,** Genetic determinants of virus susceptibility: epidemiologic implications of murine models, *Epidemiol. Rev.*, 3, 115, 1981.
167. **Fenner, F. and Ratcliffe, F. N.,** *Myxomatosis*, Cambridge University Press, Cambridge, Mass., 1965, 379.
168. **Fenner, F.,** Evolution in action: myxomatosis in the Australian wild rabbit, in *Tropics in the Study of Life. The Bio Source Book*, Kramer, A., Ed., Harper & Row, New York, 1971, 463.
169. **Scherer, W. F., Moyer, J. T., Izumi, T., Gresser, I., and McCown, J.,** Ecologic studies of Japanese encephalitis virus in Japan. VI. Swine infection, *Am. J. Trop. Med. Hyg.*, 8, 698, 1959.
170. **Lord, R. D., Work, T. H., Coleman, P. H., and Johnston, J. G., Jr.,** Virological studies of avian hosts in the Houston epidemic of St. Louis encephalitis, 1964, *Am. J. Trop. Med. Hyg.*, 22, 662, 1973.
171. **Lord, R. D., Calisher, C. H., and Doughty, W. P.,** Assessment of bird involvement in three urban St. Louis encephalitis epidemics, *Am. J. Epidemiol.*, 99, 364, 1974.
172. **Meegan, J. M. and Shope, R. E.,** Emerging concepts on Rift Valley fever, in *Perspectives in Virology XI*, Alan R. Liss, New York, 1981, 267.
173. **Thomson, G., Gainaru, M., Lewis, A., Biggs, H., Nevill, E., van der Pypekamp, H., Gerber, L., Esterhuysen, J., Bengis, R., Bezuidenhout, D., and Condy, J.,** The relationship between African swine fever virus, the wart hog and *Ornithodoros* species in southern Africa, in *African Swine Fever*, Rep, No. EUR 8466 EN, Commission of the European Communities, Paris, 1983, 85.
174. **Mishin, A. V. and Gerasimova, E. N.,** Epidemiological characteristics of tick-borne encephalitis in the Udmurt Autonomous Soviet Republic, *Med. Parazitol. Parazit. Bol.*, 28, 137, 1959.

Chapter 11

HUMAN FACTORS IN ARBOVIRUS ECOLOGY AND CONTROL

Frederick L. Dunn

TABLE OF CONTENTS

I. INTRODUCTION

Many chapters in this volume take note, to some extent, of human factors that influence arbovirus ecology, viral transmission to man, and prevention or control of the arthropod-borne viral diseases. Most epidemiologists are, at least to a degree, students of human behavior, and most epidemiologists would agree that their discipline is partly a social science.[1] Epidemiological analysis normally requires some consideration of socioeconomic and behavioral risk factors: for example, those related to occupation, gender, religious preference, income, or ethnicity. Those concerned with control of disease must also concern themselves with human factors that interfere with or contribute favorably to effective control. This chapter, therefore, recapitulates and pulls together some of the points made by others elsewhere in this volume and in the general literature on arboviral infection and disease. It broadly surveys the kinds of behavioral, cultural, social, psychological, and economic factors that need to be kept in mind in epidemiological research on these viral agents and in control of the diseases with which they are associated.

Any search of the vast array of publications on arbovirus epidemiology and control will uncover scattered references to human factors, often anecdotal or impressionistic observations, rarely supported by formal behavioral investigations. Audy's effort[2] some 15 years ago to survey what was then known about human behavior and vector control is indicative of the problem; he was forced to write in very general (although also very engaging) terms and to illustrate most of his points with anecdotes provided by friends and colleagues. Gillet[3] has written in a somewhat similar vein. In the years since 1972 much ecologically oriented arbovirus research of outstanding quality has been reported. Good recent examples include the studies by Morris and colleagues[4] of eastern equine encephalomyelitis (EEE) in upstate New York, and of *Aedes simpsoni* and *Ae. africanus* by Bown and Bang[5,6] in Nigeria. The human population at risk is usually mentioned, or even briefly characterized, in studies of this kind, but nothing has really changed since the time of Audy's review. Human factors continue to be noted only in passing, usually as incidental, casually collected information that is entered in the introductory or discussion section of the publication. It is evident that sociomedical scientists (principally sociologists, anthropologists, economists, psychologists, and geographers) have scarcely begun to contribute to this literature; certainly none seem to have attempted comprehensive field work on arboviral diseases. Thus, arbovirus research and control is another subject area within the communicable disease field which is biomedically mature but quite underdeveloped on the sociomedical side. Some of the issues that help to explain this situation have been discussed elsewhere by this writer, in contributions on other categories of communicable disease.[1,7-9]

The examination of the arbovirus literature upon which this chapter is based has clearly indicated that there are outstanding sociomedical research needs:

1. To attract some participation by medical social scientists in arbovirus research and control
2. To encourage epidemiologists working on these diseases to increase their sensitivity to and awareness of sociomedical issues and to train themselves in the basic techniques of the social-behavioral sciences
3. Because it is evident that research on the sociocultural, economic, and psychological consequences of arbovirus disease has barely begun, to encourage work on these aspects to provide a more rational basis for policymaking, planning, and allocation of resources for arbovirus disease prevention and control.

For a detailed "checklist" of sociomedical research topics — topics for the most part equally relevant in arbovirus research — the reader is referred to a recent article in the proceedings of a workshop on trachoma research and intervention.[10]

Table 1
HEALTH-RELATED HUMAN BEHAVIOR AND ARBOVIRUS INFECTION OR DISEASE

I. Behavior of those at risk ("insiders"— residents, members of the community)
  A. Behavior that enhances health or maintains current health status (of individual, group, society)
    1. Deliberate behavior (actions taken to have some impact on arbovirus transmission or control)
    2. Nondeliberate behavior (actions taken with no consideration of their possible impact on arbovirus transmission or control)
  B. Behavior (deliberate, nondeliberate) that contributes to ill-health, or that may cause death (actions that lower "level" of health of individual, group, society, i.e., by perpetuating or stepping up arbovirus transmission. or by impeding control)
II. Behavior of those ("outsiders") not at risk whose intervention may affect arbovirus transmission or control in community or in any location where humans may be at risk
  A. Behavior that enhances health or maintains current health status
    1. Deliberate behavior
    2. Nondeliberate behavior
  B. Behavior (deliberate, nondeliberate) that contributes to ill-health or that may cause death

II. HEALTH-RELATED HUMAN BEHAVIOR AND ARBOVIRUS TRANSMISSION AND CONTROL

In examining human behavior in relation to health (in this case, health as it may be affected by arboviruses), it is helpful to divide the subject into its eight components, as summarized in Table 1.[1,7-9] The discussion follows this scheme with illustrations from the arbovirus literature.

A. Behavior of Those at Risk

Arbovirus epidemiology has given much more attention to behavioral, cultural, and socioeconomic determinants relating to those at risk than to those who attempt to intervene, as outsiders, to prevent disease or control transmission. As Table 1 indicates, at risk behavior must be examined with due regard to intention or motivation, i.e., with the understanding that some action may be undertaken by people in the hope or expectation that these will have an impact on arbovirus transmission or control. Many more types of behavior, however, can usually be identified which are not recognized locally as having any bearing on transmission or control.

Each of these at risk behaviors must, secondly, be assessed with regard to their final effect on health status; some behaviors are detrimental, some affect health favorably, and some are neutral or simply ensure that the health of the individual (or group or society) is maintained at an acceptable (or perhaps unacceptable) level.

The most obvious form of deliberate health promotive behavior is that usually described as personal protective. Personal protective behavior in the case of vector-borne disease begins with the efforts of individauls or small groups (e.g., families) to minimize biting contact, principally by use of sleeping nets, house screening, mosquito coils, smoky fires, application of domestic insecticides, skin repellents, head nets, long-sleeved shirts, and other protective clothing. In traditional societies, of course, these actions are usually undertaken to reduce the nuisance problem — the problem of mosquitoes or other biting vectors as pests or irritants.[11] In many endemic countries today, however, thanks to the media and public health education, personal protection is recognized by the general public to have disease-specific value, e.g., in reducing the incidence of dengue (DEN). Waterman and colleagues[12] have recently shown how significant a variable such as window and door screening can be in relation to DEN antibody prevalence. Most forms of personal protective behavior, however, have been little studied in relation to arbovirus transmission.

Table 2
PREVENTION AND CONTROL: SOME REPRESENTATIVE ARBOVIRUS INFECTIONS

	Arbovirus	Vaccination for humans	Vector control (intervention and/or community based)	Personal protective behavior (to avoid or minimize biting contact)
I.	YF virus	+	+	+
	JE virus	+	+	+
II.	EEE virus	± (for lab personnel; horses)	+	+
	WEE virus	± (for lab personnel; horses)	+	+
	VEE virus	± (for lab personnel; horses)	+	+
III.	RVF virus	± (for lab personnel; livestock)	+	(avoidance of aerosol and contact transmission; mosquito transmission to man)
IV.	DEN virus	–[a]	+	+
	WN virus	–	+	+
	CHIK virus	–[a]	+	+
	RR virus	–	+	+
	SLE virus	–	+	+
	Phlebotomus (sandfly) fever viruses	–	+	+
V.	KF virus	± (for lab personnel)	±	+
VI.	Oropouche virus	–	–	± (It may be difficult or impossible to avoid *Culicoides* biting)

[a] Experimental vaccine under development/evaluation.

A second type of personal protective behavior is acceptance of vaccination; for the most part, in arbovirus protection, this means only yellow fever (YF) or Japanese encephalitis (JE) vaccination (Table 2). Many studies have been published on public acceptance or rejection of vaccination (especially for influenza, poliomyelitis, smallpox, and measles), but little in this regard seems to have been written about YF. Acceptance of yellow fever immunization is, on the one hand, mandatory for certain categories of international travellers, for most military personnel, and for other special groups; on the other hand, for those resident in endemic areas, choice (i.e., to accept or reject vaccination) is not generally enforced by legislation or other means. In some endemic countries the role of the media must surely be significant in encouraging YF vaccination, but studies of this topic seem not to have been undertaken or reported.

In addition to personal protection many people in areas of endemic arbovirus infection make some effort to control vector breeding and thus to reduce the risk of biting contact. Again, as in personal protection, these actions are commonly taken as a reaction to the nuisance of biting pests. Traditional approaches to control of vector breeding are manifest in many societies in peridomestic vegetation clearing and regular sweeping of the bare ground in the vicinity of the home. For many decades, however, public health workers have also encouraged householders to eliminate mosquito breeding, especially in domestic water containers, usually through health education campaigns.[13] Recently, the principle of local (householder) responsibility for vector control was widely acknowledged; programs depending on community participation in mosquito control are now under development, as, for example, in Thailand.[14] These Thai efforts to encourage *Ae. aegypti* control focus on the development of a spirit of active, voluntary participation in long-term, continuing programs.[15] In Singapore, on the other hand, the householder or shop owner is faced with legal sanctions (fines) if *Aedes* breeding is found in the course of routine surveillance activities.[13] The "integrated" program of *Aedes* control in Singapore, depending on education, surveillance, and law enforcement, exemplifies a relatively passive, only partially voluntary approach to control; this strategy is rather different to that implied by "community participation" as the term is used today by the World Health Organization, i.e., to refer to voluntary action.

Thus far we have considered only deliberate behavior, but nondeliberate actions, too, can help to reduce the chances of arbovirus transmission for those at risk. Environmental modifications undertaken for agricultural purposes obviously can change the potential for vector breeding. Environmental change may also interfere with arbovirus transmission when it results in changes in zoonotic host populations, either in their distribution or density or both. The arbovirus literature, however, emphasizes the detrimental rather than the beneficial effects (and cases) of environmental modification; the emphasis on negative consequences is understandable, but it is unfortunate that almost no research attention is directed to benefits of certain types of change.

Other kinds of change (nondeliberate behavioral change) have been recognized as beneficial, even as protective. For example, recent studies of arboviral encephalitis in California have demonstrated that the low incidence of human cases of western equine encephalomyelitis (WEE) and St. Louis encephalitis (SLE) in recent years is clearly associated with increased use of household air conditioning and television. Survey research by Gahlinger and associates[16] has shown that people tend to remain indoors during summer evenings in the Central Valley: air conditioner and television utilization times were found to correspond closely to the feeding times of the vector, *Culex tarsalis*. This interesting study is almost unique in the arbovirus literature; very little work along these lines has been reported for other arthropod-borne viral infections.

In contrast to the rather sparse record for beneficial effects of behavior by those at risk, accounts of harmful, even dangerous behavior are to be found for almost every arboviral infection or disease. Most of this kind of behavior, however, is nondeliberate; the people at risk are not aware of the hazards to which their activities (and underlying socio-economic and cultural factors) expose them. In Mysore, India it is the practices associated with cattle husbandry that place humans at risk of exposure to infected *Haemaphysalis* ticks and thus to Kyasanur forest disease.[17] In endemic YF areas it is the forest worker, the hunter, and swamp forest fisherman whose occupation places him at special risk. It has now been established, at least in Peninsular Malaysia, that a similar potential exists for transmission of DEN viruses to forest visitors. Rudnick[18] has confirmed that a forest cycle exists involving DEN viruses, forest primates, and canopy-dwelling *Aedes* mosquitoes (maintenance vectors); Gubler and others[19] have now presented evidence for a somewhat similar maintenance vector situation, also involving an *Aedes* species, in rural areas of Puerto Rico. Also in Puerto Rico, low socioeconomic status and the possession of a wood-constructed house are significantly associated with DEN incidence.[12] In Africa Rift Valley fever (RVF) virus is often transmitted to man by aerosol or direct contact. Close contact with sick and dying animals, often housed in the family compound and often slaughtered to salvage the meat, may substantially increase the risk of transmission.[20]

These actions bring people into contact with vectors and thus with the viruses, but other actions at the local level serve instead to foster vector breeding and to increase vector density. Here again, the published record is extensive. Much has been written, for example, about *Aedes* breeding in peridomestic environments, chiefly in man-made or man-modified water containers: pots, storage jars, tanks, old tires, discarded tin cans, bamboo fencing, coconut husks, etc.[21,22] It was estimated a few years ago that the city of Bangkok had about 800,000 containers of water capable of supporting the development of *Ae. aegypti;* almost 75% of the larval breeding was found to occur in large earthenware jars that store household water, but flower pots and kitchen ant traps were also significant sources of mosquitoes.[23] Similarly, in California, *Cx. tarsalis* breeding can occur in many man-made sites: standing water in the fields, swimming pools, bird baths, tin cans, and abandoned tires. SLE virus transmission in southern California recently resulted from *Cx. tarsalis* breeding in sewage drainage from a broken pipe in a septic tank leach field.[24] In a chapter on man's role in changing patterns of arbovirus infections (a rare paper in this field!) Stanley[25] lists some of the substantial changes relating especially to water use that have contributed to increased *Culex* breeding and transmission of Murray Valley encephalitis (MVE) virus in western and northern Australia: creation of a large man-made lake, development of irrigation agriculture, development of an iron ore industry, and as a potential problem, planning for large-scale hydroelectric development.

All of the above refers to unintentional actions and modifications, but deliberate risk-taking must be mentioned as well. Road construction in the Amazon or canal construction in Panama has carried with it the well-known risk of exposure to the YF virus. Tourists may or may not arm themselves with vaccinations and take other protective precautions while satisfying their travel ambitions, even in places known to be hazardous at the times of their visits. Military action, especially in forested regions, is associated with special risk in some parts of the world, hence the long tradition of military interest in and support of research on arbovirus infections.

B. Behavior by Outsiders: Intervention and Other Effects

In this second major division of health-related behavior we can review actions taken by all of those who have responsibilities or take it upon themselves to "intervene" at any level: clinical care for the sick individual, preventive care for the individual or group, and any approach to control of arbovirus transmission. We must also take note of those interventions that have detrimental effects on health status, and of all those behaviors that have unintended effects, favorable to health or otherwise.

On the positive, health-enhancing side of the ledger we can note, first, vaccine research and all of the factors in the human vaccination sequence including manufacture, distribution, storage, marketing, advertising, health education, organization of campaigns, legislation, and international travel regulations. Similar lists might be drawn up for equine and other livestock vaccine programs. On the negative side, at each step in one of these sequences, potential disruptions can arise as a consequence of human errors or misjudgments. Monath[26] illustrates this in his recent review of the status of YF vaccine research and use, noting some of the problems that have arisen and potential problems for the future:

> "Complacency with regard to the worldwide situation of yellow fever is widespread in the biomedical community, largely because of the availability of 17D vaccine; yet international and national health authorities recognize many limitations. Among these are the antiquated and cumbersome methods of present-day vaccine manufacture (growth of the virus in eggs), the limited capability for increased production, the thermal lability of the live vaccine, and the neurotropic potential limiting use in very young children. The total annual production on a worldwide basis is only about 15 million doses, an insufficient number should unforeseen disaster strike (such as the introduction and spread of the virus in Asia, where it has not occurred)."

Other major categories of interventive behavior include medical and veterinary clinical care; vector control in all its complexity; legislative aids to vector control, as employed, for example, in Singapore;[13] control directed to vertebrate hosts other than man; public health education and media reporting; surveillance; and monitoring. Any one of these topics can be broken down and described in fine detail, with specification of many necessary actions. A good example of this detailing can be found in Reeves'[27] 1967 review of factors that influence the probability of viral encephalitis epidemics in North America. Many factors in Reeves' extensive tabulation (e.g., relating to viruses, vectors, vertebrate hosts, the human population, and the environment) can be monitored in an effective surveillance program such as that now in force in California.[28]

Not only immunization but each of the other intervention strategies has its negative side as well. As Monath[26] has pointed out, in reviewing the status of YF, relaxed surveillance, poor reporting, and senescence of vector control programs have considerably increased the danger of outbreaks, both in Africa and the Americas.

In addition to deliberate intervention, many nondeliberate actions by outsiders have the potential to affect arbovirus transmission. Favorable effects (e.g., landscape modification that decreases vector breeding and reduces biting contact with humans or livestock) are seldom reported in the literature. Unfavorable effects (contributing to increased transmission) are, however, well documented. (At times it may be difficult to decide whether these effects are due to the actions of insiders — those at risk — or outsiders, certainly true in Stanley's MVE examples, cited above, of man-made changes in the environment.[25]) Unfavorable effects include increases in vector breeding potential resulting from urban growth and large-scale water development or agricultural schemes.[29,30] Scherer[30] reminds us that fields of grain can provide the food to sustain large populations of birds which can function as amplifying hosts of encephalitis viruses. Other unintended effects include the transport of vector mosquitoes from country to country by international aircraft. Goh and colleagues,[31] having studied this problem in Singapore, concluded that the solution there lies not in routine aircraft disinsecting but in continuation of the present effective system of vector surveillance and control at the international airport.

III. CONSEQUENCES OF ARBOVIRUS INFECTION AND DISEASE

It is now recognized that sociomedical research relating to any disease must go beyond risk factors (and beyond the behavior of those at risk and of those who would intervene) to include, in the broadest sense, the consequences of diseasses. "In the broadest sense" is inserted to emphasize that calculations of consequences ought not to be limited to economic, i.e., monetary, losses, and costs, although these are naturally of special interest to those in planning and policymaking who must also administer budgets.

Table 3 lists the major kinds of losses and costs that must be considered in measuring the impact of arboviral infection or disease and in allocating scarce resources for prevention, control, and clinical or veterinary care. Note that the list begins with social and psychological considerations. No systematic study of these factors has been undertaken for any one of the arboviral diseases although comments are to be found here and there in the epidemiological literature and for YF, in the work of historians and novelists. It is easy to be complacent about YF, as Monath[26] reminds us, but disaster could strike, particularly if the virus were to be introduced and disseminated in Asia. We would be forcibly reminded of the horrendous psychosocial and biological impact of the disease on human societies in centuries past. From the 17th century, for some 250 years, YF epidemics occurred with great frequency, especially in the Caribbean region. The U.S. was invaded some 26 times in the 18th century and epidemics struck North America 37 times in the 19th century.[25] The losses and costs of this epidemic disease have never been totaled, but they were obviously enormous. Public reaction, pressures, fears, and panic were documented in the newspapers of those years. It is striking,

Table 3
THE CONSEQUENCES OF ARBOVIRUS INFECTION AND DISEASE: LOSSES AND COSTS

I. Psychological and social consequences of infection and disease in the community: public fear, even panic, at times of epidemic disease; impact of sickness or death (psychological and social "loss") on family and community

II. Economic losses
- Temporary or permanent loss of wages or work capacity due to sickness or death — a loss for individual, household, community
- Losses associated with equine or other livestock sickness and death

III. The economic costs of medical and veterinary care

IV. The economic costs of prevention, protection, and control
- Costs of personal, household, and community protection against vector biting contact: netting, repellents, screening, chemicals, house construction, etc.
- Costs of prevention and control: virological research; vaccine research, development, and distribution; field research — entomology, epidemiology, etc.; surveillance, monitoring; vector control programs, both interventive and in support of community action; educational programs; media; legislation; administration

in reading some of those accounts, to see how similar the issues and concerns were then to those expressed today in contemporary press accounts of hemorrhagic DEN outbreaks in Southeast Asia. A comprehensive study of Asian media reporting of DEN since the 1950s would be welcome and instructive.

Calculations of "economic losses" for the arboviral diseases have been attempted infrequently. Accounts of epidemics and epizootics do, of courses, provide data on human, equine, and other livestock morbidity and mortality, but these losses are rarely translated into loss of productive capacity, wages, "horsepower". household income, and so forth. We are left again with impressions. Sabattini and others,[32] for example, refer to "significant" economic losses in Argentina resulting from recurrent epizootics of equine encephalitis. Similar losses have occurred in North America in the past, especially in the Midwest and in the West with the development of major irrigation projects early in this century. Schwabe[17] notes that horse mortality assumed sufficient proportions in the West to encourage farmers to replace horses by tractors.

Estimates of economic losses are essential to the process of cost-benefit calculation as a potential tool in disease prevention or control. If these estimates are not available, as is usually the case, then it will be impossible to arrive at a measure of "benefit" (which is derived from the reduction or elimination of economic, psychosocial, and demographic losses as well as the reduction or elimination of certain costs, e.g., for clinical care). While it is probably easier to assign monetary values to the kinds of costs listed in Section III and IV of Table 3, in practice the gathering of these data is certain to be difficult. As a result the whole cost-benefit subject remains essentially theoretical in the arbovirus field. This is, in fact, the case for most communicable diseases, even for malaria and schistosomiasis (despite the recent efforts of a few health economists). The appeal of the cost-benefit approach is naturally in the possibility that it may strengthen a case or provide a more persuasive rationale for public expenditure on prevention or control; thus, it is really a planning, policy, or "political" instrument.

In practice, then, informal, very rough estimates of losses and costs are routinely used in public health planning and budgetary decision making. For the arboviral diseases, at least, this situation reflects the scarcity of data and the scarcity of persons in the field with the training, interest, and encouragement (from their seniors) that are needed to gather such data.

IV. REMARKS

This brief survey of sociomedical factors demonstrates how little we really know, except in very general terms, about human behavior and the determinants of behavior in relation to any of the arbovirus diseases of man and his domestic animals. As for so many other communicable diseases the greatest amount of time and energy in ecological and epidemiological fieldwork has been devoted to studies of viruses, vectors, and vertebrate hosts other than man. Human studies have focused on populations at risk (scarcely at all on those who intervene) and very few studies have gone beyond enumeration of a few characteristics of the population, usually socioeconomic, rarely behavioral. Beyond this, what is in the literature of a sociomedical nature is largely anecdote.

This situation is scarcely surprising because arbovirus research has depended chiefly on the efforts of biologists — virologists, immunologists, entomologists, vertebrate zoologists — and of other medical and veterinary medical investigators — clinicians, veterinarians, and medical and veterinary epidemiologists — who have generally had little training in the social-behavioral sciences or contact with social scientists.

It was noted in the introduction, however, that epidemiology is partly a social science, and most epidemiologists today will acknowledge that their work should take into consideration behavioral and socieconomic determinants. Can we expect change in the direction of a greater fusion of biomedical and sociomedical interests, methodology, and research techniques? There are favorable signs. The World Health Organization through the Social and Economic Research component of the Special Programme for Research and Training in Tropical Diseases (TDR) has been encouraging research in this direction for nearly 10 years, although admittedly with a focus only on the six diseases. The TDR initiatives, extending also to substantial support of training and institutional strengthening in epidemiology and the sociomedical sciences, have certainly helped to suggest new possibilities for collaborative research on other major communicable disease problems. Biomedical investigators, epidemiologists, and social scientists have joined together, in fact, just in the last few years, in work on diarrheal diseases; the chlamydial diseases, especially trachoma; genital herpes; AIDS; and planning is proceeding for collaborative studies of DEN in Puerto Rico, studies intended to provide a basis for long-term, community-based control.[33] The Gahlinger study[16] on arboviral encephalitis risk exemplifies the possibilities for drawing on the techniques of several disciplines in an investigation of epidemiologically relevant human behavior.

Although Monath's[26] recent review of YF concludes with a call for molecular biological research, his comment on breakdowns and potential problems in control is in itself a sufficient reminder of the need for new (sociomedical) investments in arbovirus research and control.

REFERENCES

1. **Dunn. F. L. and Janes, C. R.,** Introduction: medical anthropology and epidemiology, in *Medical Anthropology and Epidemiology*, Janes, C. R., Stall, R., and Gifford, S., Eds., Reidel, Dordrecht, 1986, chap. 1.
2. **Audy, J. R.,** Aspects of human behavior interfering with vector control, in *Vector Control and the Recrudescence of Vector-Borne Diseases*, Pan American Health Organization Sci. Publ. No. 238, PAHO, Washington, D.C., 1972, 67.
3. **Gillet, J. D.,** Mosquito-borne disease: a strategy for the future, *Sci. Progr. Oxford*, 62, 395, 1975.
4. **Morris, C. D., Corey, M. E., Emord, D. E., and Howard, J. J.,** Epizootiology of eastern equine encephalomyelitis virus in upstate New York, USA. I. Introduction, demography and natural environment of an endemic focus, *J. Med. Entomol.* 17, 442, 1980.

5. **Bown, D. N. and Bang, Y. H.,** Ecological studies on *Aedes simpsoni* (Diptera:Culicidae) in southeastern Nigeria, *J. Med. Entomol.*, 17, 367, 1980.
6. **Bang, Y. H., Bown, D. N., and Arata, A. A.,** Ecological studies on *Aedes africanus* (Diptera:Culicidae) and associated species in southeastern Nigeria, *J. Med. Entomol.*, 17, 411, 1980.
7. **Dunn, F. L.,** Human behavioural factors in the epidemiology and control of *Wuchereria* and *Brugia* infections, *Bull. Public Health Soc. Malaysia*, 10, 34, 1976.
8. **Dunn, F. L.,** Behavioural aspects of the control of parasitic diseases, *Bull. WHO*, 57, 499, 1979.
9. **Dunn, F. L.,** Social determinants in tropical disease, in *Tropical and Geographical Medicine*, Warren, K. S. and Mahmoud, A. A. F., Eds., McGraw-Hill, New York, 1984, chap. 125.
10. **Dunn, F. L.,** Sociomedical contributions to trachoma research and intervention, *Rev. Infect. Dis.*, 7, 783, 1985.
11. **Dobbins, J. G. and Else, J. G.,** Knowledge, attitudes, and practices related to control of dengue haemorrhagic fever in an urban Malay kampung, *Southeast Asian J. Trop. Med. Public Health*. 6, 120, 1975.
12. **Waterman, S. H., Novak, R. J., Sather, G. E., Bailey, R. E., Rios, I., and Gubler, D. J.,** Dengue transmission in two Puerto Rican communities in 1982. *Am J. Trop. Med. Hyg.*, 34, 625, 1985.
13. **Chan, Y. C., Chan, K. L., and Ho, B. C.,** Integrated control of *Aedes* species in Singapore: public health education, law enforcement and surveillance, in *Vector Control in Southeast Asia*, Chan, Y. C., Chan, K. L., and Ho, B. C., Eds., SEAMEO, Singapore, 1974, 89.
14. **Phanthumachinda, B., Phanurai, P., Samutrapongse, W., and Charoensook, O.,** Studies on community participation in *Aedes aegypti* control at Phanus Nikhom district, Chonburi province, Thailand, *Mosq. Borne Dis. Bull.*, 2, 1, 1985.
15. **Dunn, F. L.,** Human behavioural factors in mosquito vector control, *Southeast Asian J. Trop. Med. Public Health*. 14, 86, 1983.
16. **Gahlinger, P. M., Reeves, W. C., and Milby, M. M.,** Air conditioning and television as protective factors in arboviral encephalitis risk, *Am. J. Trop. Med. Hyg.*, 35, 601, 1986.
17. **Schwabe, C. W.,** *Veterinary Medicine and Human Health*, 3rd ed., Williams & Wilkins, Baltimore, 1984, 347, 380.
18. **Rudnick, A.,** Ecology of dengue virus, *Asian J. Infect. Dis.*, 2, 156, 1978.
19. **Gubler, D. J., Novak, R. J., Vergne, E., Colon., N. A., Velez, M., Fowler, J.,** *Aedes (Gymnometopa) mediovittatus* (Diptera:Culicidae), a potential maintenance vector of dengue viruses in Puerto Rico, *J. Med. Entomol.*, 22, 469, 1985.
20. **Tesh, R. B.,** Undifferentiated arboviral fevers, in *Tropical and Geographical Medicine*, Warren, K. S. and Mahmoud, A. A. F., Eds., McGraw-Hill, New York, 1984, chap. 70.
21. **Charoensook, O., Sethaputra, S., Singklang, K., Yawwa, T., Purahong, S., and Suwankiri, P.,** Prevalence of *Aedes* mosquito in big cement jars and rain water tanks, *Mosq. Borne Dis. Bull.*, 2(Abstr.) 78, 1986.
22. **Halstead, S. B.,** Dengue, in *Tropical and Geographical Medicine*, Warren, K. S. and Mahmoud, A. A. F., Eds., McGraw-Hill, New York, 1984, chap. 69.
23. **Morrison, P.,** Review of *Disease and Urbanization*, Clegg, E. J. and Garlick, J. P., Eds., *Sci. Am.*, 244, 48, 1981.
24. **Anon.,** Arboviral encephalitis surveillance update, *California Morbidity*, weekly report from the Infectious Disease Branch, no. 33, Department of Health Services Sacramento, 23 August 1985.
25. **Stanley, N. F.,** Man's role in changing patterns of arbovirus infections, in *Changing Disease Patterns and Human Behavior*, Stanley, N. F. and Joske, R. A., Eds., Academic Press, New York, 1980, chap. 9.
26. **Monath, T. P.,** Glad tidings from yellow fever research, *Science*, 229, 734, 1985.
27. **Reeves, W. C.,** Factors that influence the probability of epidemics of western equine, St. Louis and California encephalitis in California, *Vector News*, 14, 13, 1967.
28. **Perlman, D.,** Central Valley breeding ground. Weather cycle speeds mosquito season, *San Francisco Chronicle*, 19 June 1986.
29. **Monath, T. P.,** Yellow fever, in *Tropical and Geographical Medicine*, Warren, K. S. and Mahmoud, A. A. F., Eds., McGraw-Hill, New York, 1984, chap. 68.
30. **Scherer, W. F.,** Arboviral encephalitis, in *Tropical and Geographical Medicine*, Warren, K. S. and Mahmoud, A. A. F., Eds., McGraw-Hill, New York, 1984, chap. 71.
31. **Goh, K. T., Ng, S. K., and Kumarapathy, S.,** Disease-bearing insects brought in by international aircraft into Singapore, *Southeast Asian J. Trop. Med. Public Health*. 16, 49, 1985.
32. **Sabattini, M. S., Monath, T. P., Mitchell, C. J., Daffner, J. F., Bowen, G. S., Pauli, R., and Contigiani, M. S.,** Arbovirus investigations in Argentina, 1977—1980. I. Historical aspects and description of study sites. *Am. J. Trop. Med. Hyg.*, 34, 937, 1985.
33. **Koss. J.,** personal communication, 1986.

Chapter 12

EPIDEMIOLOGICAL PRINCIPLES APPLIED TO ARBOVIRUS DISEASES

Michael B. Gregg

TABLE OF CONTENTS

I. INTRODUCTION

The purpose of this chapter is to introduce the reader to the basic principles of epidemiology, the methods used, and the application of these methods to a practical field setting. Although the discussion focuses on the practice of epidemiology in relation to arbovirus diseases, the principles and methods apply equally to other infectious diseases and noninfectious diseases as well. Epidemiologic analyses range from the most simple, almost intuitive, procedures to highly complex functions requiring extensive knowledge of biostatistics as well. The thrust of this chapter is on the practical aspects of an epidemiologic investigation of arbovirus disease emphasizing surveillance techniques, investigative procedures in the field, and appropriate analytic methods.

II. BACKGROUND AND DEFINITIONS

The word epidemiology, derived from the Greek *epi* (on or upon), *demos* (people), and *logos* (the study or knowledge of), has been defined many ways. However, all definitions include two basic elements: the study of *human health in groups of people*. A usable definition of epidemiology, therefore, might be the study of the distribution and dynamics of health events in human populations. Epidemiologists study groups of people rather than the single individual, yet they may need to examine patients and perform laboratory tests like the practicing physician. However, epidemiologists study populations using special techniques to ultimately focus their attention on prevention and control of disease rather than diagnosis and treatment of the individual patient.

What, then, are the specific purposes of epidemiology? Primarily the epidemiologist tries to determine the etiology of disease, the source(s) of the agent(s) responsible, the mode(s) of transmission, who is at risk of becoming ill, and what exposures predispose to disease. The answers to one or more of these questions may provide a basis for prevention and control. In other words, epidemiologists attempt to explain disease occurrence by studying the character of the agent, the human host, and the nature of the environment in which they interact. Clearly, these purposes apply to investigations of arbovirus diseases. Indeed, historically, extensive field and laboratory investigations of agent, host populations, and environmental determinants have attempted to define reservoirs, modes of spread, and human risk factors related to arbovirus infection.

To help define and understand the forces that produce disease in populations, epidemiologists generally perform three very basic functions: they count cases, determine rates, and then compare rates. For instance, in an investigation of a dengue (DEN) epidemic, the epidemiologist would first determine the number of ill persons and then calculate illness rates in various subpopulations in the affected community. Finally, the investigator would compare rates of illness between certain exposed and nonexposed populations to help explain why DEN affected some persons but not others.

Although these functions represent in simplistic fashion exactly what epidemiologists do, they actually reflect two basic kinds of epidemiologic study: descriptive and analytic epidemiology. In the first instance, information is collected that describes the setting in which the disease occurred, i.e., the time, the place, and the person. The duration of disease occurrence, where disease was acquired or recognized, and the characteristics of the ill people are then the *descriptive* aspects of any epidemiologic investigation. Often, simply by knowing these facts and the etiologic diagnosis, one can determine the source, mode of spread of the agent, and the group most at risk of having disease. Common sense often provides this information and little or no further investigation is required.

However, in some instances, the agent, its reservoir, its mode of spread, or the risk factors are obscure, and only by more refined *analytic* epidemiologic techniques will the true picture emerge. The analytic techniques prove most useful for posing and testing hypotheses to determine what places human populations at risk and what specific exposures predispose most frequently to infection.

Since the primary purposes of epidemiology are to prevent and control disease, what, then, are the primary tasks of the practicing epidemiologist to achieve these goals? First, there must be an established system of continual data collection, specific and sensitive enough to provide rapid reliable information. The system should be designed to define disease trends and to alert health officials to real or potential health problems. In the present parlance, the term *epidemiological surveillance* encompasses this task. Next, having recognized an outbreak, an epidemic, or a need to analyze data more carefully or rapidly, the epidemiologist must examine further the circumstances surrounding this problem and may have to perform a field investigation. Finally (and as a part of any field investigation), the epidemiologist must perform appropriate analyses of the data, draw defensible inferences from them, and make recommendations for prevention and control of disease. The next sections of this chapter address each of these three tasks in some detail.

III. EPIDEMIOLOGICAL SURVEILLANCE

Although there is no universally agreed-upon definition of epidemiological surveillance, for our purposes, it can be defined as the systematic, continued, careful watchfulness over relevant health data with appropriate analyses and inferences, and the rapid dissemination of this information to those who need to know. Intrinsic in the entire concept of epidemiological surveillance is rapid and appropriate action for disease prevention and control. The basic functions of epidemiological surveillance are included in the definition, namely data collection, analyses, and dissemination.

The purposes of epidemiological surveillance most broadly are to provide a data base to be applied in the rational prevention and control of disease. However, there are more specific purposes that should be recognized. Epidemiological surveillance provides essential data to recognize and to define health problems, to determine specific objectives, to establish priorities, to determine strategies, to evaluate prevention and control methods, and to suggest areas for further research. Therefore, epidemiological surveillance not only gives epidemiologists information regarding what the health problem is, but helps in their efforts to implement prevention and control measures.

Many sources of health-related data exist that epidemiologists may develop or use in any epidemiological surveillance system. Following are brief descriptions of some basic data systems most applicable to human arbovirus disease surveillance.

A. Mortality Data

Virtually all health jurisdictions in all countries have systems of counting deaths in their populations. Unfortunately, mortality statistics often suffer from lack of specificity, sensitivity, and particularly timeliness. Considerable variation may exist in the quality of mortality statistics from community to community, and from country to country, making comparisons difficult, if not impossible. Inherent delays in reporting, registration, analyses, and publication of mortality statistics limit their usefulness in rapid detection of infectious disease trends.

The practicality of using mortality statistics as an assessment of arbovirus diseases will vary tremendously by time, location, and disease. The St. Louis encephalitis (SLE) epidemic of 1964 in Houston was first suspected when the local health officer examined three death certificates in mid August.[1] Investigation later identified the first death as caused by SLE and confirmed cases with onset in late June of that year. Ultimately, the investigation found 160 confirmed cases. Conscious, active, ongoing examination of death certificates, therefore, can be useful in outbreak recognition, particularly in arbovirus diseases with high case-fatality ratios and in the most susceptible age groups.

B. Morbidity Data

Morbidity data can often provide much more useful, timely, and particularly sensitive systems of epidemiological surveillance. Again, most health jurisdictions have at least rudimentary systems of morbidity reporting for communicable diseases. Depending upon the degree of reporting by the practicing physician or health-care provider, such systems may give the epidemiologist extremely important and useful information. Sources of morbidity data will usually include city, county, state, or provincial data bases that list cases of communicable diseases often by week, sometimes biweekly, and occasionally by month. By careful, regular analyses of these data sources, the epidemiologist frequently can acquire the basic elements of descriptive epidemiology, namely the time and place of occurrence of disease and perhaps the age and gender of patients. Other useful data sources include schools, factories, or large businesses where absentee data are routinely collected. In the developing world, "fever clinics" or dispensaries catering to special age groups or health-care problems may be useful. Some countries have included in their reporting systems an "unusual event" category which encourages physicians to report any unusual case, cluster of disease, or death.

Hospital- or clinic-based morbidity systems have been extremely useful in arbovirus disease surveillance. For example, prior to the expected Venezuelan equine encephalitis (VEE) epidemic in Texas in 1971, three levels of surveillance were established.[2] Twenty-eight large community hospitals in central and southern Texas were contacted and asked to report each day on persons hospitalized with suspected VEE. Concomitantly, five county hospitals in the southernmost part of Texas, where risk was believed to be the greatest, not only reported on hospitalized patients, but on the number of patients seen daily in their outpatient clinics with symptoms compatible with that of VEE. Private practitioners also contributed to the surveillance effort.

All three systems of surveillance documented a simultaneous rise in the number of patients with suspected VEE during the first 2 weeks of July 1971, and a subsequent fall in the latter part of the month. Cases of VEE confirmed by virus isolation also peaked in mid July, concomitant with the other parameters of disease surveillance. Hospital-based surveillance systems for SLE have been equally useful in some of the large epidemics in the past two decades.[3-6]

Another less direct, but perhaps more specific, system of encephalitis surveillance was used during the SLE epidemic in Chicago in 1974.[5] Besides the intensive day-by-day tabulation of suspected cases seen at over 100 hospitals in greater Chicago, surveillance included registering the number of lumbar punctures performed with evidence of pleocytosis.

Intelligent use of simple, logical, clinical syndromes for case finding will often provide a rough index of disease occurrence for ongoing surveillance or during an epidemic investigation. Syndromes such as fever with rash or fever with eye pain and headache may help define DEN cases. The encephalitides may be seen with a febrile headache, obvious meningeal irritation, or frank encephalopathy — any or all of which can be counted. Fever with headache, myalgia, and nausea and vomiting was very useful in identifying suspect cases of VEE in 1971.[1] The rule of thumb, particularly where physicians and diagnostic resources are scarce or unavailable, is to establish a simple, workable, reproducible case-counting system that can be applied by paramedical help. A system that is very sensitive is superior to one that is too specific. The primary purpose of most surveillance systems is to detect, not to diagnose.

C. Single-Case Investigation

Because many arbovirus diseases infect without overt disease, an imperative exists to investigate single cases of presumed arbovirus disease in a community. These single cases should be considered as sentinel health events and followed up immediately. Just as a single case of paralytic poliomyelitis represents 100 to 200 other cases of mild to nonexistent disease elsewhere in the community, one full-blown case of classic SLE or DEN represents tens if not hundreds of other cases as yet unrecognized and/or unreported. The alert, well-trained physician with a background in epidemiology may often be the first to identify extensive arbovirus disease, as was the case in Dallas in 1966.[3]

D. Surveys

One of the most useful tools for epidemiologic assessment in infectious diseases and particularly those caused by arboviruses is the serological survey. Most frequently performed during or after the recognition of an epidemic, serological surveys may be the only means to make a firm etiologic diagnosis. Ongoing, yearly serologic studies of high-risk populations may also provide useful information, particularly by helping to define the actual incidence of disease on a year-by-year basis (see later discussion).

E. Knowledge of Arthropod and Vertebrate Vector Species

An important adjunct to or surrogate for human arbovirus disease surveillance is the monitoring of nonhuman vertebrate hosts and vector species. Humans usually represent an incidental or dead-end host, generally have a high rate of inapparent to apparent infections, and contribute insignificantly to the zoonotic cycle. Since illness and death among susceptible animal species almost always precede recognition of human disease, epidemiological surveillance of appropriate zoonotic species should be strongly considered in areas where epizootics are likely to occur and where human surveillance is poor or nonexistent. Ideally, both human and animal surveillance should be maintained.

In the case of viruses which are amplified in cycles involving animal species which have clinically silent infections (e.g., SLE virus in birds), surveillance of inapparent infections by tests for antibody or viremia may provide early warning of an impending human epidemic.

Collection and testing of vector species (usually mosquitoes) and their vertebrate hosts have been useful in documenting the presence of arboviruses in the environment. Although very essential for on-going, intensive field studies and predictive surveillance of certain diseases, this practice requires a major commitment of resources. For practical public health application, however, limited resources are often best allocated to improve existing human and, if possible, other clinically susceptible vertebrate mammalian surveillance.

F. Demographic and Environmental Factors

As with all epidemiologic investigations, it is essential that the basic demographic characteristics of the population at risk are known, i.e., the number, age, and sex distribution of the population. Without these data, no rates can be determined. In some instances these data are not readily available and have to be acquired during the field investigation itself. This has been particularly true in rural parts of developing countries where census data do not exist. However time-consuming and expensive to acquire, demographic data are indispensable; without them, valid comparisons and identification of high-risk populations are impossible.

Environmental determinants such as temperature, rainfall, humidity, and other climatic variables are basic essentials almost always needed in arbovirus disease investigations.

IV. INVESTIGATIVE TECHNIQUES

A. Background

The following section describes how to perform a field investigation of an arbovirus disease and what important administrative and public health realities are faced by field epidemiologists. The scenario consists of a request for epidemiologic assistance from a local health jurisdiction to a state, provincial, or federal health department. Also underlying this section is the theme of quick action, establishing priorities, and performing responsibly. Every effort should be made to collect and analyze data in the field and make recommendations before departure to home base. With the recent availability of computers and easy data-handling equipment, there is a strong tendency to take the data back home and perform the analyses there. Local health officials often view such action as a lack of commitment and interest, but equally as important, the patient or vector populations are no longer easily available for study. The total commitment to the field investigation, the urgency, and the impetus to investigate are frequently lost once the investigating team has left the scene.

B. Early Recognition of Disease and Response

In most instances, a state or regional health officer will be informed of an unusual number of cases of arbovirus disease from local health officials, hospitals, practicing physicians, or occasionally the news media. The local health officer, particularly in the case of arbovirus disease, probably will not have an etiologic diagnosis, and may only suspect an outbreak is in progress. However, the consultant epidemiologist should try to acquire as much information as possible in regard to the number of cases, the time and the place of occurrence, and any demographic information that is available. Depending upon the size and sophistication of the local health department, relatively extensive information may already have been collected which may be very useful in planning the field investigation.

At this juncture it is important to know why help is requested. Possibilities include simply a need for more professional help to perform or complete the investigation, to share the responsibility of the investigation with others more experienced and knowledgeable, or, most often, to acquire added professional expertise. Sometimes legal and ethical issues may complicate the investigation, and the investigator should be alert to such possibilities. No matter what the motivation behind the request for assistance, there must be an official basis for such a request and local permission for a field investigation.

C. Preparation for the Field Investigation

Successful field investigations, particularly of arbovirus disease outbreaks, depend entirely upon the collaborative efforts of clinicians, laboratory workers, epidemiologists, entomologists, mammalogists, veterinarians, and others working under a single director. As soon as the epidemic or outbreak is identified, a multidisciplinary team should be formed so that equal and timely scientific input from all disciplines can be obtained. This is particularly true for those professionals providing laboratory support who, unfortunately, are occasionally consulted and requested to provide service during the investigation or even after the investigation has taken place. Every effort should be made in advance to discuss the overall plan of the field investigation with everyone concerned, including specific scientific questions to be answered.

Depending upon the nature and extent of the field investigation, statistical support will likely be needed, particularly if large populations are to be studied and surveys are contemplated. Some consideration should also be given to including among the team an information specialist who can meet regularly with the news media to inform them of the findings and to relieve the investigators of the burden of the interruptions caused by news media personnel. Particularly useful in this regard is to identify a single spokesman for the investigative team who will meet the news media as often as necessary.

There are other useful guidelines and basic administrative activities that can materially improve the preparations for and the execution of the field investigation. These include: (1) appointing a team leader and a deputy who will be primarily responsible for the field investigaiton in collaboration or cooperation with the local health officials, (2) determining when and how communication will be established between the field team and the supervisors at "home base", and (3) making sure that the investigating team meets with local health officials immediately upon arrival in the field.

The investigator needs to identify as soon as possible who are the local health department counterparts to those on the investigative team — specifically to identify and make contact with the local team leader, the local laboratory director, entomologist, veterinarian, etc., if they are available. The team leaders should strongly consider meeting with local political officials to inform them of their presence and their proposed activities. The extra day or two required to accomplish these meetings and discussions may prove invaluable later in the investigation and may save considerably more time and effort than initially appreciated.

It is also very useful to have the investigative team leader identify a local health official who will speak for the entire field team. In general, the visiting investigators should avoid contact with the press. That responsibility should be left to the local health officials who are most familiar with the political and scientific realities in their jurisdiction.

Lastly, and sometimes of great importance, the investigator needs to determine whether scientific publication following the investigation is likely. If so, the authors, and particularly the senior authors, of any publication need to be identified as soon as possible. It is much easier to make this determination at the beginning of the investigation than at the end.

V. THE FIELD INVESTIGATION

The following discussion outlines ten steps or procedures that are normally included in any field investigation. The proposed order appears logical, but clearly some of the steps may be taken in different order or may be done simultaneously. There is no necessarily right or wrong way, for each investigation is unique, as are the epidemiologists who perform the investigations. However, in general, the data collected, the analyses, the inferences, and the ultimate prevention and control measures will likely be similar regardless of the order of activities and the epidemiologists involved.

Except in the rare instance where prospective studies of large populations are underway, it will take at least 1 to 2 weeks before arbovirus disease epidemics are recognized. This means that the investigative team arrives after the epidemic has started and, therefore, will be performing a study after the fact. Many illnesses and critical events will have already taken place, and the epidemiologist will have to look retrospectively at what probably happened. S/he will, then, rely heavily upon the memory and recollections of physicians, health officers, and other local persons to reconstruct the likely events which led to the beginning of the epidemic. This means that one is dependent upon data that are often imprecise and frustratingly fragmentary. Nevertheless, the experienced investigator will collect as much data as possible and interpret them accordingly.

A. Determine the Existence of an Epidemic

In general, local health officials know when more cases of a specific disease are occurring than would normally be expected. Health department records, particularly of conditions such as encephalitis and aseptic meningitis, are usually assessed regularly, and fluctuations in patterns of these syndromes become apparent quite readily. Unfortunately, the nonspecific nature of the encephalitides and DEN, as examples, and the sophistication needed in the laboratory to make an etiologic diagnosis hamper the sensitivity and specificity of routine disease surveillance. However, if there are no *a priori* reasons why the reporting system has changed or been altered, local health records should basically be accepted as a reflection of the disease burden. Other sources of morbidity data may need to supplement the routine reporting system, such as school, factory, or industry absentee records. Sometimes quick telephone surveys of key practitioners or outpatient clinics in local hospitals will identify or substantiate other reports of epidemic disease occurring in the area. In the developing world, local "holy men" or religious leaders may know of cases or deaths if regular data are not collected. Even contact with those who bury the dead may be useful.

B. Confirm the Diagnosis

For many infectious diseases, particularly bacterial, confirmation of the disease in question can be relatively simple and rapid. With other diseases, the clinical syndrome is virtually pathognomonic, such as measles, chicken pox, and the like. However, for arbovirus diseases, clinical confirmation is usually very difficult because of the protean manifestations of the disease syndrome and/or because of the need for specific laboratory confirmation. Very few local laboratories and only modest numbers of state or provincial microbiological laboratories are equipped to isolate and identify arboviruses. Elsewhere in the world, particularly in the developing countries, only national laboratories may be capable of isolating and identifying viruses. Although serological testing is easier to perform, specific reagents and sensitive tests are only available at national or reference laboratories.

Confirmation of diagnosis is uniquely difficult among the arbovirus diseases and consequently great attention and effort must be directed toward case finding, detailed clinical investigation, and collection of appropriate diagnostic specimens. The urgency here is establishing a close-knit, timely reporting system so that patients can be seen immediately by a qualified clinician, clinical criteria can be established, and the most reasonable clinical diagnosis can be made immediately. A presumptive etiologic diagnosis can be made much more convincingly with five or ten cases exquisitely studied than with 50 to 75 poorly documented cases of a similar syndrome. The team leader should make every effort to personally see a representative sample of the most severely ill, not only to confirm the diagnosis, but to secure the proper credibility for purposes of reporting to higher health authorities.

C. Determine the Number of Cases

After having examined several representative cases, the epidemiologist must create a workable case definition so that case finding can begin. Objective criteria for case definition are the best, such as fever, pleocytosis, rash, jaundice, vomiting, or central nervous system signs. The important point, however, is to create a reasonably accepted, simple, and applicable case definition recognizing that some cases will be missed and some noncases will be included. During the height of the field investigation, a practical balance between specificity and sensitivity in case definitions must be achieved. Ultimately, if laboratory specimens are collected and more detailed clinical information can be made available, one can refine the case definition for more detailed analyses. Particularly useful in field investigations has been classification of cases as confirmed, probable, possible, or unlikely. This helps immeasurably early in the study and may allow for some extremely useful comparisons later on.

There are several factors that may help in determining the level of sensitivity and specificity in a field investigation:

1. What is the apparent to inapparent clinical case ratio?
2. Are there important obvious pathognomonic or clinically suggestive signs and symptoms of the disease?
3. What virus isolation, identification, and serologic techniques are available that are easy, practical, and reliable?
4. How accessible are the patients or those at risk; how feasible will recontacting be

following the initial investigation for further questions, examination, or serologic testing?

5. Can the case definitions be applied easily and consistently by persons other than the current investigating team in the event that long-term follow-up is necessary?
6. Must all patients be identified during the initial investigation, or would only those seen by physicians or hospitals be sufficient?

These and other considerations may play an important role in how cases will be defined and how intensive case investigation will be. However, no matter what criteria are used, a case definition must be determined and applied without bias to all persons under investigation. If the case definition needs later revision, the resulting definitions should be applied objectively and equally to every person.

Having established a workable case definition, the epidemiologist must count cases using whatever method seems appropriate for the particular situation. If the reporting systems are deemed reasonably reflective and constant, the traditional method may be used. However, efforts are usually directed toward finding as many cases as possible by contacting physicians, hospitals, free clinics, and by performing a variety of surveys. The news media may be called upon to help in identifying cases and encouraging ill people to seek medical assistance.

Regardless of the method used to find cases, simple counting alone will not be adequate for proper analysis. Basic demographic information such as age, sex, occupation, date of illness, and residence must also be obtained. Since the primary motivation of an investigation is to document the risk factors and exposures unique to those who are ill, the epidemiologist must acquire appropriate additional information which can hopefully identify and quantitate exposures. With arbovirus disease in particular, the investigator should seriously consider acquiring information such as duration of residence in the community and in the house or dwelling; history of exposure to arthropods; the characteristics of the dwelling, e.g., screens, mosquito netting, etc.; history of vaccination; number of residents in the dwelling; daily habits of residents of the dwelling (number of hours inside and outside the home, the time of day); environmental features of areas of possible exposure; history of recent travel; the use of protective clothing; and detailed history of previous residences and possible occupational exposure.

D. Orient the Data in Terms of Time, Place, and Person

As soon as the majority of cases have been identified and appropriate demographic and epidemiologic information has been tabulated, the descriptive aspects of the investigation should be performed. This is the time to characterize the epidemic or outbreak in terms of when people became ill, where they were when they became ill (or their place of residence), and what characteristics the ill population has. This process is termed descriptive epidemiology and should be performed as soon as possible in the investigation. The primary purpose of examining these data is to generate hypotheses to explain possible risk factors and specific exposures. A tendency may be to wait until all cases have been identified before this first analysis is done. The epidemiologist should avoid waiting for the very last case to be reported, but should characterize the epidemic as soon as possible. Not infrequently, early analyses of cases will provide useful information regarding who is primarily at risk and where further investigative efforts of humans, nonhuman vertebrates, and vectors should be directed.

1. Time

A simple graphic presentation of cases by date of onset often provides extremely valuable information to the epidemiologist. Known as an "epidemic curve", this pictorial representation helps investigators to comprehend the magnitude and duration of the outbreak and

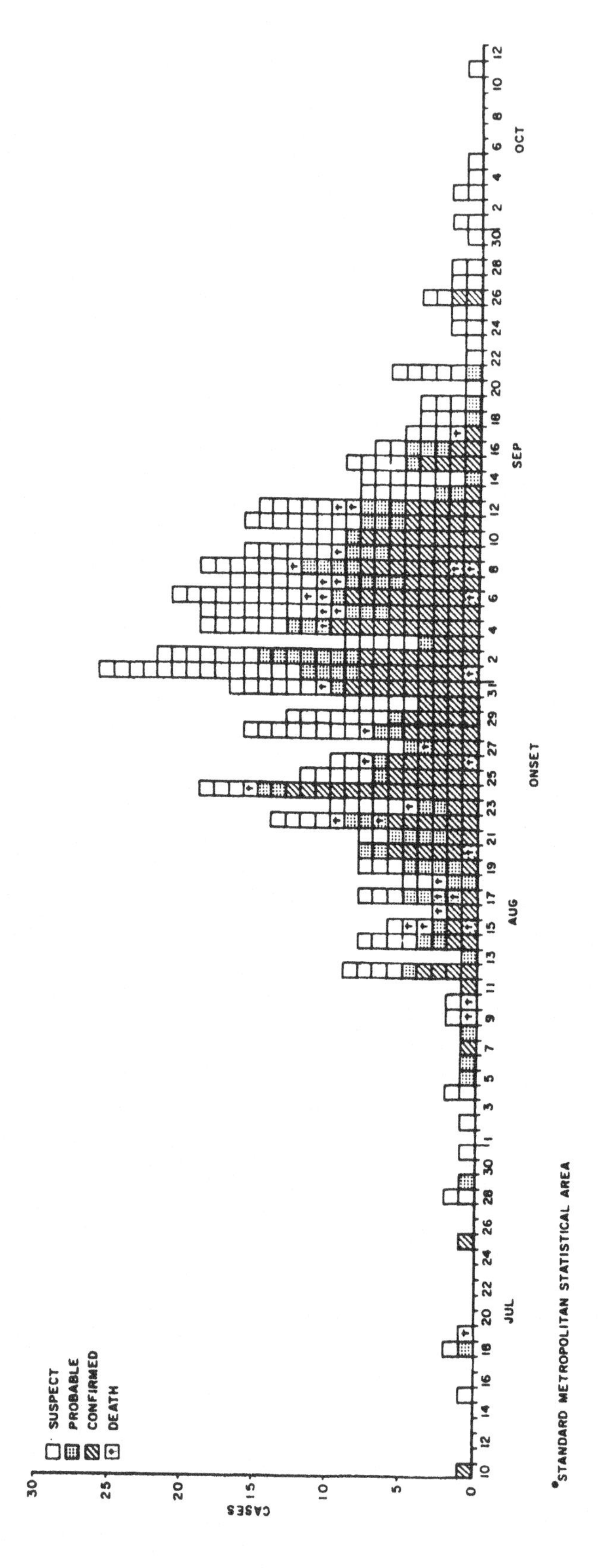

FIGURE 1. SLE cases, Chicago, Standard metropolitan statistical area, July 10—October 12, 1975.

may indicate the mode of transmission (Figure 1). A remarkable amount of information can be inferred from an epidemic curve, if the incubation period is known and if dates of onset(s) of disease are reasonably accurate. For arbovirus diseases in general, epidemic curves tend to build up relatively slowly over several weeks, reach a peak spanning several weeks, and slowly decline, depending upon what prevention and control efforts have been applied. These more leisurely appearing epidemic curves result from the inherent qualities of arbovirus disease transmission, which include such variables as vector density, amplifying hosts, extrinsic incubation period within vectors, variable climatic factors, the lack of person-to-person spread, and the usual absence of specific common sources of infection similar to those often seen in other infectious deseases.

The epidemic curve may occasionally allow the epidemiologist to predict how long the epidemic is likely to last and can often serve as a useful means of communication to the nonepidemiologist administrators or senior health officials. Figure 1 represents a recent extensive outbreak of SLE in Chicago in 1975. Investigators reported 269 serologically confirmed or probable cases and 256 clinically suspected cases were identified. As can be seen during the month of July, there were scattered occurrences of probable and confirmed cases, but it was not until the second week of August that substantial numbers of cases began to occur. An early peak of cases is seen in the fourth week of August and again in the first week of September. Cases declined fairly rapidly over the succeeding 2 weeks in September. The last confirmed cases occurred during the last week of that month.

2. Place

In some infectious disease outbreaks, the place of acquisition of disease can be quite important. This is particularly true for arbovirus diseases, primarily because of the critical nature of the proximity of the hosts to the vectors of disease. Frequently, simply by locating confirmed or probable cases of arbovirus disease, the epidemiologist can pinpoint the areas of highest risk. Such identification can direct the investigating team to focus their case-finding or vector-collection efforts to specific areas. For example, when persons with SLE were identified by place of residence in an outbreak in Corpus Christi, Tex., most lived in upper middle class neighborhoods and infection was presumably related to the presence of storm drains that offered ideal breeding grounds for *Culex vectors.*[4] A similar analysis in 1966 in Dallas showed that most patients lived near the Trinity River that runs through the city.[3]

3. Person

Lastly, the investigator should examine the character of the patients themselves by such attributes as age, sex, race, occupation, or any other characteristic that might be unique to those who have become ill. Since arbovirus diseases require contact with arthropods for disease transmission, host characteristics such as occupation, outdoor exposure, and daily activities may determine risk of exposure. Several large outbreaks of Rift Valley fever have preferentially affected certain populations such as farmers, farm laborers, and veterinarians.[7,8] The list of human characteristics is virtually endless. However, knowing the source of the agent and its usual mode of spread, the epidemiologist should be able to collect pertinent patient-related information to help generate hypotheses of disease risk.

E. Determine Who is at Risk of Becoming Ill

At this juncture in the investigation, the epidemiologist should know fairly specifically how many people are ill, when and where they became ill, and what comprises their general characteristics. Usually there will be a relatively specific working diagnosis. These data often provide enough information to feel confident as to how and why the epidemic started.

For example, these data often strongly suggest that only a certain population in a particular community exposed to a specific vector are at greatest risk of becoming ill, or that because of high concentrations of a vector species, large portions of a county or province are likely to be exposed to this vector. Under these circumstances, little in the way of analytic epidemiology needs to be applied to determine the high-risk populations. However, there may be times when the population at risk is not so apparent because of low numbers of reported cases, questionable clinical diagnoses, or lack of clear-cut information on vector populations. Under these circumstances, in order to focus on high-risk groups for later investigation, the field team may need to perform door-to-door surveys, hospital record reviews, or even serosurveys to define the high-risk groups. Rates of suspected illness are then compared area by area to determine any significant differences.

F. Develop an Hypothesis

Develop an hypothesis that explains the specific exposure that caused the disease and test this hypothesis by appropiate statistical methods. This is the next step in the field investigation and it is generally the most challenging and requires more analytic skills than any other part of the investigative process. It is hoped that at this juncture the epidemiologists feel comfortable with the diagnosis and the identification of those at greatest risk of acquiring infection. However, the field team still must determine what factors, particularly host-related and environmental, produced or allowed the greatest risk of exposure. With arbovirus disease, the source and mode of transmission are usually well known; only the environmental and host factors relating to specific exposure remain to be determined. These exposures, then, must be defined, measured, and compared between ill persons and healthy controls.

The investigation of Waterman et al.[9] is an excellent example of a careful epidemiologic analysis of the 1982 Puerto Rico DEN epidemic. In 1982, Puerto Rico experienced a large outbreak of type 1 DEN, and serological, entomological, and environmental surveys were undertaken in two Puerto Rican communities. In an effort to determine exposures to *Ae. aegypti* most likely to produce disease, the investigators carefully and systematically assessed certain environmental variables. The hypotheses were that some of these variables, more than others, predisposed to exposure and subsequent infection. Therefore, comparisons of certain environmental conditions between seropositive and seronegative individuals were made and statistically analyzed. Wood-constructed housing, tree height, shade, and screens were predictors of antibody prevalence among the affected population. Although the nature of the study was, of necessity, retrospective and could not establish proof in the true sense of the word, the epidemiological inferences and subsequent conclusions were logical and compatible with a known history of DEN infection.

This analytic phase of the field investigation challenges the epidemiologist the most. The clinical, laboratory, and epidemiologic evidence must be viewed carefully, and a single, plausible hypothesis of exposure(s) must be generated to fully explain the epidemic. If exposure histories for the ill and well (infected or noninfected) are not significantly different, a new hypothesis must be developed. This may require more environmental investigation or resurveying the population under study.

G. Compare the Hypothesis with the Established Facts

Having now determined the probable exposure of disease by epidemiologic and statistical inference, the investigators must fit the hypothesis to the clinical, laboratory, and other epidemiologic facts. In other words, do the proposed exposure, mode of spread, and population affected fit well with the known facts of the disease? In the rare field investigation when the disease is unknown, the epidemiologist will clearly find it more difficult to fit the hypothesis to the natural history of the disease in question. In such a situation, all that can be hoped is that the clinical, laboratory, and epidemiologic findings and events portray a coherent, plausible, and physiologically sound series of findings and events that make sense.

H. Plan a More Systematic Study

Following completion of the field investigation there may be a need to find more cases, to better define the extent of the epidemic, or to evaluate a new laboratory test or case-finding method. The investigative team may want to perform more detailed and carefully executed studies to improve the quality of the data and answer particular questions. The most important reasons to perform such studies are usually to improve the sensitivity and specificity of the case definition and establish more accurately the true number of persons at risk, i.e., to improve the quality of the numerators and denominators. For example, serosurveys coupled with a more complete clinical history may sharpen the accuracy of the case count and define more clearly those truly at risk for developing disease. Moreover, repeat interviews with patients with confirmed disease may allow for rough quantitation of degrees of exposure or dose responses — useful information in understanding the pathogenesis of many diseases.

I. Preparation of a Written Report

There is most often a responsibility for the investigative team to prepare some kind of written report of their findings. There are several important reasons to do so:

1. To create a document for public health action. Many times decisions to control or prevent further disease will only be made on written documentation of the investigative findings.
2. To document performance or output. There probably is no better way to justify programs of arbovirus disease activity than to create written documentation of field work.
3. To create a record for possible medical/legal issues. Objective, honest, detailed reporting of the facts of and the inferences from a field investigation can be invaluable to health officers, physicians, and constituents alike. The public is best served if the facts are available to all.
4. To improve the quality of epidemiological practice. The actual assembly and analysis of investigative findings often identify data that should have been collected and methods that need improvement. Writing epidemiology teaches epidemiology.

J. Execute Control and Prevention Measures

It is not the purpose of this chapter to discuss methods of arbovirus disease control. However, the major force behind all field investigations is control and prevention of further disease and death.

VI. ANALYTIC METHODS

In this last section, an attempt is made to describe some of the more common analytical tools used in field investigations of arbovirus disease. These discussions cover only the highlights and key points for emphasis. Standard texts of epidemiology and biostatistics elaborate more fully and should be referred to for the detailed methods and procedures.

A. Case-Control and Cohort Studies

As referred to previously, the primary purpose of any epidemiologic study is to determine why certain individuals develop disease. In order to answer this question, hypotheses of exposure are generated and rates are compared. The epidemiologist selectively compares rates of disease in various populations in hopes of uncovering what risk factors and specific exposures may be responsible for illness.

Because of their practical application to arbovirus disease investigation, two basic kinds of epidemiologic study are discussed: case-control and cohort studies. Cross-selectional studies, sometimes useful in arboviral disease evaluation, are not described nor are experimental epidemiologic studies.

1. Case-Control Studies

The first kind of study, and the one most frequently performed in arbovirus disease investigations, is that of the case-control or retrospective study. In this instance, the epidemiologist identifies cases of disease that have already occurred and compares their characteristics and/or exposure history to noncases or controls. If the exposure rates of the two groups are different, statistical analyses are made to determine if the difference was likely to have occurred by chance alone, or if it represents a likely causal association between exposure and illness. If the observed difference occurs, for example, only once in 200 similar analyses, the epidemiologist would likely conclude there is a probable causal relationship between that particular exposure and the development of infection.

The logic of case-control studies is applied frequently without regard to statistical tests when it is intuitively clear that study populations are ill and others are not after certain exposures. However, further analyses of *specific* exposures may be necessary in order to more clearly define the exact host or environmental factors responsible for disease. For example, in the previous section on investigative techniques, the analysis by Waterman and associates was a case-control study. These investigators compared infected and noninfected persons in regard to specific environmental exposures and showed a statistically significant different series of exposure rates. The inference was then drawn that these specific exposures materially contributed to acquiring infection.

Case-control studies are usually relatively simple and inexpensive to perform and frequently can be completed in several days to weeks. Also, they are ideally suited to study rare diseases because one can collect many cases that have occurred over a long time period and analyze them quite quickly.

However, in other studies, particularly ones involving large groups of people or entire communities, methodological problems arise particularly as they relate to selection of control populations. Cases and controls should be as comparable as possible in all respects other than the risk or exposure factor under study. Studies drawn from special groups such as hospital patients or particular occupational groups may not be suitable for extrapolation to the population at large. Case ascertainment, or the ability to recognize all cases, may also play an important role in performing valid, objective case-control studies. These and other methodological concerns must be considered when implementing a case-control study. The epidemiology literature is replete with articles — even entire books — on the subject. Suffice it to say, however, that carefully designed and well-performed case-control studies have served and will continue to serve the epidemiologist well.

2. Cohort Studies

The other basic epidemiologic study is called the cohort, or prospective, study. Whereas in case-control studies the event (the disease) has already taken place, in a cohort study, the event has not yet occurred, and the basic approach is one of following individuals or cohorts over time to determine if specific, known exposures are producing disease. Such studies usually require long periods of observation, often years, and considerable resources, but are traditionally considered less biased and more scientifically appealing since data are not collected after the fact. However, there are still methodologic problems that must be considered. Again, comparability of exposed and nonexposed groups may pose problems. Another important consideration is the expected frequency of disease in the exposed and nonexposed populations. With rare diseases, cohort studies may not detect enough cases to allow for valid statistical comparisons at the end of observation. Also, during prolonged prospective studies, members of both exposed and unexposed populations may be lost to

Table 1
POPULATION, DEATHS, AND CRUDE DEATH RATES (CDR) IN TWO FLORIDA COUNTIES, 1960

	1960 population	1960 deaths	CDR 1000 population
Pinellas County	374,665	5,726	15.3
Dade County	935,047	8,322	8.9

From *Vital Statistics of the United States*, 1960.

follow-up or diagnoses may not remain standardized. Successful cohort studies depend heavily on known specific exposure over time. For arbovirus diseases, such close observation may be very labor intensive and expensive.

Probably the best known noninfectious disease example of a cohort study was conduced by Sir Richard Doll,[10] who attempted to determine the effect of cigarette smoking on the development of lung cancer. After 20 years, physician-smokers had a statistically greater risk than physician nonsmokers of developing lung cancer. It was concluded that cigarette smoking was a major cause of lung cancer. Unfortunately, there are few good examples of cohort studies in the field of arbovirus disease. Most long-term studies have consisted of periodic serosurveys, which measure infection rates rather than specific exposure factors. However, in a recent cohort type study by Sangkawibha, Thai children were studied over time for the development of DEN shock syndrome (DSS).[11] The study was able to show that previous exposure to DEN-1 predisposed to the development of DSS following subsequent DEN-2 infection.

B. Age-Adjusted Rates

Emphasis has been placed throughout this chapter on the concept of determining rates of illness or exposures as the most appropriate way of determining the impact of any given factor under study. Only by determining rates can the epidemiologist make fair and scientifically objective comparisons between study populations. Despite this need to determine rates for appropriate comparative purposes, sometimes comparison of disease or death rates between different populations reveals differences that are considerably misleading. For example, as a part of a large field investigation it may be necessary to compare two (or more) study populations. However, if these populations are dissimilar with respect to age, sex, race, vaccination status, or other factors, one may have to "adjust" these rates so that fair comparisons can be made. Most often adjustments are made because of age differences in the study groups. This is particularly true when rates within the populations vary with age and when the study groups differ appreciably in their age composition. A definition of an "age-adjusted" rate is an estimated overall rate that would have occurred in a selected *standard* population had the age-specific rates, which did occur in the study population, prevailed.

By way of brief example, refer to Table 1. One can see that the crude death rate for Pinellas Country, Fla., in 1960 was almost twice as large as that of Dade County. Table 2 shows the age distribution of both counties; Dade had a younger population than Pinellas. The experienced epidemiologist needs rates to compare and thus determines age-specific death rates for both counties (Table 3). It can now be seen that although the crude death rate was much higher in Pinellas, the age-specific death rates are generally higher in Dade County. This seemingly paradoxical finding can be explained simply by an unequal distribution of persons in the various age groups that are compared.

Table 2
DISTRIBUTION OF THE POPULATION BY AGE IN TWO FLORIDA COUNTIES, 1960

Age group (years)	Pinellas County		Dade County	
	Population	Distribution (%)	Population	Distribution (%)
<1	5,674	1.5	18,819	2.0
1—4	22,167	5.9	74,554	8.0
5—14	51,932	13.9	162,633	17.4
15—24	32,565	8.7	108,310	11.6
25—34	33,877	9.0	124,938	13.4
35—44	41,633	11.1	140,768	15.0
45—54	41,670	11.1	118.013	12.6
55—64	51,985	13.9	93,058	9.9
65—74	65,783	17.6	67,994	7.3
75+	27,379	7.3	25,960	2.8
Total	374,665	100.0	935,047	100.0

From *Vital Statistics of the United States*, 1960.

Table 3
POPULATION, DEATHS, AND AGE-SPECIFIC DEATH RATES IN TWO FLORIDA COUNTIES, 1960

Age group (years)	Pinellas County			Dade County		
	1960 population	1960 deaths	Rate/1000	1960 population	1960 deaths	Rate/1000
<1	5,674	160	28.2	18,819	542	28.8
1—4	22,167	30	1.4	74,554	86	1.2
5—14	51,932	30	0.6	162,633	60	0.4
15—24	32,565	26	0.8	108,310	131	1.2
25—34	33,877	47	1.4	124,938	213	1.7
35—44	41,633	124	3.0	140,768	458	3.3
45—54	41,670	320	7.7	118,013	973	8.2
55—64	51,985	829	15.9	93,058	1557	16.7
65—74	65,783	1901	28.9	67,994	2113	31.1
75+	27,379	2259	82.5	25,960	2189	84.3
Total	374,665	5726	15.3	935,047	8322	8.9

From *Vital Statistics of the United States*, 1960.

An easy way to "adjust" for these differences is shown in Table 4. Here, the populations of both counties have been pooled to create a standard population and the age-specific death rates seen in Table 3 have been applied to the pooled numbers. This has created a new "putative" number of deaths for each age group. When added and then divided by the total populations for each county, two new adjusted death rates appear: 10.4/1000 population and 10.9/1000 for Pinellas and Dade counties, respectively. The nearly double crude death rate in Pinellas can now be explained by the much older age distribution of that population.

Age adjusting can be performed using other population standards and can be applied to morbidity data as well as deaths. However, the basic purpose is the same: maximizing comparability by minimizing inequities in age distributions.

Table 4
MORTALITY IN TWO FLORIDA COUNTIES, 1960

Age group (years)	1960 population			Age-specific death rates/1000		Putative deaths in pooled population using county age-specific rates	
	Pinellas	Dade	Pooled	Pinellas	Dade	Pinellas	Dade
<1	5,674	18,819	24,493	28.2	28.8	691	705
1—4	22,167	74,554	96,721	1.4	1.2	135	116
5—14	51,932	162,633	214,565	0.6	0.4	129	86
15—24	32,565	108,310	140,875	0.8	1.2	113	169
25—34	33,877	124,938	158,815	1.4	1.7	222	270
35—44	41,633	140,768	182,401	3.0	3.3	547	602
45—54	41,670	118,013	159,683	7.7	8.2	1230	1309
55—64	51,985	93,058	145,043	15.9	16.7	2306	2422
65—74	65,783	67,994	133,777	28.9	31.1	3866	4160
75 +	27,379	25,960	53,339	82.5	84.3	4400	4496
Total	374,665	935,047	1,309,712	15.3	8.9	13,639	14,335
Adjusted death rates						10.4	10.9

Note: Direct methods of adjustment using pooled population as standard.

C. Serosurveys

One of the most common epidemiologic tools applied in arbovirus disease investigation is the collection of blood specimens for antibody testing. Initially, specimens are usually collected for diagnostic purposes, but subsequently sera may be collected to (1) define the cumulative infection experience of a population to a given virus, (2) estimate the ratio of apparent to inapparent infection rates, (3) define environmental and host-related factors underlying exposure, and (4) estimate reinfection rates.

Detection of specific antibodies to a single infectious agent at a single point in time defines the antibody prevalence to that particular viral agent. This antibody prevalence reflects the cumulative experience, both past and present, with that specific viral agent — presuming lifelong presence of antibody. Testing of paired acute and convalescent sera for demonstration of a rise in antibody titer or the use of a serologic assay applied to single samples that measure only very recent infection (e.g., immunoglobulin M enzyme-linked immunosorbent assay [IgM ELISA] or complement-fixation [CF] test) provide an estimate of the *incidence* of infection during the epidemic period. Coupled with good clinical histories from a field investigation, serologic tests for recent infection can provide information on inapparent to apparent infection rates and also materially supplement and refine acute morbidity reporting. Serum specimens collected before an anticipated epidemic or before contact with a virus may permit determination of protective antibody levels to that virus after challenge, or, in the case of immunopathologic diseases such as DSS, relationships between previous infection and development of severe disease. Reinfection rates can also be determined by collecting paired sera before and after exposure.

Regardless of purpose, rigorous standards of sampling must be applied to the test population. There are several basic sampling methods applicable to serosurveys that may be used depending upon what questions are to be answered.

1. The random sample. Under ideal circumstances, a random sample of the population under study will provide the most useful and statistically powerful results. However,

simple random sampling of individuals is often highly impractical, expensive, and time consuming. The investigators, essentially by definition, have to know the location of the entire population under study for random sampling to be applied. A small population, on the other hand, of perhaps up to 1000 or 1500, in a small village whose population has been or can easily be counted, is very suitable for random sampling. By assigning a number to each individual and by selecting the numbers randomly, the serosurvey can be completed with relative ease.

2. Multistage sampling. When larger populations are to be studied, the sampling method most effectively is divided into several stages of selection. As an example, a city can be divided into census tracts or districts, blocks of housing units, and ultimately dwellings. Depending upon the information to be obtained and the statistical confidence desired, random samplings of each of these clusters is applied. When the blocks have been identified, there can be a simple random selection of housing units in the blocks and either all or a selected segment of each household may be interviewed and bled. Obviously the more levels or groups there are to sample, the less statistical strength there is to the results. This is usually compensated for by increasing the number of census tracts or blocks to enhance confidence in the results. In such a sampling scheme, a specific segment of the population can be targeted for serological surveys, such as the very young or perhaps women in their reproductive years. When such selectivity is applied to a household with more than one member present in that subset, random selection is preferred and strengthens the significance of the results.
3. Stratification sampling. When sampling of large and particularly diverse geographical areas is contemplated, it is highly desirable to subdivide the entire area into several strata by geographic, climatic, or meteorologic characteristics to ensure that these markedly different areas are represented in the survey. Then a sample can be selected from each stratum in proportion to its population size or other characteristic deemed important. This system of stratification guarantees representation from different geophysical areas of the entire population studied.

In all sampling of this kind, the most essential point to bear in mind is that there must be a probability of selection of geographical areas and individuals, i.e., no single geographical unit or individual should be excluded from the possibility of being selected.

REFERENCES

1. **Luby, J. P., Miller, G., Gardner, P., Pigford, C. A., Henderson, B. E., and Eddins, D.,** Epidemiology of St. Louis encephalitis in Houston, Texas, 1964, *Am. J. Epidemiol.*, 86, 584, 1967.
2. **Gregg, M.,** Human disease: USA, in *Venezuelan Encephalitis*, PAHO/WHO Sci. Publ. No. 243, Pan American Health Organization/World Health Organization, Washington, D.C., 1972, 255.
3. **Hopkins, C. C., Hollinger, F. B., Johnson, R. F., Dewlett, H. J., Newhouse, V. F., and Chamberlain, R. W.,** The epidemiology of St. Louis encephalitis in Houston, Texas, 1966, *Am. J. Epidemiol.*, 102, 1, 1975.
4. **Willams, K. H., Hollinger, F. B., Metzger, W. R., Hopkins, C. C., and Chamberlain, R. W.,** The epidemiology of St. Louis encephalitis in Corpus Chisti, Texas, 1966, *Am. J. Epidemiol.*, 102, 16, 1975.
5. **Zweighaft, R. M., Rasmussen, C., Brolnitsky, O., and Lashof, J. C.,** St. Louis encephalitis: the Chicago experience, *Am. J. Trop. Med. Hyg.*, 28, 114, 1979.
6. **Powell, K. E. and Blakey, D. L.,** St. Louis encephalitis. The 1975 epidemic in Mississippi, *JAMA*, 237(21), 2294, 1977.
7. **McIntosh, B. M., Russell, D., dos Santos, I., and Gear, J. H.,** Rift Valley fever in humans in South Africa, *S. Afr. Med. J.*, 58, 803, 1980.

8. **Hoogstraal, H., Meegan, J. M., Khalil, G. M., and Adham, F. K.,** The Rift Valley fever epizootic in Egypt 1977—78. II. Ecological and entomological studies, *Trans. R. Soc. Trop. Med. Hyg.*, 73, 624, 1979.
9. **Waterman, S. H., Novak, R. J., Sather, G. E., Bailey, R. E., Rios, I., and Gubler, D. J.,** Dengue transmission in two Puerto Rican communities in 1982, *Am. J. Trop. Med. Hyg.*, 34, 625, 1985.
10. **Doll, R. and Peto, R.,** Mortality in relation to smoking: 20 years' observations on male British doctors, *Br. Med. J.*, 2, 1525, 1976.
11. **Sangkawibha, N., Rojanasuphot, S., Ahandrik, S., Viriyapongse, S., Jatanasen, S., Salitul, V., Phanthumachinda, B., and Halstead, S. B.,** Risk factors in dengue shock syndrome: a prospective epidemiologic study in Rayong, Thailand. I. The 1980 outbreak, *Am. J. Epidemiol.* 120, 653, 1984.

Chapter 13

QUANTITATIVE MODELS OF ARBOVIRUS INFECTION

Dana A. Focks

TABLE OF CONTENTS

Models of infectious disease are systems that describe the dynamics of a communicable disease and its spread. They mimic the course of infection by incorporating and interrelating the key elements of the particular system. Models are necessarily simplifications of the real world and much of their utility comes from their ability to simulate it with far less than a complete knowledge.

I. INTRODUCTION

The purposes of modeling are varied but all stem from the benefit of integrating what is known about a particular system in a rigorous fashion that is testable. As a minimum, they are a convenient summary of what is known about a particular system. Such efforts highlight areas where additional or better information is necessary. They give insight into the dynamics of a system which is the product of multiple, nonlinear, and complex interrelationships. In an investigation to determine cause and effect, the usual approach is of repeated experiments involving manipulation of independent variables with observation of affected dependent variables. This is clearly inappropriate to the study of epidemics. A natural course around this difficulty is the attempt to develop models that fit the behavior of a particular system and then study the impact of various perturbations such as control measures or the basic parameters of the system itself such as the daily survival of the vector. In the context of quantitative models of arbovirus infection, perhaps the greatest impetus to their creation is the need to study the effects of various control strategies such as sanitation, vaccination, and vector control. Models such as those of Cvjetanovic et al.[1] on the dynamics of acute bacterial infections are good examples. They were designed for public health workers as aids in the evaluation and optimization of disease control efforts where resources were limited. That these models have an applied orientation does not detract from their use in more basic studies on the systems themselves.

In his book on the mathematical theory of infectious disease, Norman T. J. Bailey[2] indicated that it was a Swiss physican and mathematician in 1760 who first applied mathematical modeling to the study of the dynamics of an infectious disease; his goal was to affect public policy. Using differential equations, Daniel Bernoulli investigated the possible merits of preventive inoculation (or variolation) for the control of smallpox. Subsequent progress in mathematical modeling came slowly, initially because of an inadequate understanding of the basis of infectious disease and later on, due more to the lack of mathematical techniques. By 1957, there were only approximately 100 references devoted principally to the subject of the mathematical treatment of the dynamics of infectious diseases. Ten years later, that number had doubled. By 1975, the number of papers had increased to more than 500, and today, the rate of increase is faster than exponential.[2] Work today is greatly aided by a substantial wealth of knowledge on the mathematics of infectious disease and the computer which allows simulation of complex systems and statistical parameter estimation.[3-8]

One might suppose then that for every important communicable disease, there would be a considerable history of modeling efforts. The actual situation is quite different. In a review on the use of epidemiological models, Becker[9] indicated that of 75 papers published between 1974 and 1979, only ten contained any actual epidemiological data or dealt with computational techniques aimed at relating models to actual data. The rest dealt with the study of the mathematics of epidemic models. Becker concluded that the vast bulk of published work on epidemiological models is aimed merely at enriching the body of (albeit, potentially useful) theoritical knowledge of the behavior of untested mathamatical models of epidemics.

This largely theoretical orientation, however, has produced some very important contributions to the understanding of the dynamics of some systems of communicable disease.

First is the demonstration that changes in transmission rates can sometimes be the result of random chance fluctuations alone and are not necessarily due to changes in virulence or seasonal abundance of vector, etc. Considered more important is the epidemic threshold theorem which "...essentially says that in large populations there will be either minor epidemics or major epidemics with hardly any epidemics of a size between these two extremes."[9] This theorem predicts that under certain conditions, the immunization of a proportion of the susceptible population will result in the probability of small epidemics becoming essentially one and specifies the level of immunization required to eliminate large epidemics. It also indicates the existence of a critical community size, above which a disease tends to maintain itself and below which it tends to die out. In cases where a replenishment of susceptibles occurs, the theorem explains recurrent epidemics and, if the replenishment occurs at a constant rate as in the case of new births, it can explain apparent cycles of incidence without reference to periodicity in any of the system's other components such as seasonal changes in the density or daily survival of an insect vector.

As powerful as purely mathematical models are, they are not without some rather severe limitations. In order to be mathematically tractable, it is commonly necessary to introduce several simplifying assumptions at the expense of realism. This allows a rigorous solution of the mathematics, but it is a solution to an idealized system that frequently does not adequately mimic the real world. In fact, dynamic models, e.g., a system described by a set of differential equations that can be solved analytically and give practical results, are not very common.[10]

Today, an extensive body of techniques exists which permits the description and simulation of complicated systems. Since the techniques do not attempt analytical solutions, the constraints in the creation of a set of equations or statements of relationships are relaxed, and the model-builder may develop an account of his system as realistic, comprehensive, and/or detailed he deems appropriate and is able. There is no shortage of good texts on the techniques of computer simulation[10-14] and no attempt will be made to introduce it here other than to present some examples. It is the author's experience that once motivated by an appreciation for the power and utility of modeling as an adjunct to the study of a particular system, it is not very difficult to learn the techniques.

The history of the quantitative analysis of vector-borne disease began with Sir Ronald Ross who developed the first quantitative models of malaria transmission[15,16] with derivations of them providing the basic framework for our understanding of all vector-mediated transmission. Basically, Ross recognized that (on a per time basis) the number of new malarial infections was the product of the number of current infections, the number of mosquitoes that feed on humans, the probability of a mosquito surviving the extrinsic incubation period (EIP), the number of blood meals on man after the EIP, and the proportion of blood meals taken on uninfected humans. From this fundamental statement was derived the relationship between vectorial capacity (the mean number of potentially infective bites per host per day) and the density of the vector and the associated probability of feeding on any particular day and its predilection for the host's blood, the daily survival of the vector, the EIP, and the size of the host population and the proportion of hosts that are susceptible. This expression "describes a fundamental property of all vector transmission systems, and thus will be found buried superficially or deeply in most attempts to model such systems."[17] Malaria models have subsequently been expanded and improved in numerous studies.[18-28] Some of the recent models have been used to direct and evaluate control operations[29,30] and serve as examples of what may be accomplished with similar efforts directed toward a quantitative understanding of the dynamics of arbovirus infection cycles.

II. QUANTITATIVE ARBOVIRUS MODELS

There are relatively few published works of any type on quantitative models describing the dynamics of arbovirus infection. In fact, the literature on modeling malaria is more extensive than that for arboviruses. Dietz[31] has provided an excellent survey of the classical mathematics of arbovirus infection; topics include the derivation of the general vector-borne epidemic, the epidemic threshold, critical population size for maintenance of the virus, spatial spread of epidemics, models for age-specific prevalence of infection and disease, the mathematical theory of epidemic control via vaccination and vector control, and the periodicity of outbreaks. Elements of the theory presented in his paper will be found in any treatment of the dynamics of arbovirus infection; however, for the reasons cited previously, it is difficult to adequately describe the complexity of any real arbovirus system with a set of mathematical equations alone.

The following is a brief description of some models selected so as to provide the non-modeler with a feel for two commonly used modeling approaches. The models bridge the gap between a rigorous and formal mathematical treatment of a necessarily simplified system and the complexity of a real system.

A. Dynamic Life Table Models

A common goal of many arbovirus models has been to investigate the periodic recurrence or the endemic persistence of the virus. DeMoor and Steffens[32] used computer simulation to investigate the potential role of various parameters affecting the endemicity of chikungunya virus in southern Africa in a hypothetical population of 10,000 monkeys and baboons over a 30-year period. Their premise was that the simplest explanation for the maintenance of the virus was a continual transmission cycle which varied in intensity in response to seasonal changes in the vertebrate host and mosquito vector populations.

Potentially key variables included in the model were, for the mosquito, daily survival (Sa) and EIP (which were varied to reflect seasonal changes in the vector population and potential), daily biting probability (DBP), and the absolute number of mosquitoes. Vertebrate variables included the birth and death rates, duration of viremia, mortality due to infection, initial proportion of immunes, number of mosquito bites per day, and absolute numbers.

The model, written in *FORTRAN*, used the computer essentially as an accounting device, keeping track on a daily basis of the number of mosquitoes that have bitten an infected host on the current day, the number of mosquitoes infected for 1, 2, etc. days but not yet infectious, and the number of mosquitoes which were infectious. Each day the survival of each mosquito was determined by selecting a random number between 0 and 1; if the number was less than Sa, the mosquito is added to the next age classification. A number greater than Sa indicated the death of the particular mosquito. The number of newly infected mosquitoes was simulated by generating for each infectious host a Poisson distributed number (with rho being the average number of bites per host per day). In a similar fashion, the computer maintained the numbers of susceptible hosts, hosts being infected on the current day, hosts incubating or circulating virus for each day of incubation or viremia, and immune hosts. These numbers were also updated on a daily basis using random numbers to determine if a mosquito bit a susceptible host. Once a month, random numbers were again used to select hosts for death with new hosts being added to maintain a stable population.

Simulation runs with this simple model indicated that it was not necessary to hypothesize periodic reintroduction to explain the maintenance of the virus as there were many reasonable combinations of values for the variables that would result in the indefinite persistence of virus. A considerable range of values for DBP were consistent with the long-term persistence of the virus. The existence of an endemic state was seen to be very sensitive to Sa; changing Sa from 0.89 to 0.94 altered the course of the disease from loss of the virus to a stable endemic state. Using a (realistic) turnover rate of approximately one third of the animals per year, a minimum community size of about 4000 animals was required to maintain the endemic state when other variables were set to values favoring endemicity. If community size was increased to 10,000, a lower turnover rate (the result of decreasing the death rate) was still consistent with virus persistence.

The model also provided insight into the ramifications of the stochastic nature of transmission under certain conditions. When the conditions of the model were set to a very stable endemic state, the starting point for a series of random numbers used to determine if a particular animal was bitten or a particular mosquito died on a particular day, etc., was not seen to alter the endemic state. On the other hand, when conditions were set to produce a marginally stable state, i.e., characterized by large swings in the infection rates of host and vector, the choice of the starting point for the sequence of random numbers played a determining role in how long the virus persisted. This result is analogous to the role of chance events in determining the outcome of an unstable endemic state. They concluded that the persistance of the virus is more dependent upon the dynamics of the vector that those of the host population.

This particular model was selected to demonstrate the usefulness of a very simple model based on a limited amount of data. A chief advantage of this type of model is that it is easily extended to include an accounting of all life stages of the vector, host, environment, etc. This enables the simulation of a degree of complexity limited only by the available data. For example, in a model which included all life stages of the vector, the impact of adulticiding may be studied by setting Sa temporarily to a value indicating insecticide mortality. This could have an immediate as well as a delayed impact on transmission as the loss of oviposition may result in lower larval and subsequent adult populations. A descriptive name for this type of comprehensive treatment is a dynamic life table model. Finally, the approach has an additional advantage of being readily understandable to one who is not mathematically or statistically inclined.

A recently published work by Kay et al.[33] is an exciting example of using computer-based simulation to investigate epidemiological phenomenon that cannot be studied by any other means. The goal was to (1) retrospectively study past epidemics of Murray Valley encephalitis in southern Australia to determine the length of the rural amplification phase of the virus in birds prior to spillover into man and (2) evaluate the hypothesis that outbreaks are the result of virus introduction via migratory birds from northern tropical Australia where the virus is endemic. Similar to the previous example, the computer was used to keep track of the disease status of hypothetical vector and host populations through time using estimates of the various parameters from the literature.

In addition to giving insight into the relative sensitivity of the system to certain variables such as mosquito survival, length of EIP, etc., the study suggested that Murray Valley virus is likely maintained in an endemic state without human cases in the southern temperate regions of Australia throughout the year and that periodic reintroduction from the north is unlikely. The type of analysis used here is probably applicable to any type of vector-borne disease and this paper will be important reading for anyone considering a systems approach to their virus. The work by Scott et al.[35] is another example of the utility of computer simulation of arbovirus transmission that is worthy of study.

B. Statistical Models

The next example utilizes several techniques common to statistical modeling. While the methods vary from simple curve fitting via regression to more complicated methods such as time series analysis, the principal goal is to identify and quantify relationships between one or more sets of dependent and independent variables. A common use of statistical models is simply the prediction of future values of a dependent variable based on the previous behavior of the dependent variable and/or current/past values of other independent variables. Statistical models differ fundamentally from the simulation of a complete system where a major goal is the description of the interaction of the underlying mechanisms. Historically, time series techniques were used in econometric studies to derive (black box) relationships between variables presumed to afford information regarding the future value of stocks. These techniques, while perhaps more elegant statistically and mathematically than the simulations described above, are several steps removed from the biological realism afforded by computer-based simulation.

The following example, while dealing only with the population dynamics of a vector, was selected to illustrate the flexibility possible with some novel statistical techniques. Portier et al.[34] used time series analysis to relate an abundance index of the mosquito population weekly oviposition rates to meteorological parameters (minimum, maximum, and average temperatures, rainfall, and relative humidity) and past values of the dependent variable.

Briefly, a time series is nothing more than a set of regularly spaced observations of some process such as the weekly egg laying rate of *Aedes aegypti,* the weekly number of cases of dengue (DEN) in a city, or the temperature and rainfall at the local airport. Several statistical functions can be derived for these data that are of interest. The auto-covariance function provides a measure of correspondence or correlation between observations of a single variable separated temporally by increasing periods of time. Generally, the correspondence declines with increasing separation in time, i.e., high levels of adult mosquitoes or DEN cases this week will frequently be associated with high levels during the preceding week, but knowledge of mosquito abundance or case numbers 6 or 10 weeks ago may be expected to shed less light on the expected present situation. Similarly, high levels n weeks ago may be associated with correspondingly high current levels because of an n-week development or transmission period. The cross-covariance function, similar to the auto-covariance function, provides information on the correspondence between two time series such as oviposition rates and temperature, etc. Both of these functions help identify variables that correlate with the dependent variable, provide insight into their relative importance, and thereby indicate those which should be included in subsequent statistical models.

A widely used forecasting technique that makes estimates of future values of some variable based solely on past observations of the same variable is called an autoregressive integrated moving average (ARIMA) model. A transfer function forecasting model (TFFM) attempts to predict the same thing by reference to the present or past values of one or more independent variables.

After examining the auto- and cross-correlation functions, an initial TFFM model was developed for a 2-year series of data from New Orleans which expressed the current rate of oviposition as a function of nothing more than past values of the minimum temperature. It was surprising that after accounting for the effect of this variable, the effects of other temperature variables, relative humidity, and rainfall were not significantly related to oviposition nor to the residuals (observed-predicted) of this initial model. It seems that during the two years studied, the key factor driving the dynamics of *Ae. aegypti* in New Orleans was strictly temperature with the other factors being adequate so as not to limit populations. We noted that over short periods of time, this simple model tended to error in a consistent fashion; if the prediction was low last week, it was likely to also be low during the next few weeks. As a result, we added to the TFFM model the last week's residual as an ARIMA component. This model was superior to the TFFM version and did well when the relationship between temperature and oviposition held (spring through autumn), but was inadequate during the winter when oviposition remained zero independent of just how low temperatures went. A function was then developed via regression which predicted the probability of observing oviposition next week based on previous temperatures. If the probability of oviposition was above a critical value, the model predicted the rate of oviposition based on the TFFM/ARIMA model; if the probability was below the critical value, an estimate of zero was given. Based on nothing more than minimum temperature, this hybrid model explained more than 90% of the variation in a subsequent third year's data used for validation.

With the computer, extremely powerful statistical techniques are available that have rarely been applied to the task of relating large quantities of data on the many independent variables known to influence mosquito abundance, activity, or arbovirus transmission. In contrast to simulation models, statistical models are appropriate when large amounts of data of the types mentioned here are to be examined. In terms of shortcomings, it is important to realize that a statistical model will remain adequate only so long as the system producing the data used to create the equations remains unchanged. Additionally, the model cannot be used to investigate the effect of any variable not originally in the model. We cannot use the *Ae. aegypti* model for San Juan nor can it be used to investigate the impact of adulticiding or source reduction in New Orleans or the role of rainfall in a year of below normal precipitation. A practical impediment to statistical modeling is the requirement for a statistician on the modeling team.

III. CONCLUSIONS

There is a wealth of data on arbovirus systems that could be productively analyzed with quantitative models. This work would be an adjunct to present research, summarizing available data, identifying variables where more precise estimates were required, and providing insight into the dynamics and control of infection not afforded by the armchair models many have developed for their systems. A multidisciplined approach may result in accomplishing an analysis otherwise not attempted due to a lack of expertise in the model-building area. The real or perceived lack of data should not deter initial efforts.

REFERENCES

1. **Cvjetanovic, B., Grab, B., and Uemura, K.,** *Dynamics of Acute Bacterial Diseases,* World Health Organization, Geneva, 1978, 143.
2. **Bailey, N. T. J.,** *The Mathematical Theory of Infectious Diseases and Its Application,* 2nd ed., Griffin, London, 1975, 413.
3. **Dietz, K.,** Overall population patterns in the transmission cycle of infectious disease agents, in *Population Biology of Infectious Diseases,* Anderson, R. M. and May, R. M., Eds., Springer-Verlag, Berlin, 1982, 87.
4. **Anderson, R. M. and May, R. M.,** Transmission dynamics and control of infectious disease agents, in *Population Biology of Infectious Diseases,* Anderson, R. M. and May, R. M., Eds., Springer-Verlag, Berlin, 1982, 149.
5. **Muench, H.,** *Catalytic Models in Epidemiology,* Harvard University Press, Cambridge, Mass., 1959, 110.
6. **Bartlett, M. S.,** *Stochastic Population Models in Ecology and Epidemiology,* John Wiley & Sons, New York, 1960, 90.
7. **Wickwire, K.,** Mathematical models for the control of pests and infectious diseases: a survey, *Pop. Biol.,* 11, 182, 1977.
8. **Mollison, D.,** Spatial contact models for ecological and epidemic spread, *J. R. Stat. Soc. Ser. B,* 39, 283, 1977.
9. **Becker, N.,** The uses of epidemic models, *Biometrica,* 35, 295, 1979.
10. **Gordon, G.,** *System Simulation,* Prentice-Hall, Englewood Cliffs, N. J., 1969, 303 .
11. **Naylor, T. H., Balintfy, J. L., Burdick, D. S., and Chu, K.,** *Computer Simulation Techniques,* John Wiley & Sons, New York, 1966, 352.
12. **Iyengar, S. S.,** *Computer Modeling of Complex Biological Systems,* CRC Press, Boca Raton, Fla., 1984, 142.
13. **Patten, B. C.,** *Systems Analysis and Simulation in Ecology,* Vol. 1 and 2, Academic Press, New York, 1971, 1, 198.
14. **deWit, C. T. and Goudriaan, J.,** *Simulation of Ecological Processes,* Center for Agricultural Publishing and Documentation, Wageningen. The Netherlands, 1978, 175.

15. **Ross, R.,** *Report on the Prevention of Malaria in Mauritius,* Waterlow, London, 1908.
16. **Ross, R.,** *The Prevention of Malaria* (with Addendum on the Theory of Happenings), 2nd ed., John Murray, London, 1911.
17. **Fine, P. E.,** Epidemiological principles of vector-mediated transmission, in *Vectors of Disease Agents,* McKelvey, J. J., Jr., Eldridge, B. F., and Maramorosch, K., Eds., Praeger, New York, 1981, 229.
18. **Lotka, A. J.,** Contributions to the analysis of malaria epidemiology. I. General part, *Am. J. Hyg.,* 3(Suppl. 1), 121, 1923.
19. **Davidson, G. and Draper, C. C.,** Field studies on some of the basic factors concerned in the transmission of malaria, *Trans. R. Soc. Trop. Med. Hyg.,* 47, 552, 1953.
20. **MacDonald, G.,** *The Epidemiology and Control of Malaria,* Oxford University Press, London, 1957, 201.
21. **Garrett-Jones, C. and Shidrawi, G. R.,** Malaria vectorial capacity of a population of *Anopheles gambiae, Bull WHO,* 40, 531, 1969.
22. **MacDonald, G., Cuellar, C. B., and Foll, C. V.,** The dynamics of malaria, *Bull. WHO,* 38, 743, 1968.
23. **Cuellar, C. B.,** A theoretical model of the dynamics of an *Anopheles gambiae* population under challenge with eggs giving rise to sterile males, *Bull. WHO,* 40, 204, 1969.
24. **Fine, P. E. M.,** Superinfection — a problem in formulating a problem: an historical critique of MacDonald's theory, *Trop. Dis. Bull.,* 72, 475, 1975.
25. **Fine, P. E. M.,** Vectors and vertical transmission — an epidomiologic perspective, *Ann. N. Y. Acad. Sci.,* 266, 173, 1975.
26. **Frost, W. H.,** Some conceptions of epidemics in general, *Am. J. Epidemiol.,* 103, 141, 1976.
27. **Haile, D. G. and Weidhaas, D. E.,** Computer simulation of mosquito populations *(Anopheles albimanus)* for comparing the effectiveness of control techniques, *J. Med. Entomol.,* 13, 553, 1977.
28. **Molineaux, L. and Gramiccia, G.,** *The Garki Project: Research on the Epidemiology and Control of Malaria in the Sudan Savanna of West Africa,* World Health Organization, Geneva, 1980, 311.
29. **Dietz, K., Molineaux, L., and Thomas, A.,** A malaria model tested in the African savannah, *Bull. WHO,* 50, 347, 1974.
30. **Dietz, K., Molineaux, L., and Thomas, A.,** Further epidemiological evaluation of a malaria model, *Bull. WHO,* 56, 565, 1978.
31. **Dietz, K.,** Transmission and control of arboviruses, in *Proc. Soc. Industrial Appl. Math. Conference on Epidemiology,* Ludwig, D. and Coope, J., Eds., Society for Industrial and Applied Mathematics, Philadelphia, 1975, 104.
32. **DeMoor, P. P. and Steffens, F. E.,** A computer-simulated model of an arthropod-borne virus transmission cycle, with special reference to chikungunya virus, *Trans. R. Soc. Trop. Med Hyg.,* 64, 927, 1970.
33. **Kay, B. H., Saul, A. J., and McCullagh, A.,** A mathematical model for the rural amplification of Murray Valley encephalitis virus in southern Australia, *Am. J. Epidemiol.,* 125, 690, 1987.
34. **Portier, K. M., Focks, D. A., and Lai, P. Y.,** A threshold-transfer function model of *Aedes aegypti* oviposition data from New Orleans, Louisiana, presented at International Society of Ecological Modeling, Gainesville, Fla., August 11—15, 1985.
35. **Scott, T. W., McLean, R. G., Francy, D. B., and Card, C. S.,** A simulation model for the vector-host transmission system of a mosquito-borne avian virus, Turlock (Bunyaviridae), *J. Med. Entomol.,* 20, 625, 1983.

INDEX

A

C

D

E

Q

R

S

T

U

V

W

Y

Z